Phase-Locked Loop Engineering Handbook for Integrated Circuits

DISCLAIMER OF WARRANTY

For a listing of recent titles in the *Artech House Microwave Library,* turn to the back of this book.

Phase-Locked Loop Engineering Handbook for Integrated Circuits

Stanley Goldman

ARTECH
HOUSE

BOSTON | LONDON
artechhouse.com

Library of Congress Cataloging-in-Publication Data
A catalog record for this book is available from the U.S. Library of Congress.

British Library Cataloguing in Publication Data
A catalogue record for this book is available from the British Library.

ISBN-13: 978-1-59693-154-1

Cover design by Igor Valdman

© 2007 ARTECH HOUSE, INC.
685 Canton Street
Norwood, MA 02062

10 9 8 7 6 5 4 3 2 1

To Yolanda, Grace, and John

Contents

CHAPTER 3

CHAPTER 8

CHAPTER 9

Preface

This book is intended to guide readers on phase-locked loop (PLL) theory and show practical applications, designs, simulations, and testing techniques. The book emphasizes monolithic PLLs because of the growth in this area; however, the techniques also apply to other PLLs. The majority of the material was generated from presentations to customers, seminars, and work assignments at Texas Instruments (TI).

Current PLL reference books emphasize system theory and circuits that can be built in rack-mounted modules or soldered into printed circuit boards (PCBs). Most of these techniques cannot be done in integrated circuits (ICs). Integrated circuits require a lot more attention to transistor-level design detail than is feasible using an IC process. This book balances the necessary system-level background with the transistor-level design detail. The amount of design detail presented in this book is unavailable in other current references.

The difficulty of PLLs has caused many years to be spent collecting and studying a wide variety of books and articles to learn PLL concepts and develop PLL analysis, simulation, and testing techniques. However, the information, solved practical problems, SPICE listings, simulation techniques, and test setups in this book will accelerate the learning process for readers. Easy access in one reference book will allow readers to solve practical PLL problems, design PLLs, conceptualize new PLLs, simulate phase locks loops, and test and troubleshoot PLLs immediately.

What makes PLLs difficult? PLLs are used as pipe cleaners for breaking simulation tools. Cadence and Synopsys use a PLL to show the limitations of their tools. The failure mode of the simulation tool is to show a working PLL when it does not function in silicon. In addition, the PLL has long simulation times because the high-frequency voltage-controlled oscillator (VCO) clock of 1 GHz requires small time steps and the low reference frequency of 1 MHz requires hundreds of microseconds of time for the simulation to settle. Because of these simulation problems, there is not even close to 100% coverage of PLL performance. There is always a gap. The question is how big a gap one leaves and how to minimize this gap. Next, one simulation tool cannot measure all the characteristics and design requirements of the PLL. Multiple methodologies and developing one's own simulation tools add to the difficulty. Laboratory measurements require specialized equipment and measurements that have state-of-the-art accuracy.

A PLL engineer must have a breadth of knowledge in many disciplines covering almost all of electronics. Familiarity with Z transforms, La Place transforms, Fourier transforms, differential equations, numerical analysis, the C programming language, digital circuits, analog circuits, RF circuits, control systems, and communications are the recommended background areas. Very few engineers have this background. Finally, to quote Behzad Razavi, the design of PLLs cannot be described easily using a straight top-down or bottom-up approach because each level of abstraction entails issues strongly related to other levels as well (Razavi, B., *Monolithic Phase-Locked Loops and Clock Recovery Circuits*, New York: IEEE Press, 1996).

The limit of 50 MHz for crystal-oscillator frequencies, the demand for higher-speed integrated circuits, and the integration of all functions to create a system on an integrated circuit has created an unprecedented demand for monolithic PLL circuits. Microprocessors, digital signal processors, microcontrollers, and telecommunications integrated circuits are demanding clock speeds from 200 to 4,000 MHz. PLL frequencies multiply 10–50-MHz crystal-oscillator frequencies up to these required output frequency ranges. Current system designers are demanding more functions (e.g., process, control, memory) to be integrated on one integrated circuit (systems-on-a-chip concept) in order to meet consumer demands for higher-powered and lower-cost electronic products. Because of these demands, monolithic PLLs have a different design emphasis than the traditional PLL designs of the past. New designs must provide maximum isolation from other circuits on the chip, use low power, occupy small die area, use small-value capacitors, and not use inductors. Strategies for testing PLLs on a chip also present new challenges because, in many cases, only the input reference frequency and output frequency pins are available, and integrated circuit testers require functional tests that take only a few milliseconds to complete.

Traditional PLL design relied on different chip functions to make a PLL that was connected together on a PCB. Understanding the details of many of these devices was not necessary. Today, IC designers must know the transistor-level details of VCOs, dividers, phase detectors, charge pumps, operational amplifiers, lock detectors, and crystal-oscillator designs. This book gives the details of these designs and gives SPICE listings so that readers can more completely understand them by running their own simulations. In addition, the SPICE simulations show methods for verifying the performance of individual functions within the PLL. This book presents a practical guide for analyzing, designing, simulating, testing, and troubleshooting monolithic PLLs.

Design engineers, system engineers, and test engineers are the audience for this book. They will most need and use the information in this book. System engineers will learn the hardware constraints on system performance, which will lead to better functional divisions between chips, better floor planning in the chips, and fewer redesigns because of better ranking of requirements.

Design engineers will learn about requirement trade-offs, synthesis of loop components, PLL requirements, and analysis and troubleshooting of PLL problems. Consequently, higher-performing loops will be designed that can be easily tested on a tester.

Test engineers will be able to understand PLL requirements and the test procedures necessary to verify PLL performance according to those requirements. These engineers will be able to understand PLL requirements and more fully test all functions on a PLL.

The unique features of this book will make it an essential resource for engineers. My methods of PLL design, hierarchical SPICE listings, and detailed design of phase detectors, dividers, and oscillators will make it especially valuable to IC design engineers. The engineering community has a high interest in these analysis techniques.

Almost all of the equations were developed and analyzed using Mathcad. These verified equations were directly copied into the book, which enhances the accuracy of the equations by avoiding typos and other possible errors. In addition, some of the plots from Mathcad were also included.

Finally, the SPICE listings, which are used to analyze and understand PLL concepts, is another unique feature that is included on the CD for the book. Years of learning PLL concepts and developing PLL analysis techniques will be shortened. The SPICE simulations will accelerate the learning process for the readers and allow them to solve practical problems immediately.

Book Summary

This book has nine chapters and two appendixes. Chapters 1 through 5 discuss the basic tools that are required to do PLL analysis and specify PLL requirements. Chapters 6 through 9 extend the basics to practical simulation, application, and testing problems.

Chapter 1 gives an overview of PLLs, presents background information to understand how a loop operates, and gives reference material for further study of loops. Key signals of interest within a PLL and the major components are discussed so that engineers can immediately identify performance by monitoring these signals or major components on the test bench or in simulations. Key design requirements are presented so that engineers will be able to identify the important requirements among the many trivial ones that can be found in statements of work or specifications. A history of PLLs is presented to show the progression from the first implementations with large racks of equipment to today's tiny monolithic versions.

Applications of PLLs are presented to show the many applications in modern systems. In addition, the reader needs to understand the differences in specification requirements for these applications because the design differences in the PLL literature are driven by these application differences. Good design engineers must understand how their circuits will be used in order to provide the optimum circuit for the customer. Not everything the customer desires is in the specification, and the ranking of the importance of specifications must be determined.

Chapter 2 covers system analysis of PLLs. Control-systems theory and analysis of PLLs is covered in this chapter. Bode plot analysis of PLLs is studied so that stability requirements can be designed. The relationship between step response and stability is studied. Finally, error tracking of a PLL to various stimuli is studied. Understanding control systems helps an engineer build stable and robust loops.

Chapter 3 discusses system requirements for a PLL. Noise basics, phase noise, time-domain response, acquisition, jitter, and spurious signals are the system requirements that are studied in detail. First, noise basics are discussed in order to get a foundation for analyzing noise requirements and their relationship to circuit design. In this section, the sources of noise, a discription of each noise type, active noise models, equivalent input noise, noise figure, and the trade-offs between bipolar and Complementary metal oxide semiconductor (CMOS) transistors will be discussed. Understanding these topics will help readers decipher noise specifications, understand the physical limitations of the circuit design, and design low-noise circuits.

Next, frequency modulation (FM) theory and its relationship to phase noise are reviewed. The phase-noise characteristics of oscillators are discussed, and a model is developed. The relationships in this chapter have been discussed in several articles and seminars. The VCO and reference phase-noise combination in a PLL are discussed. Understanding this relationship helps engineers design low-noise loops. Phase noise is a measure of the stability of the loop. Lower-phase-noise designs allow the loop to be used in higher-precision applications (e.g., high-quality audio products).

Linear time-domain responses are studied to give an engineer the understanding of the relationship between loop variables and a faster switching time. Almost all PLL responses can be modeled by these classic equations. Next, these equations can be used to study the relationship between loop gain and tracking error. Reducing tracking error allows the output edges to more closely track the reference input edges, which can be critical in clock recovery and resynching circuits. In addition, these equations can be used to determine the natural frequency and damping factor by adjusting these parameters to fit the measured data.

Nonlinear acquisition is studied to give an engineer an understanding of loop responses to various external stimulus. Understanding these responses will help explain loop responses that seem to be inconsistant. The second-order, nonlinear, ordinary differential equation is derived. Then, the equation is simplified and converted to a difference equation to make it easier to program. Finally, responses inside and outside the separatrix are studied.

The causes of jitter, the effect of jitter on the PLL, the relationship of phase noise to jitter, and the relationship of spurious signals to jitter are covered. Jitter reduces the timing margin in digital circuits, which reduces the speed at which digital circuits can operate. Understanding the relationship of PLL parameters to jitter will allow the design of faster digital circuits. For synchronizing circuits, additional jitter increases the bit-error rate of the interface.

Spurious signals are studied because these signals cause jitter and reduce the precision of the output frequency. The intermodulation products of a mixer are studied because many spurious signals result from some nonlinear transfer function that performs the mixing operation. Understanding the mixing operation allows engineers to design methods for suppressing these products. Spurious signals at the output from the reference frequency are among the largest spurious signals in a loop. Feed-through from the phase detector output to the VCO tune line is the biggest contributor. Understanding this feed-through mechanism will help engineers design lower-jitter and lower-noise PLLs.

Chapter 4 discusses the design of the individual components in a PLL. Detailed designs of programmable dividers, VCOs, crystal oscillators, phase detectors, lock detectors, and acquisition aids are studied. Design techniques for programmable dividers are discussed to prevent a PLL from hanging up on one of the power rails.

VCO performance accounts for more of the PLL's characteristics than any other component. Design trade-offs and techniques for single-ended ring, differential ring, multivibrator, and LC resonant VCOs are discussed.

Crystal reference oscillators are also being incorporated on integrated circuits with the PLLs. Consequently, crystal-oscillator design techniques are covered. Maintaining oscillation and startup time are the main concerns of IC designers in crystal-oscillator design. Analysis of stability and computing startup time are presented.

In Chapter 5, phase detectors are studied because understanding how phase detectors work is one of the major keys to understanding how PLLs work. First, the phase/frequency detector is studied because of its wide use in the industry. The operation of the phase/frequency detector, the frequency-difference response, the phase-difference response, the generation of the phase detector transfer function, and the distortion zone are covered.

Many systems use the lock detector to reset the system. This is a disastrous change to the operation of most systems. In a PLL, a reset can start the loop operating at a very low frequency (or no output); it then reacquires lock at the normally much higher output operating frequency. Consequently, a small phase shift in a PLL that would marginally affect the system can cause a huge disruption in the operation of the system if a reset occurs. The key to lock detection is to detect behavior that shows the PLL is broken. The quadrature phase detector, time-window edge comparison, the tune-voltage window comparator, and cycle-slip detection are the lock-detection methods that are covered.

Open-loop sweep, closed-loop sweep, and discriminator-aided acquisitions are the methods that are covered. The phase/frequency detector uses discriminator-aided acquisition and is the most popular choice. Clock recovery circuits cannot use phase/frequency detectors. Consequently, these circuits require an acquisition aid. Understanding the design details and trade-offs in these components is critical to designing monolithic PLLs.

Chapter 6 presents loop-compensation synthesis. The synthesis of loop-compensation components in the loop are application dependent. Consequently, several examples of the most popular compensation schemes are presented to give an engineer an intuitive feel that allows him or her to extend these methods to designs that are not presented. These examples include passive, active, charge pump, sampling delay, optimum phase-noise, low active filter noise, and minimum capacitance design.

Chapter 7 discusses test measurements of PLLs. Measurements of PLLs are critical in verifying the performance of the PLL. Many of the tests can only be done on the bench because the length of test time (greater than tens of milliseconds) makes testing on the tester in production impractical. Understanding how a PLL is tested can help the designer incorporate testing aids into the PLL.

This chapter discusses several phase-noise measuring methods. An explanation of the advantages and disadvantages of each method and several phase-noise plots of sources and measuring equipment aids the reader in selecting the appropriate phase-noise method. Step-response measurements are discussed. Results from step-response measurements verify damping and bandwidth design goals. Jitter and spurious-signal measurements are also discussed. Understanding these issues will help design lower-noise loops. Finally, this chapter presents troubleshooting techniques for PLLs. The testing and troubleshooting of PLLs in integrated circuits presents unique problems. In order to determine which tests need to be run on a particular design, the reader must understand how to troubleshoot various problems that occur in a PLL.

Chapter 8 covers simulation techniques for PLLs. The types of PLL simulators vary from equation solutions, to behavioral-model solutions, and to SPICE transistor-level solutions. The choice of simulation depends on the speed and accuracy of the results. For instance, accurate SPICE simulations can take several weeks to run when looking at acquisition times. This chapter will help the reader make an informed choice as to which method to use.

Chapter 9 presents PLL applications and extensions. PLLs are used to generate frequencies, to do clock recovery of a signal, and to resynchronize signals. Design solutions will be studied to illustrate solutions for particular design problems. This will familiarize readers with the various design approaches, which they can then extrapolate to their particular solutions.

In clock recovery, many systems transmit or receive data without any additional timing reference. To an ever-increasing extent, communication links use digital formats to transmit information synchronously in a continuous, uniform pulse stream. Clock-recovery techniques are covered because every digital communication link uses a PLL in the reception of this information.

Communication systems, radar systems, data-acquisition systems, and test equipment convert analog signals to digital words for digital processing. Digital processing assumes equally spaced time samples; however, actual samples from analog-to-digital (A/D) converters have slightly unequal time spacing. Phase noise from the oscillator that generates the sampling clock produces this change in time spacing. In digital processing, equally spaced time samples are critical to system performance. The phase noise of the sampling clock affects the dynamic range of the A/D converter by adding noise. The limitations of the conversion process will be covered so that adjustments can be made to PLL designs that can overcome these limits.

As digital processes reduce the cost of circuit functions and design tools to handle large-scale digital circuits, design in the digital domain becomes easier. Consequently, it has created a demand for all-digital PLLs (ADPLL). ADPLLs will be covered to improve the reader's ability to recognize the correspondence between the fundamentals and structures that look very different from the basic PLL.

The variety of design topics covered illustrates solutions to particular design problems. This will familiarize the reader with the various design approaches so that they can extrapolate them to their particular solutions.

The appendixes contain a glossary of terms and symbols and other reference material. The glossary defines and identifies terms, letter symbols, and abbreviations. The IEEE standards were followed to generate the definition of these terms and the symbols.

Finally, TI transistor-level processes are not included (it is best to substitute MOSIS transistor models). The circuits in this book do not reflect TI's implementation of any product, and there is no guarantee that they will work for anyone's application or that the simulations will work. References to software tool vendors and test equipment manufacturers do not represent the current performances of these tools and are included only to show the methodology that would be used to evaluate these tools.

Acknowledgments

I am grateful to Texas Instruments, Inc., for creating a working environment in which this work could be performed and for allowing me to publish this manuscript. I would like to thank all my coworkers at Texas Instruments, whose interaction, effort, material, and support made this book possible. I appreciate the moral support of many of my friends and relatives while I was writing this book. A special thanks to Ming Chang and Baher Haroun, whose support made this book possible.

CHAPTER 1

Getting Started with PLLs

A phase-locked loop (PLL) helps keep the electronic world orderly. In a television set, one PLL keeps the head at the top of the screen and the feet at the bottom of the screen. Another PLL makes the color green remain green and the color red remain red.

This chapter presents background information to understand how a loop operates and gives reference material for further study of loops. Key signals of interest within a PLL and the major components are discussed so that we can immediately identify performance by monitoring these signals or major components on the test bench or in simulations. Key design requirements are presented so that we will be able to identify the important requirements from among the many trivial ones that can be found in statements of work or specifications. A history of PLL is presented to show the progression from the first implementations with large racks of equipment to today's tiny monolithic versions.

Applications of PLLs are presented to show the many applications in modern systems. We will need to understand the differences in specification requirements for these applications because the design differences in the PLL literature are driven by these application differences. Good design engineers must understand how their circuits will be used in order to provide the optimum circuit for the customer. Not everything the customer desires is in the specification, and the ranking of the importance of specifications must be determined.

1.1 Definition and Operation

More precisely, a PLL synchronizes the output phase and frequency of a controllable oscillator to match the output phase and frequency of a reference oscillator. Ideally, the steady-state condition will show a zero difference in phase and frequency between the controlled oscillator output and the reference output.

The simplest PLL consists of four basic functional blocks:

1. Voltage-controlled oscillator (VCO);
2. Phase detector (PD or PFD);
3. Loop filter;
4. Feedback divider (which equals 1 for the simplest case).

To maintain synchronization, a phase comparison (by a phase detector) of the outputs of the reference and controllable oscillators generates an error signal that is processed by the loop filter to control the controllable oscillator for minimum phase error. An increase in phase error produces a control signal that changes the controllable oscillator to decrease the phase error and vice versa. Consequently, the loop tracks changes in the phase and frequency of the reference oscillator.

To be more specific about the operation of the loop, a phase comparison by the phase detector occurs for each rising edge of the reference oscillator. The output of the phase detector produces a pulsed error voltage that has a pulse width equal to the difference in phase of the two signals. The loop-filter processes the pulsed error voltage (average) to produce a dc voltage for controlling the VCO. Figure 1.1 shows the ideal transfer function for a phase detector with a 5-MHz reference frequency.

Equation (1.1) shows the mathematical description for the ideal phase detector:

$$V_{pdavg} = K_d \theta_e \tag{1.1}$$

where

K_d = Phase detector gain (V/rad);

θ_e = Phase error (rad).

From the control voltage, the VCO slightly changes the frequency in a direction that reduces the phase difference. Figure 1.2 shows the ideal tune-voltage-versus-frequency curve transfer function for a VCO. Equation (1.2) shows a mathematical description of the ideal VCO transfer function:

$$\omega_{out} = \omega_{off} + K_v V_{tune} \tag{1.2}$$

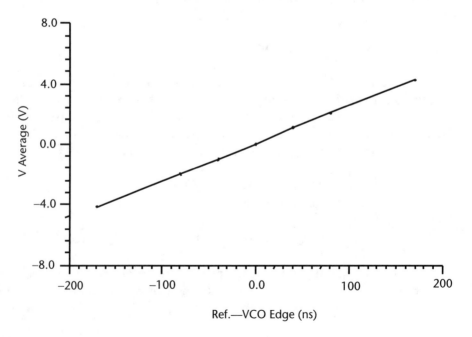

Figure 1.1 Ideal phase detector transfer function.

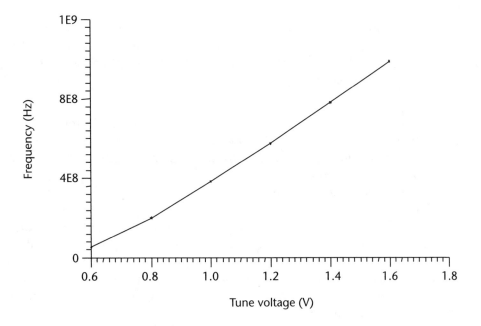

Figure 1.2   Ideal transfer function for a VCO (frequency versus tune voltage).

where

$$\omega_{out} = \Delta\theta_{out}/\Delta t$$

ω_{off} = offset frequency of the VCO, frequency with 0V on the tune line, (rad/s);

K_v = VCO gain, slope of the VCO transfer function (rad/s/V).

The next cycle begins again with a phase detector comparison with the reference-clock rising edge. This cycle repeats for each reference-oscillator period until the phase difference is minimized.

From a network-analysis point of view, the phase detector is a transducer that converts a frequency difference to a voltage. The loop-filter processes the output voltage of the phase detector and produces the control voltage to the VCO. The VCO is another transducer that converts the processed voltage from the loop filter to frequency. The output frequency is then fed back to the phase detector for comparison with the input frequency. Consequently, network functions between the output of the phase detector and the input of the VCO are expressed in terms of voltage. Network functions from the VCO to the phase detector are expressed in terms of phase or its derivative frequency.

Key signals of interest within a PLL include:

1. Input frequency or reference frequency;
2. Output frequency;
3. Tune voltage or current input to the VCO or CCO;

4. Comparison of the positive edge of the reference input signal to the phase detector with the positive edge of the fed-back signal from the VCO to the phase detector, with the positive edge of the reference input as the trigger source (phase error).

When the PLL is locked, the output frequency should follow the input frequency. The ratio of the output frequency to the input frequency should be unity. The tune voltage or current input to the VCO should also vary smoothly with the changing input frequency. Finally, the positive edge of the fed-back signal from the VCO should smoothly track the positive edge of the reference input signal with changing input frequency.

A PLL has several states of operation. Initially, the loop begins in the *unlocked* state, which occurs at the moment when power is applied to the PLL. After a transition period, a *locked* state arrives, in which the frequency of the VCO equals the average frequency of the input signal, and each input cycle has only one cycle of VCO output. The transition from the unlocked to the locked state defines the *acquisition* response for a PLL.

Figure 1.3 shows a transient response of a PLL acquiring lock. The loop starts from a high-frequency, unlocked state and finishes in a locked state. In the unlocked state, multiple cycle slips in phase occur. The maximum pulse width of the phase

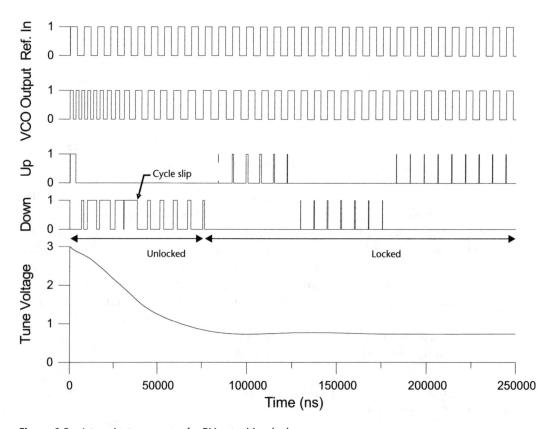

Figure 1.3 A transient response of a PLL acquiring lock.

detector, followed by a minimum pulse width at the next phase comparison, defines the occurrence of a cycle slip as shown in the figure.

One of the most common applications of a PLL is the multiplication of the reference frequency. A frequency divider placed in the feedback path of the loop between the VCO output and the phase detector input multiplies the reference frequency by the frequency-divide ratio. This multiplication occurs because, in order for the frequency after the frequency divider to be equal to the reference frequency, the VCO frequency at the input of the frequency divider must be at the reference frequency times the divide ratio. For example, with a frequency divider that has a divide ratio of 2, the frequency of the signal at the output of the VCO will be twice that of the reference.

Key design considerations for monolithic applications include:

- Timing jitter;
- Loop stability;
- Noise immunity;
- Architecture.

Low timing jitter for a loop allows the widest variation in digital timing for the logic in the core. High loop stability and high noise immunity provide a loop that stays locked in a harsh electronic environment. An unlocked loop (worst-case failure mode) causes accumulated phase and frequency errors that can result in system failures. PLL architecture decisions affect the system requirements that will be met. Single PLL, multiple PLL, direct digital synthesis with a PLL, multivibrator VCO, ring oscillator VCO, phase/frequency detector, and XOR phase detector are some example architecture choices.

1.2 Phase-Lock Loop Literature

1.2.1 Books

Several noteworthy introductory books have been published on PLLs. I will critique the ones that I am most familiar with. Floyd Gardner's book [1] stresses physical understanding of basic PLL characteristics. He uses mathematical descriptions of PLLs and avoids long mathematical derivations. Roland Best's book [2] provides explanations of PLL behavior with an easy-to-understand methodology that includes analogies of PLLs to mechanical systems. He covers analog, digital, all-digital, and software PLLs. A computer-simulation program is provided that helps readers look at PLL performance. Behzad Razavi's book [3] provides a survey of IEEE articles that cover the basics, building blocks, modeling, simulation, and monolithic implementation of PLLs and carrier-recovery circuits. Vadim Man-assewitsch's book [4] emphasizes the application of PLLs to frequency synthesis. He covers system issues of frequency synthesizer architectures, grounding and shielding, troubleshooting techniques, atomic reference sources, and the need to understand requirements. In addition, he covers circuit detail for the major components in PLLs and synthesis of the loop filter. Other noteworthy books include [5–7].

1.2.2 Articles

Several introductory articles have also been published. The articles authored by Andrzej Przedpelski are an excellent introduction to PLLs. They are:

- "Analyze, Don't Estimate Phase-Lock-Loop," *Electronic Design*, Vol. 26, No. 10, May 10, 1978;
- "Optimize Phase-Lock Loops to Meet Your Needs—Or Determine Why You Can't," *Electronic Design*, Vol. 26, No. 19, September 13, 1978;
- "Suppress Phase-Lock-Loop Sidebands without Introducing Instability," *Electronic Design*, Vol. 27, No. 19, September 13, 1979;
- "Programmable Calculator Computes PLL Noise, Stability," *Electronic Design*, Vol. 29, No. 7, March 31, 1981.

In addition, Gardner's article is extremely useful reading:

- "Charge-Pump Phase-Lock Loops," *IEEE Trans. Comm.*, Vol. COM-28, November 1980, pp. 1849–1858.

The reader should not be deceived by the initial simplicity of PLLs. Understanding PLLs involves a broad range of technical disciplines. Consequently, concentrating efforts on the immediate problem will help save time and reduce frustration. For instance, if the reader is troubleshooting an existing design, he should concentrate on system-response theory, understanding PLL requirements, grounding and shielding, measurement techniques, and troubleshooting techniques. If the reader is designing a PLL, he should concentrate on system architecture, understanding PLL requirements, feedback theory, detailed component design, and grounding and shielding.

1.2.3 Background Books

Full understanding and analysis of PLLs requires proficiency in several engineering subjects. Familiarity with Z-transforms, La Place transforms, Fourier transforms, differential equations, numerical analysis, the C programming language, digital circuits, analog circuits, radio frequency (RF) circuits, control systems, and communications is the recommended background. The following textbooks cover most of the minimum recommended background:

- *Feedback Control Systems* by Charles L. Phillips and Royce D. Harbor;
- *The Fast Fourier Transform* by E. Oran Brigham;
- *Network Analysis* by Van Valkenburg;
- *Analysis and Design of Analog Integrated Circuits* by Paul R. Gray and Robert G. Meyer;
- *Principles of CMOS VLSI Design* by Neil H. E. Weste and Kamran Eshraghian;
- *Principles of Communication Systems* by H. Taub and D. L. Schilling.

1.2.4   Web Sites

Besides books and articles, the Internet provides several locations to find information on PLLs:

- Texas Instruments (TI), high-performance PLLs: http://www.ti.com/sc/docs/products/msp/clock/pll/overview.htm;
- National Semiconductor wireless Web site: http://www.national.com/appinfo/wireless;
- Frequency response analysis and design tutorials: http://me.www.ecn.purdue.edu/~me475/ctm/freq/freq.html;
- Chip directory: http://icat.snu.ac.kr/chipdir/f/pll.htm;
- PLL fundamentals (minicircuits): www.minicircuits.com/appnote/vco15-10.pdf;
- Monolithic complementary metal oxide semiconductor (CMOS) RF transceiver (Berkeley): http://kabuki.eecs.berkeley.edu/rf/rf.html;
- Analog integrated circuit (IC) design, Dr. Hellums of the University of Texas, Dallas: www.utdallas.edu/~hellums;
- PLLs, Stan Goldman: http://home.comcast.net/~sgold_1.

1.3   Loop Classifications

Figure 1.4 shows functional block diagrams for PLL classes. At the top, a pure analog PLL (APLL) uses an analog multiplier for a phase detector [4]. Next, a DLL does not generate a frequency different from [3]. This type corresponds to a digital phase lock loop (DPLL) in some ASIC libraries. Next, the DPLL uses a digital phase/frequency comparator, digital dividers, and analog loop filter and VCO. It is really a hybrid analog/digital loop [2]. This type corresponds to an APLL in some ASIC libraries. The all-digital PLL (ADPLL) does not use any analog components (e.g., resistors, capacitors) [2]. Finally, a software PLL consists of code in a DSP.

1.4   Example Applications

Applications of PLLs include:

1. Frequency multiplier by multiplying the frequency of the reference oscillator;
2. Modulator by adding the modulating signal to the phase error;
3. Demodulator by tracking the changes in modulation to the reference input;
4. Coherent receiver by operating as a narrowband, tunable filter to track the carrier frequency;
5. Data synchronizer by operating as a narrowband tunable filter to recover the clock.

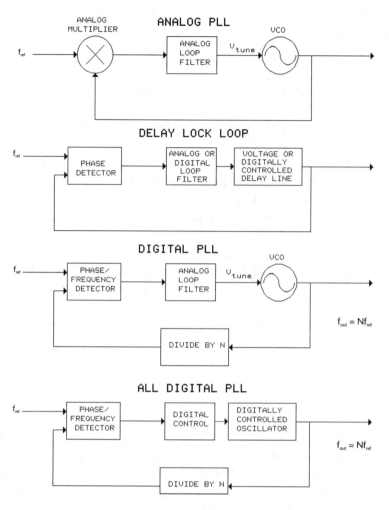

Figure 1.4 Functional block diagrams for PLL classes.

1.4.1 History

From a historical perspective, De Bellescize in 1932 published one of the earliest descriptions of phase lock, which treated the synchronous reception of radio signals (synchronous or homodyne receiver). This receiver consisted of a local oscillator, a mixer, and an audio amplifier. Adjusting the oscillator using phase-lock techniques to the incoming carrier frequency down-converted the incoming signal to 0 Hz at the output of the mixer. Mixing to 0 Hz demodulates the input signal at the output of the mixer. The audio amplifier increases the signal strength to the speaker. For various reasons, the simple synchronous receiver has never been used extensively.

The synchronization of horizontal and vertical scan in television produced the first widespread use of phase lock. A pulse transmitted with the video information signals the start of each line and the start of each interlaced half-frame of a television picture. To reconstruct a scan raster on the TV tube, the pulses, after being stripped off the video, synchronize a pair of free-running relaxation oscillators to trigger a pair of single sweep generators. Consequently, the sweep generators continue to operate even if synchronization is absent. Noise can cause starting time jitter and occasional misfiring far out of phase. Horizontal jitter reduces horizontal resolution and causes vertical lines to have a ragged appearance. Vertical jitter causes an apparent vertical movement of the picture. Phase locking examines the phase between each oscillator and many of its sync pulses and adjusts oscillator frequency so that the average phase discrepancy is small. Because a PLL averages the effects of many pulses, an occasional noise pulse does not disrupt the raster scan.

From earlier applications to those in use today, PLLs have spread to computers, Doppler radar, satellite communications, cellular phones, and telecommunications. In this next section, we will look at the basic operation of these applications. This will be an quick, informal, and nonnumeric overview. The operation of the PLLs in these applications will raise several key issues that need to be answered in studying PLLs.

1.4.2   Doppler Radar

Figure 1.5 graphically depicts a Doppler radar system that uses a PLL. This is a transmit-and-receive application. The Doppler frequency shift gives the information needed to determine vehicle velocity, and the delay time from the object gives the distance to the object. A PLL (synthesizer) provides a low-noise source to transmit off an object and a low-noise reference frequency to determine the Doppler shift. The field of view has a small radar return signal in front of a large radar return signal. Phase noise from the PLL can mask the presence of a small moving target next to a large stationery target.

A PLL has several key requirements for Doppler radar applications. The loop requires low phase noise to maximize resolution between the size of objects. High output frequency also helps in resolving objects. An accurate time base yields accuracy in determining the distance to an object.

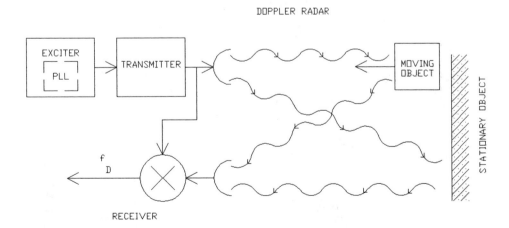

Figure 1.5   Graphical depiction of a PLL application in a Doppler radar system.

1.4.3 Satellite Communications

Figure 1.6 graphically depicts a PLL application in a satellite communications system. Ground station #1 transmits a signal to the processing satellite. The processing satellite demodulates it, remodulates it, and frequency-shifts it to produce a downlink to ground station #2. PLLs are used to frequency-translate the various signals to the proper intermediate frequencies (IFs) in order to downconvert to the demodulator and upconvert to the output transmit frequencies.

A PLL has several key requirements for satellite communications applications. High phase noise from the PLLs will mask out a weak channel when it is surrounded by strong channels. Fast switching speed from the PLL may be needed when switching between channels.

1.4.4 Cellular Phones

Figure 1.7 shows a diagram of a PLL application in a cellular phone system. A mobile telephone communicates with a cell site that connects to the public telephone system. PLLs are used to frequency-translate the IF up to the 869–894-MHz transmit frequency and downconvert the 824–849-MHz receive frequency to the IF (45 MHz) in the cell site. In the mobile telephone, PLLs are used to frequency-translate the IF up to 824 to 849 MHz for transmission and to downconvert the received 869–894-MHz signal to the IF for reception.

A PLL has several key requirements for cellular phone applications. For all applications, the handset must generate high frequencies with low power in order to increase the time before it is necessary to recharge the battery. The PLL consumes

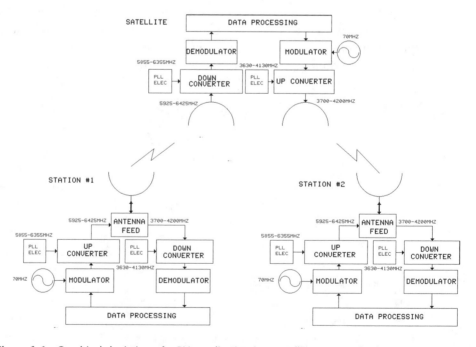

Figure 1.6 Graphical depiction of a PLL application in a satellite communications system.

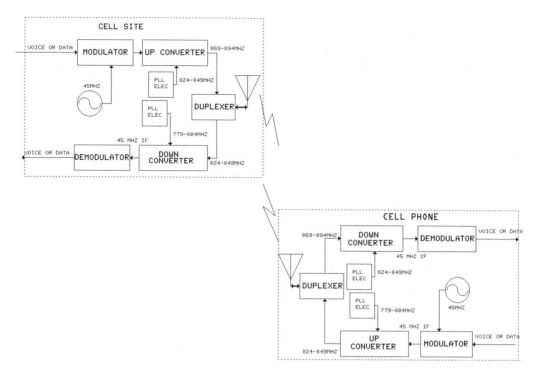

Figure 1.7   Diagram of a PLL application in a cellular phone system.

among the largest amounts of power because it has high-frequency circuitry. For nonspread-spectrum formats, high phase noise from the PLLs will mask out a weak channel when it is surrounded by strong channels. In spread-spectrum formats, high phase noise will limit the number of channels that can be on at the same time. Fast switching speed from a PLL is needed for spread-spectrum formats in order to hop between frequencies in a wide-enough band to make the spread spectrum effective.

1.4.5   Telecommunications Systems

Figure 1.8 shows a diagram of a PLL application in a telecommunications system. PLLs are used at the head end to frequency-translate cable, digital video, and telephony downstream in several 6-MHz channels from 50 to 750 MHz. At the set-top box in the house, a loop downconverts the signals to a processor that sends a video signal to a TV and telephony to a phone or computer. For interactive video and telephony, the set-top box frequency-translates the signal to the upstream central office in the 10–50-MHz band. At the head end, PLLs translate the interactive video and telephony for processing.

A PLL has several key requirements for telecommunications applications. Low phase noise is required to prevent excess noise from getting into adjacent channels and to maximize picture quality. Fast acquisition speed is required to lock a data packet preamble quickly. Narrow bandwidths are required to reject data transitions.

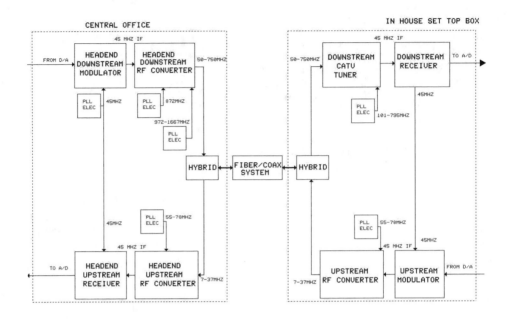

Figure 1.8   Diagram of a PLL application in a telecommunications system.

To summarize, this chapter presented background information to understand how a loop operates and gave reference material for further study of loops. Key signals of interest within a PLL and the major components were identified to help monitor these signals or major components on the test bench or in simulations. Key design requirements were identified so that we will be able to identify these requirements from the trivial many that can be found in statements of work or specifications. A history of PLLs was presented to show the progression from the first implementations with large racks of equipment to today's tiny monolithic versions. Finally, applications of PLLs were presented that showed the many applications in modern systems. This background material will provide help in understanding the control-systems analysis in the next chapter and provides references to which readers can turn for further explanation of PLLs or to refresh their knowledge of the fundamentals necessary for understanding the material in this book.

Questions

1.1 Name the three major components in a PLL.

1.2 Explain how a PLL operates.

1.3 What are the four key signals of interest in a PLL?

1.4 What are the three operational states of a PLL?

1.5 What are the four key design considerations for monolithic applications?

1.6 What are the four PLL classes?

1.7 What are the four application functions of PLL?

1.8 Name some applications of PLLs.

References

[1] Gardner, F., *Phaselock Techniques*, New York: Wiley Interscience, 1979.

[2] Best, R., *Phase-Locked Loops: Design, Simulation, and Applications*, 3rd ed., New York: McGraw-Hill, 1997.

[3] Razavi, B., *Monolithic Phase-Locked Loops and Clock Recovery Circuits*, New York: IEEE Press, 1996.

[4] Manassewitsch, V., *Frequency Synthesizers: Theory and Design*, 3rd ed., New York: Wiley Interscience, 1987.

[5] Crawford, J., *Frequency Synthesizer Design Handbook*, Norwood, MA: Artech House, 1994.

[6] Egan, W., *Frequency Synthesis by Phase Lock*, New York: Wiley Interscience, 1981.

[7] Rohde, U., *Digital PLL Frequency Synthesizers: Theory and Design*, Englewood Cliffs, NJ: Prentice-Hall, 1983.

CHAPTER 2

System Analysis

Control-systems theory and analysis of PLLs is covered in this chapter. This chapter could have been combined with the Chapter 3 on system requirements; however, this chapter was separated to emphasize the importance of system analysis. In addition, system analysis gives a deeper understanding of how a PLL works, building upon the material covered in Chapter 1. A general system concept of how PLLs work will help in understanding the following chapters.

First, VCO and phase detector mathematical relationships are discussed. Then, Bode plot analysis of PLLs is studied so that stability requirements can be designed. The relationship between step response and stability is studied in order to recognize the stability of a loop and its measured step response. Error tracking of a PLL to various stimuli is studied so that the correct loop architecture can be selected to meet tracking requirements. Finally, a simple charge pump synthesis example of a loop filter is studied to begin to understand the relationship between system analysis and component-value synthesis to meet the system requirements. Understanding control systems helps an engineer build stable and robust loops.

2.1 VCO Mathematical Description

We begin with a mathematical description of a VCO so that it can be used in the analysis of the loop. Understanding this relationship will help with the linear analysis of the loop. We will study the mathematical description of a phase-modulated signal because that is what occurs in a VCO. Consequently, (2.1) gives a mathematical description of a phase-modulated signal, from [1, p. 117] and modified with VCO terminology:

$$V_o(t) = V_a \cos\left(\omega_c t + K_v \int V_{tune}(t) \, dt\right) \tag{2.1}$$

The output voltage amplitude is ignored in PLL analysis, and we concentrate our attention on the argument of the cosine function, which is the time variation of phase. Differentiating the terms in the argument produces (2.2) for the instantaneous frequency:

$$\omega = \frac{d}{dt}\left(\omega_c t + K_v \int V_{tune}(t) \, dt\right) = \omega_c + K_v V_{tune}(t) \tag{2.2}$$

Next, we want to analyze the loop as a small change from the lock condition. Consequently, the center frequency of the VCO, ω_c, must equal the input reference frequency, ω_{ref}. Then, our analysis is only concerned with modulation of the VCO away from that center frequency, which is the deviation of instantaneous frequency from that locked condition. Equation (2.3) gives the deviation of instantaneous frequency from the locked frequency by subtracting the center frequency from (2.2):

$$f_o = \frac{\omega - \omega_c t}{2\pi} = \frac{K_v V_{tune}(t)}{2\pi} \tag{2.3}$$

Equation (2.4) gives the instantaneous phase, which is the argument of the cosine function in (2.1):

$$\theta_{oi} = \omega_c t + K_v \int V_{tune}(t)\, dt \tag{2.4}$$

Equation (2.5) gives the deviation of instantaneous phase from the lock condition by the same procedure that was done for (2.3), followed by taking the La Place transform:

$$\theta_o = \theta - \omega_c t = K_v \int V_{tune}(t)\, dt = K_v \frac{1}{s} V_{tune}(s) \tag{2.5}$$

Equation (2.5) shows the origin of the $1/s$ term that will be used in the mathematical model of the loop, and it shows the origin of this ideal integrator. This is the relationship that will be used in the linear analysis.

2.2 Phase Detector Mathematical Relationship

Next, we study the mathematical relationship of the phase detector. A phase detector is nothing more than a simple analog multiplier, which is a mixer. Consequently, we mix two signals to show the resulting relationship. First, we have to describe these two input signals mathematically. Equations (2.6) and (2.7) mathematically describe the general signal inputs to the mixer:

$$V_1(t) = V_{p1} \cos(\omega_{rf} t + \theta_e) \tag{2.6}$$

where

$V_1(t)$ = source 1 signal;

V_{p1} = maximum amplitude of source 1 (V);

ω_{rf} = angular frequency of a signal at the radio frequency (RF) port of the mixer (rad/s);

$\omega_{rf} = 2\pi f_{rf}$

θ_e = phase-error difference between signal 1 and 2 (rad);

t = time variable (sec).

$$V_2(t) = V_{p2} \cos(\omega_{lo} t) \tag{2.7}$$

where

$V_2(t)$ = source 2 signal;

V_{p2} = maximum amplitude of source 2 (V);

ω_{lo} = angular frequency of a signal at the local oscillator (LO) port of a mixer (rad/s).

Mixing (analog multiplication) (2.6) and (2.7) produces (2.8):

$$V_1(t) V_2(t) = V_{p1} V_{p2} \cos(\omega_{rf} t + \theta_e) \cos(\omega_{lo} t) \tag{2.8}$$

Using the trigonometric identity for products of a trigonometric function produces (2.9):

$$V_1(t) V_2(t) = V_{p1} V_{p2}\, 0.5[\cos(\omega_{rf} t - \omega_{lo} t + \theta_e) + \cos(\omega_{rf} t + \omega_{lo} t + \theta_e)] \tag{2.9}$$

where

$V_1(t) V_2(t)$ = mixing process.

Eliminating the high-frequency product with a lowpass filter yields (2.10):

$$V_1(t) V_2(t) = V_{p1} V_{p2}\, 0.5[\cos(\omega_{rf} t - \omega_{lo} t + \theta_e)] \tag{2.10}$$
$$= V_{pbeat} \cos(\omega_{beat} t + \theta_e)$$

where

$\omega_{beat} = \omega_{rf} - \omega_{lo}$ for $\omega_{rf} > \omega_{lo}$;

$V_{pbeat} = V_{p1} V_{p2} \times 0.5 \times$ mixer losses;

= the resulting voltage level after mixing (V).

The derivative of (2.10) calculates the incremental phase slope. For the mixer operating with a 0-beat frequency ($\omega_{beat} = 0$, which is dc) and 90° phase shift, the derivative of (2.10) produces (2.11) and (2.12):

$$V_{pds} = \frac{d}{d\theta_e} V_{pbeat} \cos(\theta_e) = V_{pbeat} \sin(\theta_e) \tag{2.11}$$

where

$V_{pds}(\phi)$ = phase detector phase slope (V).

$$V_{pds}(\theta_e) = V_{pbeat} \sin(\theta_e) \tag{2.12}$$

For a phase error θ_e equal to $\pi/2$ rad (quadrature) in (2.12), the phase slope V_{pds} equals the peak resulting voltage V_{pbeat} and the gain of the phase detector (V/rad), $K_d = V_{pbeat}$. For θ_e equal to 0 rad in (2.12), K_d equals 0 V/rad. This shows that maximum phase sensitivity occurs for a 90° phase difference between the input signals, while a minimum phase sensitivity of 0 occurs for a phase difference of 0°. With a 0-Hz beat frequency in (2.10), adjusting the phase to 90° phase difference produces 0V at the intermediate frequency (IF) port of the mixer and gives maximum phase sensitivity for a measurement. Adjusting the phase to 0° phase difference produces a maximum voltage and gives minimum phase sensitivity for a measurement. The terms LO, RF, and IF are from receiver terminology, where the LO is the oscillator in the receiver, the RF is the signal from the antenna, and the IF is the down-converted frequency received, which, in this case, is baseband (dc).

Figure 2.1 shows the sinusoidal phase detector transfer function to a phase error θ_e with a 5-MHz reference frequency. This figure illustrates that phase can be given in the units of time (20 ns for one period), cycles, phase in degrees, or phase in radians. To prevent errors, it is simplest to do everything in radians and at the end convert to the units of interest. Furthermore, Figure 2.1 shows that the phase detector slope is maximized at $\pi/2$ and that is assumed to be the operating point for the linear analysis when the loop is locked. Notice that the phase range is restricted to ±90° and that the other points on the negative slope of the curve cannot be locked conditions. One reason is that the negative slope turns the negative feedback into positive feedback.

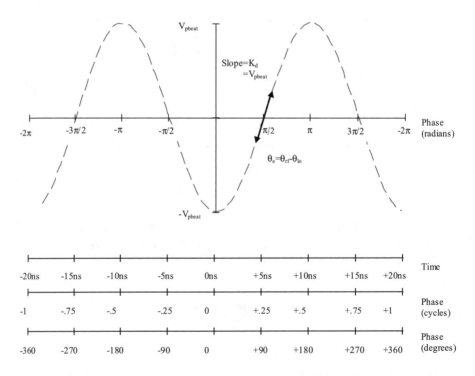

Figure 2.1 The sinusoidal phase detector transfer function and the different measures of phase error.

2.3   PLL Transfer Function and Control-Systems Theory

With the mathematical description of the VCO and phase detector completed, we can now analyze the control systems of the PLL. Figure 2.2 shows the general control-systems block diagram for a PLL.

The following control-system equation, which results from evaluating Figure 2.2, shows the general equation for a closed-loop transfer function [2]:

$$\frac{C_o}{R_i}(s) = \frac{G(s)}{1 \pm G(s)H(s)} \tag{2.13}$$

where

$$- = \text{for positive feedback;}$$

$$+ = \text{for negative feedback;}$$

$$\frac{C_o}{R_i} = \text{the closed-loop transfer function;}$$

$$G(s) = \text{forward transfer function;}$$

$$H(s) = \text{feedback transfer function;}$$

$$G(s)H(s) = \text{open-loop transfer function;}$$

$$G(s)H(s) = \text{ratio of 1 and angle of } 0° \text{ for positive feedback and } 180°$$
$$\text{for negative feedback; these are the conditions for oscillation.}$$

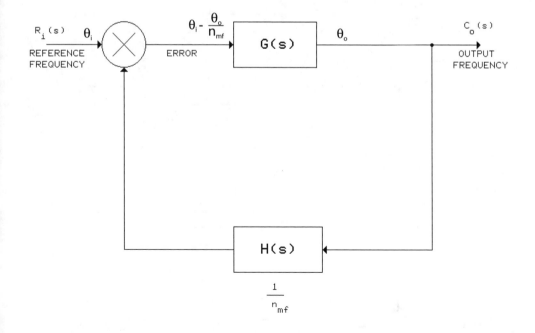

Figure 2.2   General PLL block diagram.

Equation (2.14) expresses the closed-loop transfer function for a PLL from the above general equation for a control system in terms of phase, which is the control variable in a PLL [3]:

$$\frac{\theta_o}{\theta_i}(s) = \frac{G(s)}{1 + G(s)H(s)} \qquad (2.14)$$

where

θ_o = output phase (rad);

θ_i = input phase (rad).

The open-loop gain $G(s)H(s)$ term determines the stability of a PLL. If $G(s)H(s)$ ever becomes equal to -1, then (2.14) becomes unstable. Therefore, the circuit components in the $G(s)H(s)$ function determine the stability of the PLL. Consequently, the parameters that define the $G(s)H(s)$ function must be chosen carefully. Besides stability, the design of the control system determines the step response and error tracking of the system.

Figure 2.3 shows the frequency step responses for various phase margins and the significant difference in effect on the response of the loop. The reference frequency for the loop is 32 kHz. The horizontal axis is time in seconds, and the vertical axis is frequency in hertz. This figure shows the effects of the stability

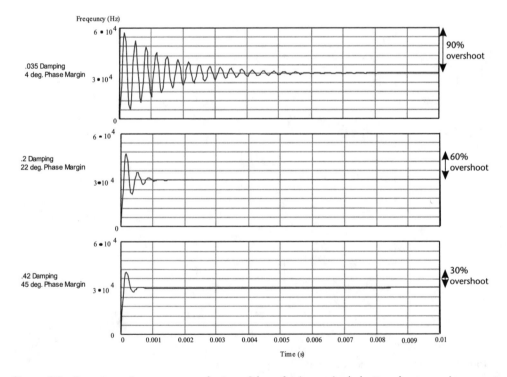

Figure 2.3 Frequency step responses of a type 2 loop for increasingly better phase margins.

margin on loop performance. Lower stability produces undesirable ringing and longer settling time. A damping factor of 0.42 has 1 peak and 1 undershoot. A damping factor of 0.2 has 3 peaks, and a damping factor of 0.035 has more than 18 peaks. Look at how much the first overshoot peak changes with the damping factor, and for 0.035 damping factor, it is 100%. Later on, this characteristic will be used to help us determine stability.

Next, Figure 2.4 shows a type 2 loop error-tracking response for step, ramp, and parabolic phase stimulus. The horizontal axis is normalized time to ω_n, and the vertical axis is normalized magnitude. The input stimulus is shown in dashed lines. The loop tracking response is a solid line. The loop has a damping factor of 0.1, and the error is amplified in the ramp and parabolic responses so that the tracking error can be easily identified. Notice for a type 2 loop, the parabolic response has a constant error that lags behind the input stimulus.

Figure 2.5 adds transfer functions for the phase detector, loop filter, VCO, and programmable divider to the control-systems block in Figure 2.2. These mathematical relationships allow control-systems analysis to be done. From studying Figure 2.5, the component transfer functions that are contained in the open-loop gain can be identified as shown in (2.15):

$$G(s)H(s) = \text{(phase detector gain)(filter transfer function)} \qquad (2.15)$$

$$\text{(VCO transfer function)(divider transfer function)}$$

Figure 2.6 shows a type 2 second-order loop with an active loop compensation that will be used to illustrate the analysis technique. This circuit configuration is generally used in frequency-generation applications. Substituting component transfer functions into (2.15) from Figure 2.6 produces (2.16):

$$G(s)H(s) = \frac{K_d K_v}{n_{mf} C R_1} \left(\frac{1}{s^2}\right)(sCR_2 + 1) \qquad (2.16)$$

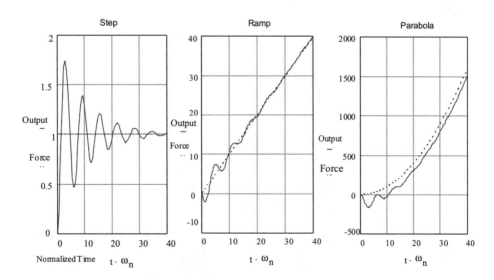

Figure 2.4   Error tracking of a type 2 loop for step, ramp, and phase stimulus.

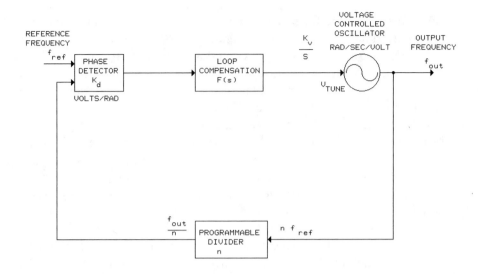

Figure 2.5   PLL component block diagram.

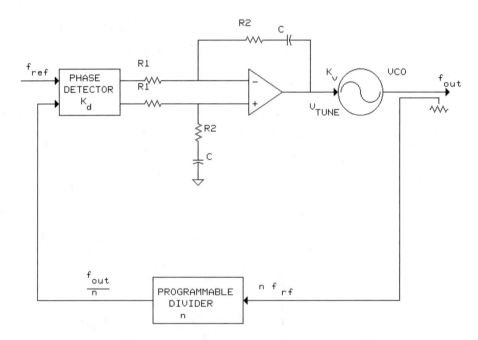

Figure 2.6   Type 2 PLL component block diagram with active loop compensation.

Equation (2.16) computes the open-loop-gain function. Substituting (2.16) into (2.14) produces (2.17), which computes the closed-loop transfer function:

$$\frac{G(s)}{1 + G(s)H(s)} = \frac{\dfrac{K_d K_v}{CR_1}(sCR_2 + 1)}{s^2 + s\left(\dfrac{K_d K_v}{n_{mf}} \dfrac{R_2}{R_1}\right) + \dfrac{K_d K_v}{n_{mf}CR_1}} \qquad (2.17)$$

where

K_d = phase detector gain (V/rad);

K_v = VCO transfer function gain constant (rad/s/V);

n_{mf} = integer divider value;

n_{mf} = loop frequency multiplication factor;

n_{mf} = output frequency/input frequency;

   C = capacitor in the feedback path of the operational amplifier (F);

R_1 = resistor at the negative input terminal of the operational amplifier (Ω);

R_2 = resistor in the feedback path of the operational amplifier (Ω).

2.4   Error Tracking

After all transients have died, remaining steady-state error tracking provides another measure of the performance of a feedback-control system. This parameter measures the amount of glue that keeps the loop locked. Tight tracking means it is more difficult for a disturbance to unlock the loop. Loose tracking means a slight disturbance can cause the loop to lose lock. To understand this parameter, we have to start by studying the control-system error function and apply the final-value theorem. Equation (2.18) computes control-system error:

$$E(s) = \frac{R(s)}{1 + G(s)H(s)} \tag{2.18}$$

   Applying the final-value theorem of La Place transforms to the error equation with the input forcing function computes the steady-state error. Equation (2.19) shows the final-value theorem that needs to be applied to compute steady-state error:

$$\lim_{t \to \infty} \epsilon(t) = \lim_{s \to 0} sE(s) \tag{2.19}$$

For a stable feedback-control system, (2.20) computes the steady-state error:

$$E_{ss} = \lim_{s \to 0} sE(s) \tag{2.20}$$

The amount of remaining error for step, ramp, and parabolic forcing functions determines the difference in performance for different PLL circuits. Table 2.1 analytically presents these functions.

   Table 2.2 shows the resulting steady-state error for type 1 and 2 PLL control systems with step, ramp, and parabolic input forcing functions [4]. The remaining steady-state error also distinguishes the responses for classification by type of a

Table 2.1 System Forcing Functions

Function	Mathematical Expression (Time Domain)	La Place Transform (Frequency Domain)
Step	A	A/s
Ramp	vt	v/s^2
Parabolic	$1/2at^2$	a/s^3

control system as shown in Table 2.2. A type 0 system has no poles at the origin, a type 1 system has one pole at the origin, a type 2 system has two poles at the origin, and so on. The number of poles at the origin in the open-loop gain determines the tracking performance of a PLL in the face of increasingly disturbing forcing functions, as shown in Table 2.2. The highest polynomial exponent in the denominator of the closed loop defines the order of the loop. A type 2, order 2 PLL means there are two integrators at the origin, and both of those integrators contribute to the order of the loop in the polynomial in the denominator. A type 2, order 3 PLL means there are two integrators at the origin, and another pole (usually a high-frequency pole) adds an additional order to the polynomial in the denominator. The biggest effect that these higher-order poles have on tracking is their reduction of the damping factor.

The next section analyzes the effects of the circuit components in the $G(s)H(s)$ function on stability. Equation (2.16), the open-loop transfer function, is organized into terms that can be expressed easily by using Bode plot graphic techniques. The first term computes a gain constant; the other terms compute Bode plot functions of frequency. The two poles at the origin compute two integrators, which force the error function closer to zero with decreasing frequency. Equation (2.16) is from [5, 6].

2.5 Type 2, Second-Order Active Loop–to–Servo Terminology

The most common designs of a PLL depend on the loop performance requirements of loop bandwidth, output frequency (n_{mf} · reference frequency), and stability (damping factor). The loop performance requirements and (2.16) (open-loop transfer function) and (2.17) (closed-loop transfer function) determine the circuit values. To ease computations, (2.17) is converted to (2.21) using servo terminology [7]:

$$\frac{G(s)}{1 + G(s)H(s)} = \frac{n_{mf}(\omega_n)^2\left(s\dfrac{2\zeta}{\omega_n} + 1\right)}{s^2 + s(2\zeta\omega_n) + (\omega_n)^2} \tag{2.21}$$

Equation (2.21) allows classical second-order differential equation solutions to be applied. Consequently, decisions about desired step responses and error tracking can be eased because of familiarity with these solutions. In addition, these solutions include solving the quadratic equation for the roots of the denominator. With servo terminology, the solution has a simple format where the roots are $\omega_n[-\zeta \pm (\zeta^2 - 1)^{0.5}]$. Finally, (2.21) gives a compact mathematical description of

Table 2.2 Residual Phase Errors by Loop Type and Forcing Function [4]

System Type	Step Input	Ramp Input	Parabolic Input
1	$\lim_{s \to 0} \dfrac{A}{s} s \dfrac{s}{s + K_v K_d} = 0$	$\lim_{s \to 0} \dfrac{v}{s^2} s \dfrac{s}{s + K_v K_d} = \dfrac{v}{K_v K_d}$	$\lim_{s \to 0} \dfrac{a}{s^3} s \dfrac{s}{s + K_v K_d} = \infty$
1	$\lim_{s \to 0} \dfrac{A}{s} s \dfrac{s^2}{s^2 + s(2\zeta\omega_n) + (\omega_n)^2} = 0$	$\lim_{s \to 0} \dfrac{v}{s^2} s \dfrac{s^2}{s^2 + s(2\zeta\omega_n) + (\omega_n)^2} = 0$	$\lim_{s \to 0} \dfrac{a}{s^3} s \dfrac{s^2}{s^2 + s(2\zeta\omega_n) + (\omega_n)^2} = \dfrac{a}{(\omega_n)^2}$

the loop with only three variables (ω_n, n_{mf}, and ζ) versus the six variables in (2.17).

Equating (2.17) (closed-loop transfer function with component variables) and (2.21) (closed-loop transfer function with servo terminology) produces (2.22) and (2.23) for the servo terms of natural angular frequency and damping factor [8]:

$$\omega_n = \sqrt{\frac{K_d K_v}{R_1 C n_{mf}}} \qquad (2.22)$$

$$\zeta = \frac{R_2 C}{2} \sqrt{\frac{K_d K_v}{R_1 C n_{mf}}} \qquad (2.23)$$

The natural angular frequency determines the switching rise and fall time (10%–90%) response, and the damping factor determines the amount of stability or, in other words, the margin from the point of instability.

2.6 Loop Stability: Bode Plot Analysis

The phase response of the open-loop gain $G(s)H(s)$ determines the stability of a PLL. If the phase angle equals 180°, and the magnitude equals unity, then oscillations will occur. The frequency point where the magnitude of the open-loop transfer function equals unity defines the 0-dB crossover frequency. The 0-dB crossover frequency approximately equals the closed-loop natural frequency of the PLL for damping factors less than 0.6.

The phase-angle value where the magnitude of the open-loop transfer function equals unity determines stability. From this intersection, two stability margins can be defined from the Bode plot of the open-loop gain in Figure 2.7. First, the number of phase-angle degrees greater than −180° (i.e., instability point) defines the *phase margin*. Next, the amount of gain increase to cause the unity gain crossover frequency to increase to a −180° phase angle defines the *gain margin*. Comfortable stability margins are greater than 30° for phase and 10 dB for gain [4].

Servo terminology relates damping factor and natural frequency to the open-loop-gain Bode plot by (2.24) and (2.25) [14]:

$$\theta_{\text{margin}} = \text{atan}\left(2\zeta\sqrt{2\zeta^2 + \sqrt{4\zeta^4 + 1}}\right) \qquad (2.24)$$

$$f_x = \frac{\omega_n}{2\pi}\sqrt{2\zeta^2 + \sqrt{4\zeta^4 + 1}} \qquad (2.25)$$

These equations allow design trade-offs to be evaluated between time response and stability margins.

First, let's study the relationship of phase margin to damping factor. Equation (2.24) allows phase margin to be computed from damping factor, but the reverse is more challenging. Consequently, let's plot the relationship. Figure 2.8 plots the phase margin versus damping factor in (2.24) to show the relationship of the two

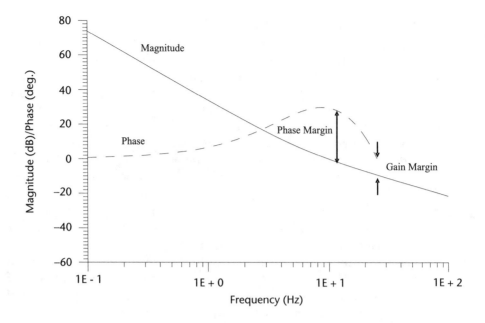

Figure 2.7   Bode plot showing phase and gain margin.

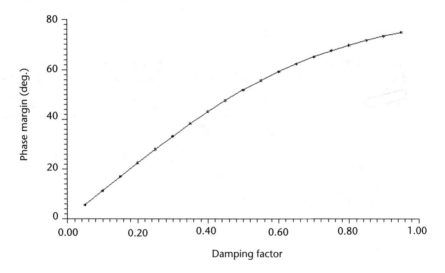

Figure 2.8   Relationship of phase margin to damping factor.

parameters. The vertical axis is phase margin in degrees and the horizontal axis is damping factor, which is a ratio with no units.

A damping factor of 0 corresponds to a 0° phase margin. A damping factor close to 1 has approximately 80° phase margin. Damping factors greater than 1 are not usually seen because other factors (e.g., higher-frequency poles) subtract enough from the phase margin to prevent these higher damping factors. Also, you can see an exponential approach to 90° with greater damping factors changing the phase margin only slightly. Finally, for phase margins less than 40° 100 × damping factor ≈ the phase margin.

Now, let's study the relationship of natural frequency to 0-dB crossover frequency. Plotting the equation makes it easier to understand this relationship. Figure 2.9 plots the natural frequency normalized to a 0-dB-crossover-frequency-point-versus-damping-factor curve in (2.35). For damping factors less than 0.5, the natural angular frequency almost equals the 0-dB crossover frequency. For a damping factor approximately equal to 0.9, the 0-dB crossover frequency is two times the natural frequency. Also, this figure shows that the frequency of oscillation for a damping factor equal to 0 defines the natural frequency of the loop in (2.21).

Figure 2.10 graphically shows the relationship between natural frequency, 0-dB crossover point, and the 3-dB bandwidth. All of these points can be described by the damping factor and the natural frequency, which makes a compact mathematical description for the PLL that can be used in complicated analysis. For the same natural frequency, adjusting the value of the damping factor toward zero moves the zero location to higher frequencies and the 0-dB crossover point to the natural frequency. Adjusting the damping factor to higher values moves the zero frequency location to lower frequencies and moves the 0-dB crossover point to higher frequencies. For the same damping factor, adjusting the natural frequency to higher or lower frequencies moves the zero location point, the 0-dB frequency crossover point, and the 3-dB bandwidth relative to the natural frequency. Consequently, the phase margin remains the same.

2.7 Loop Stability: Root-Locus Analysis

Root locus is another method for studying stability. The denominator of the closed-loop transfer function, which is the characteristic equation, is set to 0 and solved

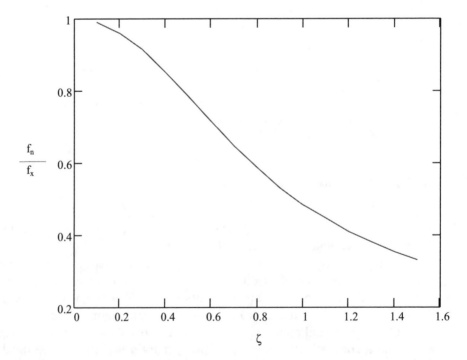

Figure 2.9 Plot of natural frequency over 0-dB crossover-point ratio versus damping factor.

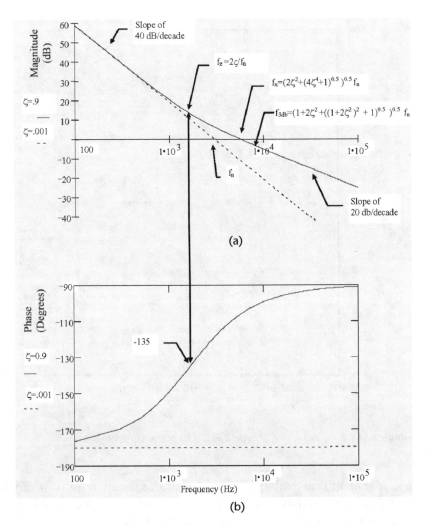

Figure 2.10 Relationship between natural frequency, 0-dB crossover point, and 3-dB bandwidth with (a) magnitude and (b) phase plots.

for its roots. Varying the VCO gain from zero to infinity and plotting the roots produces a root-locus plot as shown in Figure 2.11 for a type 2 loop. Poles that go into the right half-plane have unstable operating conditions. In Figure 2.11, the loop is stable for all gains because it is a type 2, second-order loop. If another high-frequency pole were added, then there would be a stability problem, and there would be conditions where the poles would go into the right-hand plane.

Besides stability, the location of the poles in the root locus determine the time-domain response. By observing the pole locations, we can interpret the time-domain response. The contours of constant ω_n are circles in the s plane. The contours of the constant damping ratio are straight lines through the origin. The contours of constant damping $\zeta\omega_n$ are straight lines parallel to the $j\omega$ axis. Constant damping $\zeta\omega_n$ is the exponent multiplied by time in the time-domain solution for a damped sinusoid. Faster damping occurs at minus infinity on the negative real axis, and slower damping occurs toward zero. Contours of constant frequency ω_n sqr$(1 - \zeta^2)$ are

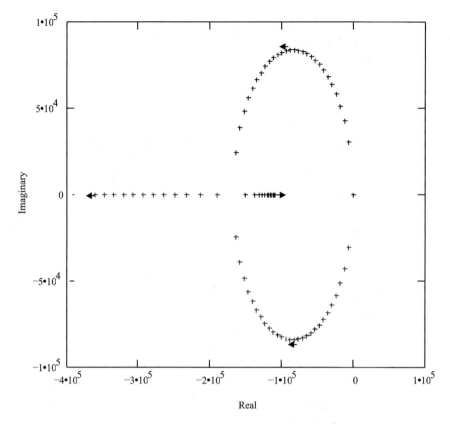

Figure 2.11   Root-locus plot of type 2 PLL with active compensation.

lines parallel to the real axis. The constant frequency term is the argument of the cosine or sine function in the time-domain solution for a damped sinusoid. Higher frequency occurs for larger values on the $j\omega$ axis.

Now that we have reviewed root locus, let's interpret the meaning of the PLL plot in Figure 2.11. Both poles originate at the origin for low gain. With increasing gain, the poles become complex and form a circle with the circle centered at $-1/\tau_2$ ($\tau_2 = R_2 C$) with a $1/\tau_2$ radius. Damping is small for small gain and increases with increasing gain. With sufficiently high gain, the locus eventually returns to the negative real axis, and the loop is overdamped. One branch of the locus terminates at the finite zero; the other terminates at infinity.

A disadvantage of root locus is that no direct laboratory measurements measure the pole locations. Furthermore, the circuits do not allow large enough variations of the VCO gain (from zero to infinity or even a large number) without losing lock. Consequently, the root locus is derived from theoretical data and does not lend itself to verification, while Bode plot analysis does.

2.8   Charge Pump Synthesis Example of Loop-Component Values

The following is a charge pump example to illustrate one of the synthesis techniques with one of the most common architectures in integrated circuits. First, the loop-compensation architecture is selected. In this case, charge pump compensation, as shown in Figure 2.12, is selected because the application is for an integrated circuit.

Next, the limitations of the components in this PLL architecture are explored. The VCO gain, the divide ratio, the charge pump gain, and the reference frequency in Table 2.3 are usually given by the components chosen or by specifications. Consequently, they either cannot be adjusted easily or have a narrow adjustment range.

Next, Table 2.4 shows the control-system specifications requirements that give the desired response for stability and switching speed. The natural frequency should be selected to be at least a factor of 10 less than the reference frequency to avoid sampling-delay effects. Higher bandwidths can be achieved but not with the approach that is being used.

Consequently, the capacitor and resistor values are all that remain to synthesize the loop compensation. With divide ratio, damping factor, and natural frequency selected, and given K_p and K_v, (2.26) and (2.27) compute the final C and R_2 component values:

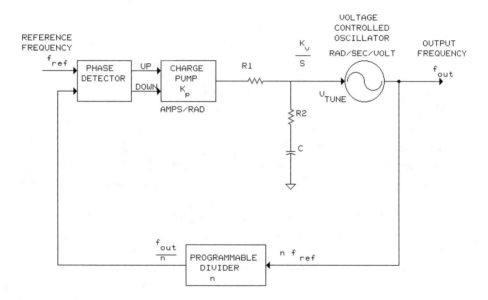

Figure 2.12 Charge pump compensation PLL.

Table 2.3 Charge Pump Example Given Component Values

f_{ref} (Hz)	N	K_v (rad/s/V)	K_p (A/rad)
80E6	1	400E6	0.0034

Table 2.4
Desired Control-
System Parameters

ζ	f_n
0.9	425 kHz

$$C = \frac{K_v K_p}{2\pi(\omega_n)^2 N} \tag{2.26}$$

$$R_2 = 2\frac{2\zeta\omega_n \pi N}{K_v K_p} \tag{2.27}$$

Table 2.5 shows the calculated values. An additional capacitor (C_2) across the VCO tune line to ground is used for additional filtering of reference sidebands, and its value is usually one-tenth the value of C.

The resistor and capacitor values have some practical limitations that can require compromises in performance in order to get practical values. For instance, capacitor values above 1 microfarad are electrolytic. Electrolytic capacitors are not recommended for filters. Therefore, a 1-microfarad capacitor or less will make the circuit component nonelectrolytic and insure excellent frequency-response characteristics. In the above example, all the values are realizable.

Discussion of other approaches and architectures is beyond the scope of this section and will be taken up in Chapter 6; however, this example does show the tight relationship between design approach and system specifications that requires the designer to identify the most important specifications from the trivial many.

2.9 Summary

VCO and phase detector mathematical relationships were discussed. Then, Bode plot analysis of PLLs was studied. The relationship between step response and stability was established. Error tracking of a PLL to various stimuli was studied. Finally, a simple charge pump synthesis example of a loop filter was done.

This chapter made several important points about control systems in PLLs. First, stability and bandwidth can be determined by frequency step response. Next, the type of loop determines how closely the loop tracks the input and the loop's sensitivity to disturbances. Next, synthesis of loop-component values can be done from servo terminology variables, which reduces the complexity of the mathematics. Finally, Bode plot analysis can be used to determine stability. Understanding these concepts about control systems helps an engineer build stable and robust loops. This chapter introduced the relationship between system analysis and component-value synthesis in meeting system requirements. The next chapter will study more system requirements that effect component selection and component-value synthesis.

Table 2.5 Component Values Calculated from the Charge Pump Example

C	R_2	C_2
3E–8	22	0.3E–8

Questions

2.1 For a damping factor of 0.707, what kind of response (overdamped, underdamped, or critically damped) will occur for a frequency step and why?

2.2 What phase margins are recommended in the literature for a stable loop?

2.3 What do *type* and *order* mean in a PLL?

2.4 What type of loop would you recommend to track a sweeping input frequency?

2.5 What conditions must be met to cause oscillation in a PLL?

2.6 What three variables are necessary to describe a loop?

2.7 What three design variables specify a PLL?

2.8 What term describes the frequency of oscillation with a damping factor of zero?

2.9 What are the input and output variables of a PLL?

2.10 Name two methods for determining the stability of a PLL?

2.11 What is wrong with the relationship of the following equation: $\theta_o/\theta_i = \omega_o/\omega_i$?

2.12 For a 32-kHz-input to 16.384-MHz-output PLL with charge pump compensation, and ignoring sampling-delay effects, compute the component values R_2 and C for a natural frequency of 300 Hz and a damping factor of 0.5, given 400E6-rad/s/V VCO gain, $N = 512$, and a 3.4-ma/rad phase detector/charge pump gain.

2.13 For a 10-MHz-input to 180-MHz-output PLL with charge pump compensation, and ignoring sampling-delay effects, compute the component values R_2 and C for a natural frequency of 145.1 kHz and a damping factor of 2, given 1.88E9-rad/s/V VCO gain, $N = 18$, and a 5-μa/rad phase detector/charge pump gain.

2.14 For a 10-MHz-input to 40-MHz-output PLL with charge pump compensation, and ignoring sampling-delay effects, compute the component values R_2 and C for a natural frequency of 1.125 MHz and a damping factor of 1.061, given 6.283E9-rad/s/V VCO gain, $N = 4$, and a 20-μa/rad phase detector/charge pump gain.

2.15 For a 2-MHz-input to 128-MHz-output PLL with charge pump compensation, and ignoring sampling-delay effects, compute the natural frequency, damping factor, and 0-dB crossover frequency point, given 300-MHz/V VCO gain, a 800-μa/rad phase detector/charge pump gain, and the component values $R_1 = 0$, $R_2 = 10$, and $C = 1,000$ nF.

References

[1] Taub, H., and D. L. Schilling, *Principles of Communication Systems*, New York: McGraw-Hill, 1971.

[2] Goldman, S. J., *Phase Noise Analysis in Radar Systems*, New York: Wiley Interscience, 1989.

[3] Przedpelski, A. B., "Analyze, Don't Estimate, Phase Lock Loop Performance," *Electronic Design*, May 10, 1978.

[4] Manassewitsch, V., *Frequency Synthesizers: Theory and Design*, 2nd ed., New York: Wiley Interscience, 1987, pp. 237–293.

[5] Gardner, F. M., *Phaselock Techniques*, New York: Wiley Interscience, 1966, pp. 7–16.

[6] Gorski-Popiel, J., *Frequency Synthesis: Techniques and Applications*, New York: IEEE Press, 1975, pp. 32–39.

[7] Robins, W. P., *Phase Noise in Signal Sources*, London, U.K.: Peter Peregrinus, 1982.

[8] "Phase-Frequency Detector MC4344-MC4044," MTTL Complex Functions, Motorola Semiconductor Products, Inc., Data Sheet, 1973.

CHAPTER 3

System Requirements

This chapter discusses system requirements for a PLL. Noise basics, phase noise, time-domain response, acquisition, jitter, and spurious signals are the system requirements that are studied in detail. In the previous chapter, stability and feedback analysis were discussed, which are also system requirements; however, they are in a chapter by themselves because of their importance and because these requirements are fundamental to understanding how a PLL works.

First, noise basics are discussed in order to lay a foundation for analyzing noise requirements and their relationship to circuit design. Next, phase-noise relationships in oscillators are discussed to give background for analyzing noise and jitter in oscillators. Linear time-domain responses are studied to give an engineer an understanding of the relationship between loop variables, faster switching time, and tracking error. In addition, studying these responses allows damping factor and natural frequency to be determined from measured or simulated step responses. Next, nonlinear acquisition is studied to provide understanding of loop responses to various external stimuli. Understanding these responses will help in understanding loop responses that seem to be inconsistent. Then, the relationships of PLL parameters to jitter are studied to show how to improve timing margin in digital circuits. For synchronizing circuits, reducing jitter will improve the bit-error rate (BER) of the interface. Finally, spurious signals are studied because these signals cause jitter and reduce the precision of the output frequency. Spurious signals at the output from the reference frequency are the largest spurious signals in a loop. Feed-through from the phase detector output to the VCO tune line is the biggest contributor. Understanding this feed-through mechanism will help in designing lower-jitter and lower-noise PLLs.

3.1 Noise Basics

In this section, the sources of noise, a description of each noise type, active noise models, equivalent input noise, noise figure, and the trade-offs between bipolar and CMOS transistors will be discussed. Understanding these topics will help the reader decipher noise specifications, understand the physical limitations of the circuit design, and design low-noise circuits.

In this discussion, small voltage and current fluctuations generated within the electrical device produce noise. This discussion does not include the coupling of man-made signals to the circuit. Noise represents the lower limit to the size of an electrical signal. Oscillators upconvert noise to the oscillation frequency, which puts a limit on the quality of the oscillator. Consequently, achieving a high signal-to-noise ratio (SNR) will produce a desirable low-jitter and low-noise oscillator. Designing a low-noise amplifier for the gain element in an oscillator will result in a low-noise oscillator.

The analysis procedure is to find the gain and bandwidth, then the equivalent input noise. Output noise then can be calculated from these components. In general, analyzing the output noise level all together is too complicated because too many variables make it easy to get confused.

Let's review some statistics of noise. In many cases, noise has a Gaussian amplitude distribution. For thermal noise, voltage models the amplitude distribution. This noise voltage has a probability-density function $p(V)$. Let's assume that the noise rides on a dc voltage V_{dc}. Multiplying the probability-density function by the band of voltage of interest $p(V)dV$ gives the probability that the voltage lies between values V and $(V + dV)$ at any time. Given σ for the standard deviation of the Gaussian distribution, the voltage amplitude lies between the limits $V_{dc} \pm \sigma$ 68% of the time. The mean-square value of the noise voltage minus its dc average $(V - V_{dc})$ defines the variance, σ^2. Theoretically, the noise amplitude can have positive or negative values approaching infinity; however, the probability falls off very quickly as amplitude increases. Consequently, $\pm 3\sigma$ gives an effective limit to the noise amplitude. The noise signal lies within these limits 99.7% of the time.

3.1.1   Sources of Noise

3.1.1.1   Thermal

Random thermal motion of electrons in conductors produces thermal noise. Temperature directly affects the level of thermal noise. Direct current does not affect thermal noise. Resistors, resistive wire, radiation resistance of antennas, loudspeakers, and microphones produce thermal noise [1]. A series voltage generator with resistor R can represent thermal noise, as shown by (3.1):

$$(V_{nstd})^2 = 4kTR\,\Delta f \tag{3.1}$$

where

k = Boltzmann's constant;

T = temperature in degrees Kelvin;

Δf = bandwidth.

Alternatively, thermal noise can be represented by a parallel current source with the resistor R, the Norton equivalent, as shown by (3.2):

$$(i_{nstd})^2 = 4kT1/R\,\Delta f \tag{3.2}$$

These equations show noise spectral densities that are independent of frequencies. Consequently, they are white-noise sources up to 10^{13} Hz and have Gaussian amplitude distributions. The $4kT$ term has the units of energy/Hz. Tracking energy in noise calculations minimizes mistakes in calculations and derivations because of the conservation of energy.

Equations (3.1) and (3.2) show that the noise voltage and current have a mean-square value that is directly proportional to the bandwidth Δf (in hertz) of the measurement. Thus, a noise-current or noise-voltage spectral density $(V_{nstd})2/\Delta f$ (with units of square volts per hertz) has a constant level as a function of frequency. White noise has this characteristic of a flat spectrum.

The randomness of noisy signal makes the instantaneous value of the waveform unpredictable at any instant. Consequently, the only information available for use in circuit calculations concerns the mean-square value of the signal, which is given by (3.2).

The circuit that has the noise source determines the bandwidth Δf in (3.2) [1]. The noise contribution of a low-noise figure amplifier to an input signal with high SNR is minimal; however, a high attenuation of a signal in a system chain of devices could bring the amplified input noise level down close to the thermal noise floor, where the next amplifier in the chain will give a significant contribution.

The thermal noise floor provides a reference point for many noise calculations. It represents the best performance that can be achieved. How close you are to the noise floor determines whether you should continue making improvements and how successful you are likely to be in making these improvements. Consequently, this level should be studied closely. For $k = 1.381 \times 10^{-23}$ and $T = 300K$, the $4kT$ term equals 1.66×10^{-20} W/Hz. On a log scale, this term equals -197 dBW/Hz [$10 \log(4kT)$] and -168 dBm/Hz [$10 \log(4kT\,1,000)$].

In almost all cases, one of the input noise sources is the equivalent source resistance. From network theory, a maximum amount of power can be delivered to a resistive load. The load must be equal to the source resistance to transfer the maximum power. Equation (3.3) shows the transfer function for the maximum available power condition:

$$P_{avl} = \frac{(|E_g|)^2}{4R_l} \tag{3.3}$$

where

E_g = voltage source in the network.

This equation shows the available power that can be delivered to the load R_l from an equal source resistance. This same transfer function can be applied to a resistive noise source. From (3.3) and (3.1), (3.4) shows the available noise power to be delivered to the load R of equal noise-source resistance [2]:

$$P_{navl} = \frac{4kTR}{4R} = kT \tag{3.4}$$

Consequently, the available noise power to be delivered to the load equals -174 dBm/Hz [$10 \log(4kT\,1,000/4$] or -168 dBm/Hz $- 6$ dB). This is the power level that would be measured by a power meter and is the method that noise-figure meters use to measure noise accurately.

3.1.1.2 Shot

Direct current flow produces shot noise in diodes and bipolar transistors. Studying a diode and the carrier concentrations in the device in forward bias shows the origin of shot noise. An electric field E exists in the depletion region and a voltage $(\psi_o - V_f)$ exists between the p-type and n-type regions, where ψ_o is the built-in potential and V_f is the forward bias on the diode. Holes from the p region and electrons from the n region with sufficient energy to overcome the potential barrier at the junction produce the forward current I_D in the diode. After crossing the junction, the carriers diffuse away as minority carriers. The carrier passage across the junction has a random time interval. These events depend on the carrier's having sufficient energy and velocity directed toward the junction. Thus, external current I appears to be a steady current. However, the current contains a large number of random, independent current pulses [1].

A series of random, independent current pulses with average value I_D produces the following mean-square noise current as shown by (3.5):

$$(i_{nstd})^2 = 2qI_D\Delta f \qquad\qquad (3.5)$$

The noise current has a mean-square value that is directly proportional to the bandwidth Δf (in hertz). Thus, the noise-current spectral density $(i_{nstd})^2/\Delta f$ (with units of square amperes per hertz) has a constant level as a function of frequency. Equation (3.5) holds true until the frequency becomes comparable to $1/\tau$, where τ is the carrier transit time through the depletion region. For most practical electronic devices, τ has an extremely small value. Consequently, the accuracy of (3.5) holds up well into the gigahertz region [1].

3.1.1.3 Flicker

All active devices, as well as some discrete passive elements, such as carbon resistors, produce flicker noise. A flow of direct current produces flicker noise. The literature has various origins of flicker noise. However, in bipolar transistors, traps associated with contamination and crystal defects in the emitter-base depletion layer cause flicker noise. The traps randomly capture and release carriers. The time constants from this process give rise to a noise signal with energy concentrated at low frequencies [1]. Equation (3.6) mathematically describes the spectral density of flicker noise:

$$(i_{nstd})^2 = K_1 \frac{I^a}{f^b}\Delta f \qquad\qquad (3.6)$$

where

Δf = bandwidth at frequency f;

I = direct current;

K_1 = constant for a particular device;

a = constant in the range of 0.5 to 2;

b = constant of about unity.

For $b = 1$, the noise spectral density has a $1/f$ frequency dependence (hence, the alternative name $1/f$ noise). Low frequencies have the highest levels of flicker noise. In devices exhibiting high flicker noise levels, this noise source may dominate the device noise into the megahertz frequency range.

For low-noise, low-frequency integrated circuits, external metal film resistors (no flicker noise) should be used to carry direct current [1]; however, a carbon resistor can be used with zero dc current flow because a zero direct current through the resistor produces no flicker noise (thermal noise always exists in the resistor).

In Sections 3.1.1.1 and 3.1.1.2, we saw that shot and thermal noise signals have well-defined mean-square values that can be expressed in terms of current flow, resistance, and a number of well-known physical constants. By contrast, the mean-square value of a flicker noise signal, as given by (3.6), contains an unknown constant K_1. This constant varies by orders of magnitude from one device type to the next. Furthermore, this constant can vary widely for different transistors or integrated circuits from the same processed wafer. Contamination and crystal imperfections, which are factors that can vary randomly, even on the same silicon slice, produce this wide variation in flicker noise. However, from experiments, measurements of K_1 from a number of devices for a given process do predict average, or typical, flicker noise performance for integrated circuits.

Finally, measurements of flicker noise show a non-Gaussian distribution because of the nonflat spectral density. This is a significant characteristic because most circuits and systems function under the assumption that noise is Gaussian.

3.1.1.4   Burst

Although not fully understood, the presence of heavy-metal ion contamination relates to the presence of burst noise. For instance, gold-doped devices show very high levels of burst noise. Some integrated circuits and discrete transistors produce low-frequency burst noise. An oscilloscope trace of this type of noise shows bursts of noise (hence, the name) on two or more discrete levels. The noise pulses have a repetition rate usually in the audio frequency range (a few kilohertz or less). Sending this noise into a loudspeaker produces a "popping" sound. Consequently, this noise has been called "popcorn noise" [1]. Equation (3.7) shows the spectral density of burst noise:

$$(i_{nstd})^2 = K_2 \frac{I^c}{I + \left(\dfrac{f}{f_c}\right)^2} \Delta f \tag{3.7}$$

where

K_2 = constant for a particular device;

I = direct current;

c = constant in the range 0.5 to 2;

f_c = particular frequency for a given noise process.

The spectrum plot of burst noise typically has a hump. At higher frequencies, the noise spectrum falls as $1/f^2$. Burst noise processes often occur with multiple time constants. Consequently, multiple humps can occur in the spectrum. Also, flicker noise usually occurs with the burst noise. Therefore, Figure 3.1 shows the composite low-frequency noise spectrum that often appears. As with flicker noise, factor K_2 for burst noise varies considerably and must be determined experimentally. Finally, burst noise also has a non-Gaussian amplitude distribution [1].

3.1.1.5 Avalanche

Zener, or avalanche, breakdown in a *pn* junction produces avalanche noise. In avalanche breakdown, holes and electrons in the depletion region of a reverse-biased *pn* junction acquire sufficient energy to collide with silicon atoms. This collision creates hole-electron pairs. This process produces a random series of large noise spikes. Like flicker noise, avalanche noise requires a dc flow.

The avalanche noise process produces noise levels much greater than shot noise for the same current. This greater noise level occurs because a single carrier can start an avalanching process that results in the production of a current burst containing many carriers moving together. The summation of the number of random bursts of this type gives the higher total noise level. Since the use of Zener diodes commonly causes avalanche noise problems, avoid using these devices with low-noise circuits.

Figure 3.2 models the noise of the Zener diode. This figure shows a series voltage generator $(v_{nstd})^2$, a series resistor R, and a series dc voltage source V_z. The dc voltage V_z models the breakdown voltage of the diode. The series resistance

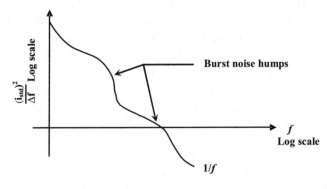

Figure 3.1 Low-frequency noise spectrum shows the presence of burst and flicker noise. (*From:* [1]. © 1993 Wiley Interscience. Reprinted with permission.)

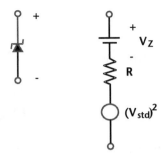

Figure 3.2 The symbol and noise model for a Zener diode. (*From:* [1]. © 1993 Wiley Interscience. Reprinted with permission.)

R has a typical value of 10Ω to 100Ω. The magnitude of $(v_{nstd})^2$ depends on the device structure and the uniformity of the silicon crystal, both of which are difficult to predict. A typical measured value is $(v_{nstd})^2/\Delta f \sim 10^{-14}$ V^2/Hz at a dc Zener current of 0.5 mA. This value equates to the thermal noise voltage in a 600-kΩ resistor. Consequently, this noise level completely overwhelms any thermal noise in the series R in the model. Avalanche noise has an approximately flat spectral density; however, it generally has a non-Gaussian amplitude distribution [1].

3.1.2 Noise Models

3.1.2.1 Diode

Figure 3.3 shows the noise model for a diode. A current generator shunting r_d, as shown in Figure 3.3, represents the low-frequency, small-signal equivalent circuit of the diode for the effects of shot noise. This noise signal has random phase and a mean-square magnitude. Thus, the arrow in the current source in Figure 3.3 has no meaning and only identifies the generator as a current source. For this discussion, we will follow this practice for dealing with independent noise generators.

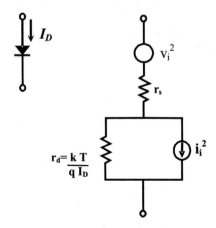

Figure 3.3 Symbol and noise model for a diode. (*From:* [1]. © 1993 Wiley Interscience. Reprinted with permission.)

A series resistance r_s completes the diode model as shown in Figure 3.3. The resistor r_s comes from the resistivity of the silicon. Consequently, a series thermal noise model for the resistor is added to the diode model as mathematically modeled by (3.8):

$$(v_{nstds})^2 = 4kTr_s\Delta f \qquad (3.8)$$

From experiments, a current generator in shunt with r_d models the effects of flicker noise. Rather than adding another current source symbol to the model, it has been combined with the shot noise generator, as shown by (3.9) [1, 3]:

$$(i_{nstd})^2 = (2qI_D\,\Delta f) + K\frac{(I_D)^a}{f}\Delta f \qquad (3.9)$$

3.1.2.2 Bipolar

Figure 3.4 shows the full small-signal equivalent circuit, including noise for the bipolar transistor. In the forward-active region, minority carriers diffuse and drift across the base region to be collected at the collector-base junction. The field in the collector-base depletion region accelerates the entry of minority carriers into this region to the collector. The diffusing (or drifting) carriers arrive at the collector-base junction at purely random time intervals. Thus, the transistor collector current consists of a series of random current pulses. Consequently, collector current I_c generates shot noise. A current generator $(i_{cstd})^2$ from collector to emitter represents the shot noise as shown in the equivalent circuit of Figure 3.4. Equation (3.10) represents the model:

$$(i_{cstd})^2 = 2qI_C\Delta f \qquad (3.10)$$

Noise voltage generator $(v_{bstd})^2$ represents the thermal noise of the transistor base resistor r_b, which is a physical resistor in the silicon. Equation (3.11) represents the noise source:

$$(v_{bstd})^2 = 4kTr_b\Delta f \qquad (3.11)$$

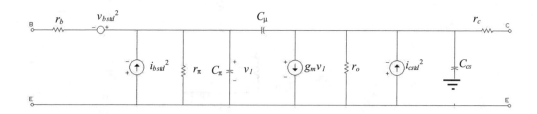

Figure 3.4 Small-signal equivalent circuit with noise generators for a bipolar transistor. (*From:* [1]. © 1993 Wiley Interscience. Reprinted with permission.)

Collector series resistor r_c also has thermal noise, and it is in series with the collector node; however, the high-impedance of the collector node makes this noise negligible. Consequently, it is usually not included in the model. Resistors r_π and r_o are modeling resistors that mathematically model the response of the transistor. Consequently, they do not have thermal noise.

Recombination in the base and base-emitter depletion regions and carrier injection from the base into the emitter produce base current I_B in the transistor. These are independent random processes. Consequently, they have shot noise. The noise-current generator $(i_{bstd})^2$ in Figure 3.4 represents the shot noise from the base current. Furthermore, from experiments, current generators across the internal base-emitter junction represent the flicker noise and burst noise in a bipolar transistor. Consequently, they have been conveniently combined with the shot noise generator $(i_{bstd})^2$. Equation (3.12) shows the combination of these noise sources:

$$(i_{bstd})^2 = 2qI_B\Delta f + K_1\frac{(I_B)^a}{f}\Delta f + K_2\frac{(I_B)^c}{1 + \left(\dfrac{f}{f_c}\right)^2}\Delta f \qquad (3.12)$$

Since the noise generators arise from separate, independent physical mechanisms, all the noise sources are independent of each other and have mean-square values. For V_{ce} 5V below the breakdown voltage BV_{ceo}, bipolar transistors have negligible avalanche noise. Consequently, this source of noise will be neglected. Finally, this equivalent circuit is valid for both *npn* and *pnp* transistors. For *pnp* devices, the magnitudes of I_B and I_c are used in the above equations [1, 3].

3.1.2.3   MOS

The structure of field-effect transistors (FET) has a resistive channel joining source and drain. The gate-source voltage controls the drain current by modulating the resistive channel. The resistive channel material exhibits thermal noise, which is the major source of noise in FETs (both JFET and MOSFET).

A noise-current generator $(i_{dstd})^2$ from drain to source in the FET small-signal equivalent circuit of Figure 3.5 represents this noise source. Also a drain-source current generator represents flicker noise in the FET. Consequently, summing these two lumps them into one noise generator $(i_{dstd})^2$. In addition, the gate leakage

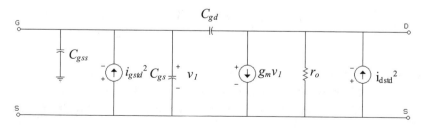

Figure 3.5   Small-signal equivalent circuit with noise generators for an MOS transistor. (*From:* [1]. © 1993 Wiley Interscience. Reprinted with permission.)

current generates shot noise. The element $(i_{gstd})^2$ in Figure 3.5 represents this noise source; however, this gate shot noise source usually has a very small level. Only connecting a very large driving-source impedance to the FET gate will make this shot noise level significant [1, 3]. Equations (3.13) and (3.14) model the independent generators in Figure 3.5:

$$\frac{(i_{gstd})^2}{\Delta f} = 2qI_G \tag{3.13}$$

$$\frac{(i_{dstd})^2}{\Delta f} = \left[4kT\left(\frac{2}{3}g_m\right)\right] + \frac{KI_D^a}{f} \tag{3.14}$$

where

I_G = gate leakage current;

I_D = drain bias current;

K = flicker noise constant for a given device;

a = constant between 0.5 and 2;

g_m = device transconductance at the operating point.

3.1.3 Equivalent Input Noise

Any active device, junction transistor, or FET has a certain amount of noise v_i (referred to the input) that is independent of source impedance. Another noise component, i_i, directly depends on the source impedance. The sqr(v_i^2) represents the spectral density of the equivalent short-circuit noise-voltage generator at the input of the MOSFET at a specified frequency and bandwidth.

Measuring the device noise output with the input ac shorted determines the equivalent input noise voltage. Dividing the output noise voltage by the circuit gain determines the input referred noise. Studying the measurement of equivalent noise voltage gives us a deeper understanding the concept. Figure 3.6 shows the test setup to measure the equivalent input noise voltage. For practical measurement purposes, setting R_g = 100Ω makes an input ac short as illustrated in Figure 3.6. The following procedure can be used to make the equivalent noise voltage measurement:

1. Set signal generator for 100 mV.
2. Then, set V_{gs} = 0.1 (100Ω/10$^6\Omega$) = 10^{-5} = 10 μV.
3. Set total gain to 1,000.
4. Voltage meter at the output now reads 10 mV.
5. Short out the 100Ω resistor.
6. Voltage meter now reads 1 mV for every microvolt of noise.

Bipolar transistors use a similar procedure, except base current is supplied to obtain the desired collector current [4].

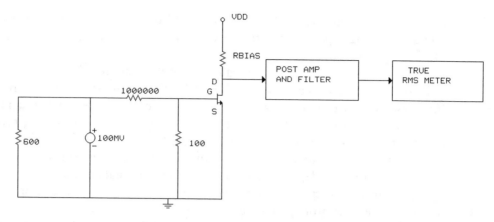

Figure 3.6 Test setup to measure equivalent input noise voltage.

The sqr(i_i^2) represents the spectral density of the equivalent open-circuit noise-current generator at the input of the transistor at a specified frequency and bandwidth. Dividing the output noise (with the input ac open-circuited) by the circuit gain and input ac impedance determines the equivalent input noise current. Figure 3.7 shows the test setup to measure equivalent noise current. For practical measurement purposes, setting $R_g = 1,000$ MΩ makes an input ac open as illustrated in Figure 3.7. The following procedure can be used to make the equivalent noise-current measurement:

1. Set signal generator for 10 mV with S_1 in position 1.
2. Set total gain to 10.
3. Voltage meter now reads 100 mV.
4. Switch S_1 to position 2.
5. Meter reads 1 mV for each 10^{-13}A (0.1 pA) of noise.

Noise voltage and current depend on frequency. In MOSFETs, the former increases while the latter decreases with decreasing frequency. The noise sources

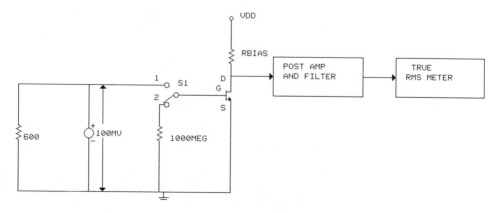

Figure 3.7 Test setup to measure the equivalent input current.

are relatively independent of drain current and drain-to-source voltage, which is unlike the strong dependence of the noise source on collector current in a transistor [4].

Figure 3.8 shows the basic equivalent input noise-source circuit. The situation in Figure 3.8 represents a noiseless MOSFET with internal-noise sources removed (the noiseless device) and with a noise voltage v_i^2 and current generator i_i^2 connected at the input. This representation holds valid for any source impedance, provided that the two noise generators are not correlated.

We cannot just assume that the two noise generators are independent in general because they are both dependent on the same set of original noise sources; however, a large number of practical circuits have a negligible amount of correlation between the two noise generators. Consequently, these effects may be neglected. Furthermore, if either equivalent input generator v_i^2 or i_i^2 dominates, the correlation may be neglected. This method of representation will be shown to be extremely useful.

The following discussion explains the need for both an equivalent input noise-voltage generator and an equivalent input noise-current generator to represent the noise performance of the circuit for any source resistance. Let's consider the extreme cases of source resistance R_s equal to zero or infinity. If $R_s = 0$, i_i^2 in Figure 3.8 is shorted out. Since the original circuit will still show output noise in general, we need an equivalent input noise voltage v_i^2 to represent this behavior. Similarly, if $R_s = \infty$, v_i^2 in Figure 3.8 cannot produce output noise, and i_i^2 represents the noise performance of the original noisy network. For finite values of R_s, both v_i^2 and i_i^2 contribute to the equivalent input noise of the circuit.

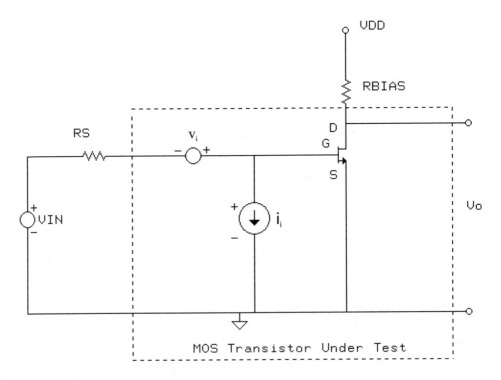

Figure 3.8 Equivalent input noise-source circuit concept.

From schematics of circuits, shorting and opening the inputs to the original noisy and the equivalent input circuit, as shown in Figure 3.8, determines the values of the equivalent input generators. First, short-circuiting the input of both circuits and equating the output noise in each case calculates v_i^2. Open-circuiting the input of each circuit and equating the output noise in each case computes i_i^2. This will now be done for the bipolar transistor and the MOS transistor [1].

Now the equations for the equivalent input noise sources for bipolar and MOS transistors are shown. Equations (3.15) and (3.16) give the equivalent input noise for a bipolar transistor:

$$\frac{(v_{istd})^2}{\Delta f} = 4kT\left(r_b + \frac{1}{2g_m}\right) \tag{3.15}$$

$$\frac{(i_{istd})^2}{\Delta f} = \frac{2qI_C}{\beta^2} + 2qI_B + \frac{\frac{K_1}{2q}I_B^a}{f} \tag{3.16}$$

Equations (3.17) and (3.18) give the equivalent input noise for an MOS transistor:

$$\frac{(v_{istd})^2}{\Delta f} = 4kT\left(\frac{2}{3}\frac{1}{g_m}\right) + \frac{K_f}{2\mu C_{ox}^2 WLf} \tag{3.17}$$

$$\frac{(i_{istd})^2}{\Delta f} = 2qI_G \tag{3.18}$$

where

μ = mobility of electrons near the silicon surface that depends on the process (e.g., $\mu_n = 0.06$ m^2/V s);

C_{ox} = gate capacitance per unit area;

W = gate width of the transistor;

L = gate length of the transistor;

K_f = SPICE flicker noise constant.

The flicker noise constant K_f in (3.17) is different from the earlier flicker noise constant because it refers to the SPICE definition of flicker noise, which has all the noise as an output current source that is converted to the input voltage noise source by the relationship of $V_g^2 = I_d^2/g_m^2$. This K_f definition is most useful in making noise calculations. These equations help us identify the sensitivity of transistor parameters in generating noise. Converting v_i and i_i measurements to noise figure in decibels as a function of a specified source resistance gives a more meaningful and useful method of bipolar and MOS transistor noise measure.

3.1.4 Noise Figure

The noise figure (F) is a commonly used method of specifying the noise performance of a circuit or a device. Its disadvantage is that it is limited to situations where the source impedance is resistive, and this precludes its use in many applications where noise performance is important; however, it is widely used as a measure of noise performance in communication systems where the source impedance is often resistive.

The definition of noise figure (F) is the ratio of the input SNR divided by the output SNR, and F is usually expressed in decibels. The advantage of the noise-figure concept is apparent in its definition. Noise figure gives a direct measure of the SNR degradation that is caused by the circuit. For example, if the SNR at the input to a circuit is 50 dB, and the circuit noise figure is 5 dB, then the SNR at the output of the circuit is 45 dB [1]. This concept is very convenient in tracking noise degradation in every receiver branch of a complicated system.

Equation (3.19) computes the figure of merit of equivalent input noise voltage and equivalent input noise current [1].

$$F = 1 + \frac{N_v^2}{4kTR_s\Delta f} + \frac{N_i^2}{4kT\dfrac{1}{R_s}\Delta f} \tag{3.19}$$

Rearranging (3.19) and setting Δf to 1 produces (3.20) in [4, pp. 6–14]. N_v and N_i terms become voltage and current noise per square-root hertz.

$$NF = 10 \log\left[1 + \frac{N_v^2}{4kTR_s} + \frac{N_i^2}{4kT\dfrac{1}{R_s}}\right] \tag{3.20}$$

Now we can enhance the understanding of noise figure to the transistor level by substituting the equivalent input noise-source equations into the noise figure equation (3.20). First, we substitute the MOS transistor equivalent noise equations into the noise figure equations [(3.20)]. The equivalent input current is neglected because we assume that it is small. Substituting (3.17) into (3.20) produces (3.21) for the noise figure of an MOS transistor:

$$NF = 10 \log\left\{1 + \frac{\left[4kT\left(\dfrac{2}{3}\dfrac{1}{g_m}\right) + \dfrac{K_f}{2\mu C_{ox}^2 WLf}\right]}{(4kTR_s)}\right\} \tag{3.21}$$

Equation (3.21) can be rearranged to a simpler equation as shown by (3.22), where the thermal noise figure and corner flicker frequency are used:

$$NF = 10 \log\left[F\left(1 + \frac{f_l}{f}\right)\right] \tag{3.22}$$

Equation (3.22) makes it easier to model simulations and make substitutions into Leeson's phase-noise model. Equation (3.23) shows the thermal noise figure portion of (3.22) for an MOS transistor:

$$F = \frac{\frac{2}{3}\frac{1}{g_m} + R_s}{R_s} \tag{3.23}$$

Equation (3.24) shows the corner flicker frequency portion of (3.22) for an MOS transistor:

$$f_l = \frac{\frac{K_f}{2\mu C_{ox}^2 WLf}}{\left[4kT\left(\frac{2}{3}\frac{1}{g_m} + R_s\right)\right]} \tag{3.24}$$

Equation (3.25) shows the thermal noise figure portion of (3.22) for a bipolar transistor:

$$F = \frac{4kTr_b + \left(\frac{2qI_c}{\beta^2}R_s^2 + 2qI_b R_s^2\right) + 4kTR_s}{4kTR_s} \tag{3.25}$$

Equation (3.26) shows the corner flicker frequency portion of (3.22) for a bipolar transistor:

$$f_1 = \frac{K_f I_b R_s^2}{4kTr_b + \left(\frac{2qI_c}{\beta^2}R_s^2 + 2qI_b R_s^2\right) + 4kTR_s} \tag{3.26}$$

The development of (3.23) to (3.26) allows us to substitute these values into Leeson's phase-noise model of an oscillator as shown by (3.27):

$$10^{\frac{\mathscr{L}(f_m)_{amp}}{10}} = \frac{0.5F_1 kT_0}{P_{avl}}\left(1 + \frac{f_l}{f_m}\right) \tag{3.27}$$

where

$\mathcal{L}(f_m)_{amp}$ = phase noise transfer function for an amplifier (dBc/Hz);

T_0 = temperature (K);
T_0 = 290 at room temperature;

k = Boltzmann constant (mW/°K · Hz));
k = 1.38×10^{-20};

$10 \log(kT_0)$ = −174 dBm/Hz;

F_1 = noise figure ratio of amplifier in oscillator;

f_l = frequency of intersection of the amplifier's flicker noise with its constant noise figure (thermal noise floor) (Hz).

This relationship gives us a deeper understanding of the effects of transistor parameters on the noise levels generated by an oscillator. This allows us to architect oscillators and size transistors to make the trade-off between low-noise and low-power performance.

3.1.5 Bipolar Versus CMOS Noise Comparison

One of the main questions that occur to circuit designers is how we make an informed selection between bipolar and MOS transistors for a low-noise circuit. From studying the equations above, a set of curves can be generated to help make this decision. The curves show that the selection between transistors depends on the source resistance presented to the circuit.

Figure 3.9 shows two bipolar transistor noise curves for 100 Hz and 100 kHz versus source resistance. The y-axis is noise figure in decibels and the x-axis is source resistance in ohms. The top 100-Hz curve shows the effect of flicker noise, and the bottom 100-kHz curve shows the effect of thermal noise. This curve shows that at high source resistance the bipolar transistor thermal noise figure goes up

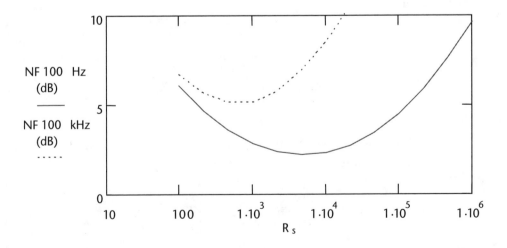

Figure 3.9   Noise figure versus source resistance for a bipolar transistor at 100 Hz and 100 kHz.

significantly. Furthermore, the flicker noise curve follows a similar pattern with even higher noise figure. Now let's look at the response of an MOS transistor.

Figure 3.10 shows MOS transistor curves for 100 Hz and 100 kHz. Again, the 100-Hz curve shows the effect of flicker noise, and the 100-kHz curve shows the effect of thermal noise. This figure shows that the noise figure goes down for high-input source resistance, which is just the opposite of the bipolar transistor. A comparison of the two figures shows high resistance (>10 kΩ) to be optimum for MOS transistors and low resistance (<10 kΩ) to be optimum for bipolar transistors.

Now let's revisit the assumption about noise current being negligible for MOS transistors in our noise figure equation development. For comparison purposes, we will first look at the noise current versus collector current for a bipolar transistor. Figure 3.11 shows that bipolar thermal and flicker noise current changes significantly with collector current. Now let's plot MOS transistor equivalent input noise versus drain current. Figure 3.12 shows that MOS transistor thermal and flicker noise current does not significantly change with drain current. Consequently, this curve verifies that we can assume noise current has no significant impact on noise figure for an MOS transistor.

Example 3.1

Let's look at an example in SPICE with a 0.3-μm gate-length transistor from a 0.22-μm process connected as a common-source amplifier with an infinite load

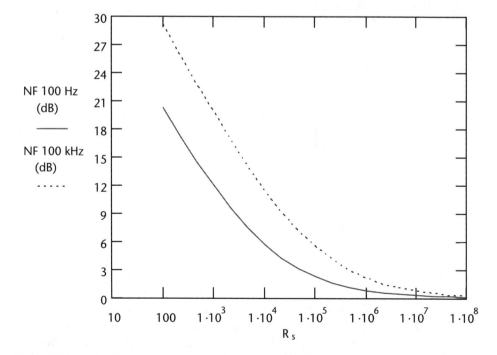

Figure 3.10 Noise figure versus source resistance for an MOS transistor at 100 Hz and 100 kHz.

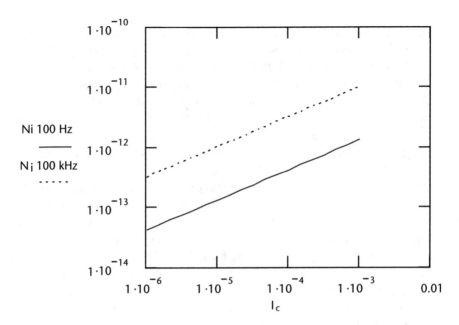

Figure 3.11 Noise current versus collector current for a bipolar transistor at 100 Hz and 100 kHz.

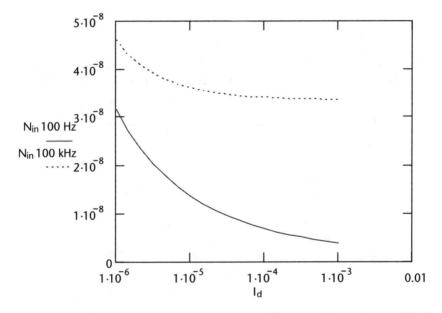

Figure 3.12 Equivalent input noise versus drain current for an MOS transistor at 100 Hz and 100 kHz.

resistance to study the effects of noise. The transistor has 80 μa of drain bias current, a 0.9V V_{gs} bias, 1V V_{ds} bias, and a 1.2-μm gate width. Figure 3.13 shows a plot of the SPICE output. The highest waveform is the gain of the amplifier, the next highest waveform is the output noise, and the lowest waveform is the equivalent input noise voltage. The difference between the input and output noise is the gain of the amplifier. The noise level is in dBV/Hz.

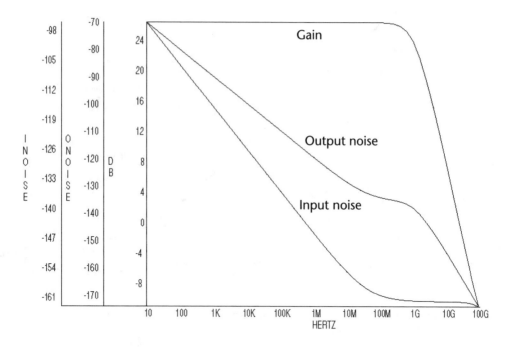

Figure 3.13 SPICE result of input and output noise and gain for an MOS transistor.

Figure 3.13 shows the relationship of equivalent input noise, output noise, and gain in decibels. The output noise equals the input noise with the addition of 26 dB of gain until approximately 1 GHz. At 1 GHz, the gain rolls off, and the output noise tracks this roll off.

Now let's vary the gate width by a factor of 10 to 12 at a width of 120 μm to see the effect on noise levels. From the equations, we should expect the flicker noise corner frequency to decrease and the thermal noise to decrease until it hits the thermal noise floor. Figure 3.14 shows the effect of increasing transistor gate width on the noise. Flicker and thermal noise are reduced by 10 dB for increasing the gate width by a factor of 10. On the last increase in gate width, the far-out thermal noise did not decrease by 10 dB. Instead, it only decreased by a few decibels due to the noise-floor limit set by the thermal noise kT. The existence of this limit and the 10-dB change in noise is a good verification of the accuracy of the simulator and the simulation test circuit.

3.2 Phase-Noise and Oscillator Theory

This section describes the basics of phase-noise analysis and applies the analysis to oscillator design. A phase-noise modeling technique is developed that will allow mathematical operations to be performed on phase-noise curves. Frequency modulation (FM) waveform theory is investigated so that the relationship between narrowband FM and single-sideband phase noise is established. Various measurement relationships to phase noise are explained so that an engineer can interpret between

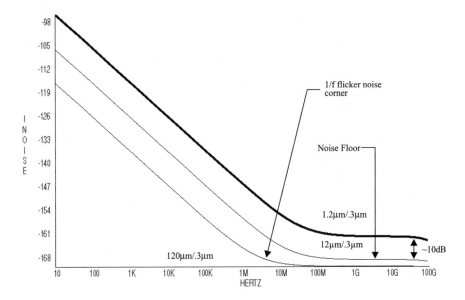

Figure 3.14   Effect of increasing transistor gate width on the noise level.

oscillator stability specifications that use these measurements. With the establishment of several phase-noise relationships, a general mathematical description for oscillators is developed. Finally, the phase-noise power slopes of an oscillator are modeled.

3.2.1   FM Theory

A study of FM theory typically begins with a mathematical model for an FM waveform. Equation (3.28) mathematically describes an FM waveform as a function of time:

$$V(t) = V_p \cos[\omega_c t + \eta \sin(\omega_m t)] \qquad (3.28)$$

where

V_p = maximum amplitude (V);

t = time variable (sec);

η = modulation index;

ω_m = angular frequency of modulation, which is usually less than the carrier frequency (rad/s);

$\omega_m = 2\pi f_m$;

ω_c = angular frequency of the carrier, usually greater than modulation frequency (rad/s);

$\omega_c = 2\pi f_c$.

The FM waveform function in (3.28) allows frequency, phase, and phase-noise relationships to be developed. The argument of the cosine function in (3.28) is the instantaneous phase. The derivative of instantaneous phase with respect to time is frequency; therefore, the study of frequency-stability analysis includes the study of instantaneous phase. The derivative of the cosine argument in (3.28) produces (3.29), which is the instantaneous frequency:

$$\omega = \frac{d}{dt}\phi$$

$$f = f_c + \frac{d}{dt}\left[\frac{\eta \sin(\omega_m t)}{2\pi}\right] \qquad (3.29)$$

$$f = f_c + \eta f_m \cos(\omega_m t)$$

where

ω = angular frequency variable (rad/s);

$\omega = 2\pi f$;

f = frequency variable (Hz).

From (3.29), the instantaneous frequency equals the carrier frequency plus an FM term. The modulation term cosinusoidally sweeps frequency with a maximum deviation from the carrier of the modulation index times the frequency of modulation. The inverse of the modulation frequency determines the sweep rate of the cosinusoidal frequency sweep. Equation (3.29) can be simulated in the laboratory by using a swept frequency source (VCO) and sinusoidally modulating the voltage tuning port. Equation (3.29) represents a sinusoidal sweep of frequency with a center frequency of f_c and a maximum frequency deviation from the carrier frequency of $f_c + \eta f_m$ and a minimum frequency deviation from the carrier frequency of $f_c - \eta f_m$. From (3.29), (3.30) defines the maximum frequency deviation:

$$f_d = \eta f_m \qquad (3.30)$$

where

f_d = maximum frequency deviation from the carrier frequency (Hz).

Rearranging (3.30) produces (3.28), which defines the modulation index [5]:

$$\eta = \frac{f_d}{f_m} \qquad (3.31)$$

3.2.2 Relationship of Phase Noise to FM

Further study of narrowband FM waveform theory shows the relationship of phase noise to an FM signal waveform. Narrowband FM theory uses trigonometry and

Bessel function identities. Equations (3.32) through (3.34) from [6] show the trigonometry identities:

$$\cos(A + B) = \cos(A)\cos(B) - \sin(A)\sin(B) \tag{3.32}$$

$$\cos(A)\cos(B) = 0.5\,\cos(A - B) + 0.5\,\cos(A + B) \tag{3.33}$$

and

$$\sin(A)\sin(B) = 0.5\,\cos(A - B) - 0.5\,\cos(A + B) \tag{3.34}$$

Equations (3.32) and (3.34) from [5] show the Bessel function identities:

$$\cos[X\sin(B)] = J_0(X) + 2J_2(X)\cos(2B) + 2J_4(X)\cos(4B) + \ldots \tag{3.35}$$

$$\sin[X\sin(B)] = 2J_0(X) + 2J_1(X)\sin(3B) + 2J_3(X)\sin(5B) + \ldots \tag{3.36}$$

Substituting (3.32) into (3.28) expands (3.28) to (3.37):

$$V(t) = +V_p\cos(\omega_c t)\cos[\eta\sin(\omega_m t)] - V_p\sin(\omega_c t)\sin[\eta\sin(\omega_m t)] \tag{3.37}$$

Equation (3.37) contains two Bessel function identities [(3.35) and (3.36)]. Substituting the two Bessel functions into (3.37) produces (3.38):

$$V(t) = V_p\left\{ J_0\,\eta\,\cos(\omega_c t) + 2\sum_{n=2}^{\infty} J_n(\eta)\cos(n\omega_m t)\cos(\omega_c t) \right. \tag{3.38}$$

$$\left. - 2\sum_{k=1}^{\infty} J_k(\eta)\sin(k\omega_m t)\sin(\omega_c t) \right\}$$

where

k = the odd harmonic integers 1, 3, 5, 7, . . . ∞;

n = the even harmonic integers 2, 4, 6, . . . ∞.

Substituting the trigonometric identities [(3.33) and (3.34)] into (3.38) reduces (3.38) to (3.39):

$$V(t) = V_p\left\{ J_0\,\eta\,\cos(\omega_c t) + \sum_{n=2}^{\infty} J_n(\eta)\cos[(\omega_c - n\omega_m)t] + \cos[(\omega_c + n\omega_m)t] \right.$$

$$\left. - \sum_{k=1}^{\infty} J_k(\eta)\cos[(\omega_c - k\omega_m)t] - \cos[(\omega_c + k\omega_m)t] \right\} \tag{3.39}$$

Equation (3.39) shows that the amount of power in each sideband depends on the modulation index and that, for a large modulation index, several sidebands appear. Figure 3.15 shows the many sidebands that appear for a large modulation index. With each harmonic, the signs alternate in Figure 3.15, which is mathematically described by (3.38).

The conservation of energy, which applies to everything in physics, applies to the total energy of FM sidebands. The Taylor series approximation for Bessel functions and a Bessel function identity are used to show the conservation of energy in FM sidebands. Equations (3.40) and (3.41) show the Taylor series approximations for the various orders of the Bessel function:

$$J_0(\eta) = 1 - \left(\frac{\eta}{2}\right)^2 \tag{3.40}$$

$$J_p(\eta) = \frac{\left(\frac{\eta}{2}\right)^p}{p!} \tag{3.41}$$

where

p = harmonic integers 1, 2, 3, 4, . . . ∞.

Equation (3.42) defines the Bessel function identity to be used in the FM sideband energy analysis [5]:

$$J_0(\eta)^2 + 2 \sum_{p=1}^{\infty} J_p(\eta)^2 = 1 \tag{3.42}$$

Equation (3.42) states that the sum of the power levels in each of the sidebands, plus the carrier, equals the total power regardless of the size of the modulation index.

An experiment in the laboratory confirms the conservation of energy in FM waveforms that is mathematically described in (3.42). This experiment connects a

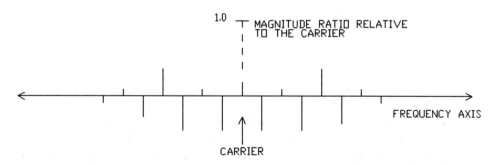

Figure 3.15 Spectral components for an FM signal with a modulation index of 5. (*From:* [5]. © 1971 McGraw-Hill. Reprinted with permission.)

source with modulation index control, modulation frequency control, and power control to a power meter and a spectrum analyzer. Leaving the power control fixed, the modulation index is adjusted to a small value that produces a carrier and an upper and lower sideband. Then, the power level is measured with the power meter. Now, the modulation index is adjusted to a larger value that produces many more sidebands, which are separated by the modulation frequency. Measuring the power again results in the same answer as for the small modulation index; however, large modulation indexes distribute the power between the carrier and sidebands.

Phase noise relates the ratio of the sideband power to carrier power at a particular modulation frequency offset from the carrier frequency. For accurate phase-noise measurements, the small modulation index condition must be satisfied because large modulation indexes affect the carrier power level. Substituting the Bessel function into (3.42) produces (3.43):

$$\text{Carrier power + Sideband power} = 1 \tag{3.43}$$

$$[J_0(\eta)]^2 + 2[J_1(\eta)]^2 = 1$$

In (3.43), the carrier-power and sideband-power levels depend on the value of the modulation index. Small modulation indexes make the effects of the second- and higher-order harmonics of the Bessel function in (3.42) insignificant [5].

A more exact expression for the single-sideband phase-noise power in a 1-Hz bandwidth can now be defined from (3.43). For a small modulation index, almost 100% of the power is in the carrier. Consequently, the carrier power equals 1 for the normalized power calculations in (3.43). This is a valid assumption for most stable oscillators. Using one-half the power of the sideband to represent a single sideband in (3.43) and dividing by the carrier power, which equals unity, produces (3.44):

$$\frac{P_{ssb}}{P_c} = \frac{[J(\eta)_1]^2}{[J(\eta)_0]^2} \tag{3.44}$$

$$\frac{P_{ssb}}{P_c} = [J(\eta)_1]^2$$

For $\eta <\!<\!<\!< 1$ and $J_0(\eta) = 1$, the single-sideband phase-noise-to-carrier ratio in a 1-Hz bandwidth equals the sideband-power-to-carrier-power ratio in (3.44). Unfortunately, (3.44) depends on the Bessel function as a function of the modulation index. To eliminate the Bessel function, which is difficult to evaluate, the Taylor series approximation for the first-order Bessel function is substituted into (3.44). This substitution produces (3.45):

$$\frac{P_{ssb}}{P_c} = \frac{\eta^2}{4}$$

$$\frac{P_{ssb}}{P_c} = \frac{(\eta_{rms})^2}{2}$$

$$\mathscr{L}(f_m) = 10 \log\left(\frac{\eta^2}{4}\right) \qquad (3.45)$$

$$= 10 \log(\eta^2) - 6 \text{ dB}$$

$$= 10 \log(\eta_{rms}^2) - 3 \text{ dB}$$

where

P_{ssb} = power level in the sideband (mW);

P_c = power level at the carrier frequency (mW);

$\dfrac{P_{ssb}}{P_c}$ = power ratio of the sideband to the carrier;

$\eta_{rms} = \dfrac{\eta}{\sqrt{2}}$;

η_{rms} = root-mean-square (RMS) of the modulation index;

$\mathscr{L}(f_m)$ = notation used for single-sideband phase noise in a 1-Hz bandwidth (dBc/Hz).

Equation (3.45) expresses the single-sideband phase-noise power in a 1-Hz bandwidth when a Taylor series approximation for the Bessel function is used. Figure 3.16 graphically shows the frequency-domain representation of single-sideband phase noise [7].

3.2.3   Different Measures of Phase Noise

Several types of phase-noise measurements have appeared in various articles. Their relationships to single-sideband phase noise are shown in (3.46) through (3.53). Equation (3.46) defines the spectral density of phase fluctuations (double-sideband phase noise):

$$S_\phi(f_m) = \eta_{rms}^2 \qquad (3.46)$$

$$S_\phi(f_m) = 2\frac{P_{ssb}}{P_c}$$

where

$S_\phi(f_m)$ = spectral density of phase fluctuations.

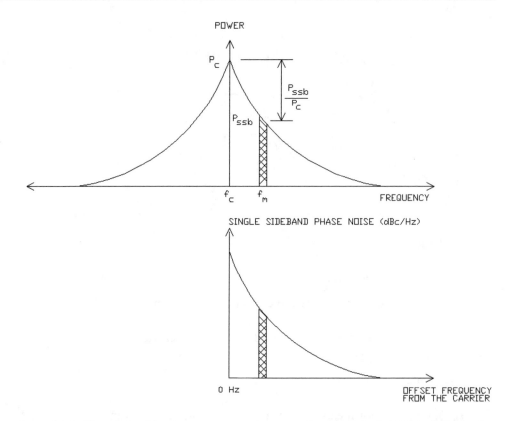

Figure 3.16 The relationship of a spectrum analyzer waveform to a single-sideband phase-noise measurement. The ratio of the single-sideband phase-noise-power level to the carrier power level in a 1-Hz bandwidth f_m Hz away from the carrier frequency defines a single-sideband measurement on a spectrum analyzer. (*From:* [7, Figure 2]. © 1979 Agilent. Reprinted with permission.)

The spectral density of phase fluctuations is related to single-sideband phase noise by a multiplication factor of 2. Equation (3.47) defines the spectral density of frequency fluctuations:

$$S_{\Delta f}(f_m) = \frac{d}{dt}(\eta_{rms})^2$$

$$S_{\Delta f}(f_m) = (f_m)^2(\eta_{rms})^2 \qquad (3.47)$$

$$S_{\Delta f}(f_m) = (f_m)^2\left[2\left(\frac{P_{ssb}}{P_c}\right)\right]$$

where

$S_{\Delta f}(f_m)$ = spectral density of frequency fluctuations.

The spectral density of frequency fluctuations is the derivative of phase fluctuations; therefore, it relates phase noise in terms of frequency changes. Equation (3.48) defines fractional spectral density of frequency fluctuations:

$$S_y(f_m) = \left(\frac{f_m}{f_c}\right)^2 \eta^2 \tag{3.48}$$

$$S_y(f_m) = \left(\frac{f_m}{f_c}\right)^2 \left[2\left(\frac{P_{ssb}}{P_c}\right)\right]$$

where

$S_y(f_m)$ = fractional spectral density of frequency fluctuations.

Equation (3.48) normalizes the frequency fluctuations to the output frequency of the oscillator for a comparison of oscillator purity at different operating frequencies.

Continuing with another measure of spectral purity, (3.49) defines residual FM:

$$\Delta f_{res} = \sqrt{2}\sqrt{\int_{f_{min}}^{f_{max}} \left(\frac{P_{ssb}}{P_c}\right)(f_m)^2\, df_m} \tag{3.49}$$

where

Δf_{res} = residual FM;

f_{min} = lowest frequency of integration;

f_{max} = highest frequency of integration.

Residual FM is the total RMS value of the frequency deviation within a specified bandwidth. This method is frequently used to describe frequency stability with a single value. Also, residual FM is a measure of the peak-to-peak deviation of the carrier as displayed on a spectrum analyzer. Finally, (3.50) defines the Allan variance:

$$\sigma_y^2(\tau) = \frac{1}{2(m-1)}\sum_{k=1}^{m-1}\left(\overline{Y(k+1)} - \overline{Y(k)}\right)^2 \tag{3.50}$$

where

$\sigma_y^2(\tau)$ = notation used for the Allan variance;

τ = measurement time period (sec);

m = total number of data points;

$\overline{Y(k)}$ = an array of the average fractional frequency difference of the kth sample, which is defined as the frequency deviation divided by the center frequency of the oscillator;

$k = \overline{Y(k)}$ array element number.

Equation (3.50) shows a time-domain measure of phase noise that calculates the standard deviation of the fractional frequency fluctuations.

So far, phase noise has been studied in the frequency domain using FM theory. However, because (3.50) shows phase noise in the time domain, phase noise in the time domain will be discussed. Several electronic systems measure an external frequency source by counting zero crossings in a time period determined by the frequency source in the system. Simulating these systems, the Allan variance measurement of phase noise uses counters to sample the fractional frequency change of the output frequency of an oscillator over a fixed measurement time and measurement bandwidth. Consequently, the Allan variance of the frequency source in the system determines the frequency-measurement accuracy of the external source. This time-domain representation of phase noise describes the amount of frequency drift in parts per million over a period of time. Many engineers are familiar with the parts per million format for the temperature drift of resistors, like the parts per million format for the measure of frequency stability of a source.

Equation (3.51) integrates phase noise that is multiplied by a windowing function to compute Allan variance from phase noise [8]:

$$\sigma_y^2(\tau) = \frac{4}{(\pi f_c \tau)^2} \int_{f_{min}}^{f_{max}} 10^{\frac{\mathcal{L}(f_m)}{10}} \sin(\pi f_m \tau)^4 \, df_m \qquad (3.51)$$

where

f_{min} = lowest frequency of integration value that is equal to 1 over the largest time value of interest or at least 0.16 Hz (Hz);

f_{max} = largest frequency of integration value that is equal to the measurement bandwidth (Hz).

The sin to the fourth term in (3.51) results from the measurement time window and multiplies the phase noise before integration. This function effectively puts a highpass filter at the inverse of half the measurement time. The more time between measurements in the time domain slides the window to lower offset frequencies in the frequency domain. Finally, care must be taken in the lower-frequency limit because too high of a phase-noise level (> −10 dBc/Hz) can blow up the phase-noise modeling equations.

Another measure of phase noise is phase error. This is not to be confused with time jitter, which is a different measure. This measure would be the peak phase error from an ideal sine wave. Equation (3.52) computes the peak phase error in radians from the phase noise:

$$\eta_e = \sqrt{4 \int_{f_{m1}}^{f_{m2}} 10^{\frac{\mathcal{L}(f_m)_{osc}}{10}} \, df_m} \qquad (3.52)$$

Another measure of phase noise is the signal-to-noise ratio (SNR). The SNR becomes important in carrier-recovery or clock-recovery loops. The modulation on the downconverting oscillator in the receiver will look like modulation on the received signal with an ideal downconverting oscillator. A low SNR on the receiving oscillator will increase the BER. Equation (3.53) computes the SNR by integrating the phase noise:

$$S/N = \int_{f_{m1}}^{f_{m2}} 10^{\frac{\mathcal{L}(f_m)_{osc}}{10}} \, df_m \tag{3.53}$$

If (3.53) refers to baseband downconversion, then (3.53) must be multiplied by a factor of 2.

3.2.4   Oscillator Design and Phase-Noise Modeling

Now, the basic equations and definitions for phase-noise analysis can be applied to the design and phase-noise analysis of oscillators. Figure 3.17 shows a general oscillator configuration that can be analyzed using feedback and control theory. Equations (3.54) through (3.56) mathematically describe the elements in the block diagram. Equation (3.54) mathematically describes the phase-noise transfer function of the amplifier in the oscillator [9–12]:

$$10^{\frac{\mathcal{L}(f_m)_{amp}}{10}} = \frac{0.5F_1 kT_0}{P_{avl}}\left(1 + \frac{f_l}{f_m}\right) \tag{3.54}$$

where

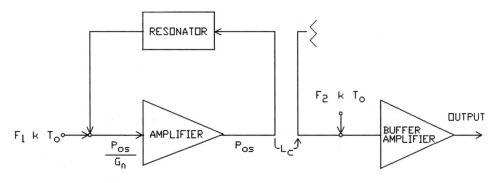

Figure 3.17   A general oscillator functional block diagram.

$\mathcal{L}(f_m)_{amp}$ = phase-noise transfer function for an amplifier (dBc/Hz);

T_0 = temperature (K);

T_0 = 290 at room temperature;

k = Boltzmann constant (mW/°K · Hz));

$k = 1.38 \times 10^{-20}$

$10 \log(kT_0) = -174$ dBm/Hz;

F_1 = noise figure ratio of amplifier in osciallator;

f_l = frequency of intersection of the amplifier's flicker noise with its constant noise figure (Hz);

P_{avl} = available input power to the amplifier, which also equals the output power divided by the amplifier gain (mW);

$P_{avl} = P_{os}/10^{G_n/10}$;

P_{os} = saturated output power of the amplifier in the oscillator (mW);

G_n = gain of the amplifier (dB).

Equation (3.54) describes the noise characteristics of the amplifier in the oscillator as a noise figure term plus a flicker noise (1/f noise) term. Equation (3.55) mathematically describes the phase-noise transfer of a resonator:

$$10^{\frac{\mathcal{L}(f_m)_{res}}{10}} = \frac{1}{1 + \dfrac{2\omega_m j Q_l}{\omega_0}} \qquad (3.55)$$

where

$\mathcal{L}(f_m)_{res}$ = phase-noise equivalent transfer function of a resonator;

ω_0 = center angular frequency of the resonator;

Q_l = loaded Q of the resonator;

ω_m = offset angular frequency from the center angular frequency of the resonator (rad/s);

$\omega_m = 2\pi f_m$.

Equation (3.55) characterizes the equivalent highpass transfer function of a resonator as described in [13]. Equation (3.56) mathematically describes the phase-noise transfer function of a coupler and a buffer amplifier:

$$10^{\frac{\mathcal{L}(f_m)_{coup}}{10}} = \frac{F_2 k T_0}{\left(\dfrac{P_{os}}{L_c}\right)} \qquad (3.56)$$

where

$\mathcal{L}(f_m)_{coup}$ = phase-noise transfer function of the coupler and buffer amplifier;

F_2 = noise figure ratio of the buffer amplifier;

L_c = coupling loss to the buffer amplifier.

Equation (3.56) describes the phase-noise effect of the coupler and buffer amplifier. The closed-loop phase noise of the oscillator is computed using feedback theory. Equation (3.57) mathematically describes positive feedback:

$$C_o/R_i(s) = \frac{G(s)}{1 \pm G(s)H(s)} \tag{3.57}$$

where

$-$ = positive feedback;

$+$ = negative feedback;

$C_o/R_i(s)$ = closed-loop transfer function;

$G(s)$ = forward transfer function;

$H(s)$ = feedback transfer function;

$G(s)H(s)$ = open-loop transfer function;

$G(s)H(s)$ = a magnitude of 1 and a phase angle of 0° for positive feedback of Barkhausen's phase criteria [14] and 180° for negative feedback; these are the conditions for oscillation.

Equation (3.57) allows all the transfer functions to be combined to form (3.58):

$$\text{Oscillator transfer function} = \frac{G(s)}{1 - G(s)H(s)} \tag{3.58}$$

$$10^{\frac{\mathcal{L}(f_m)_{osc}}{10}} = \frac{0.5F_1 kT_0}{\left(\dfrac{P_{os}}{G_n}\right)} \left(1 + \frac{f_l}{f_m}\right) \left[1 + \frac{1}{(f_m)^2}\left(\frac{B_r}{2}\right)^2\right] \frac{F_2 kT_0}{\left(\dfrac{P_{os}}{L_c}\right)}$$

where

$\mathcal{L}(f_m)_{osc}$ = single-sideband phase noise of the oscillator;

B_r = 3-dB bandwidth of the resonator (Hz).

Equation (3.58) mathematically models oscillator spectral noise-power density levels versus frequency offsets from the carrier.

Unloaded Q and insertion loss determine the bandwidth of a resonator. Unloaded Q and insertion loss are measures for the electronic performance of the resonator that appear in most resonator data sheets. Equations (3.59) and (3.60) are used to relate resonator unloaded Q and insertion loss to the 3-dB bandwidth of the resonator:

$$B_r/2 = f_c/(2Q_l) \qquad (3.59)$$

$$Q_l = Q_{ul} = \frac{Q_{ul}}{\sqrt{L_{res}}} \qquad (3.60)$$

where

f_c = output frequency of oscillator (Hz);

Q_{ul} = unloaded Q of the resonator;

L_{res} = insertion loss of the resonator.

The loaded Q value calculated in (3.60) is used in (3.58) to determine the electronic performance of the resonator in an oscillator. By computing the effect of each component in the oscillator, an engineer can select the components that will optimize the performance of an oscillator. An engineer can select a resonator and amplifier and substitute the electronic characteristics of each component into (3.58) to obtain the desired oscillator phase-noise specification.

Equation (3.58) models the phase noise of an oscillator, but it does not determine if the circuit will oscillate. A laboratory method can be devised to determine the ability of a circuit to oscillate. Figure 3.18 shows a test interconnection for a general configuration of an oscillator with the feedback loop opened. If the breakpoint of the circuit is at 50Ω, then the open-loop transfer function is easily measured on a network analyzer. For the circuit to oscillate at a particular frequency, the amplifier gain plus the resonator insertion loss must be equal to 1 and have a 0° insertion phase at that frequency to satisfy (3.57). Figure 3.19 shows a measured open-loop transfer response curve for a 590-MHz oscillator. The gain

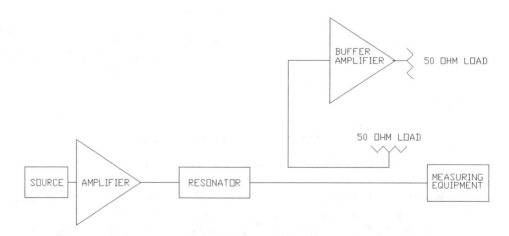

Figure 3.18   Test configuration to measure an oscillator open-loop response.

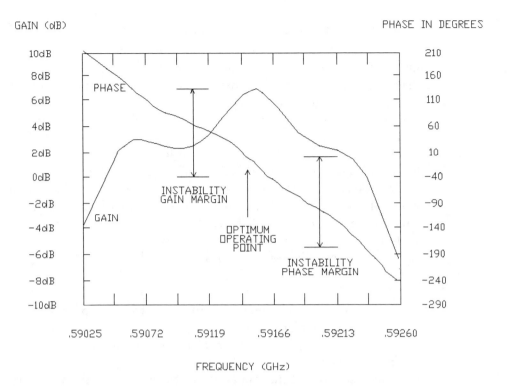

Figure 3.19   Definition of oscillator instability parameters on the measured open-loop frequency response of a 590-MHz surface acoustic wave oscillator.

in Figure 3.19 is greater than 1 at the zero-insertion phase point. The circuit will oscillate because, when the loop is closed, the measured small-signal gain is reduced by the saturation of the amplifier.

The following intuitive explanation illustrates the electronic process that insures the circuit's continued oscillatory performance. Input noise from the environment exists at the input to the amplifier in the oscillator. The noise is amplified to the output of the amplifier and filtered before reappearing at the input to the amplifier again. Larger input noise signals to the amplifier add in phase to the existing input noise signal to the amplifier, producing an even larger noise signal. The larger noise signal goes through the same amplification, filtering, and addition process until the amplifier reaches its saturation limit. The saturation limit of the amplifier reduces the gain until (3.57) is satisfied. Thus, any small-signal gain greater than unity defines the *instability gain margin*. If the insertion phase is not equal to zero at any frequency point where the gain is greater than unity, then the circuit will not oscillate. Consequently, the smallest amount of phase shift necessary to stop the circuit from oscillating defines an *instability phase margin*. Maximizing the instability gain and phase margins defines the optimum operating point for the oscillator. Figure 3.19 shows the operating frequency of the oscillator at the zero-insertion phase point. If a phase shifter is placed in series with the amplifier and resonator, then the operating frequency of the oscillator can be tuned by adjusting the phase shifter. This frequency shift occurs because the operating frequency of the oscillator tracks the frequency of the zero-insertion phase point.

3.2.5 Negative-Resistance Oscillator Model

Many oscillators are designed using a negative-resistance component to produce oscillations. Figure 3.20 shows a common negative-resistance oscillator configuration. The negative-resistance oscillator does not electrically appear to be the same as the feedback configuration of Figure 3.17.

From (3.58) the gain and insertion-loss parameters used in the analysis of the previous oscillator configuration do not model the negative-resistance oscillator; however, using reflection gain and reflection loss in place of gain and insertion loss for parameters in the general oscillator configuration analysis does model the negative-resistance oscillator. A negative-resistance component produces a reflection gain from an input signal. This reflection causes multiple reflections to occur when the total reflection gain of the circuit is greater than the reflection losses. The application of the developed oscillator analysis to negative-resistance oscillators demonstrates the versatility of the analysis approach.

3.2.6 Power Slopes of Oscillators

Various power slopes versus frequency offsets occur when plotting the spectral power density of an oscillator. Random walk FM (f^{-4}), flicker FM (f^{-3}), white FM (f^{-2}), flicker phase (f^{-1}), and white phase (f^{0}) are the names commonly used to describe the power slopes.

The power slopes and the names associated with each are plotted in Figure 3.21. The general Bode plot expression of (3.61) for a pole is a mathematical form that describes the curves in Figure 3.21:

$$P_m(f_m) = 10 \log \left\{ \left[\frac{1}{1 + \left(\frac{f_m}{f_{3p}} \right)^2} \right]^n \right\} \tag{3.61}$$

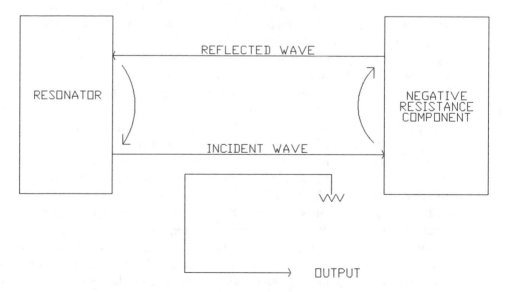

Figure 3.20 A negative-resistance oscillator functional block diagram.

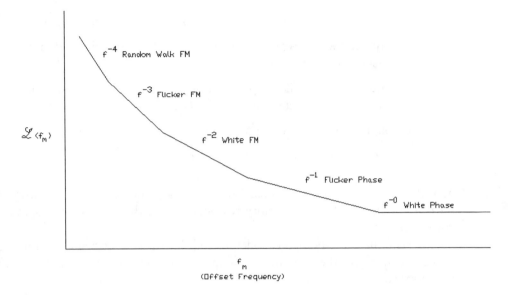

Figure 3.21   The designations of various oscillator phase-noise-power slopes.

where

$$n = \text{order of the pole that determines roll-off slope;}$$

$$f_{3p} = \text{frequency location of 3-dB point (Hz);}$$

$$P_m(f_m) = \text{power level at the offset frequency (dBc).}$$

Instead of modeling the noise curve of the oscillator with one power slope, various power slopes are used to model sections of the noise curve. Each section models a particular noise characteristic's slope. This is mathematically shown in (3.62) through (3.64):

$$\text{White FM } \mathcal{L}(f_m) = 10 \log\left\{\left[\frac{1}{1 + \left(\frac{f_m}{f_{3p}}\right)^2}\right]^1\right\} \text{ for } f_2 > f_m > f_1 \quad (3.62)$$

$$\text{Flicker phase noise FM } \mathcal{L}(f_m) = 10 \log\left\{\left[\frac{1}{1 + \left(\frac{f_m}{f_{3p}}\right)^2}\right]^{0.5}\right\} \text{ for } f_{int} > f_m > f_2$$

$$(3.63)$$

$$\text{White phase noise } \mathcal{L}(f_m) = \text{constant for } f_m > f_{int} \quad (3.64)$$

where

f_1 = frequency where white FM phase-noise characteristics begin (Hz);

f_2 = frequency where white FM phase-noise characteristics end and flicker phase-noise characteristics begin (Hz);

f_{int} = frequency where flicker phase-noise characteristics end and white phase-noise characteristics begin (Hz);

f_{3p1} = pole location for white FM (Hz);

f_{3p2} = pole location for flicker phase noise.

Modeling two or three phase-noise slopes with an equation instead of using each measured data point eases memory restrictions and the writing of programming code in the computer.

Determining the Bode plot model for the sloped sections begins by finding the corner frequency of the pole. By rearranging (3.61), the corner frequency of a pole is computed as shown by (3.65):

$$f_{3p} = \frac{f_v}{\sqrt{\dfrac{1}{\left(10^{\frac{\mathcal{L}(f_c)}{10}}\right)^{1/n} - 1}}} \qquad (3.65)$$

where

f_{3p} = frequency of pole (Hz);

f_v = frequency value at the relative power level $L(f_v)$ (Hz);

$n = 0.5$ for flicker noise (10 dB/dec);

$n = 1$ for white FM phase noise (20 dB/dec);

$n = 1.5$ for flicker FM noise (30 dB/dec).

Equation (3.65) calculates the corner frequency of a pole from the slope and one frequency and magnitude point in the region of the phase noise to be modeled. The model of phase noise defines the exponential roll-off between two offset-frequency limits that define the section. Determining the corner-frequency location of the pole of each section begins by first locating on the measured oscillator data the noise sideband power in dBc/Hz relative to the carrier at the offset-frequency point from the carrier that lies inside the oscillator noise-characteristic slope that is to be modeled. Substituting the power and frequency values into (3.65) computes the corner-frequency location of the pole for the modeled section. After modeling the power slopes for each section, the frequency of intercept with the flat phase-noise response of the oscillator must be determined. Rearranging (3.65) produces (3.66), which computes the frequency of intercept with the flat phase-noise response of the oscillator:

$$f_{int} = f_{3p2} \sqrt{\frac{1}{\left(10^{\frac{\mathscr{L}(f_m)}{10}}\right)^{1/n}} - 1} \qquad (3.66)$$

From the power level of the flat phase-noise response, (3.66) computes the frequency limit where flat phase-noise response occurs. If one cannot easily determine the power slope, an equation can be used to calculate the slope accurately. Equation (3.67) computes this slope between data points:

$$n = \frac{\mathscr{L}(f_a) - \mathscr{L}(f_b)}{10 \log \left[\dfrac{1 + \left(\dfrac{f_b}{f_{3p}}\right)^2}{1 + \left(\dfrac{f_a}{f_b}\right)^2}\right]} \qquad (3.67)$$

where

f_a = start frequency of slope (Hz);

f_b = stop frequency of slope (Hz).

Przedpelski [15] presents an alternate method for modeling phase noise. The power slopes are modeled with the general exponential expression of (3.68):

$$\mathscr{L}(f_m) = 10 \log[a(f_m)^b] \qquad (3.68)$$

The oscillator noise characteristics in (3.68) are modeled by (3.69) through (3.71):

$$\text{White FM } \mathscr{L}(f_m) = 10 \log[a_1(f_m)^2] \text{ for } f_2 > f_m > f_1 \qquad (3.69)$$

$$\text{Flicker phase noise } \mathscr{L}(f_m) = 10 \log[a_2(f_m)^1] \text{ for } f_{int} > f_m > f_2 \qquad (3.70)$$

$$\text{White phase noise } \mathscr{L}(f_m) = \text{constant for } f_m > f_{int} \qquad (3.71)$$

where

a_1 = constant for white FM noise response;

a_2 = constant for flicker phase-noise response.

Equation (3.72) calculates the multiplication constant in (3.68):

$$a = 10^{\frac{\mathscr{L}[(f_a) - 10(b \log(f_a)]}{10}} \qquad (3.72)$$

Equation (3.73) calculates the frequency of intercept with the noise floor:

$$f_{int} = \left[\frac{10^{\frac{\mathcal{L}(f_{int})}{10}}}{a} \right]^{1/b}$$

(3.73)

Finally, (3.74) calculates the required noise slope:

$$b = \frac{\mathcal{L}(f_a) - \mathcal{L}(f_b)}{10[\log(f_a) - \log(f_b)]}$$

(3.74)

The alternate method has a power slope that agrees with the slope of the oscillator spectral-noise characteristic in Figure 3.21. The Bode plot model differs with the characteristic spectral-noise slope of the oscillator in Figure 3.21 by a factor of twice the slope as shown by (3.75):

$$n = b/2$$

(3.75)

The Bode plot equation model of spectral-noise power-density slopes were programmed into the MEAS program [16]. To avoid confusion when modeling phase-noise data for the MEAS computer program, the user must remember that the phase-noise slope in the Bode plot model is two times greater than the phase-noise slopes in Figure 3.21.

3.2.7 Resonator Effects on Oscillator Phase Noise

The resonator term in (3.58) has an expression that contains the resonator bandwidth. In a spectral-noise power-density plot of an oscillator the 3-dB bandwidth of the resonator is located at an offset frequency from the carrier equal to one-half of the 3-dB bandwidth of the resonator. The relationship of the resonator bandwidth to the flicker corner frequency of the low-noise oscillator amplifier affects the spectral-noise power-density curves as shown in Figure 3.22. From Figure 3.22, the breakpoints of the measured phase-noise response of an oscillator clearly depend on the selection of the amplifier and resonator in the oscillator.

The MEAS program [16] was developed to model phase-noise characteristics of oscillators, measured data, and measuring equipment. From (3.61) through (3.63), the single-sideband phase-noise curves of an oscillator can be calculated. Equations (3.61) through (3.63) were programmed into the MEAS program. Now the effects of components on the phase noise of an oscillator can be computed. Figure 3.23 shows various single-sideband oscillator responses for various types of resonators, with the unloaded Qs shown in Table 3.1.

The amplifier characteristics of noise figure, output power, and flicker corner frequency were constant for the three designs. Figure 3.23 is an example of the analysis trade-offs that can be investigated in an oscillator design when computer models of the components in an oscillator are used.

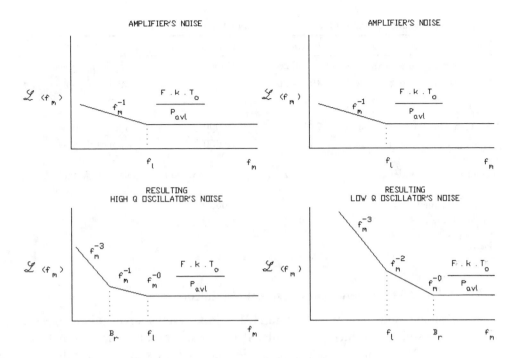

Figure 3.22   The effects of flicker noise corner frequency of an amplifier on the phase-noise-power curves for high-Q and low-Q oscillators. (*From:* [17, Figure 9]. © 1979 Agilent. Reprinted with permission.)

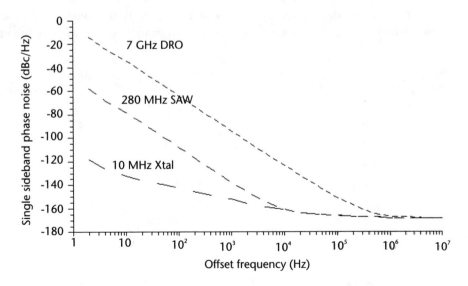

Figure 3.23   The effect of a 10-MHz crystal resonator, a 280-MHz surface acoustic wave (SAW) resonator, and a 7-GHz dielectric resonator on oscilloscope phase noise with all other parameters remaining constant.

Table 3.1 The Q and Operating Frequency of Various High-Q Resonators[a]

Type of Resonator	Frequency	Unloaded Q	Major Limitation
Quartz crystal, 3rd overtone	10 MHz	1.5×10^6	Applicable power
Quartz crystal, 7th overtone	100 MHz	90,000	Applicable power spurious modes; microphonic phase modulation
Quartz crystal, 11th overtone	280 MHz	40,000	Applicable power
Surface acoustic wave resonator on quartz	800 MHz	10,000	Applicable power
Cylindrical waveguide	>2 GHz	20,000–50,000	Size (for L and S band); microphonic phase modulation
Dielectric resonator	7 GHz	$7,000^b$	
Ceramic resonator	4–60 MHz	500–2,000	

[a]*Source:* [18, p. 11].
[b]*Source:* [19].

3.2.8 Allan Variance and Residual FM Calculations

Example Bode plot models of an HP10544B crystal source and an HP8672 microwave source produce the calculated frequency responses shown in Figures 3.24 and 3.25. Computer modeling of the sources allows mathematical operations to be performed on the spectral-noise power-density curves of the source, which allows other forms of spectral measurements to be calculated. Figure 3.26 shows computer calculations of the Allan variance by the integration method of (3.48) for an HP10544B crystal source.

Table 3.2 displays the calculation of residual FM for an HP8672 microwave source. The computer results agree with the manufacturer's performance specification sheets of the sources. The above calculations demonstrate that computer modeling of phase noise allows various mathematical operations to be performed.

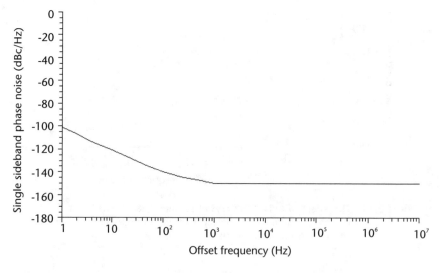

Figure 3.24 Phase-noise model response of an HP10544B 10-MHz crystal oscillator.

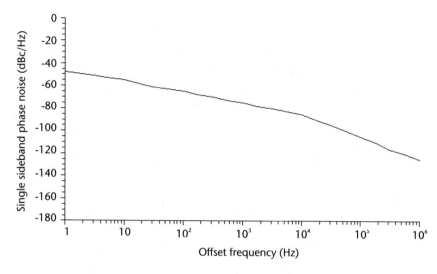

Figure 3.25 Phase-noise model response of an HP8672B microwave synthesizer operating at 10 GHz.

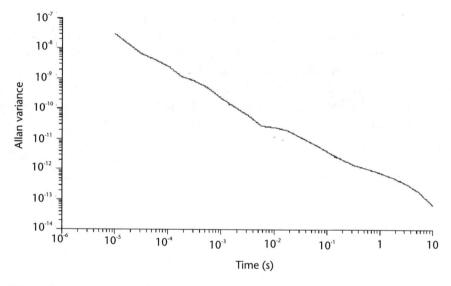

Figure 3.26 Allan variance of an HP10544B 10-MHz crystal oscillator calculated from the phase-noise model of Figure 3.24.

To summarize, this section derived phase-noise relationships from narrowband FM waveform theory. Oscillator design circuitry was explored, and modeling techniques were developed so that the effects of various oscillator components on phase noise could be evaluated. Modeling techniques of phase noise were developed based on the spectral-noise power-density characteristics of the oscillator. The Bode plot phase-noise model as an analysis tool allows mathematical operations to be performed on phase-noise curves. The Bode plot phase-noise model will be used in later chapters to analyze phase noise in PLLs and their applications.

Table 3.2 Calculation of Residual FM from the Phase-Noise Model of Figure 3.25 for an HP8672 Microwave Synthesizer at 10 GHz

Frequency	Noise	f^2	Residual FM
1	−46	0	0
1.7	−48	5	0.0082
3	−51	10	0.0168
5	−53	15	0.0311
10	−55	20	0.0559
17	−58	25	0.0998
31	−61	30	0.1777
56	−63	35	0.3162
100	−65	40	0.5623
177	−68	44	0.9999
316	−70	50	1.7783
562	−73	55	3.1623
1,000	−75	59	5.6234
1,778	−78	65	9.9999
3,162	−80	70	17.782

3.2.9 Phase Noise in PLLs

The phase noise of the VCO and the phase noise of the reference oscillator in a PLL affect the output phase noise of the PLL. An equation can be derived that computes the effects of the phase noise of the VCO and the phase noise of the reference oscillator on the output phase noise of the PLL.

Figure 3.27 shows the block diagram of the PLL for analyzing phase noise. Doing the block diagram algebra in Figure 3.27 and substituting $f_m = s$ yields (3.76):

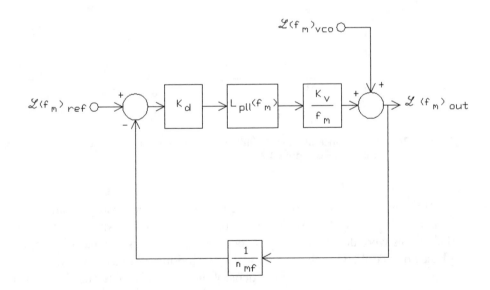

Figure 3.27 A block diagram shows the effect of each oscillator on the total phase noise of the PLL.

$$10^{\frac{\mathcal{L}(f_m)_{out}}{10}} = \left[\frac{10^{\frac{\mathcal{L}(f_m)_{out}}{10}}}{n_{mf}} + 10^{\frac{\mathcal{L}(f_m)_{ref}}{10}} \right] K_d \frac{K_v}{f_m} L_{pll}(f_m) + 10^{\frac{\mathcal{L}(f_m)_{vco}}{10}} \quad (3.76)$$

where

$\mathcal{L}(f_m)_{ref}$ = single-sideband phase-noise ratio associated with the reference oscillator (dBc/Hz);

$\mathcal{L}(f_m)_{vco}$ = single-sideband phase-noise ratio associated with the VCO (dBc/Hz);

$\mathcal{L}(f_m)_{out}$ = single-sideband phase-noise ratio as a result of the PLL (dBc/Hz);

$L_{pll}(f_m)$ = active-filter function for PLL.

Combining like terms, substituting $H(f_m) = 1/n_{mf}$, substituting $G(f_m) = K_d L_{pll}(f_m) K_v / f_m$, and solving for the output phase noise produces (3.77):

$$10^{\frac{\mathcal{L}(f_m)_{out}}{10}} = 10^{\frac{\mathcal{L}(f_m)_{ref}}{10}} G(f_m) - 10^{\frac{\mathcal{L}(f_m)_{out}}{10}} G(f_m)H(f_m) + 10^{\frac{\mathcal{L}(f_m)_{vco}}{10}}$$

$$10^{\frac{\mathcal{L}(f_m)_{out}}{10}} [1 + G(f_m)H(f_m)] = 10^{\frac{\mathcal{L}(f_m)_{ref}}{10}} G(f_m) + 10^{\frac{\mathcal{L}(f_m)_{vco}}{10}}$$

$$10^{\frac{\mathcal{L}(f_m)_{out}}{10}} = \frac{10^{\frac{\mathcal{L}(f_m)_{ref}}{10}} G(f_m)}{1 + G(f_m)H(f_m)} + \frac{10^{\frac{\mathcal{L}(f_m)_{vco}}{10}}}{1 + G(f_m)H(f_m)} \quad (3.77)$$

Equation (3.77) shows that the output phase noise of the PLL is the sum of the phase noise of the VCO and the phase noise of the reference oscillator. Inside the bandwidth of the loop, the output phase noise of the PLL has the characteristics of the phase noise of the reference oscillator because the open-loop gain $[G(f_m)H(f_m)]$ is large and cancels the phase noise of the VCO. Outside the bandwidth of the loop, the output phase noise of the PLL has the characteristics of the phase noise of the VCO because the closed-loop gain is small and cancels the phase noise of the reference oscillator.

Consider locking a 1-GHz VCO to a 10-MHz HP10544B crystal reference oscillator. Figure 3.22 shows the phase noise for the crystal oscillator. The reference oscillator has noise power of −120 dBc/Hz at 10 Hz with a 20-dB/dec slope, −140 dBc/Hz at 100 Hz with a 10-dB/dec slope, and a noise floor of −150 dBc/Hz. The VCO has −110 dBc/Hz at 100 kHz and 30-dB/dec slope and a noise floor of −158 dBc/Hz. The PLL will have a 10-kHz natural frequency and a damping factor of 2.

Figure 3.28 shows the multiplication factor responses for this loop. Equation (3.77) computes these factors. Figure 3.29 shows the phase noise of the reference oscillator, the VCO, and the resulting 1-GHz PLL. Adding the reference multiplica-

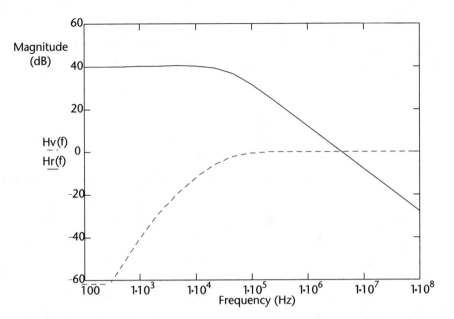

Figure 3.28 The reference oscillator and VCO phase-noise multiplication factors for a PLL with a
10-kHz natural frequency, a damping factor of 1, a 1-GHz VCO, and a 10-MHz crystal
reference oscillator.

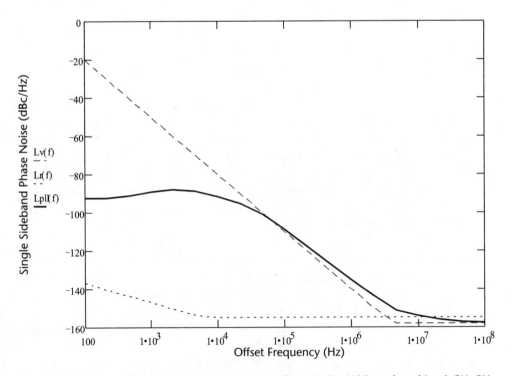

Figure 3.29 Phase noise of the 10-MHz reference oscillator, 1-GHz VCO, and resulting 1-GHz PLL.

tion factor in Figure 3.28 to the phase noise of the reference oscillator in Figure 3.29 and adding the VCO multiplication factor in Figure 3.28 to the phase noise of the VCO in Figure 3.29 evaluates (3.77) for computing the output phase noise of the PLL in Figure 3.29.

Many systems use PLLs to generate transmitter signals, clocks, or receiver LOs; consequently, phase-noise analysis of a system with a PLL requires evaluating (3.77) to determine the output phase noise of the PLL [15, 20].

3.2.9.1 Transfer Functions

To understand the effects of each component on the output phase noise, we have to determine the transfer function of each of the components. Using block diagram algebra [21] produces (3.78). Equation (3.78) computes the total transfer function from reference input to loop output.

$$L_{pll} = \frac{n_{mf}\left(K_d L_A \dfrac{K_v}{s} \dfrac{1}{n_{mf}}\right)}{1 + K_d L_A \dfrac{K_v}{s} \dfrac{1}{n_{mf}}} \tag{3.78}$$

where

L_A = loop filter transfer function.

In the numerator, the open-loop transfer function is multiplied by the multiplication ratio to get the forward transfer function. The denominator remains the same. Equation (3.79) computes the transfer function from the output of the phase detector to the loop output:

$$L_{pd} = \frac{n_{mf}\dfrac{1}{K_d}\left(K_d L_A \dfrac{K_v}{s} \dfrac{1}{n_{mf}}\right)}{1 + K_d L_A \dfrac{K_v}{s} \dfrac{1}{n_{mf}}} \tag{3.79}$$

In the numerator, the forward transfer function is divided by the gain of the phase detector. The denominator remains the same. Equation (3.80) computes the transfer function from the input to the control line to the loop output.

$$L_{tune} = \frac{n_{mf}\dfrac{1}{K_d}\dfrac{1}{L_A}\left(K_d L_A \dfrac{K_v}{s} \dfrac{1}{n_{mf}}\right)}{1 + K_d L_A \dfrac{K_v}{s} \dfrac{1}{n_{mf}}} \tag{3.80}$$

In the numerator, the forward transfer function is divided by the gain of the phase detector and the loop-compensation filter. The denominator remains the same.

Equation (3.81) computes the transfer function from the VCO output to the loop output:

$$L_{vco} = \left[\frac{n_{mf} \dfrac{1}{K_d} \dfrac{1}{L_A} \dfrac{1}{\left(\dfrac{K_v}{s}\right)} \left(K_d L_A \dfrac{K_v}{s} \dfrac{1}{n_{mf}}\right)}{1 + K_d L_A \dfrac{K_v}{s} \dfrac{1}{n_{mf}}} \right] = \left[\frac{1}{1 + K_d L_A \dfrac{K_v}{s} \dfrac{1}{n_{mf}}} \right]$$

(3.81)

In the numerator the forward transfer function is divided by the forward transfer function, which results in 1. The denominator remains the same. These equations are used to develop equations that determine the effects of noise contributions from internal and coupled sources.

3.2.9.2 Noise Contributions from Internal and Coupled Sources

In the optimum low-noise PLL design, the output phase noise should be dominated by the noise from the crystal reference oscillator and the VCO. Consequently, noise from other sources should be minimized [22].

Noise sources in a PLL are as follows:

- Divider;
- Phase detector;
- Loop filter;
- Coupled noise onto tune line;
- Coupled noise from power supply.

Equation (3.82) computes the effect of noise from the phase detector output and the input noise of the active filter:

$$\mathscr{L}_{amp}(f) = 20 \log \left[\frac{1}{2K_d} \frac{N \dfrac{(f_n)^2}{[f(i)]^2} \sqrt{1 + \left(\dfrac{2\zeta f(i)}{f_n}\right)^2}}{\sqrt{\left[1 - \left(\dfrac{f_n}{f(i)}\right)^2\right]^2 + \left(\dfrac{2\zeta f_n}{f(i)}\right)^2}} e_{amp} \right]$$

(3.82)

where

e_{amp} = equivalent input noise of an amplifier [V/sqr(Hz)].

Equation (3.83) computes the effect of divider noise on the output phase noise of the PLL, which has the same multiplication factor as the noise from the reference oscillator to the output of the PLL:

$$\mathscr{L}_{plldiv}(f) = 20 \log\left(10^{\frac{\mathscr{L}_d + H_r(f)}{20}}\right) \tag{3.83}$$

where

$H_r(f)$ = reference phase noise transfer function, which is the closed loop transfer function;

$\mathscr{L}_d$ = noise level of the divider [dBc(Hz)].

Equation (3.84) computes the effect of noise at the tune line to the output phase noise of the PLL:

$$\mathscr{L}_{tune}(f) = 20 \log\left[\frac{1}{2} \frac{\frac{K_v}{[f(i)]^1}}{\sqrt{\left[1 - \left(\frac{f_n}{f(i)}\right)^2\right]^2 + \left(\frac{2\zeta f_n}{f(i)}\right)^2}} e_{tune}\right] \tag{3.84}$$

where

e_{tune} = noise level at the tune line of the VCO [V/sqr(Hz)].

Example 3.2

Let's use an example PLL to show the phase-noise contributions of the key components in a PLL. The example PLL has an input frequency and output frequency of 2 MHz and 480 MHz, respectively. The parameters of the loop have a feedback divide ratio of $N = 240$, a loop natural frequency of $f_n = 3$ kHz, a damping factor of $\zeta = 0.9$, a flat divider phase noise of $\mathscr{L}_d = -165$ dBc/Hz, an operational amplifier noise of $e_{amp} = 30$ nV/sqr(Hz), $\mathscr{L}_{pd} = -165$ dBc/Hz, $K_d = 0.4/(2\pi)$, $K_v = 2\pi 500E6$, and $e_{tune} = 0.1$ nV/sqr(Hz).

The VCO has phase noise of $\mathscr{L}_{vco} = -82$ dBc/Hz at 1 kHz with 30-dB/dec slope and a −140-dBc/Hz noise floor. The reference crystal oscillator has phase noise of $\mathscr{L}_{ref} = -134$ dBc/Hz at 10 Hz with 5-dB/dec slope and −149-dBc/Hz noise floor.

Figure 3.30 shows the noise contributions from the reference oscillator, VCO, amplifier, phase detector, divider, and noise on the tune line after applying the transfer function equations. The figure shows that noise from the amplifier is greater than the noise from the reference oscillator and VCO. The noise from the dividers, tune line, and phase detector is less than the noise from the reference and VCO. In addition, this figure shows that the shape of the noise can be used to identify the coupling location. For instance, flat noise coupled before the tune line will have a lowpass filter shape. Flat noise coupled at the tune line will have a bandpass filter shape. Flat noise coupled at the output of the VCO will have a highpass filter shape.

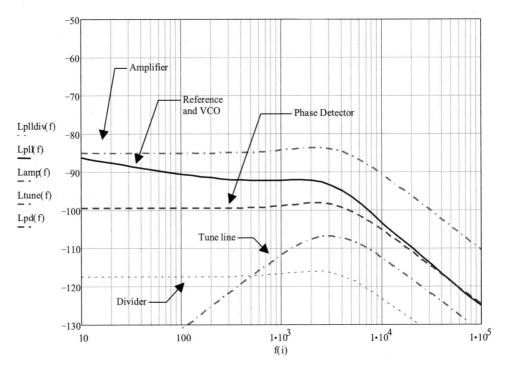

Figure 3.30 Noise contributions from the divider, amplifier, phase detector, and tune line compared with output phase noise of the PLL.

To summarize, this section developed a PLL phase-noise model. Equations were developed to model and design PLLs for low phase noise. These equations will be referred to often since PLLs are a common occurrence, affecting SNR in communication systems and the amount of jitter that occurs in digital circuits and digital signal processors.

3.3   Jitter in PLLs

The main concern for jitter in digital systems is the decrease in timing margin that it causes. Controlling jitter allows more flexibility in the timing budget.

For example, Figure 3.31 shows the effect of jitter on the edge-to-edge timing margin. A positive skew results in the final register (U_2) of the data path that leads the time of arrival of the clock signal at the initial register (U_1). Under these conditions, the maximum attainable operating frequency is decreased (worst case). Equations (3.85) through (3.87) calculate the effects of jitter on the timing margin [23]:

$$T_{T_{margin}} = T_{ckmin} - (T_{U_1} + T_{comb} + T_{U_2 setup} + T_{skew} + T_{jitter}) \qquad (3.85)$$

$$T_{T_{margin}} = 11 \text{ ns} - (3 \text{ ns} + 5 \text{ ns} + 1 \text{ ns} + .5 \text{ ns} + 1 \text{ ns}) \qquad (3.86)$$

$$T_{T_{margin}} = 0.5 \text{ ns} \qquad (3.87)$$

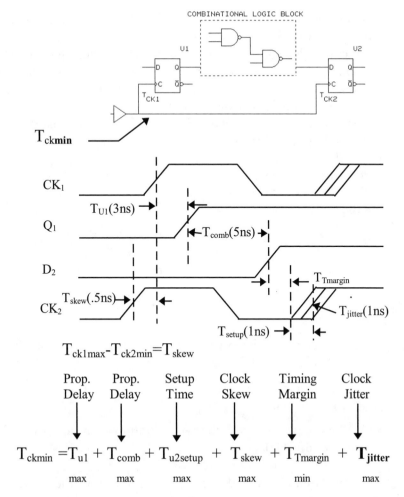

Figure 3.31 Jitter effect on edge-to-edge timing. (*From:* [23]. © 1997 Wavecrest. Reprinted with permission.)

3.3.1 Causes of Jitter

The effects of jitter have three main components in the frequency domain. These components are close-to-the-carrier phase noise, far-from-the-carrier phase noise, and spurious signals. The highest levels in silicon are spurious signals. The following are causes of jitter from spurious signals [24]:

- Data/clock dependency;
- Conducted (e.g., switching power supply, motors, relays);
- Simultaneous switching outputs (SSOs) or ground bounce;
- Crosstalk;
- Reflections;
- EMI;
- Low loop stability of $<10°$ phase margin in a control loop.

Figure 3.32 shows possible injection paths into a PLL that can cause jitter. The most sensitive spot in any PLL is the voltage-controlled oscillator's tune line because it usually has high-gain and high-impedance characteristics. Next, the power-supply lines to the VCO are usually sensitive. Recent measurements have shown that output buffer switching has been modulating PLL outputs. The most likely path is the common power-supply and ground lines between the input/output (I/O) buffers and the PLL. Figure 3.33 shows simultaneous switching of output and ground bounce. These signals can feedback into the PLL through the power-supply and ground leads and cause jitter.

The next question is, How does the PLL respond to these disturbances? Equation (3.88) shows lowpass and highpass functions to phase noise, depending on where the noise couples into the PLL:

$$10^{\frac{\mathscr{L}(f_m)_{out}}{10}} = \frac{10^{\frac{\mathscr{L}(f_m)_{ref}}{10}} G(f_m)}{1 + G(f_m)H(f_m)} + \frac{10^{\frac{\mathscr{L}(f_m)_{vco}}{10}}}{1 + G(f_m)H(f_m)} \tag{3.88}$$

where

$\mathscr{L}(f_m)_{ref}$ = single-sideband phase-noise ratio associated with the reference oscillator (dBc/Hz);

$\mathscr{L}(f_m)_{vco}$ = single-sideband phase-noise ratio associated with the VCO (dBc/Hz);

$\mathscr{L}(f_m)_{out}$ = single-sideband phase-noise ratio as a result of the PLL (dBc/Hz).

Equation (3.88) shows that the phase noise from the reference oscillator is lowpass filtered to the output of the PLL. Figure 3.34 shows the lowpass filter (LPF) transfer function for the PLL. Equation (3.88) also shows the phase noise from the VCO is highpass filtered to the output of the PLL. Figure 3.35 shows the highpass filter (HPF) transfer function for the PLL. A roll-off capacitor in the compensation filters the signal outside the 1-MHz loop bandwidth. Consequently, the function behaves like a bandpass filter.

The equation for phase noise also shows the effect of signal sources coupling into the PLL. Consequently, where signal sources couple into the PLL affects the response that occurs. Changes in the reference frequency and in the divide ratio are lowpass filter responses. A dc change in the supply to the VCO and signals coupled to the output of the VCO produce highpass filter responses. Understanding these responses is one of the keys to understanding the source of the interfering signal and troubleshooting these problems.

3.3.2   Phase-Noise Analysis on Jitter

Analysis illustrates the effects of the three major components on jitter. Varying the values of these components and calculating jitter versus measurement time shows their relationships. First, let's study the effects of phase noise on jitter.

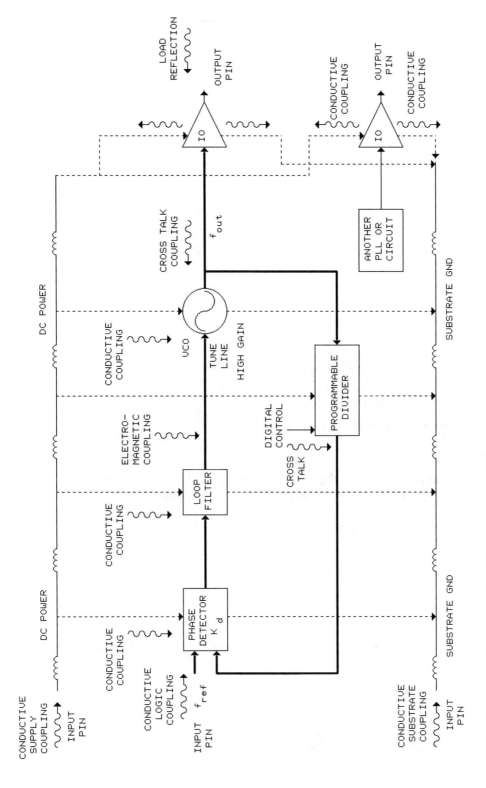

Figure 3.32 External injection paths into a PLL that can cause jitter.

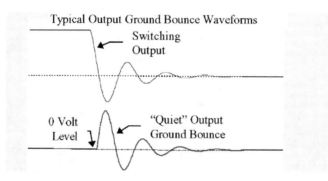

Figure 3.33   Diagram of simultaneous switching. (*After:* [23].)

The following lists the causes of jitter from far-from-the-carrier and close-to-the-carrier phase noise, which results from the physical material:

- Thermal noise (Johnson);
- Shot noise (Schotky);
- Popcorn noise;
- Flicker noise ($1/f$ noise);
- Avalanche noise.

Equation (3.89) converts phase noise to the Allan variance by integrating over the measurement bandwidth.

$$T_{jit}(T) = T \sqrt{\frac{4}{(f_c)^2 \pi^2 T^2} \int_0^{f_{max}} S_\phi(f_n) \left[\sin(\pi f_n T)\right]^4 df_n} \tag{3.89}$$

where

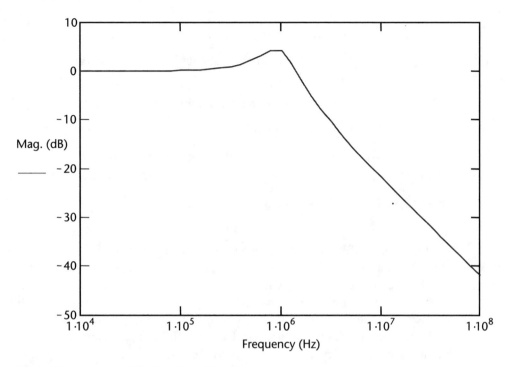

Figure 3.34   Lowpass filter function of the PLL.

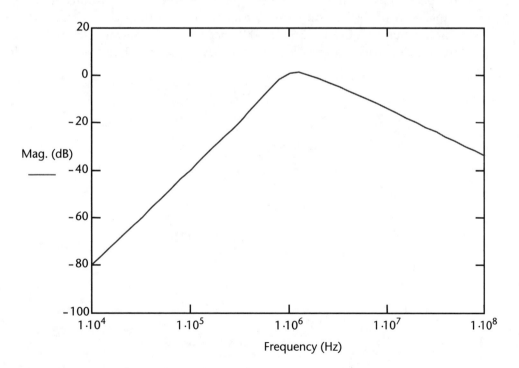

Figure 3.35   Highpass filter function for the PLL.

$S_\phi(f_n)$ = spectral density of phase fluctuations (rad^2/Hz);

f_{max} = maximum measurement frequency, analog-to-digital (A/D) bandwidth (Hz);

T = gate time of measurement (time interval between edges) (sec);

f_c = frequency of carrier (Hz);

f_n = modulation frequency of noise component (Hz);

$\mathcal{L}(f_m) = 1/2\,S_\phi(f_m)$ = spectral density of phase fluctuations to phase noise;

$\mathcal{L}(f_m)$ = single-sideband phase noise (dBc/Hz).

Equation (3.89) has two major terms in the integration: (1) spectral density of phase fluctuations, which is related to phase noise, and (2) a sin to the fourth term, which is related to the time windowing of the signal in the time domain. When we only want to look at the rising edges after 1,000 zero crossings of a 100-MHz sine wave in the time domain, convolution of this time-domain window (10 μs) to the frequency domain results in a filter (at 100 kHz), as shown in Figure 3.36. For illustration purposes, this filter for a 10-μs time interval was layed over the phase-noise plot of a PLL. Consequently, the magnitude of the filter has no meaning, and for discussion sake we will normalize it to 0 dB. As an approximation, the multiple humps for frequencies greater than 100 kHz can be modeled as a pass condition, making the filter response a highpass filter. Equation (3.89) shows that the phase noise is multiplied by this filter, and the product is integrated with the bandwidth of the measurements system. For increasing time intervals (or more skipped output VCO edges > 1,000) the filter slides to the left for lower offset frequencies, which allows lower-frequency noise to affect the jitter result. For

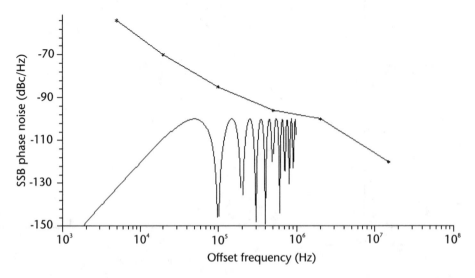

Figure 3.36 Overlayed plots of phase noise and the windowing function in the Allan variance (jitter) equation.

decreasing time intervals (or fewer skipped edges < 1,000), the filter slides to the right for higher offset frequencies, which rejects more lower-frequency energy.

Starting the filter at the maximum right-hand side in Figure 3.36 and increasing the time interval, which slides the filter curve to the left, generates Figure 3.37. Mirroring the phase noise about the y-axis in Figure 3.36 gives the general shape of the waveform in Figure 3.37. This waveform gives the jitter for all values of divide ratios after the VCO. Furthermore, the flattening out of the phase noise at 10^{-6} seconds is approximately equal to the bandwidth of the loop filter.

We can understand the trade-offs and oscillator design implications for jitter by varying the phase-noise sections in an oscillator. First, let's show the variation of the close-to-the-carrier phase noise, which can also be referred to as resonator noise. Figure 3.38 shows close-to-the-carrier phase noise, which has a 30-dB/dec slope, and the offset in phase noise is changed in 15-dB increments. This variation

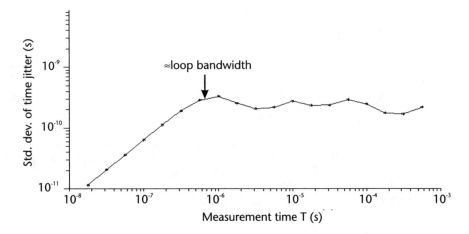

Figure 3.37 Allan variance (time jitter) versus time interval between edges (skipping edges).

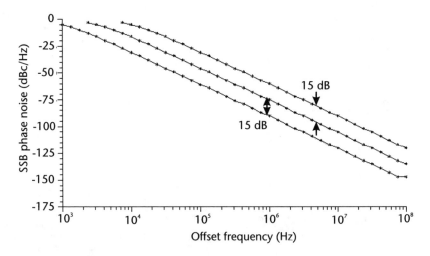

Figure 3.38 Varying resonator phase noise.

in phase noise represents changing the Q of the resonator in the oscillator or changing the corner flicker frequency of the amplifier in the oscillator.

Next, Figure 3.39 shows far-from-the-carrier phase noise, which is Gaussian white noise. The noise level is changed by 15-dB increments. This variation in phase noise represents changing the thermal or shot noise of the oscillator amplifier or changing the thermal or shot noise of the following buffer-amplifier stage. For computing the jitter from the varied inputs, a 33-MHz clock frequency is used, and the phase noise of the VCO single end ring is used as a reference point.

Now let's calculate the jitter for the three main components. Figure 3.40 shows the effect of jitter from varying the resonator noise (close-to-the-carrier phase

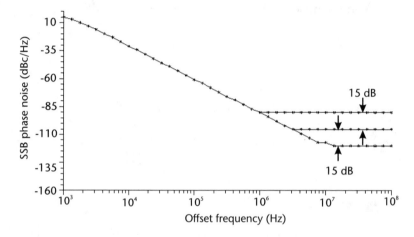

Figure 3.39   Varying far-from-the-carrier phase noise.

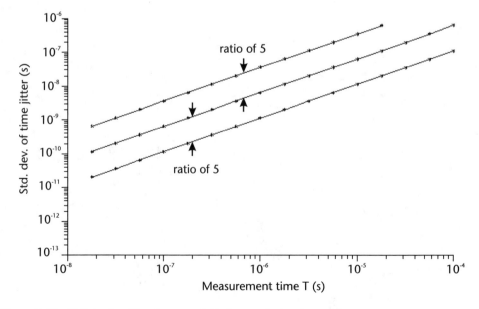

Figure 3.40   Shift in time jitter from varying close-to-the-carrier noise.

noise). The measurement time is equivalent to the gate time in a counter or the time interval in a modulation-domain analyzer. For edge-to-edge jitter in a clock, the measurement time is one cycle of the clock, or 33 ns in this case. Consequently, this figure shows that the time jitter varies from 40 ps, to 200 ps, to 1,000 ps for 15-dB increases in the resonator noise level.

Figure 3.41 shows the effect of jitter from varying the Gaussian noise (far-from-the-carrier noise). The jitter at 33 ns varies from 100 ps, to 500 ps, to 3,000 ps for 15-dB increments in noise level. The shift in time-jitter magnitude matches the 15-dB increments in noise level. Varying the far-from-the-carrier phase noise only shifts the flattened part of the time-jitter curve. These figures show two of the three major contributions to jitter. Next, the effects of spurious signals will be analyzed.

3.3.3 Analysis of Spurious Signals on Jitter

Analyses of spurious signals are derived from narrowband FM theory. Sampling of the continuous waveform changes the results. First, we will study narrowband FM theory. Later, the effects of sampled data will be shown.

3.3.3.1 Narrowband FM Theory

Equation (3.90) computes peak-to-peak jitter for a spurious signal, which is from narrowband FM theory:

$$T_{pp} = 2 \left[\frac{1}{f_{vco} - f_m 2 \cdot 10^{\frac{P_{sb}}{20}}} - \frac{1}{f_{vco}} \right] \qquad (3.90)$$

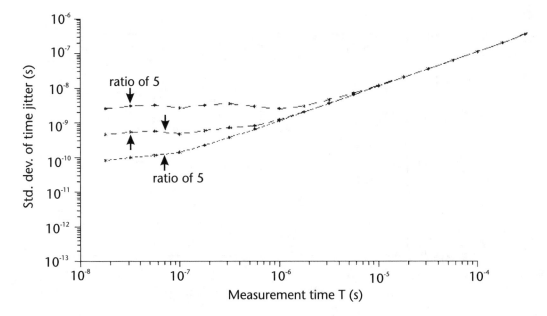

Figure 3.41 Shift in time jitter from far-from-the-carrier phase noise.

where

P_{sb} = measured single sideband of −6 dB, assumes equal amplitude modulation (AM)/phase modulation (PM) component (dB);

f_m = modulation frequency (frequency difference between sideband and carrier (Hz);

f_{vco} = VCO frequency (Hz);

T_{pp} = peak-to-peak jitter (sec).

Figure 3.42 shows variation of peak-to-peak jitter on the y-axis at 100 MHz versus FM on the x-axis and sideband level. The calculations assume the coupling factor is the same at each frequency. This figure shows that spurious signals close to the carrier have less effect on jitter than signals farther away from the carrier.

Figure 3.43 shows peak-to-peak period jitter versus FM and VCO frequency and a −30 dBc sideband level. This figure shows that higher output frequencies produce lower jitter for the same sideband level.

Rearranging (3.90) produces (3.91) for the sideband level required to maintain peak-to-peak requirements:

$$P_{sb} = 20 \log \left[\frac{\dfrac{-1}{\dfrac{T_{pp}}{2} + \dfrac{1}{f_{vco}}} + f_{vco}}{f_m 2} \right] \tag{3.91}$$

Figure 3.44 shows sideband levels to maintain desired 200-ps peak-to-peak jitter versus VCO frequency. This figure shows that higher frequencies can have higher

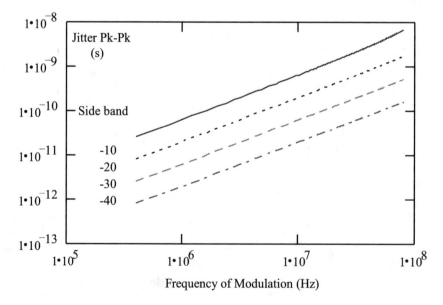

Figure 3.42   Variation of peak-to-peak jitter at 100 MHz versus FM and sideband level.

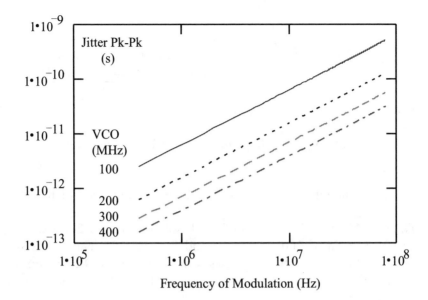

Figure 3.43 Peak-to-peak period jitter versus FM and VCO frequency.

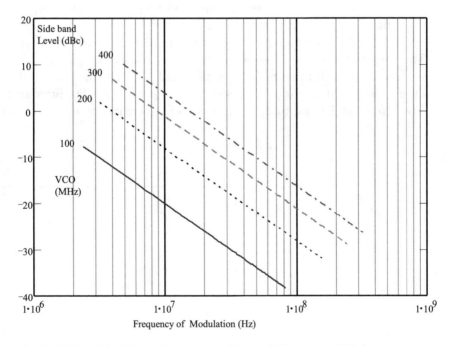

Figure 3.44 Sideband level for maintaining peak-to-peak jitter versus VCO frequency.

sideband levels and maintain the required jitter level. This figure also shows that lower sideband levels are needed for spurious signals further away from the carrier.

Equation (3.92) computes peak-to-peak jitter versus isolation on silicon and VCO gain:

$$T_{pp} = 2 \left[\frac{1}{f_{vco} - \left(V_{lpk} \cdot 10^{\frac{P_{iso}}{20}} \frac{K_v}{2\pi} \right)} - \frac{1}{f_{vco}} \right] \tag{3.92}$$

where

V_{lpk} = peak-to-peak voltage of source to be isolated (V);

K_v = VCO gain (rad/s/V);

P_{iso} = amount of isolation from voltage source to be isolated (dB).

Figure 3.45 shows jitter versus board isolation from a 2-V peak-to-peak source and VCO gain of 200, 400, 600, and 800 MHz/V. This figure shows that the amount of board isolation increases with increasing VCO gain in order to maintain the same amount of jitter. Consequently, reducing VCO gain helps reduce jitter.

3.3.3.2   Sampled Data Response for Spurious Analysis

Figure 3.46 shows the effect on RMS jitter of varying the power level and measurement time of a spurious signal. The FM signal is at 30 MHz, the carrier is at 33 MHz, and the spurious level is varied from −20 to −60 dBc in 20-dB increments. The measurement time (time between samples) has a sampled response where at some measurement times no jitter is measured (e.g., 100 ns, 1 μs, and 10 μs). The jitter at the worst case is at a measurement time of 20 ns. The jitter at 20 ns varies from 20 ps, to 200 ps, to 2,000 ps for 20-dB increases of sideband power level.

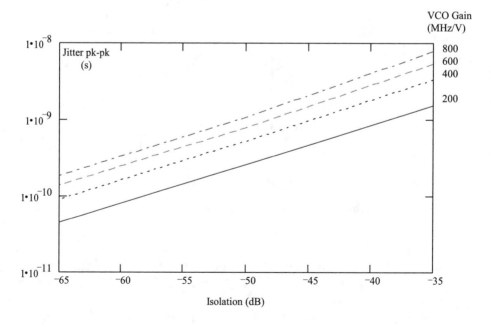

Figure 3.45   Jitter versus circuit isolation and VCO gain.

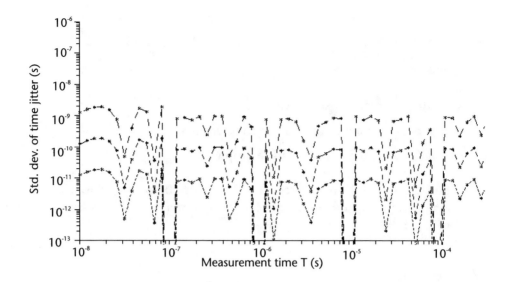

Figure 3.46   RMS time jitter versus sideband level and time measurement between samples.

3.3.4   Spurious-Noise-Reduction Techniques

Noise-reduction techniques can be implemented in two domains:

- Printed circuit board (PCB) techniques [25];
- Monolithic techniques (system-design styles for noise reduction) [26].

The monolithic circuit-design styles for noise reduction are listed in order from least to most noisy:

- Steer current and sense current (current mode analog and digital);
- Balanced current steering with small voltage switches (differential pair, differential ECL (emitter coupled logic), CML (current mode logic), or SCL (source coupled logic) [FSCL]);
- Unbalanced current steering with larger voltage switches (unbalanced ECL, CML, or CSL);
- Switch current and switch voltage (TTL);
- Switch maximum current and voltage (CMOS);
- Concentrate current and voltage transient during small time interval (CMOS).

One of the best ways to reduce noise other than using design technique is to use decoupling capacitors. Decoupling capacitors should be able to supply all the current required by an IC when it switches. The smallest-value capacitor that will do the job is the best choice because lead inductance is minimized. In many cases decoupling capacitors may be needed on the silicon to minimize supply ripple. Equation (3.93) computes the size of decoupling capacitor in order to keep the supply ripple low [25].

$$C = \frac{dI \, dt}{dV} \tag{3.93}$$

where

dI = current transient (A);

dt = slew time of the current transient (sec);

dV = desired transient voltage drop on the supply line.

For example, a 10-ma current transient with 500-ps time duration and a 50-mV maximum ripple on the supply line computes a 100-pF decoupling capacitor by using (3.93).

3.4 Time-Domain Solution

The manner that a PLL changes to an input frequency or phase shows an important aspect of PLL performance. A step change of the input is one of the most important measures of its response. Does the output reach its final value quickly or slowly? Does it overshoot? If so, how long does it take to come finally within a specified offset from the final value? How great is the phase error during a frequency step? Does it surpass the range of the phase detector? To help answer such questions, we will generate a set of curves showing time responses under various conditions.

The challenge is to find a reasonably small set of curves that covers a large number of cases of practical importance. First, we will restrict our attention to linear performance. Consequently, the response to any size input step will be proportional to that step. So, a curve of unit step response will give us the response to all step sizes. Next, we will restrict our study to type 2 loops as another device for limiting the number of curves. Let's restrict our curves to damping factors less than 1 because of the difficulty of getting phase margins greater than 80° to have damping factors greater than 1. Finally, type 2 curves will be plotted against normalized time, $\omega_n t$, for various damping factors ζ.

3.4.1 Importance of Solving for the Time-Domain Response

Studying the time-domain response of a loop gives us valuable information about the loop. First, the time-domain response tells us what input conditions will cause an unstable loop. For example, many methods sweep the VCO frequency to find lock. Time-domain-response analysis can show that sweeping too quickly will not give the loop enough time to acquire lock. Next, the time-domain response shows the accuracy with which a loop can track an input signal.

In addition, many specifications require a phase settling time or frequency settling time to within a certain accuracy. The time-domain response can determine the settling time. Without doing time-response analysis, many engineers rely on measuring the voltage on the tune line of the VCO. However, the oscilloscope does not have enough dynamic range to measure small frequency differences.

Consequently, small frequency variations go undetected. Also, the time-domain response can show the amount of residual phase or frequency error. Finally, the time-domain response to a frequency step can be used to determine PLL bandwidth and phase margin because the 10% to 90% rise time is proportional to bandwidth, and the amount of overshoot is proportional to phase margin. Consequently, time-domain simulations and laboratory data can be fit to these equations.

3.4.2 Time-Domain Solution Using La Place Transforms

In a first-order loop, the resulting transient phase errors are simple exponentials, like an RC network. The exponent equals $-K_v K_d t$. Type 1 loops have wide bandwidth and poor tracking. Consequently, they are not often used, and we will not go into detail about their performance. As a rough rule of thumb for loop orders higher than the second, one can assume that the peak error of a third-order loop in response to any of the three forcing functions (step, ramp, and parabolic) will be about the same as that of a second-order loop with the same bandwidth and similar positions of the dominant poles. The differences arise in the steady-state errors and the detailed transient behavior. Consequently, we will study a type 2 loop response because it has the most general solution.

3.4.2.1 Closed-Loop Equation for PLL

The time-domain solution begins by obtaining the closed-loop equation for a PLL. First, a circuit topology is chosen. Figure 3.45 shows a general type 2, second-order PLL circuit with active compensation in terms that can be used for analysis.

Using the variables in Figure 3.47 and solving for the closed-loop equation in phase produces (3.94):

$$\frac{\theta_o}{\theta_i} = \frac{\frac{K_d K_v}{CR_1}\left(\frac{1}{s^2}\right)(sCR_2 + 1)}{1 + \frac{K_d K_v}{n_{mf} CR_1}\left(\frac{1}{s^2}\right)(sCR_2 + 1)} \tag{3.94}$$

Executing the multiplication and division in (3.94) produces (3.95):

$$\frac{\theta_o}{\theta_i} = \frac{\frac{K_d K_v}{CR_1}\left(\frac{s}{CR_2} + 1\right)}{s^2 + s\left(\frac{K_d K_v}{n_{mf}}\frac{R_2}{R_1}\right) + \frac{K_d K_v}{n_{mf} CR_1}} \tag{3.95}$$

3.4.2.2 Definition of the Closed Loop in Terms of Second-Order Parameters

The denominator in (3.95) has the familiar general form for a linear, constant coefficient, second-order differential equation $s^2 + s2\zeta\omega_n + \omega_n^2$. Comparing the terms of (3.95) with the second-order differential equation produces an equation with servo terminology terms. Now (3.95) for the closed-loop transfer function

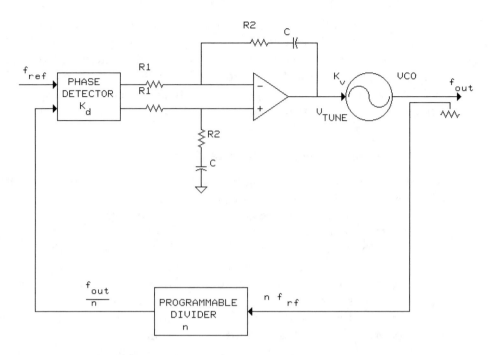

Figure 3.47 General type 2, second-order PLL circuit with active compensation.

with circuit components can be transformed into (3.96) with servo terminology format:

$$\frac{\theta_o}{\theta_i} = \frac{n_{mf}(\omega_n)^2 \left(s\dfrac{2\zeta}{\omega_n} + 1\right)}{s^2 + s(2\zeta\omega_n) + (\omega_n)^2} \tag{3.96}$$

3.4.2.3 Defining Step in Phase, Step in Frequency, and Frequency Ramp

Loops are used in many different applications. Consequently, a variety of disturbances perturb the performance of the loop. Finding the response of the loop to these disturbances determines the robustness of the loop and insures the performance of the loop under the maximum variety of disturbances. These disturbances (forcing functions) are broken down into phase step, frequency step (phase ramp), and frequency ramp (phase acceleration) [27]. Equation (3.97) defines the phase step input function from La Place transform theory:

$$\theta_{step}(s) = \frac{1}{s}\frac{\theta_o(s)}{\theta_i(s)} \tag{3.97}$$

Equation (3.98) defines the phase-ramp input function from La Place transform theory:

$$\theta_{ramp}(s) = \frac{1}{s^2}\frac{\theta_o(s)}{\theta_i(s)} \tag{3.98}$$

Equation (3.99) defines the phase-parabolic input function from La Place transform theory:

$$\theta_{par}(s) = \frac{1}{s^3} \frac{\theta_o(s)}{\theta_i(s)} \tag{3.99}$$

Now we will use these forcing functions to find the transient solutions.

3.4.2.4 Example Phase-Step Solution

Studying a time-domain solution makes it easier to solve time-domain solutions on other loops with modified equations and is a good review of taking the inverse La Place transform. First, the solution to the phase-step function is derived. Substituting (3.96) for the closed loop into (3.97) for the phase-step forcing function produces (3.100) for the phase step:

$$\theta_{step} = \frac{1}{s} \frac{n_{mf}(\omega_n)^2 \left(s\frac{2\zeta}{\omega_n} + 1 \right)}{s^2 + s(2\zeta\omega_n) + (\omega_n)^2} \tag{3.100}$$

We will do the solution for damping factors less than 1 because it is the most common case since phase margins of greater than 80° are rarely achieved in practical circuits. For damping factors less than 1, the roots in the denominator are imaginary. Consequently, the complex-conjugate format for imaginary roots can be used. Now, using the complex-conjugate roots and factoring the roots in (3.100) produces (3.101):

$$\theta_{step} = \frac{1}{s} \frac{n_{mf} \left[\dfrac{(\omega_d)^2 + \alpha^2}{z} \right] (s + z)}{(s + \alpha - j\omega_d)(s + \alpha + j\omega_d)} \tag{3.101}$$

Equations (3.102) through (3.104) show the redefined terms in (3.100) in terms of the factored roots in (3.101):

$$z = \frac{(\omega_d)^2 + \alpha^2}{2\alpha} \tag{3.102}$$

$$\alpha = \zeta\omega_n \tag{3.103}$$

$$\omega_d = \sqrt{1 - \zeta^2}\,\omega_n \tag{3.104}$$

Partial Fraction Expansion
To find the solution to (3.101), the transform expression must be broken into simpler terms before any transform can be applied. Partial fraction expansion (also known as Heaviside's Expansion Theorem) produces the simpler equation. First,

the equation is expanded by separating the original equation into a summation of simple expressions with simple roots in the denominator and unknown coefficients in the numerator. This equation is equated to the original equation. Then, for each coefficient, multiplying by the denominator of one of the coefficients, setting s to the root of the denominator, and solving for the coefficient finds the value of the coefficient. Now, applying the expansion rules to (3.101) produces (3.105) [28]:

$$\theta_{step} = \frac{K_1}{s} + \frac{K_2}{(s + \alpha - j\omega_d)} + \frac{K_3}{(s + \alpha + j\omega_d)} \tag{3.105}$$

Equating (3.101) to (3.105) gives an equation that can be used to find the coefficients K_1, K_2, and K_3. Coefficient K_3 is the complex conjugate of K_2 because the denominator is a complex conjugate. These roots are a complex-conjugate pair. Consequently, we only have to solve for K_1 and K_2 and take the complex conjugate of K_2 to have a solution.

Now we have to solve for the two coefficients in (3.105). We will start with coefficient K_1. To solve for K_1, we equate (3.101) to (3.105), multiply both sides by s, and set $s = 0$. This produces (3.106) for K_1:

$$K_1 = \frac{n_{mf}\left[\dfrac{(\omega_d)^2 + \alpha^2}{z}\right] z}{(\alpha - j\omega_d)(\alpha + j\omega_d)} \tag{3.106}$$

Further algebraic manipulation produces (3.107) through (3.109). Doing the multiplication in the numerator and denominator produces (3.107):

$$K_1 = \frac{n_{mf}[(\omega_d)^2 + \alpha^2]}{[\alpha^2 - (j\omega_d)^2]} \tag{3.107}$$

Doing the squaring operation on the imaginary part in the denominator produces (3.108):

$$K_1 = \frac{n_{mf}[(\omega_d)^2 + \alpha^2]}{(\omega_d)^2 + \alpha^2} \tag{3.108}$$

Canceling the equal numerator and denominator parts produces (3.109) for the final K_1 solution:

$$K_1 = n_{mf} \tag{3.109}$$

Next, we solve for K_2 in a similar manner as we did for K_1. To solve for K_2 requires multiplying both sides by $(s + \alpha - j\omega_d)$, setting $s = -\alpha + j\omega_d$, and substituting (3.101) into the left-hand side of (3.105). These steps produce (3.110):

$$K_2 = \frac{n_{mf}\left[\dfrac{(\omega_d)^2 + \alpha^2}{z}\right](-\alpha + j\omega_d + z)}{(-\alpha + j\omega_d)[(-\alpha - j\omega_d)(\alpha + j\omega_d)]} \tag{3.110}$$

Further algebraic simplification produces (3.111) and (3.112). Doing the multiplication in the numerator and simplifying terms produces (3.111):

$$K_2 = \frac{n_{mf}\left[\dfrac{(\omega_d)^2 + \alpha^2}{z}\right](-\alpha + j\omega_d + z)}{(-\alpha + j\omega_d)(2j\omega_d)} \tag{3.111}$$

Now we redefine some terms to make it easier to use Euler's identities, as shown by (3.112):

$$K_2 = \frac{n_{mf}K_4}{(2j\omega_d)} \tag{3.112}$$

Next, we simplify the complex variable K_4. Equation (3.113) shows the complex variable K_4 from (3.112):

$$K_4 = \frac{\left[\dfrac{(\omega_d)^2 + \alpha^2}{z}\right](-\alpha + j\omega_d + z)}{(-\alpha + j\omega_d)} \tag{3.113}$$

Multiplying the top and bottom of (3.113) by $(-\alpha - j\omega_d)$ begins the process of separating out real and complex quantities and produces (3.114):

$$K_4 = \frac{\left[\dfrac{(\omega_d)^2 + \alpha^2}{z}\right](-\alpha + j\omega_d + z)(-\alpha - j\omega_d)}{[\alpha^2 + (\omega_d)^2]} \tag{3.114}$$

Algebraic simplification on (3.114) produces (3.115) through (3.117). Canceling a term in the numerator with a term in the denominator does an algebraic simplification of (3.114) to produce (3.115):

$$K_4 = \frac{(-\alpha + j\omega_d + z)(-2\alpha - 2j\omega_d)}{z} \tag{3.115}$$

Separating out real and imaginary parts in (3.115) produces (3.116):

$$K_4 = \frac{[\alpha^2 + (\omega_d)^2] - z\alpha}{z} + j\frac{-\alpha\omega_d - (z\omega_d + \alpha\omega_d)}{z} \tag{3.116}$$

Multiplying the terms in (3.116) and simplifying produces (3.117):

$$K_4 = \frac{[\alpha^2 + (\omega_d)^2]}{z} - \alpha - j\omega_d \tag{3.117}$$

With K_4 simplified, we move on to finding the third coefficient, K_3.

La Place transform theory states that K_3 is the complex conjugate of K_2. Consequently, the complex conjugate of (3.111) produces (3.118):

$$K_3 = \frac{n_{mf}\left[\dfrac{(\omega_d)^2 + \alpha^2}{z}\right][(-\alpha - j\omega_d) + z]}{(-\alpha - j\omega_d)(-2j\omega_d)} \tag{3.118}$$

Again, we redefine some terms to produce (3.119), which make it easier to use Euler's identities:

$$K_3 = \frac{n_{mf}K_5}{(2j\omega_d)} \tag{3.119}$$

Again, we simplify the complex variable K_5. From La Place transform theory, the complex number K_5 is the complex conjugate of K_4. Performing the complex conjugate on (3.113) produces (3.120):

$$K_5 = \frac{\left[\dfrac{(\omega_d)^2 + \alpha^2}{z}\right][(-\alpha - j\omega_d) + z]}{(-\alpha - j\omega_d)} \tag{3.120}$$

Multiplying the top and bottom by $(-\alpha - j\omega_d)$ separates out the real and imaginary parts to produce (3.121):

$$K_5 = \frac{\left[\dfrac{(\omega_d)^2 + \alpha^2}{z}\right][(-\alpha - j\omega_d) + z](-\alpha + j\omega_d)}{[\alpha^2 + (\omega_d)^2]} \tag{3.121}$$

Separating (3.121) into real and imaginary parts and simplifying produces (3.122):

$$K_5 = \frac{[\alpha^2 + (\omega_d)^2] - z\alpha}{z} + j\frac{\alpha\omega_d + z\omega_d - \alpha\omega_d}{z} \tag{3.122}$$

Simplifying (3.122) produces (3.123):

$$K_5 = \left\{\frac{[\alpha^2 + (\omega_d)^2]}{z} - \alpha\right\} + j\omega_d \tag{3.123}$$

Final Simplifications
Finally, equations are simplified to match equations in the references, verifying the procedure that was taken. First, we simplify (3.117) and (3.123) to produce (3.124) and (3.125):

$$K_4 = \alpha - j\omega_d \tag{3.124}$$

$$K_5 = \alpha + j\omega_d \tag{3.125}$$

Combining (3.109), (3.112), (3.119), (3.124), and (3.125) into (3.105) and taking the inverse La Place transform from a transform table produces (3.126):

$$\theta_{step} = n_{mf}\left[1 + \exp(-\alpha t)\left(\frac{\alpha - j\omega_d}{2j\omega_d}\exp(j\omega_d t) + \frac{\alpha + j\omega_d}{-2j\omega_d}\exp(-j\omega_d t)\right)\right]$$

$$(3.126)$$

Separating out real and imaginary parts produces (3.127):

$$\theta_{step} = n_{mf}\left[1 + \frac{\exp(-\alpha t)}{\omega_d}\left(\frac{\alpha\exp(j\omega_d t)}{2j} - \frac{\omega_d\exp(j\omega_d t)}{2} + \frac{\alpha\exp(-j\omega_d t)}{-2j} - \frac{\omega_d\exp(-j\omega_d t)}{2}\right)\right]$$

$$(3.127)$$

Using Euler's identities produces (3.128):

$$\theta_{step} = n_{mf}\left\{1 + \frac{\exp(-\alpha t)}{\omega_d}[\alpha\sin(\omega_d t) - \omega_d\cos(\omega_d t)]\right\} \qquad (3.128)$$

Further simplification produces (3.129):

$$\theta_{step} = n_{mf}\left\{1 + \exp(-\alpha t)\left[\frac{\alpha}{\omega_d}\sin(\omega_d t) - \cos(\omega_d t)\right]\right\} \qquad (3.129)$$

Substituting parameters from (3.103) and (3.104) and using initial phase conditions produces (3.130), which is the final equation for a phase step response:

$$\theta_{step} = n_{mf}\left\{\theta_{ref} + \Delta\theta_{ref}\right. \qquad (3.130)$$

$$\left. - \Delta\theta_{ref}\exp(-\zeta\omega_n t)\left[\cos\left(\sqrt{1-\zeta^2}\,\omega_n t\right) - \frac{\zeta}{\sqrt{1-\zeta^2}}\sin\left(\sqrt{1-\zeta^2}\,\omega_n t\right)\right]\right\}$$

This equation agrees with [21].

Understanding this procedure to solve for the time-domain solution gives us a procedure to find the time-domain solution to any PLL circuit. This linear solution assumes that during the step response there are no cycle slips and that the phase remains within the range of the phase detector. In other words, if the phase error is greater than 2π and you are using a phase/frequency detector, then the solution will be incorrect. Consequently, this solution holds only for small phase steps ($<2\pi$ initial phase error). This solution also assumes a high-gain PLL, which has $K_v K_d \gg \omega_n$.

3.4.3 Relationship of Error Function to Closed Loop

Rather than solving for each response, [29, pp. 44–60] gives the error function equations for the time-domain responses as shown by Table 3.3. To use these

Table 3.3 Transient Phase Error of Type 2 Loop, $\theta_e(t)$ (in radians) [29, pp. 44–60]

	Phase Step ($\Delta\theta$ rad)	Frequency Step ($\Delta\omega$ rad/s)	Frequency Ramp ($\Delta\dot\omega$ rad/s^2)
$\zeta < 1$	$\Delta\theta\left(\cos\sqrt{1-\zeta^2}\,\omega_n t - \dfrac{\zeta}{\sqrt{1-\zeta^2}}\sin\sqrt{1-\zeta^2}\,\omega_n t\right)e^{-\zeta\omega_n t}$	$\dfrac{\Delta\omega}{\omega_n}\left(\dfrac{1}{\sqrt{1-\zeta^2}}\sin\sqrt{1-\zeta^2}\,\omega_n t\right)e^{-\zeta\omega_n t}$	$\dfrac{\Delta\dot\omega}{\omega_n^2} - \dfrac{\Delta\dot\omega}{\omega_n^2}\left(\cos\sqrt{1-\zeta^2}\,\omega_n t + \dfrac{\zeta}{\sqrt{1-\zeta^2}}\sin\sqrt{1-\zeta^2}\,\omega_n t\right)e^{-\zeta\omega_n t}$
$\zeta = 1$	$\Delta\theta(1 - \omega_n t)e^{-\zeta\omega_n t}$	$\dfrac{\Delta\omega}{\omega_n}(\omega_n t)e^{-\zeta\omega_n t}$	$\dfrac{\Delta\dot\omega}{\omega_n^2} - \dfrac{\Delta\dot\omega}{\omega_n^2}(1 + \omega_n t)e^{-\zeta\omega_n t}$
$\zeta > 1$	$\Delta\theta\left(\cosh\sqrt{\zeta^2-1}\,\omega_n t - \dfrac{\zeta}{\sqrt{\zeta^2-1}}\sinh\sqrt{\zeta^2-1}\,\omega_n t\right)e^{-\zeta\omega_n t}$	$\dfrac{\Delta\omega}{\omega_n}\left(\dfrac{1}{\sqrt{\zeta^2-1}}\sinh\sqrt{\zeta^2-1}\,\omega_n t\right)e^{-\zeta\omega_n t}$	$\dfrac{\Delta\dot\omega}{\omega_n^2} - \dfrac{\Delta\dot\omega}{\omega_n^2}\left(\cosh\sqrt{\zeta^2-1}\,\omega_n t + \dfrac{\zeta}{\sqrt{\zeta^2-1}}\sinh\sqrt{\zeta^2-1}\,\omega_n t\right)e^{-\zeta\omega_n t}$
	Steady-state error $= 0$	Steady-state error $= \dfrac{\Delta\omega}{K_v}$ (not included above)	Steady-state error $= \dfrac{\Delta\dot\omega t}{K_v} + \dfrac{\Delta\omega}{\omega_n^2}$ ($\Delta\dot\omega t/K_v$ not included above)

equations effectively, we will have to establish some mathematical relationships by doing block-diagram algebra. Figure 3.48 shows the general block diagram for a PLL where we will derive these relationships. From Figure 3.48, (3.131) relates the error function to the closed loop:

$$\frac{\theta_e}{\theta_i} = 1 - H\frac{\theta_o}{\theta_i} \qquad (3.131)$$

Rearranging (3.131) produces (3.132):

$$\frac{\theta_o}{\theta_i} = \frac{1 - \dfrac{\theta_e}{\theta_i}}{H} \qquad (3.132)$$

For a normalized feedback, substituting $H = 1$ (combine divide-by-n into K_v for the VCO) and an output phase-step-response variable θ_{step} produces (3.133):

$$\theta_{step} = 1 - \frac{\theta_e}{\theta_i} \qquad (3.133)$$

Equation (3.133) converts the normalized error functions from references (Table 3.3) to output phase response for a normalized phase-step input.

Because frequency is the derivative of phase, taking the derivative of the phase step response gives the frequency-impulse response, taking the derivative of the phase-ramp response gives the frequency step response, and taking the derivative of the phase-parabolic response gives the frequency-ramp response.

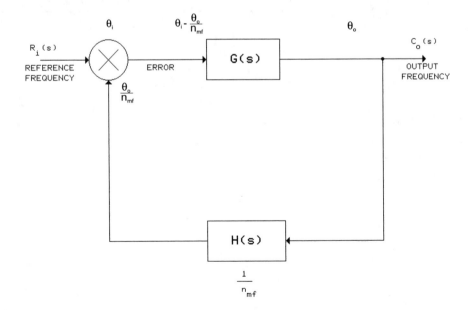

Figure 3.48   General PLL block diagram.

The derivative of the phase-ramp error equals the frequency-step error. Consequently, (3.134) computes the output frequency for a normalized frequency step:

$$f_{outn} = 1 - \frac{1}{\omega_n} \left\{ \omega_n \cos\left(\sqrt{1 - \zeta^2}\ \omega_n t\right) \exp\left(-\zeta\omega_n t\right) + \right. \tag{3.134}$$

$$\left. \left[\frac{1}{\sqrt{1 - \zeta^2}} \sin\left(\sqrt{1 - \zeta^2}\ \omega_n t\right)\right] \left(-\zeta\omega_n\right) \exp\left(-\zeta\omega_n t\right) \right\}$$

Figure 3.49 shows a family of output-frequency-step-response curves for a normalized input-frequency change with damping factors of 0.3, 0.5, 0.7, and 0.99. The first overshoot varies between approximately 50%, 30%, 20%, and 10%. Consequently, damping factor decreases with increasing overshoot. Damping factor 0.7 has a 20% overshoot and is not critically damping, as one would expect a second-order differential-equation solution to have. Next, the 10% to 90% rise time slightly changes with damping factor, and the peak of the first overshoot significantly changes with damping factor. Later, we will use this peak overshoot to help us compute ω_n and damping factor from measured or simulated step responses. Finally, Figure 3.49 can be used to determine ω_n and damping when switching time becomes an important requirement for an application.

The graph in Figure 3.49 and some of the following graphs, as well as some seen in the references, are normalized. Most of the graphs have unit step responses so the ordinate must be multiplied by the magnitude of the step that is actually used. For example, a step of 0.1 rad must multiply the y-axis by 0.1 rad to unnormalize the response on the curve. For a step of 10 Hz, a reading of 0.4 on

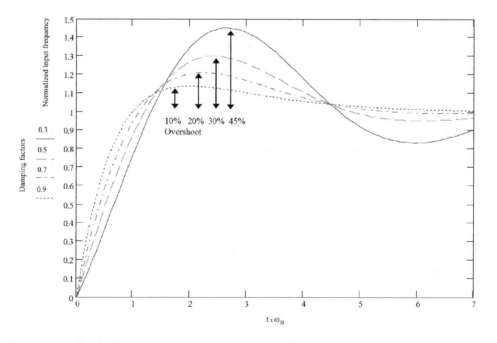

Figure 3.49 Output-frequency step response to a normalized input-frequency change.

the normalized y-axis of the response would represent 0.4×10 Hz, which equals 4 Hz. In the case of the response to an input ramp, the graph is given for an input slope of ω_n. For example, an input slope of 0.1 rad/s has a graph value at the desired x-axis time value of 0.6. Then, the unnormalized response has a 0.6×0.1 rad/s per ω_n, which has no units. For a natural radian frequency ω_n of 10 rad/s, a frequency ramp with a slope of 10 Hz/s and a y-axis normalized reading of 0.6 would represent 0.6×10 Hz/s per 10 rad/s, equaling 0.6 Hz, which is the frequency error in response to the frequency ramp. In the same manner, a parabolic input with a slope of kt must be multiplied by k rad^2/ω_n^2 to be unnormalized.

Figure 3.50 shows the envelope of a damped sinusoid versus normalized time. This figure gives a picture of how fast the loop will settle and within what frequency error over a long period of time [30]. Damping factors of 0.3, 0.5, 0.7, and 0.99 are plotted in the figure. At the normalize time of 15, the damping factor of 0.99 has the smallest value and the damping factor of 0.3 has the largest value. Consequently, this figure shows that a damping factor of 0.99 will settle most quickly. This is only a first-order approximation that conflicts with other analyses.

3.4.4 Output Responses to Unnormalized Input Steps

To unnormalize the input phase or frequency step response requires using initial conditions. Let's do the phase response first. Subtracting the phase change from initial condition to final condition, multiplying the result by the normalized error, and subtracting this result from the final phase amount produces the output phase response to an unnormalized phase step response as shown by (3.135):

$$\theta_{step} = \theta_{fc} - \frac{\theta_e}{\theta_i}\left(\theta_{fc} - \theta_{ic}\right) \tag{3.135}$$

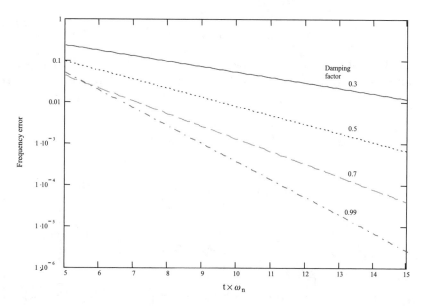

Figure 3.50 Envelope of damped sinusoid versus damping factor.

Redefining terms for delta phase change produces (3.136):

$$\theta_{step} = \theta_{ic} + \Delta\theta - \frac{\theta_e}{\theta_i}\,\Delta\theta \tag{3.136}$$

Consequently, (3.137) shows the output-frequency response to an unnormalized frequency step response of a PLL:

$$f_{step} = f_{fc} - \theta_{stepe}(f_{fc} - f_{ic}) \tag{3.137}$$

Redefining terms for the delta frequency change produces (3.138):

$$f_{step} = f_{ic} + \Delta f - \theta_{stepe}\,\Delta f \tag{3.138}$$

Equations (3.139) through (3.141) show the output phase response to unnormalized phase step ($\Delta\theta_{ref}$) responses for damping factors greater than 0 and less than 1, damping factors greater than 1, and damping factor equal to 1:

$$\theta_{step} = n_{mf}\left\{\theta_{ref} + \Delta\theta_{ref} - \Delta\theta_{ref}\exp(-\zeta\omega_n t)\right. \tag{3.139}$$

$$\left.\cdot\left[\cos\left(\sqrt{1-\zeta^2}\,\omega_n t\right) - \frac{\zeta}{\sqrt{1-\zeta^2}}\sin\left(\sqrt{1-\zeta^2}\,\omega_n t\right)\right]\right\}$$

$$\theta_{step} = n_{mf}\left\{\theta_{ref} + \Delta\theta_{ref} - \Delta\theta_{ref}\exp(-\zeta\omega_n t)\right. \tag{3.140}$$

$$\left.\cdot\left[\cosh\left(\sqrt{\zeta^2-1}\,\omega_n t\right) - \frac{\zeta}{\sqrt{\zeta^2-1}}\sinh\left(\sqrt{\zeta^2-1}\,\omega_n t\right)\right]\right\}$$

$$\theta_{step} = n_{mf}\{\theta_{ref} + \Delta\theta_{ref} - \Delta\theta_{ref}[(1-\omega_n t)\exp(-\omega_n t)]\} \tag{3.141}$$

Equations (3.142) through (3.144) show the output-frequency response to unnormalized frequency step (Δf_{ref}) responses for damping factors greater than 0 and less than 1, damping factors greater than 1, and damping factor equal to 1:

$$f_{out} = n_{mf}\left\{f_{ref} + \Delta f_{ref} - \Delta f_{ref}\exp(-\zeta\omega_n t)\right. \tag{3.142}$$

$$\left.\cdot\left[\cos\left(\sqrt{1-\zeta^2}\,\omega_n t\right) - \frac{\zeta}{\sqrt{1-\zeta^2}}\sin\left(\sqrt{1-\zeta^2}\,\omega_n t\right)\right]\right\}$$

$$f_{out} = n_{mf}\left\{f_{ref} + \Delta f_{ref} - \Delta f_{ref}\exp(-\zeta\omega_n t)\right. \tag{3.143}$$

$$\left.\cdot\left[\cosh\left(\sqrt{\zeta^2-1}\,\omega_n t\right) + \frac{-\zeta}{\sqrt{\zeta^2-1}}\sinh\left(\sqrt{\zeta^2-1}\,\omega_n t\right)\right]\right\}$$

$$f_{out} = n_{mf} \{ f_{ref} + \Delta f_{ref} - \Delta f_{ref} [(1 - \omega_n t) \exp(-\omega_n t)] \} \qquad (3.144)$$

The frequency step response of measure data can be used to compute crossover frequency and phase margin. For example, let's look at a PLL with an output frequency of 50 MHz and a unity feedback divide ratio. Figure 3.51 shows a measured frequency step from 50 to 60 MHz applied to the example PLL. Computing the 10% to 90% time and comparing with test results helps determine crossover frequency. The example shows a rise time of approximately 200 ns. Adjusting the 0-dB crossover frequency in the time-domain equation with a 0.9 damping factor for an initial guess produces a good curve fit for a 0-dB crossover frequency of 1.5 MHz. Computing percentage overshoot and comparing with test results helps determine phase margin. The example shows a 61.357-MHz overshoot. Subtracting the 60-MHz final value and dividing by the 10-MHz input frequency-step size computes a 13.57% overshoot. Adjusting the damping factor in the time-domain equation with a 0-dB crossover frequency of 1.5 MHz produces a good fit for a phase margin of 80°.

In some applications, settling time is important. With the developed equations, settling time can be solved as shown in Figure 3.52. The dotted line in Figure 3.52 shows the 60.1-MHz settling amount limit. The response in Figure 3.52 settles to within 1% of the final value at 1.4 μs.

3.4.5   Ramp Phase Solution

Equations (3.145) through (3.147) show the normalized ramp error function for damping factors greater than 0 and less than 1, damping factor equal to 0, and damping factors greater than 1:

$$\frac{\theta_e}{\theta_i} = \frac{\Delta \omega}{\omega_n} \left[\frac{1}{\sqrt{1 - \zeta^2}} \sin\left(\sqrt{1 - \zeta^2}\ \omega_n t\right) \right] \exp(-\zeta \omega_n t) \qquad (3.145)$$

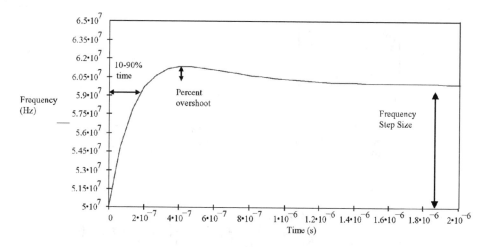

Figure 3.51   Output-frequency step of an example PLL.

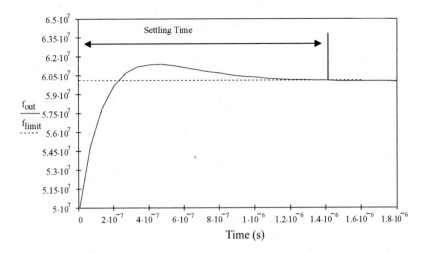

Figure 3.52 Output-frequency step response for 1% settling-time limit.

$$\frac{\theta_e}{\theta_i} = \frac{\Delta\omega}{\omega_n} \, \omega_n t \, \exp(-\omega_n t) \tag{3.146}$$

$$\frac{\theta_e}{\theta_i} = \frac{\Delta\omega}{\omega_n} \left[\frac{1}{\sqrt{\zeta^2 - 1}} \sinh\left(\sqrt{\zeta^2 - 1} \, \omega_n t\right) \right] \exp(-\zeta\omega_n t) \tag{3.147}$$

3.4.6 Parabolic Phase Solution

For a parabolic input, (3.148) through (3.150) show the normalized error functions for damping factors greater than 0 and less than 1, damping factor equal to 1, and damping factor greater than 1:

$$\frac{\theta_e}{\theta_i} = \left[\frac{\frac{d}{dt}\Delta\omega}{K_v} + \frac{\frac{d}{dt}\Delta\omega}{(\omega_n)^2} \right] \tag{3.148}$$

$$- \frac{\frac{d}{dt}\Delta\omega}{(\omega_n)^2} \left[\exp(-\zeta\omega_n t) \cos\left(\sqrt{1 - \zeta^2}\,\omega_n t\right) + \frac{\zeta}{\sqrt{1 - \zeta^2}} \sin\left(\sqrt{1 - \zeta^2}\,\omega_n t\right) \right]$$

$$\frac{\theta_e}{\theta_i} = \left[\frac{\frac{d}{dt}\Delta\omega}{K_v} t + \frac{\frac{d}{dt}\Delta\omega}{(\omega_n)^2} \right] - \frac{\frac{d}{dt}\Delta\omega}{(\omega_n)^2} \left[1 + \omega_n t \, \exp(-\omega_n t) \right] \tag{3.149}$$

$$\frac{\theta_e}{\theta_i} = \frac{\frac{d}{dt}\Delta\omega}{K_v}t + \frac{\frac{d}{dt}\Delta\omega}{(\omega_n)^2} \tag{3.150}$$

$$+ \frac{\frac{d}{dt}\Delta\omega}{(\omega_n)^2}\exp(-\zeta\omega_n t)\left\{\cosh\left(\sqrt{\zeta^2-1}\ \omega_n t\right) + \left[\frac{\zeta}{\sqrt{\zeta^2-1}}\sinh\left(\sqrt{\zeta^2-1}\ \omega_n t\right)\right]\right\}$$

Figure 3.53 shows the normalized phase-error response to parabolic phase input of ω_n^2 with an underdamped system. This response shows that the error does not go to zero. Because a type 2 loop cannot cancel out all of the parabolic phase error, at some point increasing the speed of the ramping input frequency will cause the phase error to be large enough for the loop to lose lock or not acquire lock. Equation (3.151) computes the maximum sweep rate [29]:

$$\frac{d}{dt}\omega = \frac{1}{2}(\omega_n)^2\left[1 - \left(\frac{2}{\sqrt{SNR}}\right)\right] \tag{3.151}$$

For example, with a natural frequency f_n = 1,000 and SNR = 10, the maximum sweep rate = 1.16 MHz/s

In summary, this section showed that solutions to the PLL equations can be used to understand several time-domain response characteristics of a PLL. Developing a familiarity with and insight into these responses helps with analyzing and troubleshooting many conditions in a PLL.

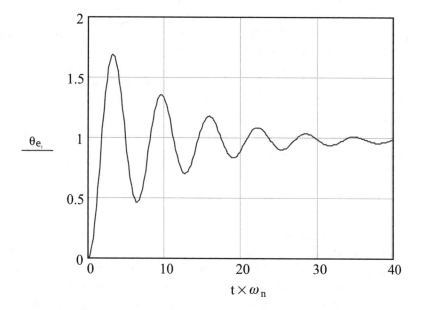

Figure 3.53   Normalized phase-error response to parabolic input change.

3.5 Acquisition of Lock

Up until now, the loop being analyzed has been assumed to be locked, and any changes to the condition of the loop have caused a small change in phase error. Of course, we seek to ensure that the loop operates this way. Consequently, we must know under what conditions the loop will not lock and under what conditions the loop will lose lock. Furthermore, we must understand how phase lock is achieved in order to decide if an additional support circuit is required and how this supporting circuit will help achieve lock.

This section discuses several key acquisition characteristics for a PLL. It shows the nonlinear equations that characterize the loop's time-domain response outside the range of the phase detector. Next, the time-domain response outside the range of the phase detector depends on initial phase and frequency conditions. We will discuss lock-in range, pull-in range, and hold-in range. Finally, we will study the derivation of the second-order, nonlinear, ordinary differential equation for a PLL. This will allow us to know clearly the approximations that are made to solve the equation and when these approximations are violated.

A loop normally starts in an unlocked state. The loop's own actions or a support circuit must bring the loop to the locked state. The process of bringing a loop into lock is called *acquisition*, which is the topic that we are going to discuss. The loop achieves *self-acquisition* by acquiring lock by itself. The loop achieves *aided-acquisition* when it is assisted by some other circuitry. The loop acquires lock slowly and unreliably with the self-acquisition process. In fact, the loop clumsily acquires lock even though it is an excellent tracking device when it is locked. Consequently, acquisition-aided circuits are commonly used. The digital phase/frequency detector has an acquisition-aided circuit included that is not discussed in depth in explanations of phase detector operations.

Several frequency ranges describe the operation of the loop outside the linear conditions that we have been discussing. The pull-in and seize ranges apply to the process of acquiring lock from the unlocked state. The hold-in range applies to the process edge of losing the locked condition.

To understand these ranges, we must define some terms. The frequency of the VCO, when the phase detector is in the center of its range, will be called the *center frequency, ω_c*. Adjusting the fixed frequency-determining elements in the oscillator or some bias that is added to the tuning voltage or to the phase detector output will vary this center frequency. We will call the difference between the input frequency and the center frequency the *mistuning*:

$$\omega_e(0) = \omega_{in} - \omega_c \qquad (3.152)$$

With the frequency variable for mistuning defined, let's define the ranges that we discussed earlier. We will begin by defining the *seize* or *lock-in range*. The mistuning frequencies where ω_{out} will immediately move to the steady-state lock frequency, ω_{in}, characterize the loop's operation with in the seize or lock-in range. In other words, $\omega_e(0)$ immediately decreases toward zero when the input switch to the next frequency is closed.

The mistuning frequencies where ω_{out} will eventually lock, but may skip cycles, sometimes many cycles, before doing so characterize the loop's operation with in the *pull-in* or *acquisition range*. The moment when the difference between the input and the output frequency causes the phase detector characteristic to travel past the phases for minimum voltage and maximum voltage of the detector define the *skipping cycle*. The mistuning frequencies where ω_{out} will maintain lock, but over which it will not necessarily acquire lock, characterize the loop's operation within the *hold-in* or *synchronization range*.

The lock-in and pull-in ranges are guaranteed maximum values. Consequently, the actions will occur regardless of initial phase; however, these actions may occur at wider mistuning frequencies if the phase happens to have a particular value.

Figure 3.54 illustrates the acquisition ranges that have just been described. We will sometimes use notation such as $\pm\omega_{PI}$ to emphasize that we are considering the deviation of ω_{in} from ω_c rather than the total range over which ω_{in} may vary, which is twice as large.

The relationship of the pull-in range greater than the lock-in range and the hold-in range greater than the pull-in range must always hold because of the definitions. These relationships apply to first-order loops and highly damped, type 2 loops, which act like type 1 loops. Type 1 loops move toward the steady state whenever it is possible to tune to it. The phase does not overshoot but moves directly toward its final value. In a first-order loop, the value of the phase error instantly relates to the value of the frequency. Consequently, a given phase error has only one frequency value. In a second-order loop, due to the filter, this does not occur. Various states can exist at a given frequency. Thus, at a given frequency, the second-order loop might move directly toward the steady state. or it might skip cycles at the same frequency, depending on the mix of state variables under each condition. In other words, the response depends on its history [30].

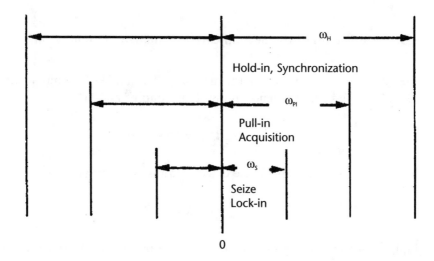

Figure 3.54 Mistuning ranges. (*From:* [30]. © 1998 Wiley Interscience. Reprinted with permission.)

An nth type loop contains n integrators, which can be ideal (such as the VCO), near-ideal (as in an active filter), or imperfect (as in a passive filter) approximations. Each integrator has a state variable of phase, frequency, frequency rate, and so forth. Locking the loop requires adjusting each of the state variables (each of the integrators) to be in close agreement with the corresponding parameters of the input signal. Therefore, we should discuss acquisition of phase, frequency, and so forth, up to n forms for an nth-type loop. (In most instances, we will limit our discussions to just two- or three-state variables because the other elements only have high-frequency poles that can be neglected.) Frequency acquisition has received the most attention; however, the other state variables can also be important. Acquisition is inherently a nonlinear phenomenon. We must use nonlinear analysis and cannot make the path easier by using linear approximations.

The nonlinear analysis begins by understanding the response of a phase detector. The periodic response of a phase detector produces a PLL response that depends on the solution of a second-order, nonlinear, ordinary differential equation. This nonlinear equation models the response of a disturbance to the PLL from known initial conditions. Studying the solution helps us understand variations in PLL responses to different stimuli. To do this analysis, we will assume a sinusoidal phase detector for the PLL.

Phase-plane trajectory analysis solves the second-order, nonlinear, ordinary differential equation that is shown by (3.153) [29, pp. 55–58; 31–34]:

$$\frac{d^2}{dt^2}\theta_e + \frac{2\zeta\omega_n}{\underbrace{\frac{2\zeta}{\omega_n}}} \sin[\theta_e(t)] + 2\zeta\omega_n \cos[\theta_e(t)]\frac{d}{dt}\theta_e = 0 \qquad (3.153)$$

To understand the nature of this equation and to unnormalize solutions, let's derive it and put it in a form that we can use to analyze it.

3.5.1 Derivation of the Second-Order, Nonlinear, Ordinary Differential Equation

The derivation begins by mathematically defining PLL components. Equation (3.154) defines the difference in input and output phase:

$$\theta_e(t) = \theta_i(t) - \theta_o(t) \qquad (3.154)$$

Next, (3.155) defines the phase detector transfer function:

$$V_d(t) = K_d \sin[\theta_e(t)] \qquad (3.155)$$

Equation (3.156) computes the derivative of the phase detector transfer function:

$$\frac{d}{dt}V_d = K_d \cos[\theta_e(t)]\frac{d}{dt}\theta_e \qquad (3.156)$$

Equation (3.157) mathematically models the VCO transfer function:

$$\frac{d}{dt}\theta_o(t) = K_v V_{tune}(t) \tag{3.157}$$

Equations (3.158) and (3.159) model the loop-filter transfer function with a La Place transform:

$$F(s) = (1 + \tau_2 s)/(\tau_1 s) \tag{3.158}$$

$$V_{tune}(s)/V_d(s) = (1 + \tau_2 s)/(\tau_1 s) \tag{3.159}$$

Rearranging terms produces (3.160):

$$\tau_1 s V_{tune}(s) = V_d(s) + \tau_2 s V_d(s) \tag{3.160}$$

Converting to the time domain produces (3.161):

$$\tau_1 \frac{d}{dt} V_{tune}(t) = V_d(t) + \tau_2 \frac{d}{dt} V_d(t) \tag{3.161}$$

With the PLL components mathematically described, we begin to make substitutions to build our equation [33].

Substituting phase detector and VCO transfer functions and the phase-error definition from (3.154) through (3.156) into (3.161) produces (3.162) [33]:

$$\tau_1 \left(\frac{d^2}{dt^2}\theta_o\right)\frac{1}{K_v} = K_d \sin[\theta_e(t)] + \tau_2 K_d \cos[\theta_e(t)]\frac{d}{dt}\theta_e \tag{3.162}$$

Multiplying (3.162) by K_v and rearranging produces (3.163):

$$K_v K_d \sin[\theta_e(t)] + \tau_2 K_v K_d \cos[\theta_e(t)]\frac{d}{dt}\theta_e = \tau_1 \frac{d^2}{dt^2}\theta_o \tag{3.163}$$

Rearranging (3.154) for output phase produces (3.164):

$$\theta_o = \theta_i - \theta_e \tag{3.164}$$

Substituting (3.164) into (3.163) for θ_o produces (3.165):

$$K_v K_d \sin[\theta_e(t)] + \tau_2 K_v K_d \cos[\theta_e(t)]\frac{d}{dt}\theta_e = \tau_1 \frac{d^2}{dt^2}(\theta_i - \theta_e) \tag{3.165}$$

Rearranging (3.165) produces (3.166):

$$\tau_1 \frac{d^2}{dt^2}\theta_e + K_v K_d \sin[\theta_e(t)] + \tau_2 K_v K_d \cos[\theta_e(t)]\frac{d}{dt}\theta_e = \tau_1 \frac{d^2}{dt^2}\theta_i$$

$$(3.166)$$

With the above equation, we can do some more substitution and rearranging to simplify the equation [33].

Next, the input signal is defined by (3.167):

$$v_i(t) = A \sin\left[(\omega_i - \omega_o)t + \Delta\theta + \frac{\pi}{2}\right]$$

$$(3.167)$$

Extracting the phase portion of the input signal from (3.167) and ignoring the amplitude (A) produces (3.168):

$$\theta_i = (\omega_i - \omega_o)t + \Delta\theta + \pi/2$$

$$(3.168)$$

Equation (3.169) computes the derivative of the phase portion of the input signal in (3.168):

$$\frac{d}{dt}\theta_i = \omega_i - \omega_o$$

$$(3.169)$$

Equation (3.170) computes the second derivative of the phase portion of the input signal:

$$\frac{d^2}{dt^2}\theta_i = 0$$

$$(3.170)$$

Substituting (3.170) into (3.166) produces (3.171):

$$\tau_1 \frac{d^2}{dt^2}\theta_e + K_v K_d \sin[\theta_e(t)] + \tau_2 K_v K_d \cos[\theta_e(t)]\frac{d}{dt}\theta_e = 0 \qquad (3.171)$$

Dividing by τ_1 produces (3.172):

$$\frac{d^2}{dt^2}\theta_e + \frac{K_v K_d}{\tau_1}\sin[\theta_e(t)] + \frac{\tau_2 K_v K_d}{\tau_1}\cos[\theta_e(t)]\frac{d}{dt}\theta_e = 0 \qquad (3.172)$$

Substituting servo parameters into (3.172) produces (3.173):

$$\frac{d^2}{dt^2}\theta_e + \omega_n^2 \sin[\theta_e(t)] + \frac{2\zeta}{\omega_n}\omega_n^2 \cos[\theta_e(t)]\frac{d}{dt}\theta_e = 0 \qquad (3.173)$$

Dividing by $2\zeta/\omega_n$ produces (3.174):

$$\frac{\dfrac{d^2}{dt^2}\,\theta_e}{\underbrace{2\zeta}_{\omega_n}} + \underbrace{\frac{\omega_n^2}{2\zeta}}_{\omega_n}\sin[\theta_e(t)] + \omega_n^2\cos[\theta_e(t)]\frac{d}{dt}\,\theta_e = 0 \tag{3.174}$$

Multiplying by $1/\omega_n^2$ produces (3.175):

$$\frac{\dfrac{d^2}{dt^2}\,\theta_e}{2\zeta\omega_n} + \underbrace{\frac{1}{2\zeta}}_{\omega_n}\sin[\theta_e(t)] + \cos[\theta_e(t)]\frac{d}{dt}\,\theta_e = 0 \tag{3.175}$$

Multiplying by $2\zeta\omega_n$ produces (3.176):

$$\frac{d^2}{dt^2}\,\theta_e + \underbrace{\frac{2\zeta\omega_n}{2\zeta}}_{\omega_n}\sin[\theta_e(t)] + 2\zeta\omega_n\cos[\theta_e(t)]\frac{d}{dt}\,\theta_e = 0 \tag{3.176}$$

This gives a basic nonlinear PLL equation that we can use.

3.5.2 Simplifying and Normalizing the Nonlinear Equation

Next, we will simplify and normalize the nonlinear equation that was derived. Letting $t = \tau/(2\zeta\omega_n)$ and substituting into (3.176) produces (3.177):

$$(2\zeta\omega_n)^2\frac{d^2}{d\tau^2}\,\theta_e + \underbrace{\frac{2\zeta\omega_n}{2\zeta}}_{\omega_n}\sin[\theta_e(t)] + (2\zeta\omega_n)^2\cos[\theta_e(t)]\frac{d}{d\tau}\,\theta_e = 0 \tag{3.177}$$

Separating the $(2\zeta\omega_n)^2$ term produces (3.178):

$$(2\zeta\omega_n)^2\left\{\frac{d^2}{d\tau^2}\,\theta_e + \frac{1}{\dfrac{2\zeta}{\omega_n}(2\zeta\omega_n)}\sin[\theta_e(t)] + \cos[\theta_e(t)]\frac{d}{d\tau}\,\theta_e\right\} = 0 \tag{3.178}$$

Simplifying (3.178) produces (3.179):

$$(2\zeta\omega_n)^2\left\{\frac{d^2}{d\tau^2}\,\theta_e + \frac{1}{(2\zeta)^2}\sin[\theta_e(t)] + \cos[\theta_e(t)]\frac{d}{d\tau}\,\theta_e\right\} = 0 \tag{3.179}$$

Equations (3.180) through (3.182) redefine terms to simplify and normalize (3.179):

$$a_p = 1/(2\zeta)^2 \tag{3.180}$$

$$d\phi = \frac{d}{d\tau}\,\theta_e \tag{3.181}$$

$$d\phi = \frac{1}{2\zeta\omega_n}\frac{d}{dt}\,\theta_e \tag{3.182}$$

Substituting (3.180) through (3.182) into (3.179) produces (3.183):

$$\frac{d}{d\tau}\frac{d}{d\tau}\,\theta_e + a_p\,\sin[\theta_e(t)] + \cos[\theta_e(t)]\frac{d}{d\tau}\,\theta_e = 0 \tag{3.183}$$

Dividing by $\frac{d}{d\tau}\,\theta_e$ produces (3.184):

$$\frac{d\left(\frac{d}{d\tau}\,\theta_e\right)}{d\theta_e} + a_p\,\frac{\sin[\theta_e(t)]}{\left(\frac{d}{d\tau}\,\theta_e\right)} + \cos[\theta_e(t)] = 0 \tag{3.184}$$

Rearranging (3.184) produces (3.185) [33]:

$$\frac{d\left(\frac{d}{d\tau}\,\theta_e\right)}{d\theta_e} = -a_p\,\frac{\sin[\theta_e(t)]}{\left(\frac{d}{d\tau}\,\theta_e\right)} - \cos[\theta_e(t)] \tag{3.185}$$

The above equation gives us the normalized nonlinear equation of a PLL, but we need to turn it into a difference equation so that we can put it into a program.

3.5.3 Difference Equation for Making the Phase-Plane Trajectory Plot

A phase-plane trajectory plot is computed by using a difference equation. A difference equation is made from (3.185). Equation (3.186) shows the normalized difference equation [32, 33]:

$$d\theta e_i = d\theta e_{i-1} + \left[-\frac{a_p\,\sin(\theta_{i-1})}{d\theta e_{i-1}} - \cos(\theta_{i-1})\right](\theta_i - \theta_{i-1}) \tag{3.186}$$

where

i = index value in the array.

Equation (3.180) is used to convert damping factor to the variable a_p in (3.186).

Figure 3.55 shows the normalized phase-plane trajectory plot from solving (3.186). The phase, which is on the horizontal axis in Figure 3.55, is the phase error of the PLL. The phase velocity, which is the vertical axis in Figure 3.55 and

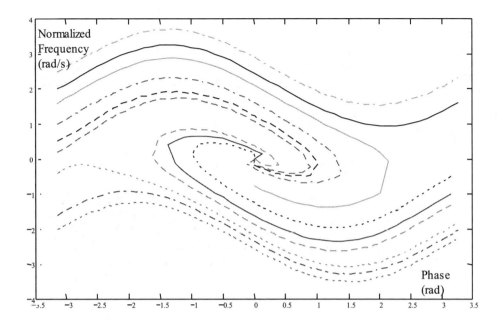

Figure 3.55 Complicated and normalized curves.

is normalized to $(2\zeta\omega_n)$, is the frequency difference between the reference and the VCO frequencies. To use the plot, the initial phase and frequency difference must be known. Pathways on the top half of the plot travel from left to right. Pathways on the bottom half of the plot travel from right to left. Finally, this curve shows an infinite number of solution pathways for the nonlinear differential equation.

Figure 3.55 shows only one 2π range of the response. An infinite number of duplicate 2π plots exist to the left and right of this plot. The point in the middle of Figure 3.55 is called a *stable focus point*. At even multiples of π are the stable points where the loop achieves lock. Figure 3.56 shows other trajectory points that can occur in a phase-plane portrait solution. At odd multiples of π are centers of instability called *saddle points*. Near these points, no matter what the direction of the trajectories may be, the magnitude of the frequency error will decrease until the saddle point is almost reached, then quickly increase again [33].

3.5.4 Unnormalized Solution

Let's find the unnormalized and simpler phase/phase velocity trajectory plot solution of a nonlinear differential equation for a PLL. To find this solution, PLL parameters (phase margin and 0-dB crossover frequency) must be converted to damping factor and natural frequency. Damping factor is calculated from the phase margin of the loop. Equation (3.187) computes phase margin from damping factor:

$$\theta_{margin} = \text{atan}\left(2\zeta\sqrt{2\zeta^2 + \sqrt{4\zeta^4 + 1}}\right)\left(\frac{180}{\pi}\right) \tag{3.187}$$

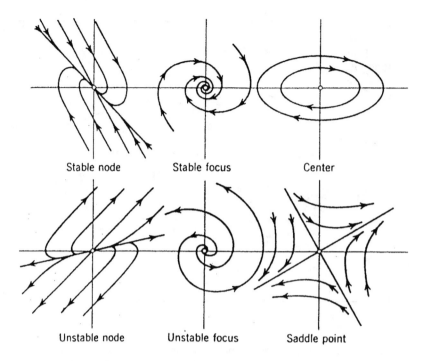

Figure 3.56 Possible trajectory points in a phase-plane solution.

We compute the damping factor by iterating the damping factor in (3.187) until it equals the phase margin. Equation (3.188) computes natural frequency from the 0-dB crossover frequency:

$$\omega_n = \frac{2\pi f_x}{\sqrt{2\zeta^2 + \sqrt{4\zeta^4 + 1}}} \tag{3.188}$$

Equation (3.189) shows the multiplicative factor $\dfrac{2\zeta\omega_n}{2\pi}$ that unnormalizes the phase-plane plot to frequency error:

$$\Delta f = d\theta e \frac{2\zeta\omega_n}{2\pi} \tag{3.189}$$

Substituting the damping factor computed in (3.187) and the natural frequency computed in (3.188) into the multiplicative factor in (3.189), then using this factor to multiply the normalized plot, produces the unnormalized plot in Figure 3.57. There are many different responses inside these waveforms as shown in Figure 3.57, but for simplicity we have left these responses off.

The separatrix curves shown in Figure 3.57 demarcate cycle-slip responses and no cycle-slip responses. A response with an initial condition inside the separatrix will not have a cycle slip and will follow the spiral track and lock at zero phase difference and zero frequency difference. A response with an initial condition outside the separatrix cycle slips in a sinusoidal-type waveform to the next 2π

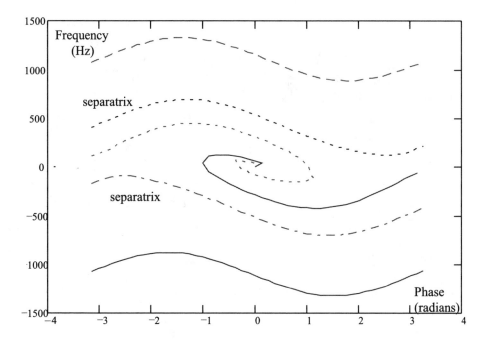

Figure 3.57   Unnormalized and simple phase/phase velocity trajectory plot.

phase plane that is identical to the one in Figure 3.57, but it takes a slightly different and lower trajectory path.

The slightly different path occurs because the end point frequency at +3.14 rad is slightly less than the beginning point frequency at −3.14 rad. Consequently, at the next 2π phase plane, a slightly different path will be taken, and will be the case for the following 2π phase planes, until at some point the separatrix is crossed. Once the separatrix is crossed, the loop locks without cycle slipping.

Outside the separatrix, there are infinite possible solutions because there is an infinite number of phase- and frequency-step initial conditions. The reacquisition times overlap making it unmanageable to do a worst-case analysis because the results will vary from a few milliseconds to infinity. Consequently, the different and varying reacquisition times that have been observed in the laboratory for the same stimulus are explained by the phase-plane portrait because the same initial frequency difference can have different initial phase conditions that causes a wide variation in the reacquisition response time.

3.5.5   Measured Step Responses Inside and Outside the Separatrix

Let's use measured data from an example PLL with a damping factor of 1 and 0-dB crossover point of 300 Hz to show the circuits response inside and outside the separatrix. Figure 3.58 shows the PLLs measured response outside the separatrix. This figure shows an initial 1-kHz frequency difference that is produced by a 0.2V step in the 5-V supply line to the VCO. The loop takes 50 ms to achieve relock. With a 1-kHz frequency difference, there are about 50 cycle slips before lock is achieved. Consequently, 50 2π phase-plane plots were crossed before the

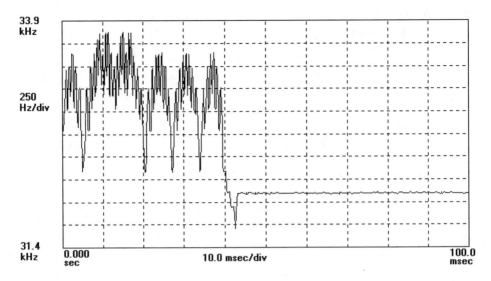

Figure 3.58   Outside-the-separatrix measured response.

separatrix was crossed and the loop locked at zero phase error and zero frequency difference.

Figure 3.59 shows the PLLs measured response inside the separatrix. A 200-Hz initial frequency step response is caused by a 100-mV step in the 5-V supply line to the VCO. This figure shows a well-behaved response that settles in a few milliseconds and does not have cycle slips.

These plots show that, outside the separatrix, the reacquisition time is unbounded and that many different responses can occur for a given frequency-step disturbance to the PLL. In addition, these plots show that, for a certain frequency-step size, the loop can take an infinite amount of time to reacquire lock.

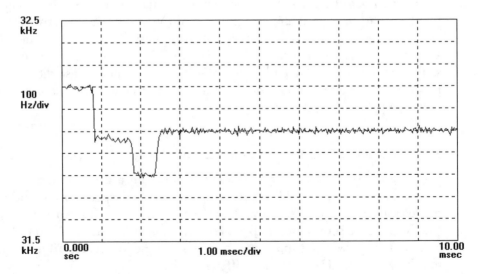

Figure 3.59   Inside-the-separatrix response.

The vertical size of the separatrix depends on the loop bandwidth. However, the amount of loop bandwidth is limited by the reference frequency (a factor of 10 less than the reference frequency is desirable). Consequently, for phase-only detectors, such as an exclusive-OR phase detector, an additional reacquisition circuit is necessary to make sure that the loop reacquires lock from any disturbance to the loop.

To summarize, this section discussed several key acquisition characteristics of a PLL. First, we discussed the lock-in range, pull-in range, and hold-in range. Then, we derived the second-order, nonlinear, ordinary differential equation for a PLL. This nonlinear equation characterized the loop's time-domain response outside the range of the phase detector. It also showed that the time-domain response outside the range of the phase detector depended on the initial phase and frequency conditions.

3.6 Spurious Signals

Any nonharmonically related signal is a spurious signal. These signals limit the electronic performance of many devices. Consequently, minimizing these signals and their amplitude levels is the goal of any good system design. Studying the generation mechanism of these signals helps us understand how to minimize their generation.

Figure 3.60 shows a spectrum analyzer that contains many interfering signals in a PLL. This figure shows harmonics of the carrier, additional phase noise from loop instability, reference sideband feed-through, and an intermodulation product. Harmonics of the carrier do not present a significant problem in many PLL designs. We have already studied loop instability and its causes. Intermodulation products

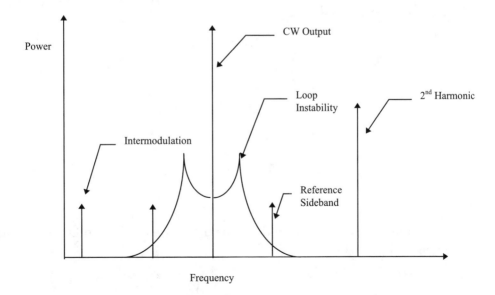

Figure 3.60 Typical intermodulation spectrum for a PLL.

are the next spurious signals that we will study. In Section 3.6.2, we will study the effects of reference sideband feed-through.

3.6.1 Intermodulation Products

In this section, equations are shown that compute the frequencies where intermodulation products will occur and the power levels that they will have. Next, some relationships between product-level and design parameters are discussed. Then, examples of upconversion and downconversion are shown.

Nonlinear circuits (e.g., diodes, I/O buffers, amplitude limiters, frequency dividers, analog multipliers) cause intermodulation products. Studying the intermodulation products of a mixer helps one understand how these products are generated. Equation (3.190) describes the mathematical relationship between the mixing of two signals, which are called the local oscillator (LO) and radio frequency (RF) signals [21, p. 62]:

$$\omega_{nm} = N_{lo}\,\omega_1 \pm M_{rf}\,\omega_2 \tag{3.190}$$

The desired output is at $M_{rf} = N_{lo} = 1$ and is in the form $\cos(\omega_2 + \omega_1)$ or $\cos(\omega_2 - \omega_1)$. All other levels are undesirable and must be reduced by filtering to the levels established by system requirements. Usually a level of −60 dBc is acceptable in most systems.

3.6.1.1 Computing Intermodulation Levels

Figure 3.61 shows a double-balanced mixer with variables defined so that intermodulation analysis can be done. Computing the level of intermodulation requires defining several terms. First, (3.191) through (3.190) define the mismatch of voltages across each diode:

$$\delta_2 = \frac{V_2}{V_1} \tag{3.191}$$

$$\delta_3 = \frac{V_3}{V_1} \tag{3.192}$$

$$\delta_4 = \frac{V_4}{V_1} \tag{3.193}$$

The differences in (3.191) through (3.193) are due to the diode capacitances and series resistances. Next, the ratio of the diode voltage drop to LO peak voltage is computed by (3.194):

$$V_f = \frac{V_F}{V_L} \tag{3.194}$$

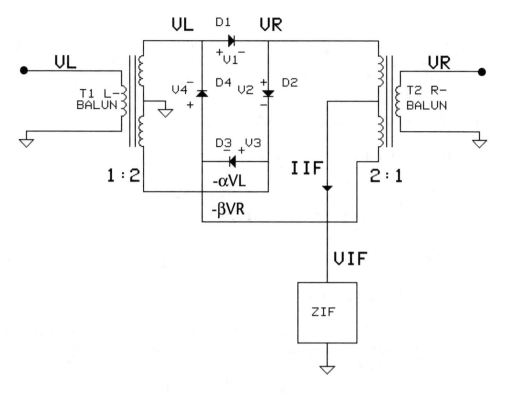

Figure 3.61 Double-balanced mixer with variable definitions.

where

V_F = forward diode voltage drop (V).

Next, Figure 3.62 shows the variable definitions for a balun so that the effects of balun isolation on intermodulation products can be analyzed. For the LO balun, (3.195) computes the isolation ratio:

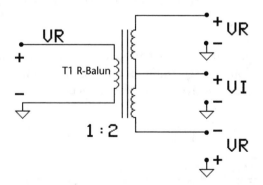

Isolation = 20 Log (V_I/V_R)=20 Log (1-β)

Figure 3.62 Balun imbalance as a function of isolation.

$$\alpha = 1 - 10^{\frac{L_{lo}}{20}} \qquad (3.195)$$

where

L_{lo} = LO port isolation (dB).

For the RF balun, (3.196) computes the isolation ratio:

$$\beta = 1 - 10^{\frac{L_{rf}}{20}} \qquad (3.196)$$

where

L_{rf} = RF port isolation (dB).

Next, coefficients for odd-by-odd products, even-by-even products, odd-by-even products, even-by-odd products, and IF products are computed by (3.197) through (3.201):

$$B_{oo} = 1 + \delta_4 + \alpha(\delta_3 + \delta_2) - |M_{rf}| [(\delta_4 + \delta_2) + \alpha(\delta_3 + \delta_2) - \beta(\delta_3 + \delta_4)] \qquad (3.197)$$

$$B_{ee} = [(-1 + \delta_4)] - \alpha(\delta_3 + \delta_2) - |M_{rf}| \{[(\delta_4 + \delta_2) - \alpha(\delta_3 - \delta_2)] + \beta(\delta_3 - \delta_4)\} \qquad (3.198)$$

$$B_{oe} = |M_{rf}| [-\delta_4 - \delta_2 + \alpha(\delta_3 + \delta_2) + \beta(\delta_4 - \delta_3)] \qquad (3.199)$$

$$B_{eo} = |M_{rf}| [\delta_4 + \delta_2 + \alpha(\delta_3 - \delta_2) - \beta(\delta_4 + \delta_3)] \qquad (3.200)$$

$$B_{if} = 1 + \delta_4 + \alpha(\delta_3 + \delta_2) - [(\delta_4 - \delta_2) + \alpha(\delta_3 + \delta_2) - \beta(\delta_3 + \delta_4)] \qquad (3.201)$$

With definitions and coefficients defined, the amplitude suppression can be computed. Equation (3.202) computes the amplitude-suppression ratio for the odd-by-odd products and the even-by-even products.

$$A_{nm1} = \frac{\Gamma\left(\frac{|N_{lo}| + |M_{rf}| - 1}{2}\right)}{\Gamma\left(\frac{|N_{lo}| - |M_{rf}| + 3}{2}\right)} \qquad (3.202)$$

$$\cdot \frac{1}{2}\left[\sin\left(\frac{|N_{lo}|\pi}{2}\right)\sin\left(\frac{|M_{rf}|\pi}{2}\right)B_{oo} + \cos\left(\frac{|N_{lo}|\pi}{2}\right)\cos\left(\frac{|M_{rf}|\pi}{2}\right)B_{ee}\right]$$

Equation (3.203) computes the amplitude suppression ratio for the odd-by-even products and the even-by-odd products.

$$A_{nm2} = \frac{\Gamma\left(\frac{|N_{lo}| + |M_{rf}|}{2}\right)}{\Gamma\left\{\frac{[(|N_{lo}| - |M_{rf}|) + 2]}{2}\right\}}$$
$$\cdot V_f\left[\sin\left(\frac{|N_{lo}|\pi}{2}\right)\cos\left(\frac{|M_{rf}|\pi}{2}\right)B_{oe} + \cos\left(\frac{|N_{lo}|\pi}{2}\right)\sin\left(\frac{|M_{rf}|\pi}{2}\right)B_{eo}\right] \quad (3.203)$$

Equation (3.204) combines the two amplitude-suppression ratios:

$$A_{nm} = \frac{A_{nm1} + A_{nm2}}{B_{if}|M_{rf}|!} \quad (3.204)$$

Equation (3.205) computes the power-level difference between the LO and RF power (dBm) applied at each port.

$$\Delta P = P_{rf} - P_{lo} \quad (3.205)$$

Finally, (3.206) computes the intermodulation suppression (dBc):

$$S_{nm} = (|M_{rf}| - 1)\Delta P + 20\log(|A_{nm}|) \quad (3.206)$$

The derivation of these parameters is based on the switching characteristics of an ideal diode. Consequently, mixing from the nonlinearity of a diode is ignored. This approximation has been addressed in the literature [35] and is justified by the close agreement between measured and calculated intermodulation suppression for small values of n and m and ΔP less than −15 dB. For $n < 8$ and $m < 4$, the predicted results are accurate enough for most system designs. For larger integer values, the predicted results are better than the actual performance.

One advantage of the formula is that the parameters can be adjusted to meet actual performance. By measuring the intermodulation suppression and adjusting the parameters in the equations, the actual performance can be modeled. Table 3.4 shows an example measurement of intermodulation suppression [36]. Adjusting the equation parameters to the values in Table 3.5 matches the data in Table 3.4. To verify the calculations of the equation, Table 3.6 shows the computed values for a 3×-2 intermodulation product.

3.6.1.2 Intermodulation Suppression

Studying the equations show several rules for suppressing intermodulation products. First, a large power difference between LO and RF suppresses intermodulation products. Next, well-balanced circuitry (high interport isolation) and diodes suppress even harmonic products.

Table 3.4 Example Measurement of
Intermodulation Suppression

LO	RF	Suppression (dBc)
1	1	0
1	2	$\Delta P - 41$
1	3	$2\Delta P - 28$
2	1	-35
2	2	$\Delta P - 39$
2	3	$2\Delta P - 44$
3	1	-10
3	2	$\Delta P - 32$
3	3	$2\Delta P - 18$
4	1	-35
4	2	$\Delta P - 39$

Table 3.5 Parameter
Values Adjusted to
Measured Suppression
Values in Table 3.4

Parameter	Value
α, β	0.7
δ_2	0.85
δ_3	0.95
δ_4	1.05
V_f	0.1

Table 3.6 Computed
Coefficients for Example
Suppression

Variable	Value
B_{IF}	3.25
B_{oe}	1.14
A_{nm}	0.026

Here are some rules for adjusting the equation parameters. First, balancing the L and R ports and matching the four diodes suppresses even-by-even products. Balancing the L port balun ($\alpha = 1$) and matching the diode across it ($\delta_3 = \delta_4$) suppresses odd-by-even products. Balancing the R port balun ($\beta = 1$) and matching the diodes across it ($\delta_2 = \delta_3$) suppresses even-by-odd products. Odd-by-odd products, especially with $m = 1$, should not be allowed inside an IF bandwidth because nothing can be done to improve their suppression without degrading the suppression of other products.

Two modes of operation exist for intermodulation products: up-converting and down-converting signals. Examples will show these mechanisms and the intermodulation products that are created.

Figure 3.63 shows the intermodulation products generated from upconversion. A constant 1.185-GHz LO upconverts the 0.45–0.63-GHz (0.54-GHz center frequency) IF to an output of 1.635 to 1.815 GHz at the RF output. In upconversion, the low-frequency input signal usually is connected to the low-frequency IF port,

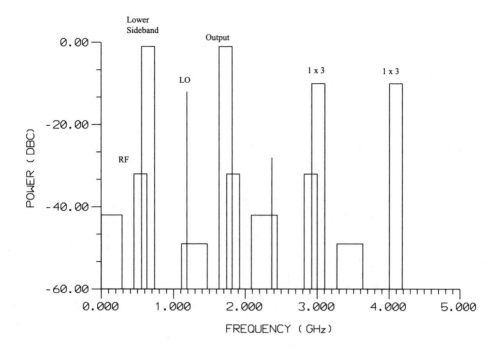

Figure 3.63 Intermodulation products from an upconversion.

and the output is taken from the high-frequency RF port. A constant line represents a signal that does not change, while a box represents a range of signals that depend on the value of the varying input frequency.

Figure 3.63 shows that the mixing products are spread all over the spectrum for just the interaction of two signals. In addition, the highest signals are the LO feed-through and the 1 × 3 product. System designers will select frequency plans so that these signals are far enough away that they can be easily filtered. If these signals are not suppressed or filtered out, the performance of an electronic device can be severely hampered.

Figure 3.64 shows the intermodulation products generated from downconversion. A constant LO of 76 MHz downconverts an RF signal of 40 to 46 MHz to an output IF signal of 30 to 36 MHz. Again, many signals are spread across the frequency spectrum from the mixing of two signals. This downconversion frequency plan has high intermodulation products from the feed-through of the RF, the LO, the 1 × 2 product, and the 1 × 3 product. The RF signal's being only 4 MHz away from the RF signal presents a difficult filtering problem. Better frequency planning by the designer would have prevented this problem.

3.6.1.3 Amplitude-to-Phase Modulation Conversion

Mixing two signals is an amplitude modulation process. So, why should we worry about it when PLL circuits are concerned with phase-modulated signals? The nonlinear circuits that mix signals together also have a transfer function component that converts AM to PM.

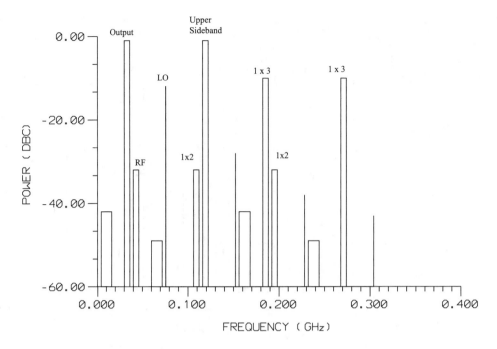

Figure 3.64   Intermodulation products from a downconversion.

Circuits with phase characteristics (or delay) dependent on the instantaneous amplitude of the input signal convert AM to PM. For example, limiters, mixers, voltage-tuned filters, and varactor multipliers represent circuits that convert AM to PM. These circuits convert envelope variations at their input to phase variations at their output. If small envelope perturbations are assumed, a conversion constant of $K°/dB$ characterizes these circuits. As an AM wave passes the converter, every decibel of variation in the envelope of the input signal produces $K°$ of peak change in the phase of the output signal.

Putting intermodulation AM waveforms into a system will produce intermodulation PM waveforms that will trash the output of a PLL or corrupt the input to a PLL by creating more, higher-level intermodulation products. Consequently, a good design approach will minimize the number of intermodulation products.

To summarize, equations were developed that compute the frequencies where intermodulation products will occur and the power levels that they will have. Next, some relationships between product level and design parameters were discussed. Then, examples of upconversion and downconversion were shown. Understanding these relationships will help us troubleshoot and prevent spurious signals in designs.

3.6.2   Minimizing the Generation of Reference Sidebands

Spurious signals in PLLs can leak to the tune line of a monolithic VCO. These signals frequency-modulate the VCO to produce unwanted sidebands at the VCO output. The tuning slope of the VCO determines the sideband levels. Monolithic VCOs have sensitive tuning slopes (>100 MHz/V). Consequently, these VCOs are very susceptible to producing high sideband levels when the VCO tune line is modulated by spurious signals from a PLL. The clock signal, which is used as reference, to the phase detector in the PLL causes the largest sidebands at the output of a PLL. Consequently, these modulated sidebands are appropriately called *reference sidebands*. Filtering alone cannot reduce the reference sidebands enough to satisfy a majority of the system requirements; however, this section shows several PLL design parameters that can be adjusted and several circuit techniques that can be used to reduce reference sidebands before adding any filters.

In a PLL, a phase detector produces a pulsed error signal by comparing the output frequency of the VCO with the reference-clock frequency input to the phase detector. The pulses are generated at the rate of the reference-clock frequency. The phase difference (or error) between the reference-clock input to the phase detector and the variable-clock input to the phase detector determines the pulse width of the signal at the output of the phase detector. The filtered and processed error-signal output of the phase detector tunes the VCO so that a minimum error signal (minimum pulse width) is produced. Without using techniques to minimize the reference-clock signal level, some amount of reference-clock signal couples to the VCO tune line. This signal modulates the VCO, which produces the reference sidebands. Many unfortunate engineers have been assigned the difficult task of removing these reference sidebands from a PLL or synthesizer after construction. Reference sidebands and other spurious sidebands should be removed at the beginning of the design phase because these signal levels can be easily minimized at this stage.

The many sources of coupling to the VCO tune line make the task of suppressing these reference sidebands difficult. Figure 3.65 identifies the coupling points and the paths of the reference clock and other spurious signals to the VCO tune line in a microwave PLL. The other spurious signals consist of power-supply-generated signals (60 cycles), logic-control-interface signals, intermediate-programmable-divider signals, reverse-output-interface signal, radiated-logic signals, any radiated signals, and many other signals too numerous to list; however, the signal level of the reference is usually higher than all the other spurious signals.

Postconstruction solutions to minimize the reference sidebands in a PLL limits an engineer to adding more filters to minimize these sidebands; however, an optimistic maximum of only 30–40-dB suppression can be achieved by a filter. Many synthesizer designs require >40-dB suppression because frequency multiplication in the synthesizer raises the reference sideband levels and other spurious sideband levels. For example, frequency multiplying a 100-MHz signal, with 40-dBc sidebands at a 10-MHz offset from the carrier, by 10 produces a 1-GHz signal with 20-dB sidebands. Consequently, a 100-MHz signal with 60-dBc sideband levels is required to meet a 40-dBc sideband specification for the 1-GHz signal.

The simple solution of using a narrow-bandwidth filter to attenuate the reference signal before the VCO tune line causes another problem. The insertion phase of a narrow-bandwidth filter that achieves 40-dB suppression of the sidebands usually causes the loop to be unstable. This instability requires widening the filter bandwidth to maintain a stable loop; however, a wider filter bandwidth degrades the original intent of sideband suppression. Consequently, this contradiction makes filtering alone inadequate for suppressing reference sidebands after a synthesizer has been constructed. The task of suppressing reference and other spurious side-

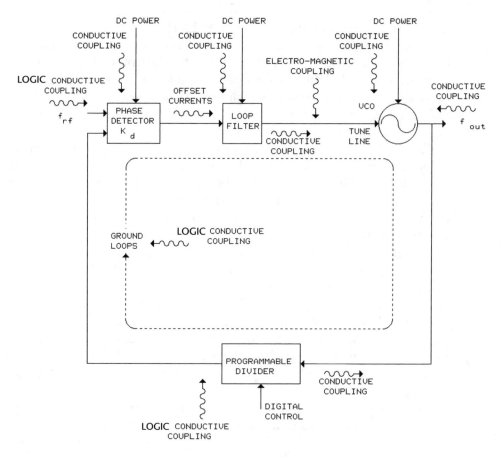

Figure 3.65   Unprotected PLL from signals coupling into the circuit.

bands should be attacked at the beginning of the design effort where these signal levels can be significantly reduced in the design stage even before adding any filters. Then, the lowered spurious-signal levels can be further reduced by filtering. This method will not only solve the problem, but it will make the filtering requirements easier.

This section concentrates on techniques to minimize reference sideband levels. However, in addition to reducing reference sidebands, many of these techniques can also be applied to reducing other spurious sidebands. The techniques that also reduce other spurious sidebands will be identified throughout the section.

Experiments have identified three major contributors to reference sideband levels. The contributors are from the operational amplifier or charge pump, the phase detector, and the chip's isolation. This section studies each effect so that adjustments can be made to suppress reference sidebands; however, these major contributors also have a common element. Each effect frequency-modulates the VCO to produce the unwanted sidebands. Reducing this modulation sensitivity equally reduces the effects of the three major contributors on the reference sideband levels. Consequently, studying the modulation mechanism first will make under-standing the three major contributors easier.

3.6.2.1 Maximizing Reference Frequency and Minimizing the VCO Tuning Slope

From narrowband FM theory, [37, 38] have generated (3.207), which relates the reference-frequency and VCO-tuning-slope parameters to reference sideband levels:

$$\sqrt{P_{sbv}} = V_{rf}\left(\frac{K_v}{2\omega_{rf}}\right) \tag{3.207}$$

where

V_{rf} = peak magnitude of reference spur on the tune line of the VCO (V);

ω_{rf} = angular modulation frequency of the reference spur (rad/s);

P_{sbv} = power ratio of the reference sideband level;

K_v = VCO gain or tuning slope (rad/s/V).

Equation (3.207) from [37, 38] computes the level of sidebands from a reference-clock signal at the tune line of the VCO. Studying this equation can yield some ways to reduce modulation sensitivity.

Plotting sideband levels using (3.207) by varying the peak modulation signal and the modulation frequency shows the sensitivity of a VCO to reference sideband signal levels on the tune line. For example, Figure 3.66 plots the sideband levels that are produced by 1-μV, 100-μV, and 1-mV peak signals for a VCO with a 100-MHz/V tuning slope. In the figure, a signal on the tune line with a 100-μV level and a 1-MHz frequency causes a 60-dB sideband. Attenuating 1-MHz signals

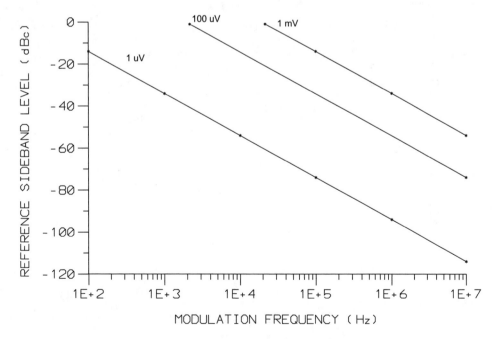

Figure 3.66 Reference sidebands versus peak modulation voltage.

to less than 100 μV on the tune line can be difficult. Figure 3.66 and the location of the reference-frequency variable in the denominator of (3.207) show that maximizing the reference frequency does indeed reduce reference sideband levels.

Plotting sideband levels using (3.207) by varying the modulation slope and the modulation frequency shows that the modulation slope of a microwave VCO determines the level of reference sidebands. For example, Figure 3.67 plots the sideband levels that are produced by a 10-μV peak signal on the tune line for a VCO with 10-MHz/V, 100-MHz/V, and 1-GHz/V tuning slopes. In the figure, a 1-MHz signal on the tune line requires a VCO with a tuning slope less than 100 MHz/V to keep reference sidebands below 60 dB. Wideband frequency operation at microwave frequencies requires a tuning slope greater than 100 MHz/V. Consequently, PLL design necessitates trading off between frequency range and reference sideband levels. In addition, Figure 3.67 and the location of the tuning-slope variable of the VCO in the numerator of (3.207) show that minimizing the tuning slope of the VCO reduces the level of reference sidebands. Consequently, (3.207) shows that adjusting the PLL design to maximize the reference frequency and to minimize the VCO tuning slope minimizes the level of the reference-frequency sidebands.

3.6.2.2 Minimizing Offset Current

Now, the contribution of the operational amplifier or charge pump to reference sideband levels can be studied. Consider the PLL of Figure 3.68, which shows an interconnect diagram of the phase/frequency detector, loop filter, and VCO in a PLL. The operational amplifier in the figure requires a dc bias at its inputs. This bias requirement produces dc offset currents in the circuit. However, dc offset currents cause a frequency error in PLLs by creating an excess dc voltage to the

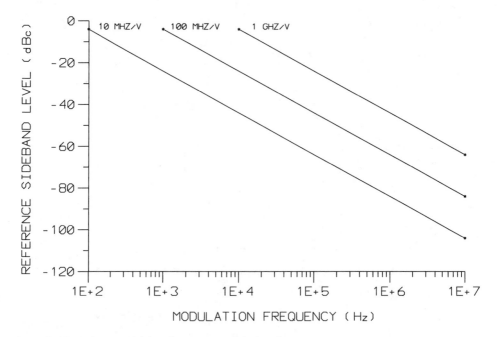

Figure 3.67   Reference sidebands versus modulation slope.

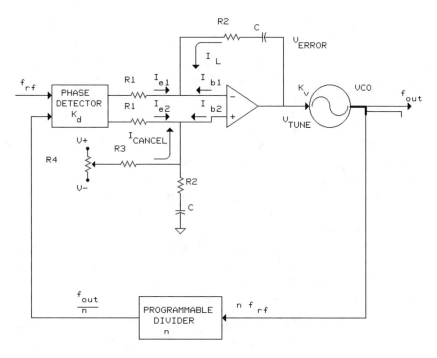

Figure 3.68 PLL with detail of loop filter using differential interconnection.

tune line of the VCO. To reduce this error, the loop generates a higher output error-signal level from the phase/frequency detector at the rate of the reference-clock frequency. This additional signal level modulates the VCO tune line to cancel the excess dc voltage that is caused by the offset currents. An increased error signal produces a wider digital pulse width at the output of the phase detector. This increased phase detector signal level increases the reference sideband levels. Consequently, minimizing the offset currents by using the differential amplifier connection in Figure 3.68 minimizes the reference sideband levels.

Next, studying an equation that relates offset currents to reference sideband levels can yield more ways of minimizing these offset current effects. For this interconnection, (3.208) from [16] computes the reference sideband level due to the effects of the offset current of the operational amplifier:

$$\sqrt{P_{sbi}} = I_b R_2 \left(\frac{K_v}{f_{rf}} \right) \tag{3.208}$$

where

I_b = operational amplifier offset bias current (A);

R_2 = feedback resistor of the operational amplifier (Ω);

P_{sbi} = power ratio of the reference sideband because of the offset bias current from the operational amplifier;

f_{rf} = reference frequency (Hz).

Equation (3.208) shows that reducing the offset current of the operational amplifier reduces the level of reference sidebands. In addition, (3.208) shows that minimizing feedback resistance reduces the level of reference sidebands; however, the lower limit of the feedback resistance depends on the current-drive-level capability of the operational amplifier. Consequently, the ideal operational amplifier to use in a PLL will have low offset-current and high current-drive capabilities to minimize the level of reference sidebands.

Finally, adding an opposing offset current in the operational amplifier circuit, as shown in Figure 3.68, will also reduce the level of reference sidebands. Adding a large potentiometer in series with a stable supply voltage will adjust out the residual operational amplifier offset current and will lower the levels of the reference sidebands. Adding the opposing offset current adjustment without reducing the operational amplifier's offset current presents an easy solution and a great temptation; however, a large opposing offset current increases the sensitivity of the circuit to larger sideband levels over temperature. A change of potentiometer resistance and supply voltage over temperature significantly increases reference sidebands that have been suppressed with a high (μa) canceling offset current.

3.6.2.3 Minimizing Phase/Frequency Detector Minimum Pulse Width

Next, the contribution of the phase/frequency detector to reference sideband levels can be studied. A phase detector produces a digital signal at the rate of the reference-clock frequency. The phase difference between the reference-clock input and the variable-clock input to the phase detector determines the pulse width of the signal at the output of the phase detector. For the $0°$ phase (phase-locked) condition, the phase detector generates the smallest possible pulse width. The simultaneous edge logic comparison of the VCO signal to the reference-clock signal causes a race condition to occur. This race condition produces a residual error signal of narrow pulses at the output of the phase/frequency detector. The speed of the phase/frequency detector directly determines the pulse width of these pulses. These pulses leak to the tune line of VCO and cause reference sidebands.

Studying an equation that relates pulse width to reference sideband level can yield ways of minimizing the reference sidebands. Calculating the Fourier series of these narrow pulses computes the signal levels at each harmonic that modulates the tune line of the VCO.

Pulse Width Reduction for Operational Amplifier Compensation
Equation (3.209) from [6] computes the Fourier series of a voltage pulse:

$$V(0) = \frac{V_{pk}\alpha}{\pi}$$

$$V(f_{rf}) = \frac{2V_{pk}}{\pi} \sin(\alpha)$$

$$V(2f_{rf}) = \frac{2V_{pk}}{2\pi} \sin(2\alpha) \tag{3.209}$$

$$\vdots$$

$$V(nf_{rf}) = \frac{2V_{pk}}{n\pi} \sin(n\alpha)$$

where

$$\alpha = \frac{t_{pw} f_{rf} \pi}{2};$$

V_{pk} = peak-to-peak voltage or voltage swing of pulse (V);

$V(n f_{rf})$ = voltage level that was produced at each harmonic (V);

n = order of harmonic or reference frequency;

t_{pw} = pulse width (sec).

Figure 3.69 graphically shows the pulsed waveform with the parameters in (3.209) identified. The condition of narrow pulses under consideration makes α in (3.209) a small number. Usually, the loop has a roll-off capacitor to filter the reference sidebands, and the higher harmonics of the reference frequency are attenuated. Consequently, the fundamental has the largest signal level. Substituting the magnitude for the fundamental from (3.209) into (3.207) and multiplying by the high-frequency transfer function of the loop filter produces (3.210):

$$\sqrt{P_{sdet}} = \left(\frac{2 V_{pk} \sin(\alpha)}{\pi} \right) \frac{R_2}{R_1} \frac{K_v}{2 \omega_{rf}} \qquad (3.210)$$

where

P_{sdet} = power ratio of the reference sideband due to the phase detector;

R_2/R_1 = high-frequency transfer function of loop filter with an operational amplifier.

Equation (3.210) computes the sideband level at the fundamental due to the minimum pulse width of the phase/frequency detector.

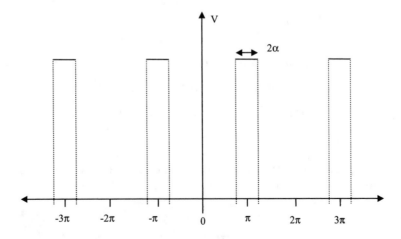

Figure 3.69 Pulsed waveform definition.

Equations (3.209) and (3.210) show that minimizing the pulse width, t_{pw}, and voltage swing, V_{pk}, of the residual error signal out of the phase/frequency detector minimizes the reference sideband levels. The speed of a phase/frequency detector directly determines the pulse width of these pulses. Consequently, a designer's selection of a process with 0.55 μm to 0.15 μm gate length for the phase/frequency detector directly determines the level of the reference sidebands. Table 3.7 presents a comparison of reference sideband performance versus the process for the phase/frequency detector with an example charge pump PLL to show the change in the reference sideband level.

Consequently, studying the narrow pulse widths of the residual error signal has shown that adjusting the PLL design to use the fastest phase/frequency detector minimizes the level of reference-frequency sidebands. For a phase/frequency detector that is made of gates, the minimum pulse width is approximately four gate delays.

Pulse-Width Reduction for Charge Pump Compensation
Now, let's look at pulse-width reduction for charge pump compensation, which will follow a similar analysis. Equation (3.209) from [6] computes the Fourier series of a current pulse:

$$I(0) = \frac{(I_{pk}\,\alpha)}{\pi}$$

$$I(f_{rf}) = \frac{(2I_{pk})}{\pi}\sin(\alpha)$$

$$I(2f_{rf}) = \frac{(2I_{pk})}{(2\pi)}\sin(\alpha) \qquad (3.211)$$

$$\vdots$$

$$I(nf_{rf}) = \frac{(2I_{pk})}{(n\pi)}\sin(\alpha)$$

where

$$\alpha = \frac{t_{pw}f_{rf}\pi}{2};$$

I_{pk} = peak-to-peak current or current swing of pulse (A);

$I(nf_{rf})$ = current level that was produced at each harmonic (A);

n = order of harmonic of reference frequency;

t_{pw} = pulse width (sec).

Substituting the magnitude for the fundamental from (3.209) into (3.207) and multiplying by the high-frequency transfer function of the loop filter produces (3.212):

$$\sqrt{P_{scp}} = \left[\left(\frac{2I_{pk}\sin(\alpha)}{\pi}\right)R_2\frac{K_v}{2\omega_{rf}}\right] \qquad (3.212)$$

Table 3.7 Reference Sideband Performance Versus Process of Phase/Frequency Detector

Effective Gate Length (μm)	Offset Voltage	Gain (V/rad)	Max. Min. Pulse Width (sec)	Reference Sideband Performance (dBc)
0.55	1.96E–03	0.816	9.99E–10	–21
0.44	2.57E–04	0.793	1.05E–09	–21
0.35	7.32E–04	0.382	1.13E–09	–21
0.25	3.37E–04	0.399	6.45E–10	–25
0.18	6.98E–04	0.286	4.32E–10	–28
0.15	8.46E–05	0.24	2.42E–10	–33

where

P_{scp} = power ratio of the reference sideband due to the phase detector/charge pump;

R_2 = high-frequency transfer function of loop filter with a charge pump.

Equation (3.212) computes the sideband level at the fundamental due to the minimum pulse width of the phase/frequency detector.

Example 3.3

What is the reference sideband level in dBc and the peak fundamental voltage on the tune line for a 4-MHz PLL with a 10-kHz reference frequency, $N = 400$, 10 MHz/V VCO gain, a phase detector with minimum pulses out of 1 ns, a charge pump peak current of 20 μa, and a charge pump resistor R_2 of 10 kΩ?

Substituting into (3.212) gives a –54-dBc sideband level and a 4-μV peak fundamental on the tune line.

3.6.2.4 Maximizing On-Chip Isolation

Finally, the contribution of chip isolation to reference sideband levels can be studied. The high tuning sensitivity of microwave VCOs, 100 MHz/V to 500 MHz/V, allows reference signals on the ground plane to modulate the VCO and produce a significant level of reference sidebands. For example, Figure 3.66 shows that a signal with a 100-μV level and 1-MHz frequency produces a 60-dB sideband on a VCO with a 100-MHz/V tuning slope. This requires >80-dB isolation from a 5-V logic clock signal to keep the reference sidebands on the VCO less than 60 dB. This sensitivity of the VCO to small signals on the ground plane and large signal levels from digital circuits in the PLL makes using good grounding and shielding techniques a requirement for minimizing the level of reference sidebands.

The digital sections for the reference-clock generator and programmable divider in the PLL usually have the highest levels of reference signals and have other spurious signals present. Consequently, these signals require the highest isolation from the VCO tune line. The leakage of these signals to the VCO tune line can be related to the desired isolation, P_{iso} (dB), and peak-to-peak voltage, V_{lpk}, by (3.213):

$$\sqrt{P_{slk}} = V_{lpk}\; 10^{\frac{P_{iso}}{20}}\; \frac{K_v}{2\omega_{rf}} \tag{3.213}$$

where

V_{lpk} = peak-to-peak voltage of worst spurious signal, which is usually is the voltage swing of the device (V);

P_{slk} = power ratio of the reference sideband due to chip isolation;

P_{iso} = chip isolation (dB).

Equation (3.213) shows that improving the isolation of the digital circuits from the VCO tune line will reduce the level of reference sidebands. For the interested reader, [37, 39] show several grounding and shielding techniques for isolating these circuits.

Isolating the loop filter from the digital circuitry and isolating the VCO circuitry from the loop filter and the digital circuitry maximizes the isolation of the VCO tune line from the digital sections. Isolation can be achieved by disconnecting the grounds on the printed circuit board, by separating the circuits with egg-crate construction techniques [37], or by separating the circuits into separate modules. Additional isolation from digital circuit signals can be achieved by using differential current mode line drivers and receivers to interface with outside digital control circuits. Figure 3.70 shows one implementation of these measures to minimize the levels of reference sidebands in a PLL.

3.6.2.5 Computing the Total Effect

Laboratory experiments have shown that the combined effects of the offset current, the phase detector, and the chip isolation all linearly add up to produce the reference sideband level. Each one contributes equally to this total. Consequently, combining (3.208) through (3.213) produces (3.214) for the total sideband level, P_{stot}, in decibels:

$$P_{stot} = 10\, \log\left(P_{sbi} + P_{sdet} + P_{slk}\right) \tag{3.214}$$

where

P_{stot} = total reference sideband level (dB).

Achieving low reference sidebands before filtering requires adjusting six parameters to their minimum effect. The common factors in the three major contributors to the reference frequency and the tuning-slope sensitivity of the VCO give these factors a slightly greater effect in reducing reference-sideband-level contributions than the other four parameters.

Example 3.4

Consider a 4-GHz PLL with a 10-MHz reference. The output of the 4-GHz VCO is divided by 400 to produce a 10-MHz signal with a 0.7-V peak-to-peak voltage

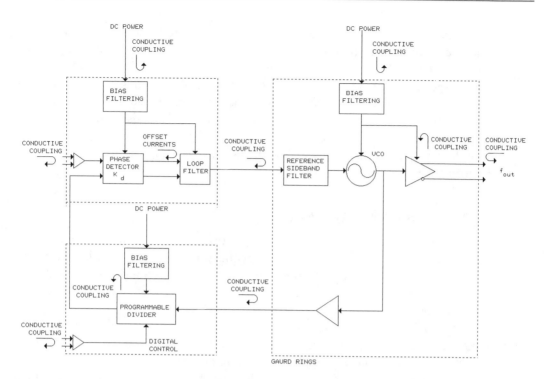

Figure 3.70 Protected PLL from coupling in an integrated circuit.

swing at the input to the phase/frequency detector. The loop filter with a 50-na offset-bias-current operational amplifier, a 1,000Ω R_2, and a 5,447Ω R_1 processes the 1-ns minimum pulse-width output of the detector. The output of the loop-filter voltage controls the 60-dB isolated tune line of the 4-GHz VCO that has a 614-Megarad/s/V tuning slope. This design produces a PLL with a phase margin of 69° and a loop bandwidth of 10 kHz. Compute the sideband levels from the offset current, the detector pulse width, and the board isolation. Then, compute the sideband level for the combination. Finally, determine the RC filter corner frequency for attenuating the reference sideband to achieve a 60-dBc sideband level.

Computing the reference sideband levels begins by evaluating (3.208), which computes a −50-dBc effect for the offset current. Evaluating (3.210) computes a −43.8-dBc effect for the phase/frequency detector. Finally, evaluating (3.213) computes a −49.1-dB effect for the PWB isolation. Combining all three effects computes a −37.7-dBc reference sideband level. A simple RC filter with a 0.5-MHz corner frequency can be added after the loop filter to further attenuate the reference sidebands by 22.3 dB. The additional filter achieves the required 60-dBc reference sideband levels and does not affect loop performance.

To summarize, this section identified parameters that are available to designers for reducing reference sideband and other spurious sideband signals. In a practical design, several of these parameters are adjusted simultaneously. Accounting for these parameters in the design will produce a more stable PLL with lower reference sideband levels and reduced filtering requirements. In addition, this section showed that the factors that affect reference sideband levels should be addressed at the beginning of the design effort. Many of these parameters require sacrificing performance specifications, changing synthesizer topology, changing digital interfaces, or changing the electronics packaging design to achieve low reference sideband levels. It is imperative that the problem of undesired sidebands be addressed at the beginning of the design effort. Trying to solve this problem by filtering after completion of the PLL design will result in a more complex circuit with poorer performance.

3.6.2.6 Filtering Reference Sidebands

Frequency synthesis always involves PLL-performance compromises. Making loop bandwidth as wide as possible reduces acquisition time and close-in noise from the VCO noise but allows high reference sidebands. Making the loop bandwidth smaller suppresses the reference sidebands but increases the acquisition time and close-in noise from the reference oscillator.

Fortunately, in many cases, the reference frequency lies considerably above the required loop bandwidth. A high ratio of reference frequency to loop bandwidth (>100) eases the amount of sideband suppression to some extent; however, heavy suppression of the undesired reference sidebands requires extra filtering. This additional filtering must be done carefully to prevent loop instability [38]. The phase margin of the PLL gives a measure of the loop stability. References recommend a phase margin greater than 30° or 45° for stable operation.

An RC, an elliptic, and active lowpass filter circuits can be used to attenuate the reference sidebands and minimize the reduction in bandwidth or VCO noise attenuation without significantly reducing the phase margin. We will do an example comparison to study the effectiveness of each filter type. All methods assume a type-2, third-order loop that meets all requirements except adequate reference-frequency sideband suppression. The reduction in magnitude at the reference sideband frequencies provides the criterion for the suppression effectiveness of the filter.

The simplest approach to filtering the reference sideband adds in series after the integrator and filter block an RC lowpass section. Making the cutoff frequency larger than the upper end of the bandwidth minimizes the effect on stability. For another method, Figure 3.71 shows a Sallen key filter [40]. This second-order, active, lowpass filter with variable damping has the most versatility. Bandwidth can be adjusted with the resistance and capacitance, and insertion phase can be adjusted with the ratio of the capacitors.

Equations (3.215) and (3.216) compute the damping factor and natural frequency for the Sallen key filter. These equations are rearranged to solve for C_1 and C_2 after selecting the resistor value and the damping factor:

$$d = \sqrt{(C_2/C_1)} \qquad (3.215)$$

$$\omega_n = \frac{1}{R\sqrt{(C_2 C_1)}} \qquad (3.216)$$

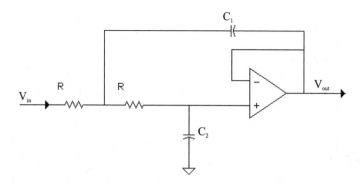

Figure 3.71 Sallen key LPF schematic.

where

ω_n = natural pole frequency of the filter;

d = filter's damping factor.

For the final sideband suppression method, Figure 3.72 shows an elliptic third-order filter. Elliptic filters have the reputation of providing the most attenuation. They have a lowpass filter response that incorporates a notch at specific frequencies. From an attenuation point of view, these should be the optimum filters.

Let's do an example comparison between an RC, active RC, and elliptic filters to reject a 33-MHz reference frequency in an example PLL. We assume that the loop has a crossover frequency of 3 MHz and want the phase margin to be 50°. Now we will synthesize the components for the filters. Setting the corner frequency for the RC filter to 4 MHz and the series resistor to 1,000Ω computes a parallel capacitor of 39 pF, which can be monolithically integrated. For the Sallen key filter, setting a resonance frequency at 30 MHz, a damping factor of 0.2, and a 39-kΩ resistor value computes a C_1 value of 10 pF and a C_2 value of 0.4 pF. From the tables for the third-order elliptic filter [41], we are given the normalized coefficients for C_1 of 0.554, C_2 of 0.0457, C_3 of 1.1521, L_1 of 1.1082, ω_2 of

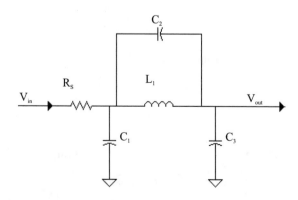

Figure 3.72 Schematic of an elliptic, third-order filter.

4.4423, and a 600Ω series resistance. Unnormalizing the coefficients computes a C_1 of 19.8 pF, C_2 of 1.63 pF, C_3 of 4.11 pF, and L_1 of 14.2 μH.

Running SPICE on all three filters allows us to compare the results. Figure 3.73 shows an overlay of the magnitude and insertion phase of the filters. The y-axis is in decibels for magnitude and degrees for insertion phase curves. From these plots, the RC filter has −20 dBc of rejection at 33 MHz but has a 40° insertion phase, which makes the loop unstable (phase margin 10°). The elliptic filter has −55 dBc of rejection at 33 MHz but also makes the loop unstable with a 10° phase margin from the 40° of insertion phase, and the 14.2 μH inductor requires the use of components external to the silicon. The Sallen-key filter has −30 dBc of rejection but only has a 10° insertion phase, which results in 40° phase margin and a stable loop.

In conclusion, the simple RC circuit has the least efficiency. It gives the least sideband attenuation and the largest phase-margin deterioration. The elliptic filter has a very high attenuation of the first sidebands with the same phase-margin deterioration as the RC circuit; however, achieving high-notch attenuation requires component tolerances that are too critical. In addition, the high inductance value will require an external silicon component. These disadvantages may be acceptable for only a few special applications. The active second-order lowpass filter, however, can be tailored to most applications. In addition, the active operational amplifier configuration allows flexibility in using other more complex filter circuits. These other configurations may offer even better solutions to sideband reduction.

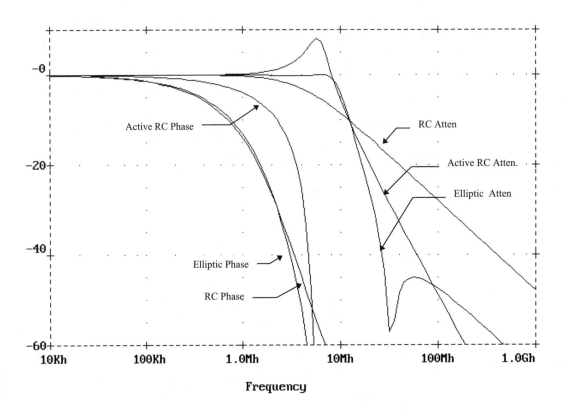

Figure 3.73 Attenuation and insertion phase for the example RC, active RC, and elliptic filters.

3.6.3 Noise-Reduction Techniques

Most integrated circuits do not produce random noise unto themselves and usually perform the single task they were designed to do. When assembled and connected to other circuits to form a system on a chip, interconnecting wires through the direct contact action of ground loops and common-mode returns generate unwanted noise. Inductive and capacitive pickup of nearby radiated fields also generate unwanted noise in the circuit. A desired signal in one circuit can be noise to another, and interferences can be produced by local circuits within the system or by equipment completely removed from and external to the system.

A newly designed system might work fine on paper or when first assembled for checkout, but when installed at its final, crowded location on ship or shore, might not perform as anticipated. Only then is it realized that the complete system has picked up much noise and hum or is itself radiating so heavily that the equipment is unusable. Costly additional effort, parts, and time must be used to locate and attempt to eliminate the causes of the noise pickup, sometimes with little success.

To avoid this unnecessary waste, cable-to-equipment interface engineering (also grounding and shielding) should be applied at the start of system planning and design. This applies to all systems, irrespective of whether the signal is low or high frequency or used in TV, telemetry, timing, ordnance, environment testing, computer, telephone, test instrumentation, or just plain communications. Each system must be considered individually since the signal frequencies and amplitudes within the system, as well as the anticipated external interference, will dictate what type of cabling and installation techniques (and grounding and shielding techniques) are to be used [42].

Electrical noise for this section has the accepted definition of being any unwanted and interfering voltage developed within, or external to, a system, which reduces the performance of that system. Interfering noise has always been a problem and in the past was usually reduced by brute-force filtering, which worked on the principle of stopping the noise after it had entered the system. This method was quite expensive but reasonably effective since signal-information voltages were low in frequency while systems were few and not too large or complex [42].

Present-day communications and data systems are continually becoming larger and more numerous, using higher information rates and frequencies in an atmosphere of expanded electrical and electronic equipment usage. The net result is ever-increasing interference and noise, creating an electronic traffic jam of major proportions. This applies equally whether low-level analog or digital pulse systems are used. Filtering is practically useless or, in some cases, completely unusable since it produces excessive deterioration of the desired pulse waveforms or inaccuracies and distortion of analog signal voltages. Obviously, noise reduction is best accomplished by simply stopping the noise before it enters the system [42].

Figure 3.74 shows coupling of noise between digital and analog circuits on a silicon substrate [26]. Switching transients of the digital circuit capacitively couple into the substrate and conduct to the analog circuit. The transient modulates the backgate diode in the analog circuit to couple into the analog circuit.

The following discussion describes how external noise is introduced into the system and the improvements that can be realized by applying good circuit isolation

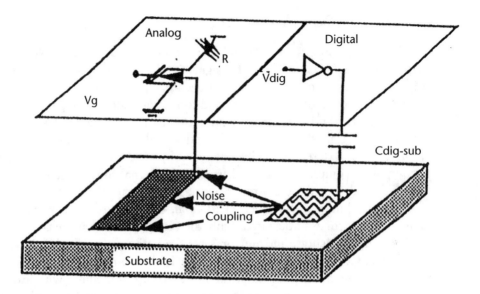

Figure 3.74 The substrate coupling noise problem. (*From:* [43]. © 1993 Carnegie Mellon University. Reprinted with permission.)

and grounding techniques. First, we will discuss methods for eliminating interference. Next, suppressing noise at the source (from other circuits on the chip), eliminating noise coupling, and reducing noise at receiver will be discussed. Finally, guidelines for controlling emissions in digital systems and external component guidelines will be discussed, which are mainly printed-circuit-board techniques. Extra emphasis will be placed on eliminating noise coupling because that is of the greatest concern on integrated circuits. Circuit techniques to separate noisy and quiet leads, avoid muxed clock signals, clean reference clocks, isolate noisy and quiet grounds and supply lines, keep the length of sensitive leads as short as possible, and operate source and load balanced to ground for very sensitive applications will be discussed.

First, here are the general methods for eliminating interference for ICs [25]:

- Grounding;
- Balancing;
- Filtering;
- Isolation;
- Separation and orientation.

Now these techniques will be applied in the following sections.

3.6.3.1 Suppressing Noise at the Source (Other Circuits on the Chip)

Several methods minimize noise generation at the source. First, enclose noise sources with guard rings or other means of isolation. Next, filter all leads leaving a noisy environment. Limit pulse rise times. Separate high-frequency digital circuits from low-frequency digital circuits. Use CMOS for low-frequency logic. Use differential

CMOS for high-frequency logic. Figure 3.75 shows different low-noise switching logic [24, 43, 44].

Finally, use separate power-down circuits to turn off individual functions in order to identify significant noise generators. Specify the maximum peak-to-peak current transient and current rise and fall times that each major function on the chip can give to the global supply and ground leads.

3.6.3.2 Eliminating Noise Coupling

Several techniques can be used to separate noisy and quiet leads. First, do not route signals through or above the PLL block. Minimize the number of input and output chip interfaces by using different architectures, eliminating off-chip filters, or combining more functions onto the chip. When low-level signal leads and noisy leads are in the same long signal path, separate them and place the ground leads between them.

Isolate the PLL from noise by carefully selecting pin inputs and outputs. Keep CMOS/TTL I/Os (high-switching current functions) as far as possible (100–200-μm spacing for maximum isolation) from the PLL. Separate the PLL connections by at least five pins slots (100–200-μm spacing for maximum isolation) from fast-switching I/O signals. For those five separating pin slots, choose signals as follows:

- Ground (G_{nd}), supply (V_{dd}) and no-switching pins—Best;
- Dedicated input pins (no-I/O)—OK;
- Output switching—Forbidden.

Figure 3.76 shows one way to separate noisy circuits from sensitive circuits. At low frequencies, CMOS logic can be used because coupling is lower. At high frequencies, a lower-noise-logic technique is used. Consequently, design functions should be separated into high- and low-frequency circuits so that guard ring spacers can be put around them and the other analog sensitive circuits.

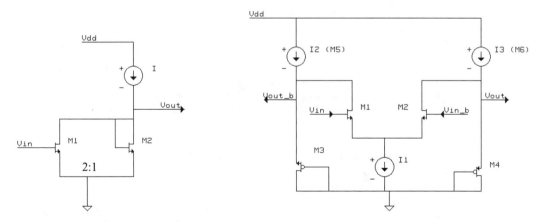

Figure 3.75 Low-noise digital logic techniques of current-steering logic and fully differential folded source-coupled logic. (*From:* [43]. © 1993 Carnegie Mellon University. Reprinted with permission.)

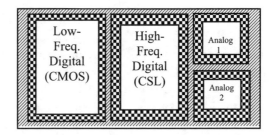

Figure 3.76   Floor plan to separate noisy circuits from sensitive circuits. (*From:* [43]. © 1993 Carnegie Mellon University. Reprinted with permission.)

Figure 3.77 shows that Nwell can improve isolation between resistors and other components by reducing the coupling capacitance to the substrate. Also this circuit structure can be used to help compute isolation [45]. With a 100-μm distance between devices, Figure 3.78 shows numerical calculations of crosstalk isolation between the different basic substrate structures and shows silicon bulk with and without buried layer, silicon on the insulator for two different substrate resistivities and with or without backside connection to ground [44]. This figure shows much higher isolation for substrate structures with higher resistivity and a few decibels of improvement with the backside connected to ground.

For low noise, we want a silicon process that gives us thick metal (low sheet resistance) to minimize interconnect resistance, a low dielectric-constant insulator to minimize parasitic capacitance, a large distance from the substrate to minimize parasitic capacitance, and silicided gates to reduce resistance.

Muxed Signals

Jitter can be greatly reduced by eliminating transmission gate multiplexors (MUXs) used for test bypass. An example with a MUX used to bypass the clock around

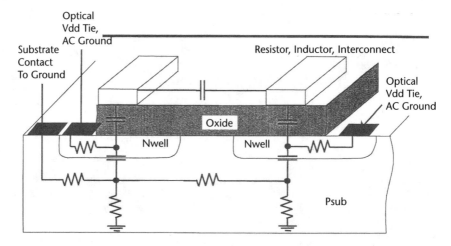

Figure 3.77   Nwell improves isolation between components. (*From:* [45]. © 2000 James P. Young. Reprinted with permission.)

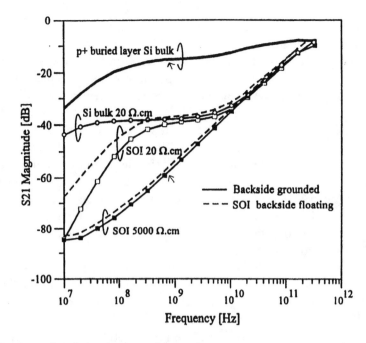

Figure 3.78 Crosstalk isolation between different basic substrate structures. (*From:* [45]. © 2000 James P. Young. Reprinted with permission.)

the PLL output shows the improvement. For example, with the transmission gate MUXs, the jitter was measured at 488 ps *p-p* with 132-ps standard deviation. Replacing the transmission gate MUXs with NAND gates reduced the jitter to 90 ps *p-p* with a standard deviation of 14 ps.

Transmission gates can be used near the PLL if they are not in a clock path. They only add jitter if two clock signals are muxed together (e.g., test clock and normal clock). The main concern is the potential risk of having both clock signals interfering or coupling with each other due to poor isolation of the inputs of the multiplexers (a problem inherently due to transmission gates).

Noise can be improved by separating supply and ground lines [46]. Several techniques can be used to separate supply and ground lines. First, use separate package pins for analog, digital, guard ring, and I/O supplies and guard rings. Do not share analog supply and ground (VDDA or GNDA are typical analog signal names) pins for a given PLL block with any other analog components in the integrated circuit, including other PLLs. Next, place respective supply bondpads close to each other to prevent ground loops. Non-PLL components in the clock path (e.g., input buffer, output buffer, other test bypass logic) running off dirty supplies can modulate delays, which adds to jitter. These components should be run off separate clean supplies, ground, and guard rings.

Avoid common ground connection between high- and low-level current cells by guard ringing them. Use a supplemental guard ring biased with a dedicated voltage between the analog and digital circuits. PLL needs to be surrounded by an N+ guard ring in Nwell that is connected to a VDD shield and a separate package pin. The PLL needs to be surrounded by a P+ guard ring that is connected to a GND shield and a separate package pin.

Other precautions can be applied to grounds. First, keep ground leads as short as possible. Next, ground low-frequency, low-level circuits at one point only; high frequencies and digital logic are exceptions. Avoid questionable or accidental grounds. Next, use low-impedance power-distribution lines. Avoid ground loops in low-frequency, low-level circuits. Finally, consider using differential amplifiers for breaking ground loops.

Noise coupling can be further improved by using distance between circuits [38]. First, leave a minimum of a 20–30-μm gap (100–200 μm for maximum isolation) between the PLL and other circuits to minimize noise coupling through the die substrate. Next, use substrate contacts to isolate analog (sensitive) and digital (noisy) circuits. Place SRAM blocks as far as possible from the PLL block within the die as SRAMs have a tendency to have high switching activity. Figure 3.79 shows good design practice in isolating substrate grounds [26]. Many of the techniques from above are used.

Noise levels can be lowered further by filtering the supply lines [46]. Voltage regulators inside the chip can be used to help filter the supply lines, but there is a voltage dropout penalty. At low modulation frequencies inside the loop bandwidth, the PLL control system reduces the effect of these signals, and regulators without the effects of bypass capacitors are not effective at high frequencies. This leaves the loop sensitive to high-frequency modulation. Consequently, circuit designs rely on using on-chip decoupling capacitors (maximum of 100 pF) to filter the high frequencies on a supply line. On-chip decoupling capacitors are more effective than off-chip decoupling capacitors because the inductance of the bond wire is eliminated. Furthermore, they circulate RF (high-frequency) currents on chip and reduce circulating currents through package parasitics. In addition, they reduce radiating currents and increase stage-to-stage isolation. The major cost of using on-chip capacitors is an increase in chip size. Also, care must be taken with decoupling

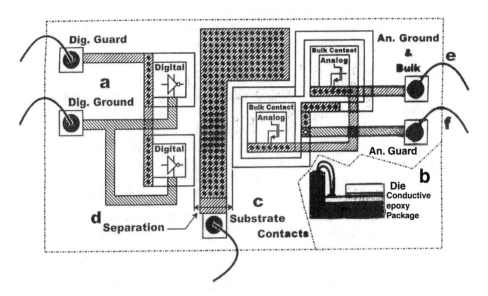

Figure 3.79 Example of good design practice in isolating grounds. (*From:* [44]. © 2006 Willy Sansen. Reprinted with permission.)

capacitors with amplifiers because resonances and feedback can occur through the supply lines. In these cases, the resonances can be dampened with a series resistor with the capacitor, active biasing, separate pins for each amplifier, or a parallel R, L, C tank [44, 45, 47]. Equation (3.217) computes the size of the capacitor to give the desired voltage ripple for the given supply-current transient conditions:

$$C = \frac{dI \, dt}{dV} \qquad (3.217)$$

where

dI = current transient (A);

dt = slew time of the current transient (sec);

dV = desired transient voltage drop on the supply line.

For a 10-ma peak current with 500-ps rise time and for a 50-mV supply ripple, (3.217) computes a 50-pF capacitor.

Figure 3.80 shows ideal ground, supply, and guard ring connections with dedicated pins for optimum grounding and isolation. Circuits should be laid out with this ideal connection in mind because connecting this way after a circuit has been laid out is extremely difficult. If pins are limited, then leave the bond pads and connect the supplies together with multiple bond wires to one pad or multiple

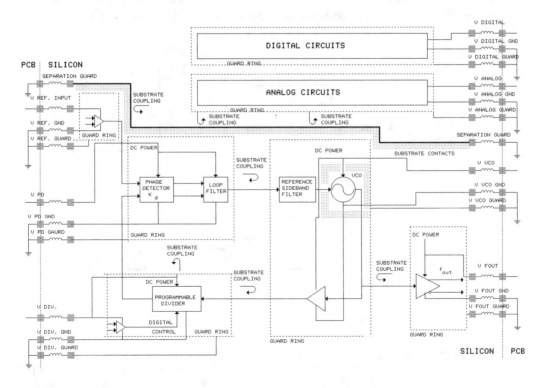

Figure 3.80 Ideal supply, ground, and guard ring routing for a PLL.

highest metal traces, as shown in Figure 3.81 [44]. The top picture shows dedicated pins to the supplies, and the bottom picture shows one way of using multiple bond wires to a pin. In this configuration, the circuit can be FIBed and bonded out to the optimum condition to see the effect of proper grounding on performance. The trace widths to the cells are increased to reduce resistance, and bond pad area is increased to reduce resistance and make room for the multiple bond wires. In addition, a VDD guard ring is used to further isolate the cells. The VDD guard ring has a dedicated bond-wire connection, and the figure shows that other connections to the VDD guard ring are not allowed.

Minimizing the lead lengths of sensitive signals minimizes coupling. First, the PLL macro should be placed at the midpoint of the die edge (away from die corners) of the ASIC, which will result in the shortest-length bond wires and lowest inductance for wire/lead-frame fingers. Next, circuitry associated to PLL should be clustered next to the PLL rather than being spread over in other regions further away from the PLL.

For very sensitive application, operate the source and balance to ground. The use of differential I/O will help to minimize the effect of switching-noise SSO (lower jitter). A terminated I/O will also have little ringing. Figure 3.82 shows applications of many of the techniques that have been discussed on an actual mixed-signal layout [44].

3.6.3.3 Reducing Noise at the Receiver (Noise Rejection)

Reduction of noise at the receiver (noise rejection) is similar to noise generation at the source. First, use only the necessary bandwidth needed for the circuit to

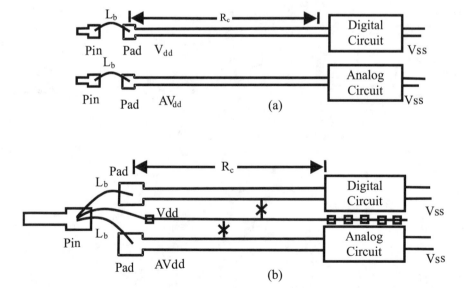

Figure 3.81 Alternative connections for bond pads and pin connections to separate out analog and digital supplies: (a) separate pins for digital and analog supply, which is preferred; and (b) single point connection for digital and analog supply with extra contacts for a supply guard ring. (*From:* [44]. © 2006 Willy Sansen. Reprinted with permission.)

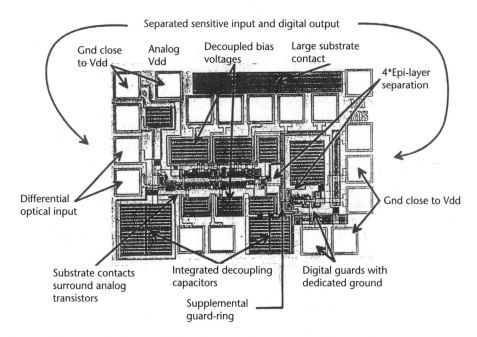

Figure 3.82 Example of many of the techniques that have been discussed. (*From:* [44]. © 2006 Willy Sansen. Reprinted with permission.)

function. Next, use frequency-selective filters when applicable. Provide proper power-supply decoupling. Enclose sensitive circuits with guard rings or use other means of isolation. Regulate the supply lines to the VCO (most sensitive) and analog circuitry in the PLL. Finally, use the low-noise-logic circuits to suppress noise at the source circuit area. Not only do these circuits minimize noise generation, but they also minimize noise pick up.

3.6.3.4 Guidelines for Controlling Emissions in Digital Systems

Circuit techniques mostly on a printed circuit board can be used to minimize emissions. First, minimize ground inductance by using a ground plane or ground grid. Next, all unused inputs on logic gates should be connected to either power or ground. Use the lowest-frequency clock and slowest rise time that will do the job. Minimize ringing on clock lines and I/O pins by matching line impedances and minimizing lead inductance. Use ferrite beads on clock lines and I/Os to slow down edges and reduce ringing on the lines.

Proper usage of decoupling capacitors also helps reduce emissions. First, locate decoupling capacitors next to each IC in the system. Next, put vias to ground and supply planes after the decoupling capacitor. Use the smallest-value decoupling capacitor that will do the job. Finally, use a bulk decoupling capacitor to recharge the individual IC decoupling capacitors (this should be 10 times the value of the decoupling capacitors [25]).

3.6.3.5 Guidelines for Using External Components

In almost every application, the PLL will have external components on the printed circuit board. We will discuss issues concerning these components. First, analog and digital supplies and analog, digital, and guard ring grounds should come from separate feed-throughs/traces from the power planes on the board. The power applied to analog supply pins should be as free from noise as possible. Digital switching noise on these pins increases jitter in the VCO. Next, use EMI filters on the power-supply pins of the PCB to help reduce noise. The ground lead of the EMI filter should be tied to system ground and *not* directly to the analog ground (GNDA) of the PLL. Otherwise, it could significantly increase the jitter.

Extra filtering in the form of ferrite beads and bypass capacitors at the VDDA pins can help to ensure that the power applied is clean. Use a bypass capacitor between the analog supply and ground (VDDA and GNDA) to keep any common-mode noise and reduce jitter. Typical values can be in the range of 0.1 μF. The bypass capacitors should be placed as close as possible to the IC and respective pins on the ASIC. Ceramic capacitors are preferred to electrolytic and tantalum capacitors because they have better high-frequency rejection. Refer to Section 3.6.3.2 for other techniques in using bypass capacitors.

3.7 Summary

In summary, this chapter discussed system requirements for a PLL. Noise basics, phase noise, time-domain response, acquisition, jitter, and spurious signals were the system requirements that were studied in detail. Noise basics were discussed and mathematical relationships were developed in order to lay a foundation for analyzing noise requirements and understanding the relationship to circuit design. Next, phase-noise relationships in oscillators were discussed and mathematical relationships were developed to give a background for analyzing noise and jitter in oscillators. Linear time-domain responses were studied to provide an understanding of the relationship between loop variables, faster switching time, and tracking error. These studies showed that damping factor and natural frequency can be determined from measured or simulated step responses.

Next, nonlinear acquisition was studied. Mathematical relationships were developed that explain loop responses that seem to be inconsistent. Next, the causes of jitter, the effect of jitter on the PLL, the relationship of phase noise to jitter, and the relationship of spurious signals to jitter were covered. Then, the relationships of PLL parameters to jitter were studied. Mathematical relationships were developed to understand how to improve timing margin in digital circuits and improve the bit-error rate of synchronizing circuits. Finally, spurious signals were studied. Mathematical relationships were developed that showed how these signals cause jitter and reduce the precision of the output frequency. Spurious signals at the output from the reference frequency is one of the largest spurious signals in a loop. Mathematical models and design techniques were developed to reduce this feedthrough from the phase detector output to the VCO tune line. Reducing this feedthrough mechanism will help in designing lower-jitter and lower-noise PLLs. With our understanding of this chapter, system requirements can be translated to circuit design of the key components in a PLL.

Questions

3.1 What are the sources of noise?

3.2 Which source of noise has $1/f$ characteristics?

3.3 What passive component should be avoided for low-noise design?

3.4 What active component should be avoided for low-noise design?

3.5 Which noise source requires that dc current be present?

3.6 What are the main advantages of equivalent-input noise sources?

3.7 What is noise figure, and why should you use it?

3.8 An amplifier chain has 60 dB of gain and a 10-dB noise figure. You are assigned to design a low-noise amplifier with 20-dB gain and 20-dB noise figure after this stage. What conditions have to occur for the noise figure of this amplifier to have an equal amount of noise contributed to its output?

3.9 Name the five spectral-density power laws and plot their responses.

3.10 Plot the five spectral power laws for an oscillator's phase noise. Assume a new power law is dominant in each decade.

3.11 Derive the closed-loop transfer function for an oscillator, using feedback theory and (3.51) through (3.54).

3.12 An amplifier has a noise figure of 6 dB, a gain of 10 dB, a flicker corner frequency of 100 kHz, and a 1-dB saturation level of +10 dBm. Calculate the oscillator's phase-noise response from 1-Hz to 10-MHz offset frequencies when a resonator with an unloaded Q of 1,000, a center frequency of 50 MHz, and an insertion loss of 12 dB is connected with the amplifier in a closed loop to form an oscillator. The oscillator amplifier's output signal is coupled by 10 dB into a buffer amplifier with a 6-dB noise figure, a 10-dB gain, and a +20-dBm, 1-dB saturation point.

3.13 An amplifier has a noise figure of 6 dB, a gain of 14 dB, a flicker corner frequency of 100 kHz, and a 1-dB saturation level of +14 dBm. Calculate the oscillator's phase-noise response from 1-Hz to 10-MHz offset frequencies when a resonator with an unloaded Q of 1,000, a center frequency of 2 GHz, and an insertion loss of 12 dB is connected with the amplifier in a closed loop to form an oscillator. The oscillator amplifier's output signal is coupled by 13 dB into a 6-dB noise-figure buffer amplifier with 10-dB gain and a 1-dB saturation point of +30 dBm.

3.14 From Question 3.13, calculate the power level at the output of the buffer amplifier.

3.15 Find the parameter of the oscillator in Question 3.13 that could be changed to reduce the phase noise of the oscillator at 1-Hz offset frequency by 30 dB without changing the oscillator's far-from-the-carrier noise floor.

3.16 Find the parameter of the oscillator in Question 3.13 that could be changed to reduce the 1-Hz offset frequency and the oscillator noise floor by 20 dB.

3.17 An amplifier has a noise figure of 6 dB, a gain of 14 dB, a flicker corner frequency of 100 kHz, and a 1-dB saturation level of +14 dBm. The amplifier's output signal is coupled by 13 dB into a buffer amplifier with a 6-dB noise figure, a 10-dB gain, and a 1-dB saturation point of +30 dBm. One resonator has a Q of 10,000, an insertion loss of 5 dB, a center frequency of 2 GHz, and an input power limit of +10 dBm, while another resonator has a Q of 1,000, an insertion loss of 10 dB, a center frequency of 2 GHz, and an input power limit of +30 dBm. Which resonator should be chosen for the lowest phase-noise response at each offset-frequency point of the oscillator? Why?

3.18 An amplifier has a noise figure of 6 dB, a gain of 14 dB, a flicker corner frequency of 100 kHz, and a 1-dB saturation level of +14 dBm. The amplifier's output signal is coupled by 13 dB into a buffer amplifier with a 6-dB noise figure, a 10-dB gain, and a 1-dB saturation point of +30 dBm. One resonator has a Q of 10,000, an insertion loss of 5 dB, a center frequency of 200 MHz, and an input power limit of +10 dBm; the tenth harmonic mode of the resonator is used in the oscillator. Another resonator has a Q of 1,000, insertion loss of 10 dB, a center frequency of 2 GHz, and an input power limit of +10 dBm. Which resonator should be chosen for the lowest far-from-the-carrier phase noise in an oscillator circuit to be made with the components described? Why?

3.19 An amplifier has a noise figure of 6 dB, a gain of 26 dB, a flicker corner frequency of 100 kHz, and a 1-dB saturation level of +10 dBm. A resonator with an unloaded Q of 1,000, a center frequency of 300 MHz, and an insertion loss of 12 dB is connected to the amplifier in a closed loop to form an oscillator. A buffer amplifier with a 6-dB noise figure and a 10-dB gain is coupled into 10 dB less power at the output of the oscillator amplifier. What is the additive noise to the oscillator's far-from-the-carrier phase-noise response when another buffer amplifier with a 10-dB noise figure is added to the circuit after the first buffer amplifier?

3.20 Why must the modulation index be small to have a valid phase-noise measurement?

3.21 Derive the solution for the constants of the power-slope line-segments equations.

3.22 Explain the relationship of insertion loss to the quality factor (Q) in a resonator. What happens when the resonator is matched?

3.23 Draw the phase-noise curve for a PLL with an input frequency of 20 MHz, an output frequency of 160 MHz, $N = 8$, $f_n = 1$ MHz, = 0.4, $K_d = 0.4/(2)$, $K_v = 2\ 500E6$, VCO phase noise of $L_{vco} = -85$ dBc/Hz at 1 MHz with 30 dB/dec slope, -140 dBc/Hz noise floor, reference phase noise of $L_{ref} = -134$ dBc/Hz at 10 Hz with 5 dB/dec slope, and a -140 dBc/Hz noise floor.

3.24 At what offset frequency does one of the following contributors dominate the loop in Question 3.23 with a phase detector phase noise of $L_{pd} = -165$ dBc/Hz, amplifier noise voltage of $e_{amp} = 600$ nV/sqr (Hz), and noise on the tune line of $e_{tune} = 0.1$ nV/sqr(Hz)?

3.25 What is the peak-to-peak jitter and percentage of the output period for a PLL with an output frequency of 10 MHz, single-sided reference sideband of −30 dBc, a reference frequency of 2 MHz, $N = 5$, and a VCO gain of 100 MHz/V?

3.26 What is the peak-to-peak jitter and percentage of the output period for a PLL with an output frequency of 1,000 MHz, single-sided reference sideband of −30 dBc, a reference frequency of 200 MHz, $N = 5$, and a VCO gain of 100 MHz/V? Compare answers in Question 3.25 and this question, and draw some conclusions about period versus jitter and percentage jitter.

3.27 What is the maximum single-sided reference sideband for a PLL with an output frequency of 100 MHz, peak-to-peak jitter desired of 100 ps, a reference frequency of 10 MHz, $N = 10$, and a VCO gain of 100 MHz/V?

3.28 What value decoupling capacitor is needed to keep current spikes with a rise time of 100 ps and a peak current of 10 ma from causing ripple on the 5V supply less than 10 mV?

3.29 For a damping factor of 0.99, a 0-dB-crossover frequency of 1.5 MHz, $N = 1$, $K_v = 100$ Mrad/s/V, $K_d = 0.4$, an initial starting frequency of 50 MHz, and a step frequency of 10 MHz, compute the natural frequency, the 10% to 90% rise time, the percentage overshoot of the first peak, and the 1% settling time.

3.30 For a damping factor of 0.5, natural frequency of 300 Hz, $N = 512$, $K_v = 400$ Mrad/s/V, $K_d = 3.4$ ma/rad, an initial starting frequency of 32 kHz, and a step frequency of 3.2 kHz, compute the 10% to 90% rise time, the percentage overshoot of the first peak, and the 1% settling time?

3.31 For a damping factor of 0.8, natural frequency of 2.924 MHz, $N = 62$, $K_v = 300$ MHz/s/v, $K_d = 0.8$ ma/rad, an initial starting frequency of 0 Hz, and a step frequency of 175 MHz, compute the 10% to 90% rise time?

3.32 For a damping factor of 0.99, natural frequency of 492 kHz, $N = 4$, $K_v = 1,000$ MHz/s/v, $K_d = 0.02$ ma/rad, an initial starting frequency of 0 Hz, and a step frequency of 100 MHz, compute the 10% to 90% rise time.

3.33 For a damping factor of 0.269, natural frequency of 279.2 kHz, $N = 18$, $K_v = 1,880$ Mrad/s/V, $K_d = 0.005$ ma/rad, an initial starting frequency of 0 Hz, and a step frequency of 10 MHz, compute the 10% to 90% rise time.

3.34 Solve the step response for a type 2, third-order loop filter with a damping factor less than 1 that has the following equation:

$$\theta_{step} = \frac{1}{s} \frac{n_{mf}(\omega_n)^2 \left(s\dfrac{2\zeta}{\omega_n} + 1\right)}{\left[s^2 + s(2\zeta\omega_n) + \omega_n^2\right]\left(s\dfrac{1}{10\omega_n} + 1\right)}$$

3.35 What is a separatrix?

3.36 If you are comparing two responses that are outside the separatrix, and the initial frequency for both responses is the same, but the initial conditions for phase are different, will the responses be different?

3.37 For an LO frequency of 972 MHz, an RF frequency that varies from 177 to 377 MHz and an output frequency of 550 to 770 MHz, what is the highest in-band spurious power level relative to the output? Assume an LO power level of 10 dBm, RF power level of –10 dBm, LO port balun imbalance of 0.3, RF port balance of 0.99, diode2 imbalance of 0.8, diode3 imbalance of 0.6, diode4 imbalance of 1.2, LO leakage factor of 0.6, LO isolation of –40 dB, RF isolation of –40 dB, and conversion loss of 8 dB.

3.38 For a block converter with an LO frequency of 927 MHz, an RF frequency that varies from 177 to 277 MHz, and an output frequency of 650 to 750 MHz, what is the highest power level excluding the LO leakage, RF leakage, and image response and products greater than 4 GHz and a maximum order $(n + m)$ of more than 20? Assume an LO power level of 10 dBm, RF power level of –10 dBm, LO port balun imbalance of 0.3, RF port balance of 0.99, diode2 imbalance of 0.8, diode3 imbalance of 0.6, diode4 imbalance of 1.2, LO leakage factor of 0.6, LO isolation of –40 dB, RF isolation of –40 dB, and conversion loss of 8 dB.

3.39 For an LO frequency of 450 MHz, an RF frequency that varies from 100 to 350 MHz and an output frequency of 550 to 770 MHz, what is the highest in-band spurious power level relative to the output? Assume an LO power level of 10 dBm, RF power level of –10 dBm, LO port balun imbalance of 0.3, RF port balance of 0.99, diode2 imbalance of 0.8, diode3 imbalance of 0.6, diode4 imbalance of 1.2, LO leakage factor of 0.6, LO isolation of –40 dB, RF isolation of –40 dB, and conversion loss of –8 dB.

3.40 For an LO frequency of 62 MHz, an RF frequency that varies from 16 to 22 MHz, and an output frequency of 40 to 46 MHz, what is the highest power level, excluding the LO leakage, RF leakage, and image response and products greater than 100 MHz and a maximum order $(n + m)$ of more than 20? Assume an LO power level of 10 dBm, RF power level of –10 dBm, LO port balun imbalance of 0.3, RF port balance of 0.99, diode2 imbalance of 0.8, diode3 imbalance of 0.6, diode4 imbalance of 1.2, LO leakage factor of 0.6, LO isolation of –40 dB, RF isolation of –40 dB, and conversion loss of 8 dB.

3.41 From Question 3.40, what is the highest power level, excluding the LO leakage, RF leakage, and image response and products greater than 400 MHz and a maximum order $(n + m)$ of more than 20?

3.42 What is the reference sideband level in dBc for a 620-MHz PLL with a 10-MHz reference frequency, $N = 62$, 1-GHz/V VCO gain, and a 1-mV peak signal on the tune line?

3.43 What is the reference sideband level in dBc for a 622-MHz PLL with a 1-MHz reference frequency, $N = 622$, 1-GHz/V VCO gain, and a 1-mV peak signal on the tune line?

3.44 What is the reference sideband level in dBc for a 622-MHz PLL with a 1-MHz reference frequency, $N = 622$, 0.1-GHz/V VCO gain, and a 1-mV peak signal on the tune line?

3.45 What is the reference sideband level in dBc for a 900-MHz PLL with a 1-MHz reference frequency, $N = 900$, 1-GHz/V VCO gain, an opamp bias current of 1 nA, and a feedback resistor R_2 of 10 kΩ and an R_1 of 100 kΩ?

3.46 What is the reference sideband level in dBc for a 900-MHz PLL with a 1-MHz reference frequency, $N = 900$, 1-GHz/V VCO gain, an opamp bias current of 10 nA, and a feedback resistor R_2 of 10 kΩ and an R_1 of 1,000 kΩ?

3.47 What is the reference sideband level in dBc for a 400-MHz PLL with a 1-MHz reference frequency, $N = 400$, 1-GHz/V VCO gain, an opamp bias current of 10 nA, and a feedback resistor R_2 of 1 kΩ and an R_1 of 100 kΩ?

3.48 What is the reference sideband level in dBc and peak fundamental voltage on the tune line for a 400-MHz PLL with a 1-MHz reference frequency, $N = 400$, 1-GHz/V VCO gain, a phase detector with minimum pulses out of 2 ns, an opamp feedback resistor R_2 of 20 kΩ, and an R_1 of 500 kΩ?

3.49 What is the reference sideband level in dBc and the peak-to-peak fundamental voltage on the tune line for a 600-MHz PLL with a 1-MHz reference frequency, $N = 400$, 1-GHz/V VCO gain, a phase detector with minimum pulses out of 0.2 ns, an opamp feedback resistor R_2 of 20 kΩ, and an R_1 of 500 kΩ?

3.50 What general methods are used for eliminating interference in ICs?

3.51 What lower-noise-logic circuit techniques can be used to minimize noise?

3.52 What is the most sensitive circuit-element-to-noise coupling in a PLL?

3.53 What are the main benefits of decoupling capacitors?

3.54 Why are decoupling capacitors necessary on silicon since the PCB has decoupling capacitors?

3.55 What is the maximum separation distance between circuits?

3.56 What circuit element should you use to reduce ringing lines on the PCB interface in order to reduce emissions?

3.57 Why do you use double guard rings?

3.58 Why should you guard ring a circuit like CMOS logic that is insensitive to noise coupling?

References

[1] Gray, P. R., and R. G. Meyer, *Analysis and Design of Analog Integrated Circuits*, New York: Wiley Interscience, 1993, pp. 715–778.

[2] "Noise Figure Primer," HP Application Note 57, January 1965.

[3] Hellums, J. R., "Physics of Noise," TI Internal Presentation, Dallas, TX, January 1999.

[4] "JFET Design and Application Manual," Teledyne Semiconductor, Mountain View CA, 1980/1981, pp. 6-12 to 6-16.

[5] Taub, H., and D. L. Schilling, *Principles of Communication Systems*, New York: McGraw-Hill, 1971, pp. 113–131.

[6]   Spiegel, M. R., *Mathematical Handbook of Formulas and Tables*, New York: McGraw-Hill, 1968.

[7]   Scherer, D., "Today's Lesson—Learn about Low-Noise Design Part I," *Microwaves*, Vol. 18, No. 4, April 1979, pp. 116–122.

[8]   Rutman, J., "Characterization of Frequency Stability: A Transfer Function Approach and Its Application to Measurements Via Filtering of Phase Noise," *IEEE Transactions on Instrumentation and Measurement*, Vol. IM-23, March 1974, pp. 40–48.

[9]   Motchenbacher, C. D., and F. C. Fitchen, *Low-Noise Electronic Design*, New York: Wiley Interscience, 1973.

[10]   Cutler, L. S., and C. L. Searle, "Some Aspects of the Theory and Measurement of Frequency Fluctuations in Frequency Standards," *Proceedings of the IEEE*, Vol. 54, No. 2, February 1966, pp. 136–154.

[11]   Hafner, E., "The Effects of Noise in Oscillators," *Proceedings of the IEEE*, Vol. 54, No. 2, February 1966, pp. 179–198.

[12]   Gerber, E. A., and R. A. Sykes, "State of the Art Quartz Crystal Units and Oscillators," *Proceedings of the IEEE*, Vol. 54, No. 2, February 1966, pp. 103–116.

[13]   Leeson, D. B., "A Simple Model of Feedback Oscillator Noise Spectrum," *Proceedings of the IEEE*, Vol. 54, February 1966, pp. 329–330.

[14]   Veeser, M. A., "ASIC Oscillator Cells Reduce System Timing Costs," *EDN*, July 20, 1995, pp. 81–86.

[15]   Przedpelski, A. B., "Programmable Calculator Computes PLL Noise, Stability," *Electronic Design*, Vol. 29, No. 7, March 31, 1981, pp. 183–191.

[16]   Goldman, S. J., *Phase Noise Analysis in Radar Systems Using Personal Computers*, New York: Wiley Interscience, 1989.

[17]   Scherer, D., "Design Principles and Measurement of Low Phase Noise RF and Microwave Sources," unpublished HP Seminar, Dallas, TX, January 1979.

[18]   Scherer, D., "Generation of Low Phase Noise Microwave Signals," unpublished HP Seminar, Dallas, TX, September 1981.

[19]   Fiedziuszkp, S. J., "Dielectric Resonator Technology Their Applications in Filters and Oscillators," unpublished Dallas IEEE Section MTT Society Workshop, April 1984.

[20]   Howe, D. A., "Frequency Domain Stability Measurements: A Tutorial Introduction," NBS Technical Note 679, March 1976.

[21]   Manassewitsch, V., *Frequency Synthesizers: Theory and Design*, 3rd ed., New York: John Wiley and Sons, 1987, pp. 260–261.

[22]   Martin, L., "Noise-Property Analysis Enhances PLL Designs," *EDN*, September 16, 1981, pp. 91–98.

[23]   "Jitter Analysis 101, a Foundation for Jitter Measurements," Wavecrest Seminar, Dallas, TX, May 1997.

[24]   Verghese, N. K., T. J. Schmerbeck, and D. J. Allstot, *Simulation Techniques and Solutions for Mixed-Signal Coupling in Integrated Circuits*, Boston, MA: Kluwer Academic Publishers, 1995.

[25]   Ott, H. W., *Noise Reduction Techniques in Electronic Systems*, New York: Wiley Interscience, 1988, pp. 288–289.

[26]   Verghese, N. K., T. J. Schmerbeck, and D. J. Allstot, *Simulation Techniques and Solutions for Mixed-Signal Coupling in Integrated Circuits*, New York: Kluwer Academic Publishers, 1995.

[27]   Best, R. E., *Phase-Locked Loops: Design, Simulation, and Applications*, 3rd ed., New York: McGraw-Hill, 1997.

[28]   Valkenburg, V., *Network Analysis*, Upper Saddle River, NJ: Prentice Hall, 1964.

[29]   Gardner, F. M., *Phaselock Techniques*, New York: John Wiley and Sons, 1979, pp. 46–50.

[30]   Egan, W. F., *Phase-Lock Basics*, New York: Wiley Interscience, 1998, pp. 105–137.

[31]   Egan, W. F., *Phase-Lock Basics*, New York: Wiley Interscience, 1988, pp. 167–170.

[32]   Truxal, J. G., *Automatic Feedback Control System Synthesis*, New York: McGraw-Hill, 1955, Chapter 11.

[33]   Viterbi, A. J., *Principles of Coherent Communications*, New York: McGraw-Hill, 1966, Chapter 3.

[34]   Blanchard, A., *Phase-Locked Loops*, New York: Wiley Interscience, 1976, Chs. 3 and 10.

[35]   Tucker, D. G., "Intermodulation Distortion in Rectifier Modulators," *Wireless Engineer*, June 1954, p. 145.

[36]   Henderson, B. C., "Reliably Predict Mixer IM Suppression," *Microwaves and RF*, November 1983, pp. 63–70, 142.

[37]   Manassewitsch, V., *Frequency Synthesizers: Theory and Design*, 2nd ed., New York: John Wiley and Sons, 1987.

[38]   "Phase-Frequency Detector MC4344-MC4044," MTTL Complex Functions, Motorola Semiconductor Products, Inc., Data Sheet, 1973.

[39]   Ott, H. W., *Noise Reduction Techniques in Electronic Systems*, New York: Wiley Interscience, 1988.

[40]   Przedpelski, A., "Suppress Phase-Lock-Loop Sidebands Without Introducing Instability," *Electronic Design*, Vol. 27, No. 19, September 19, 1979, pp. 142–144.

[41]   Williams, A., and F. Taylor, *Electronic Filter Design Handbook*, 2nd ed., New York: McGraw-Hill, 1988, pp. 11–70.

[42]   *Cable and Wiring Data Book*, Trompeter, Westlake, CA, 1991, p. 5.

[43]   Allstot, D., "Noise Considerations in Mixed Signal Design," Mixed Signal Design Center Training Class, CMU, October 22, 1993.

[44]   Sansen, W., *Analog Design Essentials,* Dordrecht, the Netherlands, Springer, 2006, Ch. 24.

[45]   Young, J. P., "RFCMOS Design Class," Besser and Associates, October 2000.

[46]   Vittoz, E. A., "The Design of High-Performance Analog Circuits on Digital CMOS Chips," *IEEE J. Solid State Circuits*, Vol. SC-20, No. 3, June 1985, pp. 657–665.

[47]   Meyer, R. G., "High Frequency IC Design Techniques," University of California, Berkeley, notes.

Appendix 3A: Single-Ended Explanation of Offset Currents

The single-ended interconnection that is shown in Figure 3A.1 provides a graphically simpler explanation of the offset current effects in a PLL. In a locked condition, the charge on the feedback capacitor remains constant, and this provides a constant tune voltage to the VCO. Consequently, zero current should flow through the feedback capacitor. A bias current from the operational amplifier draws charge away from the feedback capacitor and produces an error. The loop cancels this error by producing a wider pulse out of the phase detector, which generates a current that counteracts the bias current from the operational amplifier. This counteracting current maintains the loop in the locked condition.

Consequently, using operational amplifiers with higher-bias currents causes the loop to produce a wider pulse width out of the phase detector to maintain lock. This larger pulse width produces higher reference sidebands because the energy in the first harmonic of a pulse increases with wider pulse widths, as shown in a Fourier series of a pulse. Adding an additional current source as shown in

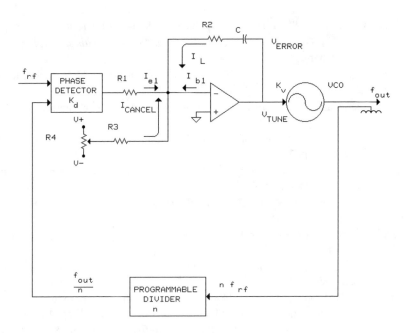

Figure 3A.1   Schematic of offset currents in a single-ended loop filter.

Figure 3A.1 is one way to cancel the bias current from the operational amplifier and reduce the pulse width out of the phase detector. The differential interconnection requires significantly less bias current. Consequently, the differential interconnection, as indicated in the text, is the preferred method.

CHAPTER 4

Components, Part 1—Dividers and Oscillators

This chapter begins to discuss the design of the individual components in a PLL. In this chapter, frequency generators are covered. Detailed designs of programmable dividers, VCOs, and crystal oscillators are studied. Design techniques of programmable dividers are discussed to prevent a PLL from hanging up on one of the power rails. Programmable dividers, pulse-swallowing counters, and fractional dividers are the types of dividers that are covered.

System-on-a-chip integration has made PLL noise immunity to supply variations extremely important because digital circuits add noise to the supply and substrate. VCO performance accounts for this noise immunity to supply variations. It also accounts for more of the other PLL characteristics than any other component in a PLL. The major output characteristics of the sensitivity of the PLL to injected noise on the supply, the time jitter of the output, the phase noise, the frequency tuning range width, non–harmonically related spurious modes, loop stability, linear modulation, linear demodulation, and output that toggles in unlocked conditions depend on the characteristics of the VCO. For these reasons, design trade-offs and techniques for a wide variety of VCOs are studied.

Single-ended ring, differential ring, multivibrator, and LC-resonant VCOs are covered. The single-ended ring is analyzed because of its simplicity and popularity. A multivibrator is analyzed because of its wide and high frequency range, because it is current controlled instead of voltage controlled, and because it easily absorbs parasitic capacitances. A negative-resistance LC resonant oscillator will be analyzed to show that a feedback path can exist from a single node connection. Understanding this concept is important in troubleshooting many applications where the oscillating feedback path is not clear.

Finally, design techniques for crystal reference oscillators will be covered because they are also being incorporated onto integrated circuits with the PLLs. Maintaining oscillation and startup time are the main concerns of IC designers in crystal oscillator design. Analysis of stability and computing startup time are presented

4.1 Dividers

A malfunction in the feedback divider from a loop transient response can cause a PLL to get lost and hang the VCO control voltage at one of the power-supply rails. Recovery from this condition requires a reset or power-up control signal to reinitialize the circuit. A feedback divider that is faster than the VCO does not require a reset or power-up control signal because it never gets lost. Consequently, the highest output frequency of the VCO in a PLL determines the speed requirement for the feedback divider.

Dividers have several different configurations in PLLs. A series of divide-by-2 flip-flop dividers, a programmable divider, a pulse-swallowing divider, or a fractional frequency divider can be used. Selection of the divider depends on the application. For one divider output, a series of divide-by-2 flip-flops will in many cases meet requirements. For multiple steps, a programmable divider will meet requirements. For high output frequency and multiple steps, a pulse-swallowing divider will meet requirements. For high output frequency, small frequency steps, and multiple steps, a fractional frequency divider will meet requirements.

A series of divide-by-2 flip-flop dividers is a very straightforward application. Consequently, we will not go into detail about this application. Usually a flip-flop divide-by-2 has a high-frequency response and does not limit PLL applications. For frequencies that are close to the process limit, special flip-flops may have to be designed. Programmable dividers are used in many applications. Therefore, we will start with this divider type to understand the use and design of frequency dividers.

4.1.1 Programmable Divider

The highest output frequency of the VCO in a PLL determines the speed requirement for the feedback divider. In many cases, this is faster than the conventional programmable divider. Design techniques of programmable dividers are discussed to prevent a PLL from hanging up on one of the power rails. This section will first discuss background theory for programmable dividers and pulse-swallowing dividers. Then, changes to the PLL will be outlined.

Figure 4.1 shows a block diagram of a conventional programmable divider. Integer divide ratios from 2 to 255 can be programmed with the data bits to the 8-bit counter. The counters count up from all logic low values to all logic high values. The counters have clock input (ck), carry out (co), count enable (ce), parallel

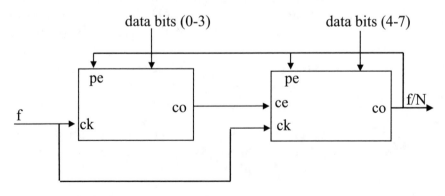

Figure 4.1 Conventional programmable divider.

load enable (pe), and data input ($p_0, p_1, \ldots p_x$) pins. The carry-out pin is normally low and goes high when the all logic high value is reached. The count-enable pin is high when counting up and low to hold the count. The parallel-load-enable pin loads data into the counters and synchronously loads data when in the logic high state.

The feedback path of the ripple carry out reduces the operating frequency of the conventional programmable divider. A maximum input frequency of up to 250 MHz for the 0.6-μm gate-length process (33C15) is achievable; however, the VCO has an output frequency as high as 1 GHz.

In most applications, a programmable divider consists of a series of flip-flops that toggle at the rising clock edge in a manner that causes frequency division of the input clock. Initially, a binary word is loaded into the flip-flops. Then, the flips flops increment states up to the terminal count (all-1s state). The terminal count is detected and causes the initial binary word to be reloaded, and the process repeats.

Figure 4.2 shows the simulated output for a divide-by-9 divider. The horizontal scale is time in nanoseconds. The clock input, the flip-flop outputs Q_0 to Q_4, and the parallel-enable signals are shown and listed in order from top to bottom. Initially, a 0111 is loaded at 10 ns. The flip-flops count up to the terminal count at 40 ns. The terminal count is detected (signal PE_1) at 42 ns. At 45 ns, the initial 0111 binary word is reloaded into the flip-flops, and the process repeats.

Figure 4.3 shows the system logic combination blocks for a synchronous 4-bit programmable frequency divider. Each bit consists of combinational logic, a positive-edge latch, and a negative-edge latch. The positive and negative-edge latches operate as a master-slave flip-flop. The combinational logic controls when

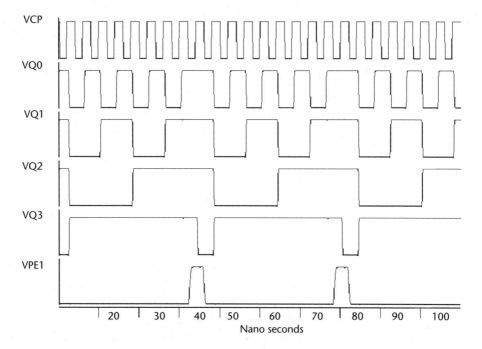

Figure 4.2 Timing diagram for an example 4-bit programmable divider.

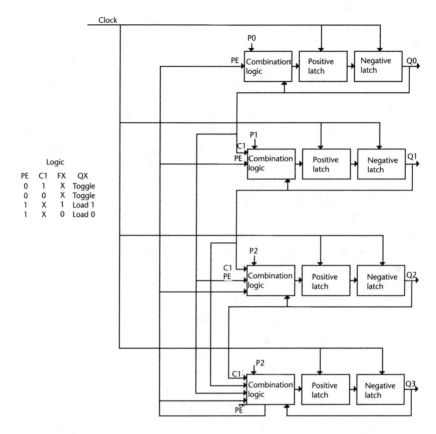

Figure 4.3 The logic combination of functions that make up a 4-bit programmable frequency divider.

the flip-flops toggle, hold, load a 1, or load a 0. The flip-flops have a carry-in signal that controls when the flip-flop toggles or holds its value. The input for the carry in comes from the LSBs of the referenced flip-flop. The final flip-flop detects the terminal count and generates the parallel-load signal for the process to repeat.

This logic has several critical delay paths: (1) the delay times from the positive and negative-edge clocks to the outputs of the positive and negative-edge latches, (2) the delay from the negative-edge latch, through the feedback path and combinational logic, to the input of the positive latch for the same flip-flop, (3) the delay from the negative-edge latch, through the combinational logic of the next MSB flip-flop, to the input of the positive-edge latch so that the count can be enabled, and (4) the delay from the negative-edge latch of the LSB flip-flop, through the combinational logic of the MSB flip-flop and the carry out, through the parallel-load-enable combination logic, to the inputs of all the positive latches.

Figure 4.4 shows a positive-edge-sensitive latch and a positive-edge-triggered register. These are the building blocks for programmable dividers. A positive-edge latch holds the data line value at the output after the positive edge of the clock and lets the output vary after the negative edge of the clock. A negative-edge latch holds the data line value at the output after the negative edge of the clock and lets the output vary after the positive edge of the clock. An edge-triggered register

Latches and registers

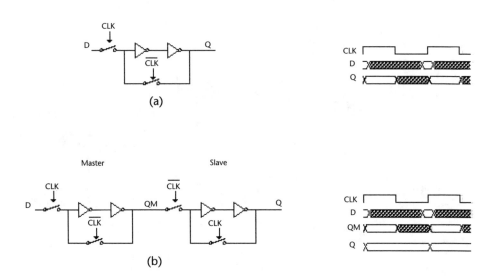

(a)

(b)

Figure 4.4   Functional logic diagram and timing diagram of (a) positive-edge-triggered latch, and (b) positive-edge-triggered register. (*From:* [1]. © 1993 AT&T. Reprinted with permission.)

combines the positive- and negative-edge-triggered latches as shown in Figure 4.4. By convention, the first latch is called the master, and the second latch is called the slave. While the clock input is low, the master negative-level sensitive latch output Q_M follows the D input, while the slave positive-latch holds the previous value. When the clock transitions from 0 to 1, the master latch ceases to sample the input and stores the D value at the time of the clock transition. The slave latch opens, passing the stored master value Q_M to the output of the slave latch Q. The D input is prevented from affecting the output because the master is disconnected from the D input. When the clock transitions from 1 to 0, the slave latch locks in the master-latch output, and the master starts sampling the input again. This sequence is shown in Figure 4.4 [1].

Figure 4.5 shows a simulation where the control line of a VCO is ramped to find the maximum operating frequency for a divide-by-2 flip-flop and a 4-bit programmable divider. The top waveform (VVCTRL) is the ramped control line to the VCO. The second waveform (VVCO_CLK) is the VCO output. The third waveform (VDIVOUT) is the output of the divide-by-2 flip-flop. The last waveform (VCO) is the ripple carry out of the 4-bit programmable divider. The figure shows that the divide-by-2 flip-flop works to 1 GHz. The programmable divider stops working at 400, 500, and 700 MHz, and at greater than 900 MHz, but at 600 and 800 MHz, it starts working again. This shows why ramping the clock frequency helps identify the first place where the circuit stops working. For operation in a PLL, the programmable divider is not useful above 400 MHz.

A fixed prescaler in front of the programmable divider as shown in Figure 4.6 is one way around the speed reduction of a conventional programmable divider.

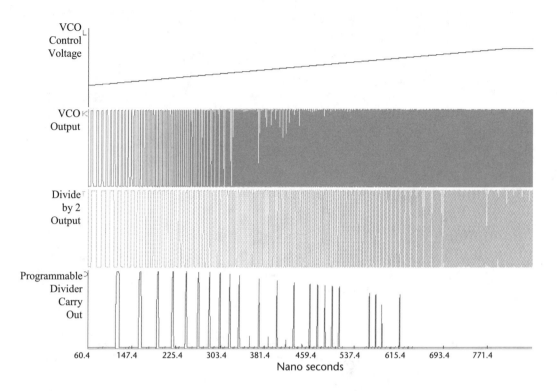

Figure 4.5   SPICE simulation of divide-by-2 flip-flop and programmable divider.

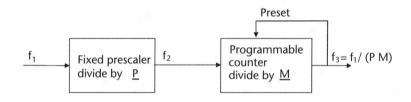

Figure 4.6   A high-speed fixed prescaler in front of a programmable divider.

A very high-speed divide-by-4 prescaler can be made to work at 1 GHz, which will match the 250-MHz programmable-divider maximum operating frequency; however, divide ratios of 8, 12, 16, ... 64 are only possible instead of integer steps. Equation (4.1) computes the possible divide-by-N values for a fixed prescaler divide ratio P and the following programmable counter with divide ratio M:

$$N = PM \tag{4.1}$$

4.1.2   Pulse Swallowing

A divider that uses a dual modulus prescaler and a swallow counter is called a *pulse-swallowing counter* [2]. A dual modulus prescaler allows integer divisions while retaining the high-speed characteristics of a fast prescaler. Selecting a divide-by-4 or -5 dual modulus prescaler will allow integer divide ratios of 12, 13, 14, ... 64 and 8, 9, 10, respectively. The selection of divide-by-4 in the prescaler makes the highest operating frequency for the swallow and programmable counters 250 MHz (1 GHz/4). This will allow operation of the loop up to 1 GHz with a reference frequency that can range from 16 MHz (1 GHz/64) to 83 MHz (1 GHz/12).

Figure 4.7 shows a simplified diagram of a pulse-swallowing counter. The dual modulus prescaler is shown as two fixed prescalers with a switch to select the output of either one. The divide ratios in the pulse-swallowing counter in Figure 4.7 are defined [2]:

U = upper (larger) divide ratio of the prescaler (usually $L + 1$);

L = lower divide ratio of the prescaler;

S = divide ratio of the swallow counter;

S = number of times the prescaler divides by U in a complete divide cycle;

M = divide ratio of the program counter;

M = total number of prescaler cycles in a complete divide cycle.

The divide cycle begins with the slowest divide ratio selected. After S pulses of f_2, the swallow counter switches to the fastest divide ratio. Later, the program counter reaches maximum count and causes a preset to occur, which causes the swallow counter to reset the dual modulus prescaler back to the slowest divide ratio and restarts the divide cycle.

Figure 4.8 shows a more detailed diagram of the pulse-swallowing counter. The divide ratio M of the program divider determines the number of f_2 pulses in one divide cycle. The swallow counter modifies the number of f_1 pulses into the prescaler that are required to produce M pulses at f_2.

The divide cycle in Figure 4.8 begins with a parallel-enable signal that synchronously presets the data. The two carry outs (COs) go to the inactive state, which ends the preset mode. The inactive state of the carry out enables the swallow divider

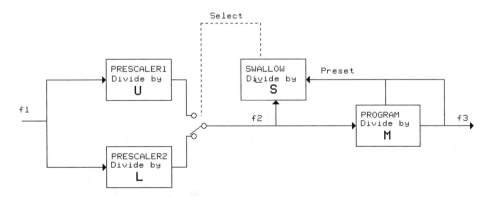

Figure 4.7 Simulated function of the swallow counter.

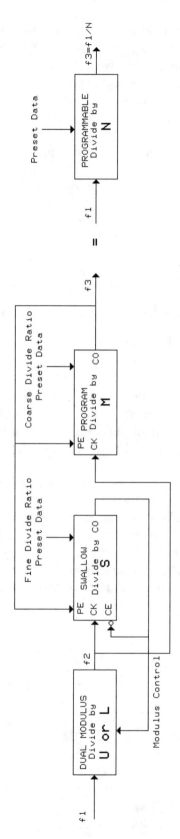

Figure 4.8 Pulse-swallowing counter and equivalent system building block.

and changes the prescaler to its highest divide ratio. The dividers start to count until the swallow divider counts to its maximum value (S pulses of f_2). At this point $S \times U$ input clock pulses have occurred. At maximum value, the carry out is enabled, which stops the swallow divider and changes the prescaler to its lowest divide ratio. The program counter continues to count up to its maximum value [$(M - S) \times L$ more input clock pulses], while the swallow counter holds its count. The carry out of the program divider enables the preset mode, and a new divide cycle begins. Equation (4.2) computes the total number of clocks counted N in the divide cycle [2]:

$$N = SU + (M - S)L \tag{4.2}$$

Substituting $L + 1 = U$ into (4.2) and rearranging produces (4.3), which computes the divide ratio N for a dual modulus prescaler:

$$N = S + LM \tag{4.3}$$

An example shows how to set the data bits to get the desired divide ratio. To divide-by-13 with $U = 5$ and $L = 4$, M would be set to 3 [integer($13/L$)], and S would be set to 1 ($13 - L \times M$).

A practical limitation on the pulse-swallowing technique is that M cannot be less than S. If M were less than S, the program counter would reach maximum count before the swallow counter. Consequently, the prescaler value would never change, and the circuit would operate like the fixed prescaler in Figure 4.6.

Let's study a possible implementation of the dual modulus prescaler in an example PLL. Figure 4.9 shows a schematic for a dual modulus prescaler (divide-by-4 or -5). The circuit consists of a 4-bit shift register that shifts a logic high to the last shift register. A logic high at the output of the last shift register synchronously reloads the shift registers with a logic high in the first register and logic lows in all the other registers. The load operation is done by switching to the parallel-load data word for the duration of the load, then switching back to the shift register connection. Then, the shift operation restarts.

Figure 4.10 shows the schematic for a 4-bit counter that is used as a programmable divider. This is based on the logic in a 74163 counter. The carry out is used to feed back to the parallel enable to create a 4-bit programmable divider.

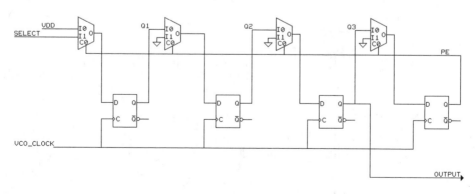

Figure 4.9 Dual modulus prescaler (divide-by-4 or -5) schematic.

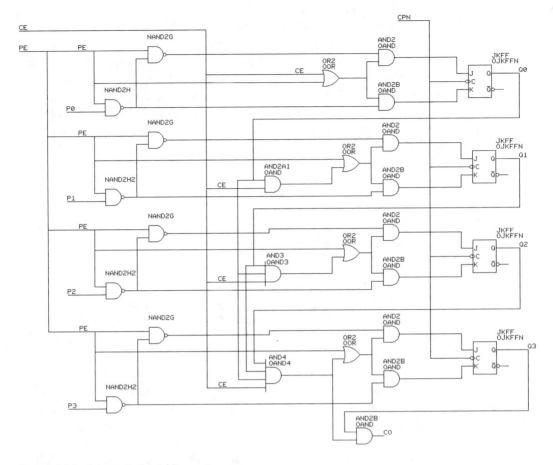

Figure 4.10 Schematic for 4-bit counter.

Figure 4.11 shows the SPICE simulation results for the pulse-swallowing counter with a divide ratio of 13 versus the ramped VCO. The first waveform shows the control voltage of the VCO. The second waveform shows the ripple carry out of the swallow counter. The last waveform shows the ripple carry out

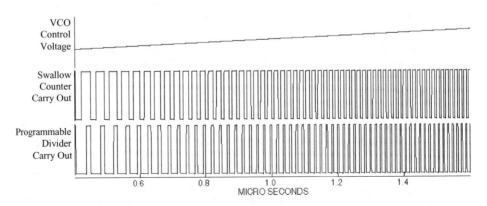

Figure 4.11 Pulse-swallowing counter simulation (divide-by-13) versus ramped VCO.

of the programmable divider. The divide ratio of 13 was used to show that the ripple carry out of the swallow counter occurs before the ripple carry out of the programmable divider and holds its value until the ripple carry out of the programmable divider goes high. In the figure, the output of the programmable divider divides by 13 until the 1-GHz output of the VCO at 1.5 μs is reached. The smooth change in the waveforms shows that no race condition was found and that the divider divides correctly.

Figure 4.12 shows a zoomed in view of the waveforms. The first waveform shows the control voltage of the VCO. The second waveform shows the output clock of the VCO. The third waveform shows the output of the dual modulus prescaler that clocks the swallow counter and the programmable counter. The fourth waveform shows the ripple carry out of the swallow counter. The last waveform shows the ripple carry out of the programmable divider. When both carry outs transition from a high to a low, the prescaler divides by five. Then, one swallow count later, the swallow counter carry out goes high, and the prescaler divides by four, until two counts later, when the program counter carry out goes to a high.

Figure 4.13 shows an example functional connection for adding the pulse-swallowing counter to a PLL. The output of the VCO connects to the feedback divider. The output of the feedback divider connects to the input of the phase detector. The output of the phase detector (PHCMP1) connects to a charge pump loop-filter combination that connects to a buffer amplifier and control line of the VCO to complete the loop. Speed-up phase detector is used to speed up the PLL response when the loop is unlocked. The speed-up phase detector senses only large phase errors and then increases the charge pump current. A divide-by-2 or -3 reference divider gives more flexibility in output-frequency combinations.

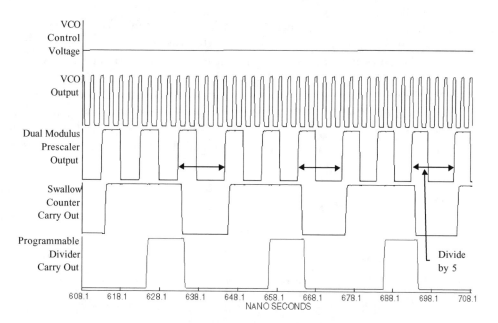

Figure 4.12 Zoomed-in view of pulse-swallowing counter simulation (divide-by-13).

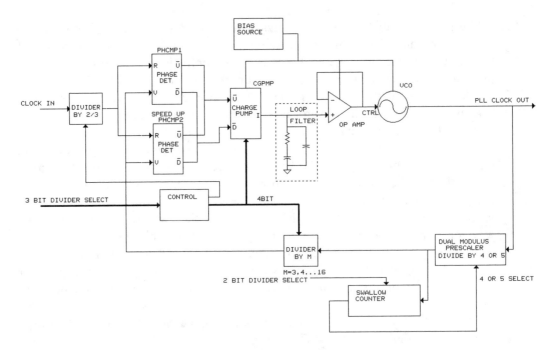

Figure 4.13　Functional configuration of a PLL with the pulse-swallowing counter.

4.1.3　Fractional Divide-by-N

In many applications, obtaining the desired output frequency requires dividing the output frequency by a large divide ratio to a low reference frequency. A low reference frequency narrows the loop bandwidth to keep the loop stable because, as a rule of thumb, the loop bandwidth needs to be less than 1/20th of the reference frequency. Narrow bandwidth makes monolithic implementation more difficult. In other applications, a large divide ratio (>1,000) multiplies the reference phase noise and internally generated noise so that the output has excessive jitter or the loop becomes unstable. Fractional divide-by-N synthesis provides an alternative [3, pp. 141–143].

Figure 4.14 shows the block diagram of a fractional divide-by-N PLL. A dual modulus prescaler, a swallow counter, a programmable counter, an accumulator, and an adder make up a fractional divide-by-N with pulse swallowing. The divide ratios in the fractional divider with pulse swallowing in Figure 4.14 are defined:

U = upper (larger) divide ratio of the prescaler (usually $L + 1$);

L = lower divide ratio of the prescaler;

S = divide ratio of the swallow counter;

　= number of times the prescaler divides by U in a complete divide cycle;

M = divide ratio of the program counter;

　= total number of prescaler cycles in a complete divide cycle;

F = fractional part.

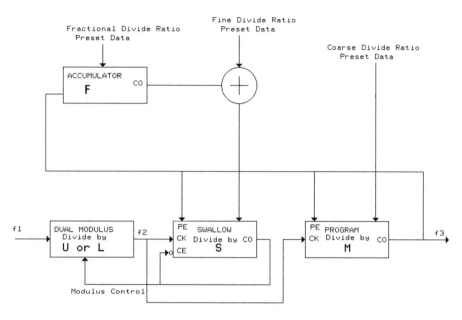

Figure 4.14 Functional block diagram of fractional divide-by-*N* with pulse swallowing.

Figure 4.15 shows the schematic of the 4-bit accumulator, the 2-bit swallow adder, and the adder subcircuit. The accumulator (sometimes referred to as a rate multiplier) consists of a 4-bit adder and a 4-bit register. The 4-bit register stores the results of the adder and feeds them back to the 4-bit adder for the next comparison at the next clock edge. An additional flip-flop is added to the carry-out bit to deglitch the output. The 2-bit adder sums the swallow data with the accumulator carry out to control the number of times a pulse is swallowed in the dual modulus prescaler (divide by 4 or divide by 5).

The divide cycle in Figure 4.14 begins with a parallel-enable signal that synchronously presets the data. The two carry outs (COs) go to the inactive state, which ends the preset mode. The inactive state of the carry out enables the swallow divider and changes the prescaler to its highest divide ratio. The dividers start to count until the swallow divider counts to its maximum value (S pulses of f_2). At this point, $S \times U$ input clock pulses have occurred. If the carry out of the accumulator is active, then the swallow count increases by 1, and $(S + 1) \times U$ clock pulses will have occurred. The carry out of the accumulator depends on the data input and is active after F number of f_3 pulses. At maximum value, the carry out is enabled, which stops the swallow divider and changes the prescaler to its lowest divide ratio. The program counter continues to count up to its maximum value [$(M - S) \times L$ more input clock pulses], while the swallow counter holds its count. The carry out of the program divider enables the preset mode, and a new divide cycle begins.

Equations (4.4) and (4.5) compute the divide ratio for a fractional divide-by-*N* with pulse swallowing:

$$N = LM + S + F \tag{4.4}$$

$$F = A/A_{tot} \tag{4.5}$$

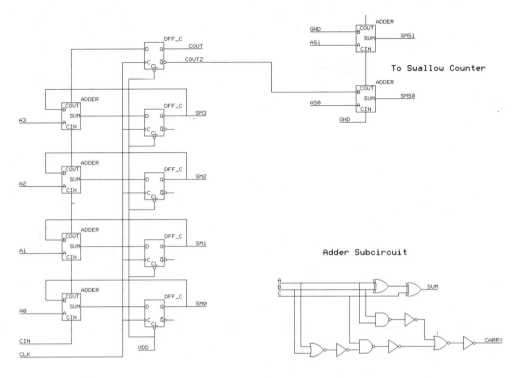

Figure 4.15 Schematic of accumulator, swallow adder, and adder subcircuit.

where

A = decimal equivalent for accumulator input data;

$A_{tot} = 2^{A_{bit}}$ maximum decimal count value of accumulator;

A_{bit} = number of bits in the accumulator.

The fractional divide-by-N technique conveniently separates the divide ratio into three parts, as shown in (4.4). These parts consist of coarse (LM), fine (S), and fractional (F) adjustments.

Let's use an example with a 4-bit accumulator to show the design and operation of the fractional divider. For 4 bits, $A_{tot} = 16$, and the fractional part steps equal 1/16. Consequently, for a minimum fractional part of 1/16, the divided-down output period changes once every 16 reference periods. For a loop bandwidth at 1/20th of the reference frequency, the jitter caused by the fractional divider will be present at the output of the PLL. Designing the loop bandwidth to be 1/200th of the reference frequency should allow the loop filter to average out the jitter from the fractional part.

The major point in fractional divider design is that only the average frequency equals the fractional divide-by-N ratio [4]. Without averaging, instantaneously following the divided output will result in frequency changes that will cause jitter at the output.

Let's continue to use an example to understand fractional dividers and add an input clock of 500 MHz and prescaler divide ratios of (4 or 5). From SPICE simulations, Figure 4.16 shows a timing diagram of the fractional divider with the divide ratio set to 12.25. The top waveform is the VCO output. The second waveform is the output of the prescaler. The third waveform is the carry out of the programmable divider. The fourth waveform is the carry out of the accumulator. The last waveform is the carry out of the swallow counter. The fractional addresses of the A_0 and A_1 bits are set to a high level. The swallow count is 0, and the programmable count is at its minimum of 3. Between 180 ns and 270 ns, the divider divides by 12 (24-ns period). At 170 ns and 270 ns, the output period of the prescaler and the output period of the programmable divider increase by 2 ns from the overflow of the accumulator at 165 ns and 265 ns. For a swallow count of 0, the carry out of the swallow counter only goes low with the carry out of the accumulator.

Processing the period in the carry-out waveform in Figure 4.16 produces Figure 4.17 for the modulation domain, which is output period versus time. Figure 4.17 shows a repetitive pattern in the carry-out signal from the programmable divider. On closer inspection the pattern can also be seen in Figure 4.16 for the prescaler output, the programmable divider carry and the accumulator carry-out signals. The pattern is one period of 26 ns and three periods of 24 ns to give an average period of 24.5 ns [average = (1 × 26 ns + 3 × 24 ns)/4 for a divide ratio of 24.5 ns/2 ns = 12.25].

Changing the swallow count from 0 to 1 and leaving the programmable count and fractional part settings alone gives a 13.25 divide ratio. From SPICE simulations, Figure 4.18 shows a timing diagram of the fractional divider with the divide ratio set to 13.25. In this case, the carry-out bit of the swallow counter gets reset

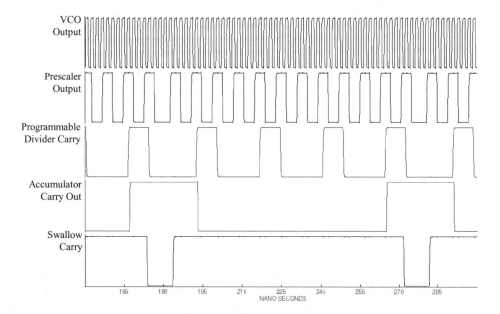

Figure 4.16 Timing diagram of SPICE simulation for fractional divide-by-12.25.

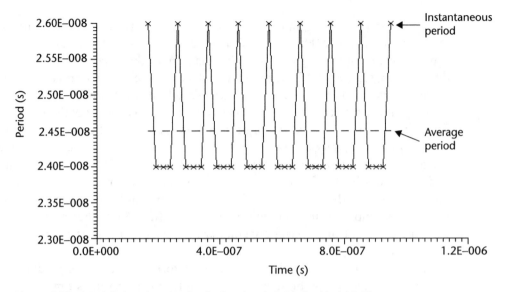

Figure 4.17 Modulation-domain plot for fractional divide-by-*N* of 12.25.

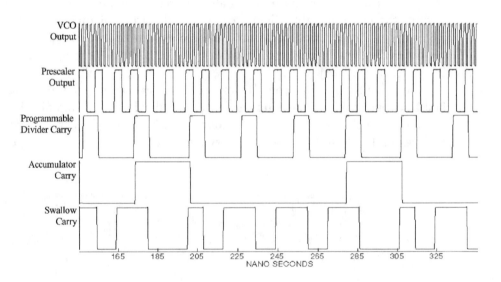

Figure 4.18 Timing diagram of SPICE simulation for fractional divide-by-13.25.

without the accumulator carry out because the divide ratio of 13 requires the modulo 2 prescaler to also swallow one period. Consequently, adding one count to the swallow count when the accumulator carry-out signal becomes active combines fractional divide with the swallow counter.

Figure 4.18 shows a repetitive pattern of four in the carry-out signal period from the programmable divider. The pattern has one period of 28 ns and three periods of 26 ns to give an average period of 26.5 ns [average = (1 × 28 ns + 3 × 26 ns)/4 for a divide ratio of 26.5 ns/2 ns = 13.25].

Studying the prescaler waveform shows the difference in operation from the previous mode (divide-by-12.25). The repetitive pattern in the prescaler has one period of 8 ns and two periods of 10 ns (28 ns), followed by three sets of one period of 8 ns, followed by one period of 10 ns, then by one period of 8 ns (26 ns), to give an average period of 26.5 ns. Consequently, an extra pulse is swallowed when the accumulator overflows.

Changing the swallow count from 1 to 2, changing the fractional part to 3/8, and leaving the programmable count alone gives a 14.375 divide ratio. From SPICE simulations, Figure 4.19 shows a timing diagram of the fractional divider with the divide ratio set to 14.375. The input clock was eliminated from the diagram because the displayed edges were too close together. In this case, the carry out does not occur in equal intervals of four reference periods.

Processing the period in the carry-out waveform in Figure 4.19 produces Figure 4.20 for the modulation domain. Figure 4.20 shows a repetitive pattern of eight periods in the carry-out signal from the programmable divider. On closer inspection, the pattern can also be seen in Figure 4.19. The programmable divider's repetitive output pattern is 30 ns, 28 ns, 30 ns, 28 ns, 28 ns, 30 ns, 28 ns, and 28 ns to give an average period of 28.75 ns [average = (5 × 28 ns + 3 × 30 ns)/8 for a divide ratio of 28.75 ns/2 ns = 14.375].

The worst-case condition for spurious is 1 swallowed pulse for every 16 input periods because this generates the lowest-frequency component (1/16 of the reference frequency). Consequently, tests on a silicon version should be run at this fractional amount to determine if the jitter is greater than the application requirements.

Figure 4.21 shows a schematic of a PLL with a fractional divide-by-N. A pulse-swallowing counter (integer divider) and accumulator with the PLL (no external components and 1-MHz loop bandwidth) allow operation at 1-GHz output with reference frequencies that range from 16 MHz (1 GHz/64) to 83 MHz (1 GHz/12). Integer steps of the divide ratio will allow 16-MHz steps. Fractional steps of the divide ratio will allow smaller steps to 1 MHz for a fractional step of 1/16th and

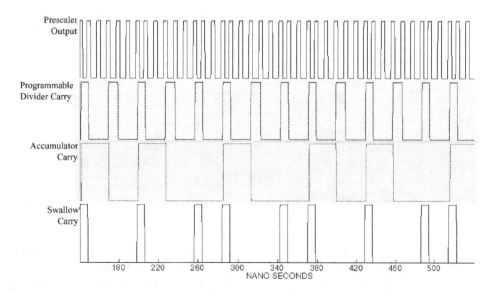

Figure 4.19 Timing diagram of SPICE simulation for fractional divide-by-14.375.

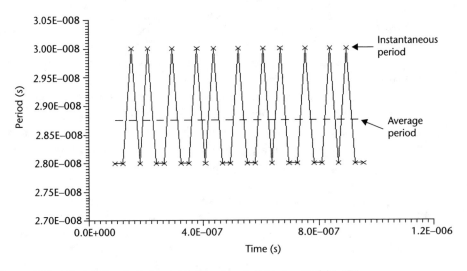

Figure 4.20 Modulation-domain plot for fractional divide-by-*N* of 14.375.

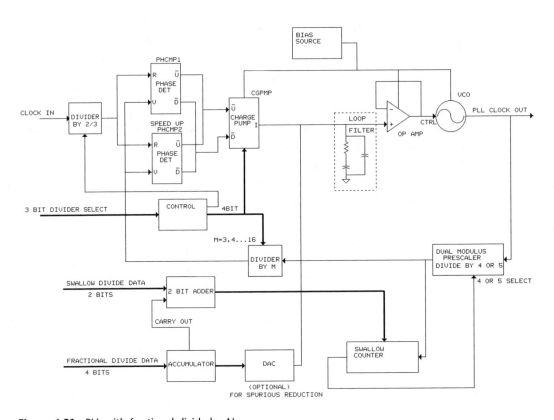

Figure 4.21 PLL with fractional divide-by-*N*.

a reference frequency of 16 MHz. The amount of additional jitter at 1 MHz (worst case) should be verified to make sure application jitter requirements are met.

Figure 4.22 shows a SPICE simulation of the fractional-*N* PLL with fractional divide-by-*N* and the *N* value set to 14.375. The top waveform shows the reference-

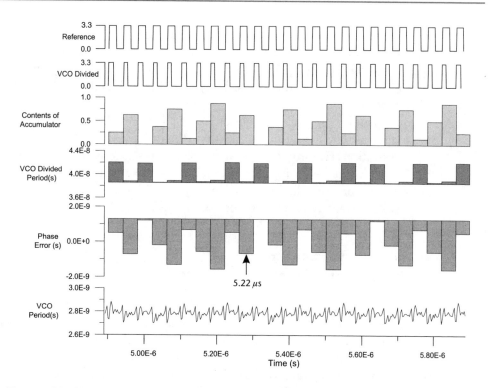

Figure 4.22 SPICE simulation of fractional divide-by-N PLL with $N = 14.375$.

clock input, the next waveform shows the divided-down output from the VCO, and the next waveform shows the fractional content in the accumulator register. The third waveform from the bottom shows the period of the VCO-divided output. The average period of the divided-down VCO output is 40 ns (25 MHz) to match the reference period. The second waveform from the bottom shows the phase error with respect to the reference input. An increase of 2 ns in period at $5.22 \mu s$ causes the phase error to change by 2 ns at the 5.22-μs X-axis value. The average of the phase error is zero. The last waveform shows the output period of the VCO. The peak-to-peak variation of the period remains within 200 ps.

In summary, fractional divide-by-N with pulse swallowing has several advantages and some disadvantages. First, this technique allows a higher reference frequency than the integer technique for synthesis of the same output frequency. Consequently, divide ratios will be lower for fractional divide-by-N. Lower divide ratios will produce lower phase noise. Finally, this technique can have additional jitter from the fractional part. However, with filtering, designs can be adjusted to minimize this effect.

4.2 Voltage-Controlled Oscillators

The VCO is the most important component in a PLL. Most of the major output characteristics of a PLL are determined by the VCO. For instance, the sensitivity of the PLL to injected noise on the supply, the time jitter of the output, the phase noise, the frequency tuning-range width, nonharmonically related spurious modes, loop stability, linear modulation, linear demodulation, and output that toggles in unlocked conditions all depend on the characteristics of the VCO. An ideal VCO would meet all of these conditions; however, practical VCOs make trade-offs in characteristics.

Consequently, varying applications require different VCOs to optimize the PLL for the application. For instance, cleanup PLLs in adaptive clock generation for audio applications require VCOs with low phase noise. Consequently, a multivibrator VCO with bipolar transistors is required because of the low flicker noise in bipolar transistors and the low phase noise in multivibrators.

Most important requirements for a monolithic VCO:

- Large frequency tuning range;
- Noise immunity to supply;
- Low noise (low jitter and low phase noise);
- No nonharmonic, spurious modes.

The tuning frequency range must accommodate variations in temperature, supply, and processing. Consequently, a large tuning range is desirable. System-on-a-chip integration has made PLL noise immunity to supply variations extremely important because digital circuits add noise to the supply and substrate. Timing accuracy and spectral purity in PLL applications impose an upper bound on the VCO jitter and phase noise.

The ideal VCO would have all these characteristics; however, these characteristics usually conflict with one another. For instance, low-noise requirements oppose wideband frequency tuning range. Consequently, obtaining wideband frequency range sacrifices low-noise characteristics.

Oscillators are generally classified as relaxation oscillators or harmonic oscillators. A relaxation oscillator switches back and forth between two stable equilibrium states. Consequently, this switching causes high harmonic content. A harmonic oscillator produces a near sinusoidal signal with good phase noise and high spectral purity. Harmonic oscillators usually use LC resonant circuits.

The following are the most common types of oscillators in an IC:

- Ring oscillator;
- Differential ring oscillator;
- Multivibrator;
- LC resonance oscillator.

Let's start by studying ring oscillators. Equation (4.6) computes the output period of a ring oscillator:

$$T_o = 2MT_d \tag{4.6}$$

where

M = number of stages;

T_d = delay for each stage.

For single-ended gates, the number of inverters must be odd. Differential stages can be even numbers, but the outputs of one of the inverters must be swapped. One round-trip around the ring produces 180° of phase shift. The total phase shift through the oscillator must be 360°, with a gain of unity to meet the criterion for oscillation. Consequently, two round-trips occur for each period of oscillation. The factor of 2 difference in phase shift is represented by the factor of 2 in (4.6).

Varying the delay term, T_d, in (4.6) varies the output frequency of the oscillator f_o ($f_o = 1/T_o$). There are two common methods for varying the delay:

- Varied capacitive loading;
- Resistive tuning (which varies drive current to the connected capacitive load).

Capacitive tuning adjusts the load capacitance to the output of the delay stage. Varying the capacitance can be done with a reverse-biased *pn* junction diode. Narrow tuning range, reduction in maximum frequency, nonlinearity in the VCO gain, and susceptibility to common-mode noise are the disadvantages of this method. Consequently, resistive tuning is preferred.

4.2.1 Operation of a Ring Oscillator

Figure 4.23 shows a schematic for a current-starved three-inverter ring oscillator. The ring oscillator works by controlling the charging and discharging of the gate capacitance of the next inverter. Decreasing the peak available charging current increases the time to charge and discharge the gate capacitance; consequently, the frequency is decreased.

Figures 4.24 and 4.25 show the three-inverter ring oscillator waveforms for a high frequency (1.4V on the PLL macro tune line) and low frequency (1.1V on

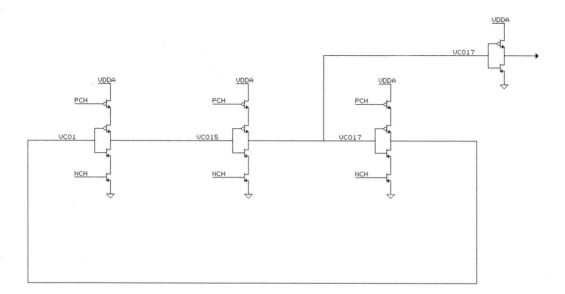

Figure 4.23 Simplified schematic of a three-inverter current-starved ring oscillator.

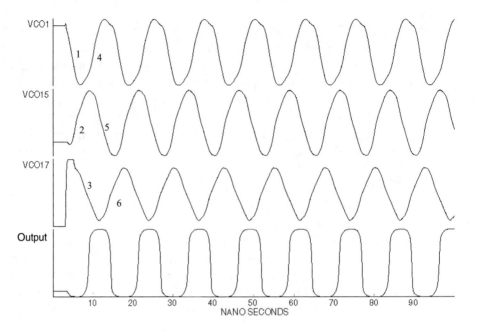

Figure 4.24   Three-inverter ring-oscillator waveforms for a high-frequency oscillation (1.4V on VCO macro tune line).

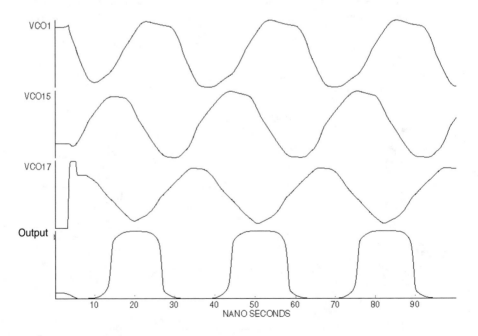

Figure 4.25   Three-inverter ring-oscillator waveforms for a low-frequency oscillation (1.1V on the VCO macro tune line).

the PLL macro tune line) at each inverter output node (VCO1, VCO15, and VCO17) and the final buffer inverter output. The simulations were done in a 0.55-μm gate-length process. These figures show the change in frequency between figures and the cascade of delay around the ring to end up in phase with the beginning node. The y-axis in the figures ranges from 0V to 5V.

In Figure 4.24, the rising edge of node VCO1 after 1.6 ns produces a falling edge at node VCO15, after 1.6 ns produces a rising edge at node VCO17, and after 1.6 ns produces a falling edge at node VCO1 to complete the cycle and half the period (4.8 ns) of the oscillation frequency. This computes an output frequency of 104.16 MHz [1/(2 × 4.8 ns)]. The numbered edges 1 through 6 in Figure 4.24 show a complete oscillation period. The slow rising edge between the inverters inside the ring oscillator adds jitter to the clock, makes it more sensitive to power supply ripple, and makes it more difficult to maintain a 50% duty cycle.

A similar cycle occurs in Figure 4.25 for tuning the VCO to a lower frequency, except the delays are 5 ns each and the period of oscillation is 30 ns. The contrast in oscillation frequencies shows the operating mechanism for the ring oscillator. The 10% to 90% point rising and falling edges are significantly different between Figures 4.24 and 4.25. In addition, the wave shape is close to a triangle wave, which we would expect for a current source driving a capacitive load (gate capacitance of the following delay inverter). The fast rising edge at node VCO17 shows the initial edge that was used to start the oscillation.

Figure 4.26 shows the current levels at one of the nodes in the ring for high and low-frequency operation. This figure shows that the low-frequency peak current (20 μa) is much lower than the high-frequency peak current (50 μa). This illustrates

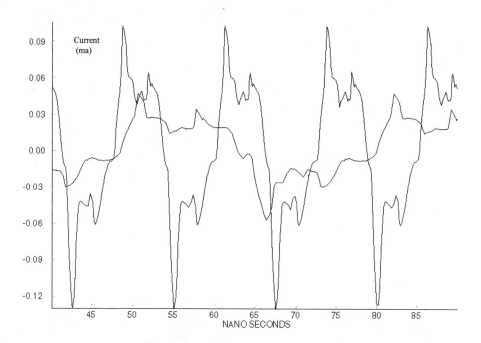

Figure 4.26   Current levels at one of the nodes in the current-starved ring oscillator for high- and low-frequency operation.

the mechanism of operation for the current-starved inverter ring oscillator. In addition, the high current spikes of (1 ma show the oscillator can cause significant high-frequency ripple in the power supply and substrate.

Figure 4.27 shows the variation of a single delay element with another delay element for loading versus tune voltage. For a voltage-control change from 0.7V to 1.5V, the delay changes from 10.4 to 2 ns. The nonlinear variation in delay with control voltage shows that the output-frequency-versus-tune-voltage curve should also be significantly nonlinear. The steepest slope in Figure 4.27 shows the tuning point that can give the widest frequency tuning range for the VCO.

Figure 4.28 shows the nonlinear (s-shaped) tuning curve for the current-starved ring oscillator. Significant variation in the tuning curve can cause moding in a PLL and narrow the nominal loop bandwidth to the point of losing lock. Computing a best fit line to the curve in the plot helps calculate the linearity by taking the ratio of the maximum derivation from the line over the full scale, which gives a linearity of 30.9%.

Figure 4.29 shows a zoomed-in view of the nonlinearity in the tuning curve by showing VCO gain versus tuning voltage. The VCO gain varies from a minimum of 20 MHz/V for a tuning voltage of 0.9V to a maximum of 150 MHz/V at 1.8V and to a minimum of 20 MHz/V at 4V. A 10-to-1 decrease in VCO gain can easily change a high-phase-margin operation point to an unstable loop. Another concern for the loop is that below 0.9V, the VCO quits oscillating. A VCO that quits oscillating in a loop can cause high-modulation moding points, and once the VCO quits oscillating, it may not restart.

For the nominal case from 0.9V to 4V on the tune line, the duty cycle varies from 47% to 54%. The duty-cycle variation is important to applications that use

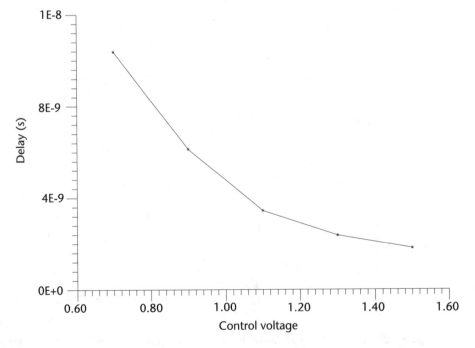

Figure 4.27   Variation of a single delay element in a ring oscillator versus tune voltage.

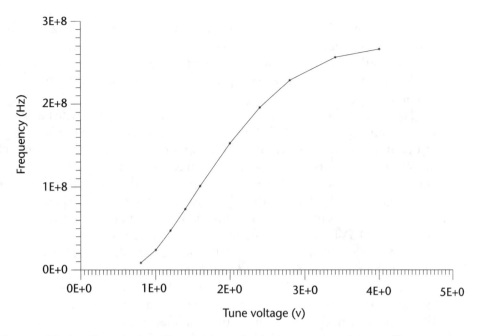

Figure 4.28 Nonlinear (*s*-shaped) tuning curve for the current-starved ring oscillator.

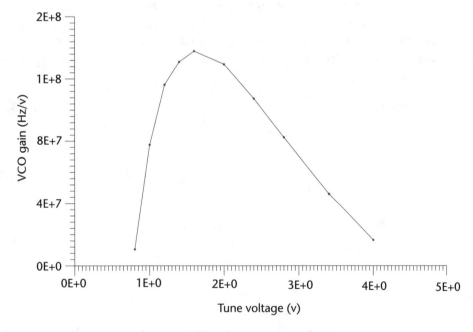

Figure 4.29 Current-starved inverter VCO gain versus tune voltage.

both clock edges. For the nominal case from 0.9V to 4V on the tune line, the supply current varies from 0.73 to 8.5 ma, the peak-to-peak current is 2.7 ma, and the current rise time varies from 30 to 0.85 ns. The supply current draw is important for low-power applications like wireless handsets. The peak-to-peak current and rise time are important in determining how much ripple will be on the supply lines. The supply ripple will vary from 0.81V to 22.95 mV for a 100 pF decoupling capacitor on silicon ($dV = dI \ dt/C$). Clearly, at the lower frequencies the external decoupling capacitance will have to take care of the large ripple on the supply. The peak-to-peak voltage swing inside the ring varies from 5 to 4.65 V_{pp}.

Figure 4.30 shows the variation in period versus an injected sine wave on the supply with a frequency of 17 MHz and 0.5 V_{pp} amplitude. Noninteger harmonics of the output frequency are selected for injection to prevent injection locking of the VCO. The PLL is usually sensitive to frequencies between 1 and 20 MHz and especially around the 0-dB crossover frequency of the loop. With the application of the injected sine wave, Figure 4.30 shows that the output frequency varies from 76.5 to 84 MHz with 0.5 V_{pp} on supply. This computes to a 15 MHz/V [(84 to 76.5 MHz)/0.5V] supply sensitivity. For a tune sensitivity of 150 MHz/V, the isolation is computed to be 20 dB [20 log(15/150)]. This also computes to a −13-dBc reference sideband level [20 log(84−76.5)/17/2].

Next, we can compute the VCO phase-noise curve from simulation of the oscillator amplifier noise. Connecting one of the delay elements up as an amplifier allows measurement of the gain of the amplifier, the thermal noise floor, and the 3-dB corner-frequency location of the beginning of flicker noise. The example has a gain of 32 dB, a thermal noise floor of −169 dBV/Hz, and a flicker corner frequency of 8 MHz for 3V on the control line. For a ring oscillator, the VCO has a Q of 1 because no energy is stored in each cycle. The charge that is stored on the gate on a rising edge is discharged back to its original amount on the falling edge. The delay element has a 5-V voltage swing at the output. Finally, the oscillator

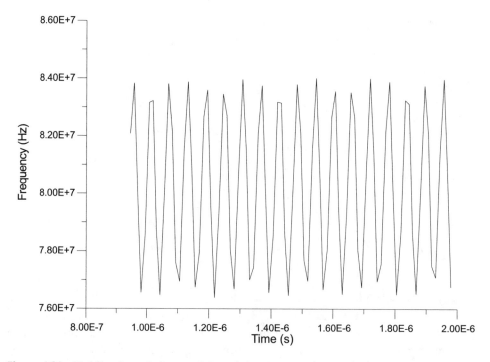

Figure 4.30   Variation in period versus injected sine wave on the supply.

has an output frequency of 80 MHz. Substituting these values into (4.7) (Leeson's equation) produces the VCO phase-noise curve in Figure 4.31:

$$\mathscr{L}(f) = \left(1 + \frac{f_l}{f}\right) \frac{10^{\frac{N_{amp}}{10}}}{\frac{V_{os}^2}{10^{\frac{G_n}{10}}}} \left[1 + \left(\frac{f_c}{2Q_l}\frac{1}{f}\right)^2\right] \tag{4.7}$$

where

f_l = flicker corner frequency (Hz);

V_{os} = output swing inside the osciallator (V);

Q_l = loaded Q of the resonator;

f_c = center frequency of the oscillator (Hz);

G_n = gain of amplifier in the oscillator (dB);

N_{amp} = equivalent input noise of the amplifier in the oscillator (dBV/Hz).

From this equation, phase-noise improvements can be determined. For example, lowering the gain of the delay element, increasing the gate width, W, and increasing the output voltage swing reduces the far-from-the-carrier phase-noise levels. Increasing the gate length, L, reduces the flicker corner frequency.

Figure 4.31 shows the calculated phase-noise curve for a current-starved ring oscillator. The bottom dashed line shows the output noise of the amplifier in the oscillator. The middle dashed line shows the noise due to the resonator Q of 1.

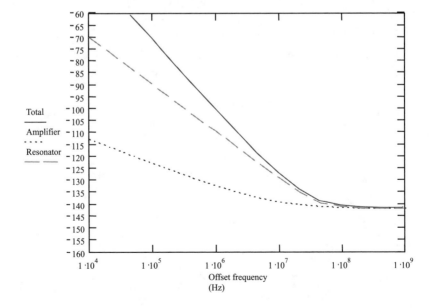

Figure 4.31 Calculated phase noise for a ring oscillator.

The top curve shows the total phase noise for the oscillator. The total phase noise curve shows a 30 dB/dec slope with a break-point change from 30 to 20 dB/dec at 8 MHz for the location of the flicker corner frequency. The close-in phase noise at 100 kHz is −70 dBc/Hz, and the far-out phase noise is −142 dBc/Hz.

Figure 4.32 shows the detailed schematic for the current-starved ring oscillator that was used in the SPICE simulations. The transistors are 50C21 process, 0.72-μm minimum gate length with 5-V supply process. The VCO consists of a control-voltage-to-current bias converter, delay elements, multiplexers, and output buffer. The first pass gate between nodes VCO2 and VCO15 is half of a multiplexer that was used to add more inverters. Consequently, for lower-frequency-operation 5, 7, 9, and 11 gates can be cascaded to give more versatility to the VCO. The second multiplexer is used to break the loop for a built in test of the ring open loop and to provide a simulator with a method for starting the oscillator.

4.2.2 Differential Ring Oscillators

Differential ring oscillators have become widely used in PLL designs. They allow quadrature outputs, reject common-mode noise, and have near 50% duty cycles. The availability of quadrature outputs and the 50% duty cycle make them desirable for clock-recovery circuits, and they have many applications in telecommunications.

The differential ring oscillator has several characteristics that are different from other oscillators. First, the differential ring oscillator allows an even number of gate delays. One of the output signals of one of the stages is swapped to achieve an even number of gates. Consequently, a quadrature output is possible, which can be important in clock-recovery circuits. The differential delay stage gives the

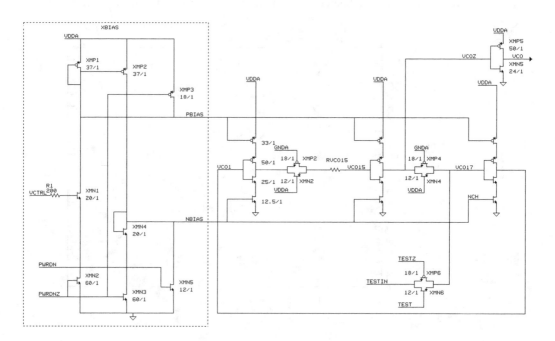

Figure 4.32 Schematic of current-starved ring oscillator.

designer a method for isolating the oscillator from substrate noise and supply noise. The differential method rejects common-mode noise, which is prevalent in ICs. Another advantage of a differential ring oscillator is the 50% duty cycle.

The differential ring oscillator also has several disadvantages. A differential delay stage increases the gate delay by 30% to 50% more than a single-ended inverter delay cell [5]. This reduces the maximum output frequency by 30% to 50%. A differential ring oscillator is noisier than a single-ended ring oscillator because the differential stage has a 6-dB noise figure penalty, and the bias current generation adds a significant amount of noise. It takes more area because the current bias generator uses a significant number of transistors. The voltage drop across the tail-current transistor makes the output have a low output voltage swing. Low voltage swing can cause edges to be skipped by the differential-to-single-ended converter that is connected to the output of the oscillator. The input and output voltages to each delay stage are not rail to rail. This makes migration to faster processes that have lower power-supply voltages (1.2V) more difficult.

Next, the oscillator can have trouble starting up because the current mirrors can get in a mode that steals all the current. In addition, V_t mismatch with minimum length gates in loads causes duty-cycle problems at low frequencies.

The schematic in Figure 4.33 shows a three-stage differential ring oscillator that operates at 1.8V in a 0.22-μm minimum gate length process. The transistor values are shown at two times the silicon size. The delay stages are connected in a noninverting mode, except for nodes VCO12 and VCO12N, which are swapped to provide 180° phase shift to meet conditions for oscillation.

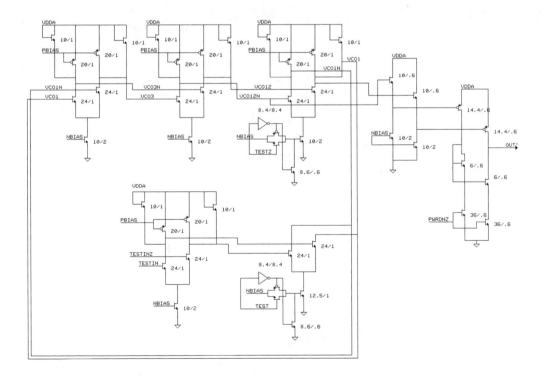

Figure 4.33 Schematic of a three-stage differential ring oscillator.

The oscillator loop is allowed to break after the second delay so that a test clock signal can be injected and in order to provide an oscillator startup path. The differential ring requires a startup circuit because it has a third state in which the current can be evenly divided in each delay element. This state is not present in a single-ended ring oscillator. Slowly ramping the power supply (over 100 μs) is one simulation test method to check for this third condition. If this condition is present, the circuit will have to detect it and reset the oscillator.

A level translator, which provides a single-ended output, follows the ring oscillator as shown on the right-hand side of Figure 4.33. The frequency is changed by adjusting the tail current, and the drive resistance of the differential delay stage. Adjusting both factors provides a broader tuning range. Diode connected p channels in parallel with the active loads of the differential amplifier provide a more symmetric load.

Figure 4.34 shows the control-voltage conversion to p channel and n channel current biases. Adjusting the tail current and drive current provides a broad frequency tune range. Current mirroring is used to balance the change in resistance between n channel and p channel transistors, which helps keep the duty cycle near 50%.

Figures 4.35 and 4.36 show the three-inverter ring oscillator waveforms for a high frequency (1.3V on the PLL macro tune line) and low frequency (1.0V on

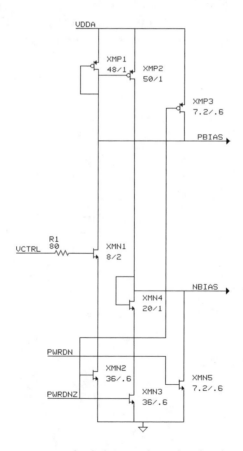

Figure 4.34   Circuit to convert control voltage to p channel and n channel bias.

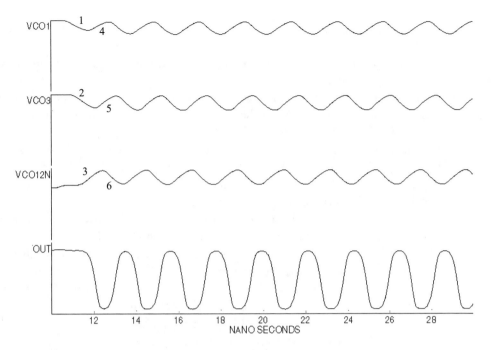

Figure 4.35   Timing for differential ring oscillator at high frequencies.

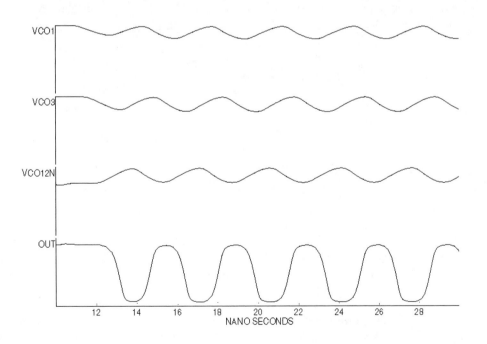

Figure 4.36   Timing for differential ring oscillator at lower frequencies.

the PLL macro tune line) at each inverter output node (VCO1, VCO3, and VCO12) and the final translation buffer output. These figures show the change in frequency between figures and the cascade of delay around the ring to end up in phase with the beginning node.

Figure 4.35 shows the waveforms for the oscillator at a frequency of 460 MHz (a 2.16-ns period with 1.3V on the tune line). The y-axis ranges from 0V to 1.8V. In the figure, the falling edge of node VCO1 after 0.36 ns produces a falling edge at VCO3, after 0.36 ns a rising edge occurs at node VCO12N, and after 0.36 ns a falling edge occurs at VCO1 to complete the cycle and half the period (1.08 ns of the 2.16-ns period) of the oscillation frequency. The numbered edges 1 through 6 in Figure 4.35 show a complete oscillation period.

Figure 4.36 shows the waveforms for the VCO with a lower output frequency of 200 MHz because of the decrease in tune voltage from 1.3V to 1.0V. A similar cycle to the previous figure occurs in Figure 4.36, except the delays are 0.83 ns each, and the period of oscillation is 5 ns (200 MHz). The contrast in oscillation frequencies shows how the ring oscillator operates. The 10% to 90% point rising and falling edges are significantly different between Figures 4.35 and 4.36. In addition, both figures show that the voltage waveforms inside the ring do not go rail to rail. This helps maintain noise immunity from the supply rails. However, the slow edge and slow swing make the phase noise worse.

Figure 4.37 shows output frequency versus tune voltage for weak, nominal, and strong conditions. These curves are linear until a tune voltage of 1V. A linear curve fit from 0.8V to 1.8V shows a linearity of 2.2%, which is much better than for the current-starved inverter. The process variation shows that the maximum output frequency varies from 500 to 1,300 MHz. The weak simulation condition has weak processed transistors, low supply voltage, and high temperature. The

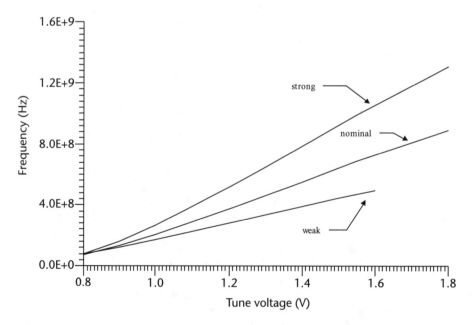

Figure 4.37 Output frequency versus tune voltage for a differential ring oscillator.

strong simulation condition has strong processed transistors, high supply voltage, and low temperature. For a robust PLL design, the following dividers must operate over these widely varying conditions in order to prevent the loop from hanging up the tune line at the power rail.

Figure 4.38 shows the nominal VCO gain. The VCO gain is flat from 1.8V to 1.0V and significantly falls off below 1.0V. The small-signal gain of each stage drops with lower control voltage. As the gain of each stage drops, the circuit will eventually fail to oscillate because the total gain around the ring at the frequency of operation will fall below unity. Consequently, for low voltages, the oscillator may not start. In addition, a 1V minimum control voltage makes the circuit not useful for low supply voltages and forces a larger VCO gain to cover a wide frequency range. Finally, the figure shows a peak gain of 840 MHz/V.

Figure 4.39 shows that a 31-MHz injected signal with 0.4V peak-to-peak amplitude produces a 36-MHz peak-to-peak waveform. Dividing the change in frequency by the variation in voltage computes a 90-MHz/V sensitivity to variations in the supply voltage. For a figure of merit, we can calculate the ratio of power-supply sensitivity to VCO gain [10 log(90/840) = −25 dB desensitivity].

Figure 4.40 shows the measured phase-noise curve for a differential ring oscillator. Measurements below −120 dBc/Hz were below the noise floor of the equipment, and for offset frequencies below 500 kHz the curve was extrapolated. The curve shows a 30-dB/dec slope without a break point. A break-point change from 30 to 20 dB/dec identifies the location of the flicker corner frequency. Consequently, with no corner flicker frequency, the flicker corner frequency is greater than 10 MHz. Thus, measuring the phase-noise curve gives us additional characteristics of the VCO [6].

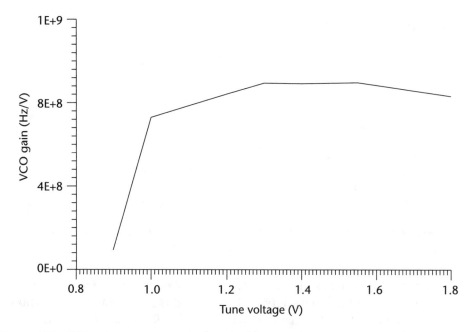

Figure 4.38   VCO gain versus tune voltage for a differential ring oscillator.

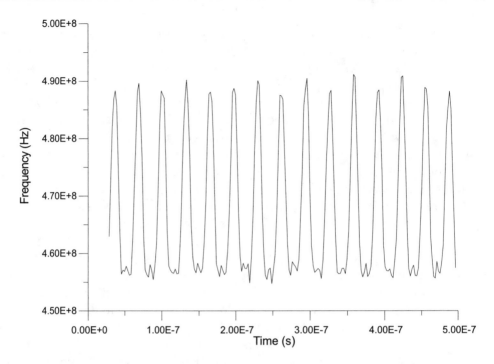

Figure 4.39   A 31-MHz signal injected on the power supply of the differential ring oscillator.

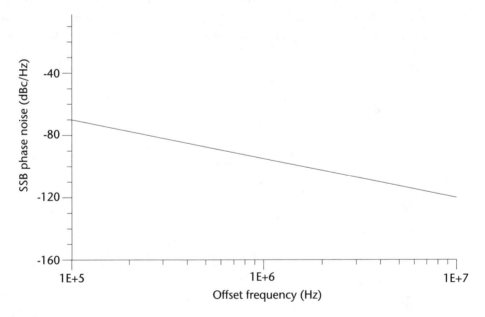

Figure 4.40   Extrapolated phase-noise curve from measured data for a differential ring oscillator.

For the nominal case from 0.8V to 1.6V on the tune line, the duty cycle varies from 0.48 to 0.52. The duty-cycle variation is important to applications that use both clock edges. For the nominal case from 0.8V to 1.6V on the tune line, the supply current varies from 0.42 to 1.78 ma, the peak-to-peak current is 0.7 ma, and the current rise time varies from 0.22 to 0.515 ns. The supply current draw is important for low-power applications like wireless handsets. The peak-to-peak current and rise time are important in determining how much ripple will be on the supply lines. The supply ripple will vary from 1.54 to 3.6 mV for a 100 pF on silicon decoupling capacitor ($dV = dI\ dt/C$). Clearly, at the lower frequencies, the external decoupling capacitance will have to take care of the large ripple on the supply. The peak-to-peak voltage swing inside the ring varies from 0.257 to 0.39 V_{pp}.

4.2.3 Multivibrators

Multivibrators are fully monolithic oscillators because the relaxation networks contain components that can be integrated. In addition, these circuits provide wide frequency range and high-speed operation.

A multivibrator consists of two parts. The first part has a constant-current source charge or discharge a capacitor network to create a voltage ramp. The second part has a voltage comparator and a latch that senses the voltage level on one of two capacitors and switches the logic to the other capacitor after a voltage threshold is crossed. Figures 4.41 and 4.42 show two different schemes to charge and discharge the capacitor in a multivibrator.

The first charging or discharging method as shown in Figure 4.41 uses a source-coupled transistor technique (modeled as switches in the figure). This technique suffers because the floating capacitor is in series with the grounded circuit parasitics, which significantly affects the performance of the circuit. Consequently, the floating capacitor value is set to a large value to avoid the effects of parasitics. The larger capacitor value, because it desensitizes the circuit to parasitics, slows down the maximum operating frequency.

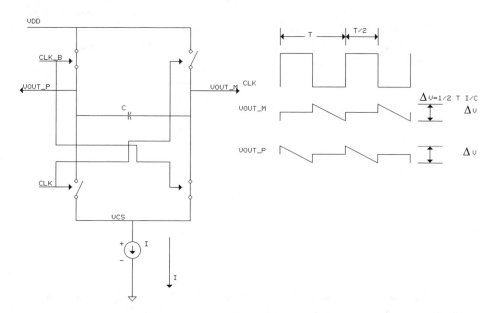

Figure 4.41 Functional diagram and timing waveforms of a series capacitor charging scheme. (*From:* [7]. © 1988 IEEE. Reprinted with permission.)

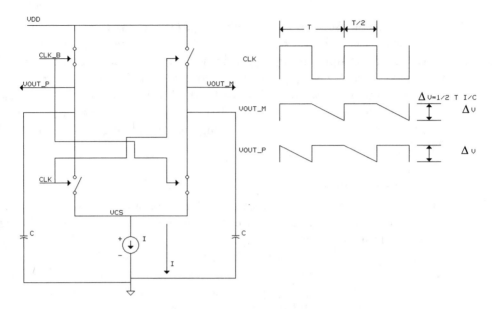

Figure 4.42   Functional diagram and timing waveforms of parallel capacitor charging scheme [5].

The second charging or discharging method as shown in Figure 4.42 uses grounded capacitors. The advantage of this technique is that the parasitics are in parallel with the timing capacitor, which allows the capacitor to go to zero for the highest output frequency with parasitic capacitance. In addition, the single terminal switching configuration at each capacitor makes the circuit insensitive to the non-ideal "ON" resistance of each switch. A potential problem is the mismatch of the capacitors. Employing symmetric and tight layout techniques avoids the mismatch.

The voltage comparator and latch portion of the multivibrator consists of two cross-coupled NOR gates. The first latch senses the voltage at the capacitor and switches upon crossing the input voltage threshold. The second latch prevents positive dc feedback because, if both nodes are high at the same time, the outputs of the nor gates are both low, and the transistors will not discharge the capacitors. The circuit will remain in a latched state.

Figure 4.43 shows a schematic of one implementation of the multivibrator. This circuit uses a 0.8-μm (50C21) minimum gate-length process. Equation (4.8) shows the formula for computing the output frequency of a multivibrator:

$$f_{vco} = \frac{1}{2\left(\dfrac{C\Delta v}{I} + \Delta t\right)} \qquad (4.8)$$

where

Δv = voltage difference from discharged state voltage to threshold voltage where comparator switches (V);

Δt = regeneration time (sec).

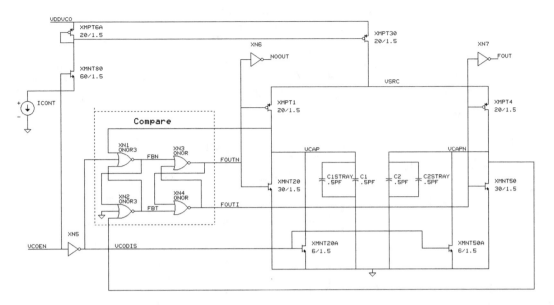

Figure 4.43   Schematic of one implementation of the multivibrator.

A multivibrator has several advantages. First, the regeneration time makes nonoverlapping clocks. This could be useful for sigma-delta modulator circuits. Next, it achieves high-speed operation because it minimizes the effects of parasitic components. The circuit minimizes complexities. Finally, the circuit uses no reactive elements other than parasitic capacitors [7].

A multivibrator also has some disadvantages. First, a possible multioscillator condition can occur because two nor gates and a CMOS inverter form a three-gate ring oscillator configuration. Selection of the logic prevents this condition from occurring. Finally, the current source generates high switching current spikes, which can cause interference with other circuits.

Figure 4.44 shows the voltage waveforms that occur in a multivibrator. The y-axis for the signals ranges from 0V to 5V. The FOUTN signal controls the charging and discharging of the in-phase capacitor (CAP), and the FOUTI signal controls the charging and discharging of the out-of-phase capacitor (CAPN). The VCAP signal shows the voltage waveform at the in-phase capacitor (CAP). The VCAPN signal shows the voltage waveform at the out-of-phase capacitor (CAPN). The ramp voltage shows the capacitor charging until a switch threshold is sensed by the logic, which then discharges the capacitor and steers the current to the other capacitor. The other capacitor behaves in the same manner, but out of phase. The peak-to-peak voltage swing of the ramp should remain constant with the change in frequency because the voltage threshold should not vary.

Figure 4.45 shows the current waveforms that occur in the multivibrator. The first waveform shows the source current, and the second waveform shows one of the steered currents into the capacitor. The steered current switches between 0 and 20 μa. The source current stays constant at approximately 22 μa. Ideally, these waveforms should be square, but practical circuits have drooping and switching spikes. The source current has switching spikes that are 6 to 10 μa. These are the

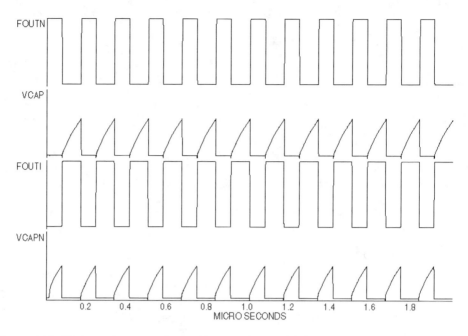

Figure 4.44 Timing voltage waveforms that occur in a multivibrator.

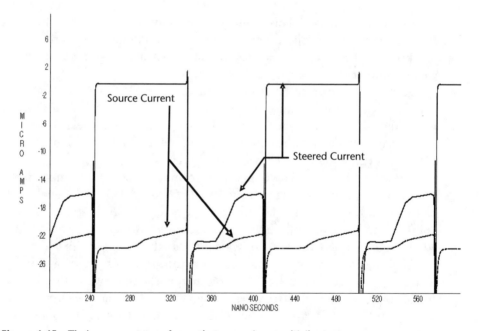

Figure 4.45 Timing current waveforms that occur in a multivibrator.

current spikes that appear on the supply lines, which can interfere with other circuits. Consequently, capacitors or voltage regulation should be used to minimize the effects of these spikes on other circuits.

Figure 4.46 shows the current-versus-frequency curve for the multivibrator for weak, nominal, and strong conditions. The frequency range varies from less than

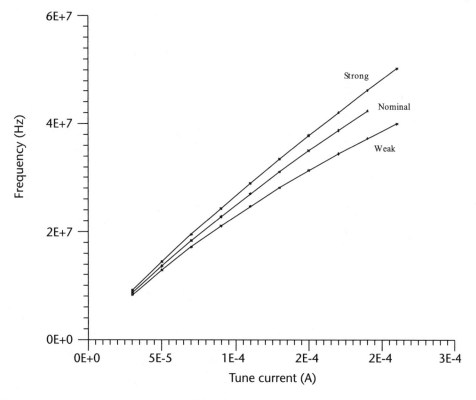

Figure 4.46   Tune current versus output frequency of a multivibrator.

10 to 40 MHz. The curves look linear and have a small variation with conditions at low frequencies and high variation with conditions at high frequencies. High linearity suppresses other modes of oscillation. Computing a best-fit line to the curve in the plot helps calculate the linearity by taking the ratio of the maximum deviation from the line over the full scale, which gives a linearity of 1.9% for the multivibrator curve.

Figure 4.47 shows gain versus tune current for the multivibrator over weak, nominal, and strong conditions. The small variation in gain on each curve shows a very linear tune curve. In addition, these curves show a small variation in gain over weak, nominal, and strong conditions (1.5E9 Hz/A to 3.0E9 Hz/A).

Figure 4.48 shows the variation of output frequency with an injected 1-MHz and $0.2\text{-}V_{p\text{-}p}$ sine wave on the power-supply line. The figure shows a sinusoidal waveform, which shows that the oscillator is linear with supply variation. In the figure, the output varies from a frequency of 10.2 to 7.5 MHz. Subtracting the peak frequency difference and dividing by the voltage variation on the supply line (2.7 MHz/0.2V) gives a 13.5-MHz/V sensitivity on the supply line. Given a jitter specification, this tells us what the tolerance needs to be on the multivibrator supply line.

Next, we can compute the VCO phase-noise curve from simulation of the oscillator amplifier noise. Connecting the multivibrator as an amplifier allows measurement of the gain of the amplifier, the thermal noise floor, and the 3-dB corner-frequency location of the beginning of flicker noise. The example has a gain

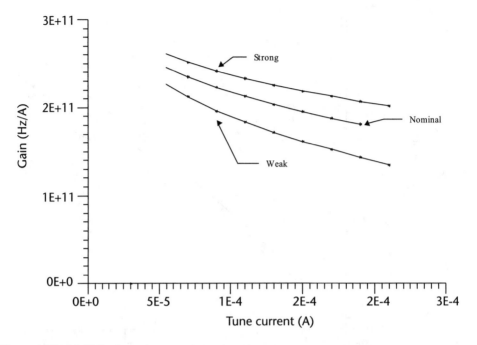

Figure 4.47   Multivibrator gain versus tune current.

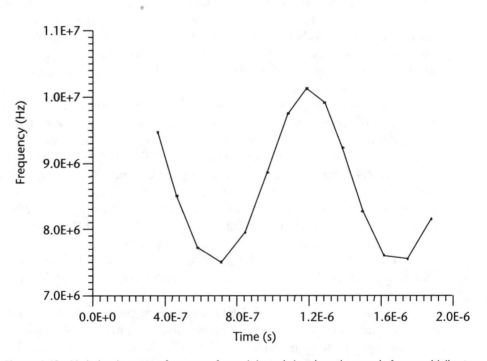

Figure 4.48   Variation in output frequency for an injected signal on the supply for a multivibrator.

of 42 dB, a thermal noise floor of −166 dBV/Hz, and a flicker corner frequency of 1.8 MHz for 3V on the control line. A multivibrator has a Q of 1 because no energy is stored in each cycle. The charge that is stored on the capacitor on a rising edge is discharged back to its original amount on the falling edge. The ramp voltage trips at 3V for oscillator voltage swing. Finally, the oscillator has an output frequency of 8 MHz. Substituting these values into (4.9) (Leeson's equation) produces the VCO phase-noise curve in Figure 4.49:

$$\mathcal{L}(f) = \left(1 + \frac{f_l}{f}\right)\frac{10^{\frac{N_{amp}}{10}}}{\dfrac{V_{os}^2}{10^{\frac{G_n}{10}}}}\left[1 + \left(\frac{f_c}{2Q_l}\frac{1}{f}\right)^2\right] \tag{4.9}$$

From this equation, phase-noise improvements can be determined. For example, lowering the gain of the delay element, increasing gate width W, and increasing the output voltage swing reduces the far-from-the-carrier phase-noise levels. Increasing gate length L reduces the flicker corner frequency.

Figure 4.49 shows the calculated phase-noise curve for a multivibrator. The bottom dashed line shows the output noise of the amplifier in the oscillator. The middle dashed line shows the noise due to the resonator Q of 1. The top curve shows the total phase noise for the oscillator. The total phase noise curve shows a 30-dB/dec slope with a break-point change from 30 to 20 dB/dec at 1.8 MHz for the location of the flicker corner frequency. The close-in phase noise at 10 kHz is −58 dBc/Hz, and the far-out phase noise is −131 dBc/Hz.

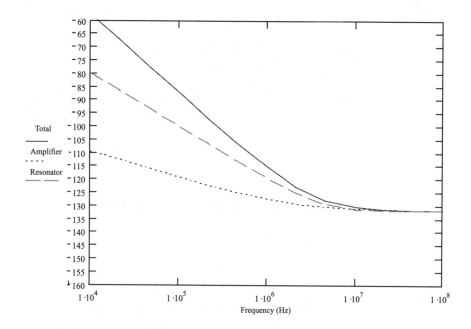

Figure 4.49 Calculated phase noise for a multivibrator.

4.2.4 LC Resonant Oscillators

The design approach base on LC resonant oscillators gives excellent results for high-speed narrowband oscillators. The external requirement for an LC tank to meet a wide tuning range limits these circuits to a small range of applications for integration [8]. Good oscillator design depends on the selection of a circuit topology, active device, and resonator and on the consideration of the coupling network.

LC resonant oscillators rely on feedback control through a resonator to generate oscillations. Table 4.1 gives the choices of resonators for wideband VCOs. For integrated circuits, there are not too many choices. A transmission-line resonator and low-frequency LC oscillators would require external components. For a fully monolithic solution, the choices are even more limited.

Figure 4.50 shows the two main feedback mechanisms for an LC oscillator. The LC feedback oscillator has a transistor amplifier that provides gain to the small signal to start and maintain the oscillation and a series LC feedback resonator that selects the frequency oscillation and provides the right amount of phase without losing positive gain to maintain oscillation.

The negative-resistance oscillator has a transistor with an inductor in the gate that provides a negative resistance to an LC resonator. The negative resistance reflects the wave presented to it so that the returning wave is larger than the

Table 4.1 The Q and Operating Frequency of Various Low-Q Resonators

Type of Resonator	Frequency	Unloaded Q	Major Limitation
LC	0.9–10 GHz	5–50	Narrow tune range
RC (ring, multivibrator)	10 kHz–2 GHz	1	High noise
Transmission line	2–10 GHz	5–50	Narrow tune range

Feedback Oscillator

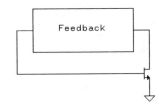

Negative Resistance Oscillator

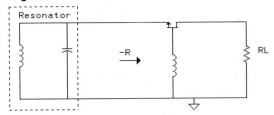

Figure 4.50 The two feedback mechanisms for LC resonant oscillators.

incident wave. The LC resonator selects the frequency and provides the right amount of phase shift without losing positive gain to maintain oscillation.

4.2.4.1   LC Feedback Oscillator

Figure 4.51 shows different LC feedback oscillator topologies. They can be derived from the basic ungrounded layout shown in the upper-left-hand corner. This figure shows a generalized ungrounded oscillator from which all the other oscillator configurations have been derived. These other configurations are achieved by selecting the grounding point (common drain, common source, and common gate) on the basic oscillator. These oscillator configurations are called Seiler, Pierce, and Colpitts. The Pierce has good out-of-band stability. Stable oscillation frequencies can be generated from low frequencies up to 1/3 f_t. The Colpitts topology has negative real-part impedance at its source from about 1/5 f_t to f_t, depending on the base-to-ground parasitic inductance. The Seiler topology with capacitive loading on its source has a negative real-part impedance at its gate over a significant range of frequencies. Negative resistance at frequencies other than the desired oscillation can cause spurious oscillation and sharply nonlinear tuning characteristics. These undesired responses can occur when a harmonic of the desired frequency crosses through a region of negative resistance.

Figure 4.52 shows an example circuit of a Pierce oscillator. Equations (4.10) through (4.13) compute the component values in the circuit:

$$L = \frac{190}{2\pi f} \tag{4.10}$$

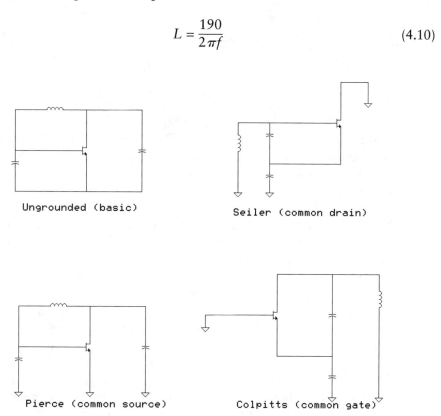

Ungrounded (basic)   Seiler (common drain)

Pierce (common source)   Colpitts (common gate)

Figure 4.51   Different LC oscillator feedback topologies.

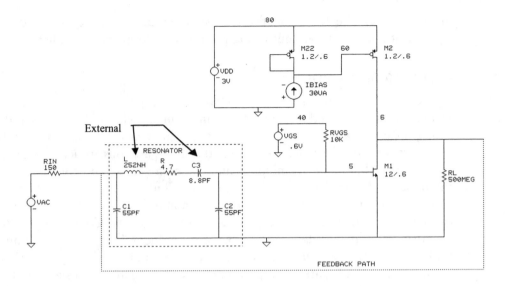

Figure 4.52 Example circuit of a Pierce oscillator.

$$C_1 = \frac{1}{48\pi f} \tag{4.11}$$

$$C_2 = C_1 \tag{4.12}$$

$$C_3 = \frac{1}{300\pi f} \tag{4.13}$$

An example will show how these equations are used. Let's design an oscillator with a 120-MHz output frequency. Equations (4.10) through (4.13) compute an L of 252 nH, C_1 of 55.26 pF, C_2 of 55.26 pF, and C_3 of 8.8 pF. The high inductance value makes it difficult for this circuit to use a monolithic inductor. A much higher oscillation frequency is required to reduce the value of the inductor so that it can be integrated.

Figure 4.53 shows the open-loop gain and phase of the example Pierce oscillator. This figure shows a >0-dB gain and 0° phase at 120 MHz. Furthermore, this figure shows an instability margin of over 40 dB at 120 MHz. Consequently, a 40-dB loss in the feedback would be necessary for the oscillator to quit. Finally, the zero phase shift occurs close to the maximum phase slope, which gives the best performance. Now, let's see how far we can tune the frequency.

Figure 4.54 shows the effect of a factor-of-10 reduction in the C_3 capacitor. The output frequency increases by 66% from 120 to 200 MHz. Internal spiral inductors have a range of 1 to 10 nH, Q of 1 to 10, 1–5-GHz frequency range, and a size of 100 to 400 μm on one side, which limits these oscillators to high frequencies [9]. In addition, there are some limitations on the variation of the capacitor, which has to maintain high Q with wide tuning range.

Monolithic integration of VCOs in a silicon-based IC requires high quality on-chip varactors. Conventionally, on-chip varactors have been implemented with *pn*

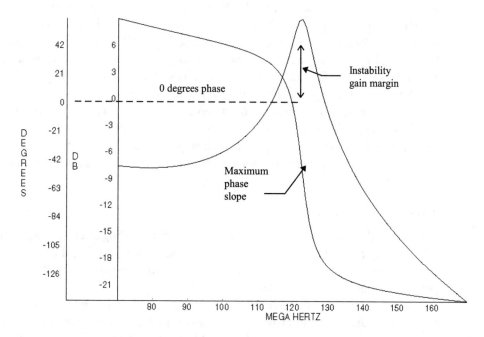

Figure 4.53 Open-loop gain and phase of the example Pierce oscillator.

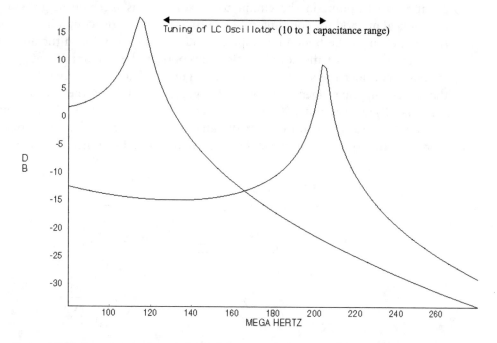

Figure 4.54 Open-loop gain and phase of the example Pierce oscillator with a factor-of-10 capacitance reduction.

junctions under reverse-bias or MOS capacitors in a depletion-inversion regime. A quality factor (Q) of less than 7 for capacitance of 1 to 10 pF at 0.9 to 2.4 GHz has been reported for pn-junction varactors. Typical MOS capacitors can achieve higher Q (14 GHz/pF) with larger capacitance per area [10]. References have shown 150% tuning for MOS capacitors with values of 0.4 to 1.4 pF at 1 to 5 GHz.

A large monolithic variable capacitor will not change by a factor of 10. Consequently, an external varactor would have to be used. Therefore, the LC Pierce oscillator has a limited tuning range of 50% to 100% and needs an external varactor and inductor. As a result of these requirements for an LC tank to meet a wide tuning range, these circuits have limited practical advantage for integration.

4.2.4.2 Output Coupling

Until now, the oscillator has been studied ignoring loading effects, which will occur when trying to couple energy out of the oscillator. Connecting the cascade output back to the input forms the oscillator circuit. No provisions have been shown to get the output from the circuit.

A number of circuit techniques provide a coupling method to get the output of the oscillator. For microwave oscillators that are not on integrated circuits, inserting a coupler or power splitter into the closed loop gives one path for the feedback and another path for the output. A low coupling value gives the greatest output power. For loops with a high-gain margin, the higher-loss coupled path closes the loop. This allows the output path to have the majority of power delivered to the load.

Another technique couples power by connecting a capacitor or inductor from an appropriate point in the circuit to the load. This mechanism gives a more economical method for coupling to the output; however, the coupling element and the load changes the Bode plot response and must be included in the analysis.

The reactance of the coupling element controls the amount of coupling. For example, consider the simple network in Figure 4.55(a). This network represents the series-coupling element and the load. Figure 4.55(b) shows the effective parallel resistance R_p and reactance X_p, which load the cascade. For small values of coupling reactance (a large capacitor or a small inductor), the reactance has little effect. The load directly connects to the circuit. Consequently, the output load has *tight*

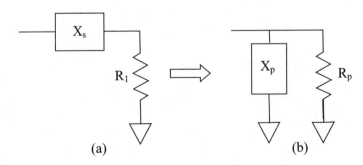

(a)   (b)

Figure 4.55 Converting (a) series coupling and loading to (b) effective parallel resistance and reactance.

coupling to the circuit. This technique works well for a high-impedance load from a buffer amplifier; however, tight coupling causes oscillator performance parameters to be heavily dependent on load characteristics and changes.

Loading from a low-impedance (50Ω) load reduces the gain margin. This reduction can cause the oscillator gain to go below unity and stop oscillations. Consequently, loading effects need to be studied to minimize them. Increasing the coupling reactance, X_S, reduces the resistive loading on the cascade by increasing R_p. Equations (4.14) through (4.17) relate the parallel and series values:

$$R_p = (R_l^2 + X_s^2)/R_l \tag{4.14}$$

$$X_p = (R_l^2 + X_s^2)/X_s \tag{4.15}$$

$$R_l = (X_p^2 R_p)/(R_p^2 + X_p^2) \tag{4.16}$$

$$X_s = (R_p^2 X_p)/(R_p^2 + X_p^2) \tag{4.17}$$

The coupling reactances in the equations above may be inductive or capacitive. A series-coupling reactance equal to the load resistance produces a parallel load resistance twice the load resistance. A coupling reactance of three times the load resistance produces a parallel load 10 times the load resistance. The amount of loading on the cascade by the coupling network depends on the impedance level at the coupled location. For example, an effective 500Ω load at a circuit location with an impedance level of 50Ω has a minimal effect on the gain margin or loaded Q; however, the same load across the resonator at a high impedance level in the oscillator may have a significant effect.

In monolithic circuits, a high impedance buffer amplifier couples out the oscillation. In this case, the gate capacitance of the buffer-amplifier transistor loads down the circuit. Consequently, this loading should be taken into account, especially for high-frequency oscillators.

4.2.4.3 Negative-Resistance Response

Understanding how a negative-resistance oscillator works is important in troubleshooting many applications where the oscillating feedback path is not clear. It is not obvious that a feedback path can exist from a single node connection; however, negative-feedback oscillator theory shows how this is possible.

A negative-resistance oscillator example illustrates how this oscillator works. The negative-resistance oscillator is connected to an external resonator on a PCB. An inductance in the gate of a transistor produces a negative resistance.

Figure 4.56 shows the SPICE schematic for the negative-resistance oscillator with the 0.72-μm gate-length process with a 5-V supply analog model for the transistor. This common-gate transistor topology is often used to build compact VCO circuits. The inductor L_2 adjusts the amount of negative resistance. The transistor and circuit values listed in Figure 4.56 are appropriate for an oscillator operating near 1 GHz. The series circuit $R_v - C_v$ represents the varactor device

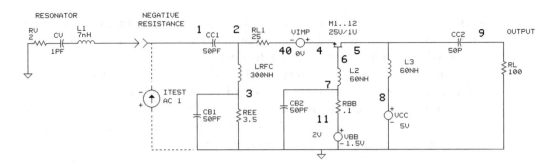

Figure 4.56 Schematic for negative-resistance oscillator.

that provides a variable capacitance for oscillator tuning. L_1 provides the inductance to complete the tank circuit on the left-hand side of the figure. The simple parallel combination of resistor R_1 and inductor L_3 represents the output circuit. This is usually an interstage or output amplifier circuit designed to meet specific conditions for load matching, harmonic rejection, or both. These more ambitious output circuits are not considered here.

Evaluating the impedance (or admittance) at the circuit plane between the negative resistance of the transistor on the right-hand side of Figure 4.56 and the LC tank on the left-hand side identifies the conditions for oscillation. The element Z_{amp} refers to the series impedance looking into the transistor source. The impedance Z_{res} refers to the complex impedance presented by the varactor and inductor. The threshold for circuit oscillation occurs when the impedance sum at the plane between the LC tank and the transistor is zero.

The sum of the real parts must be less then zero ($R_{res} + R_{neg} < 0$) and the sum of the imaginary parts must equal zero ($X_{res} + X_{amp} = 0$) to have an oscillation from negative resistance. At this point in threshold oscillation, circuit conditions are linear. So, a linear simulator like SPICE ac analysis can be utilized. While the equations must be precisely satisfied at the exact threshold of oscillation, a linear circuit analysis that finds net negative resistance over the desired frequency range will assure that the circuit is capable of oscillating over that range [11].

Figure 4.57 shows the real part of the impedance presented to the resonator from the transistor. It also shows that the resistance starts out positive at 500 MHz and goes negative at 900 MHz. This characteristic verifies the correctness of the simulation since a minus-sign error can occur easily. A plot with an error would never go negative. For real resistances less than the 2Ω in the resonator, the circuit has enough gain to oscillate. So this satisfies one of the conditions for oscillation.

Figure 4.58 shows a plot of the imaginary part for the amplifier and the resonator. At 1,150 MHz, the imaginary parts cancel out. Consequently, 1,150 MHz should be the frequency of oscillation. In addition, the simulations show that a net negative resistance is observed at various frequencies as the transistor gate circuit inductance L_2 is varied. Consequently, L_2 is varied to change the negative resistance to operate at frequencies in the desired operating range.

The previous figures have shown a negative resistance from 1 to 3 GHz. Consequently, any resonance between 1 and 3 GHz will cause this circuit either to oscillate or to have a long ringing response.

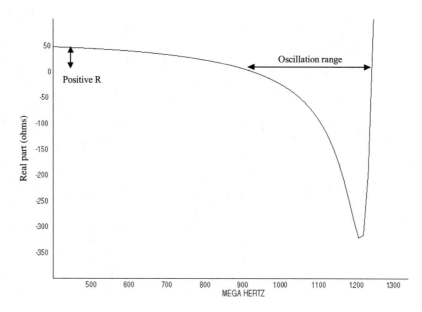

Figure 4.57   Plot of the real part of the impedance looking into the transistor and away from the resonator.

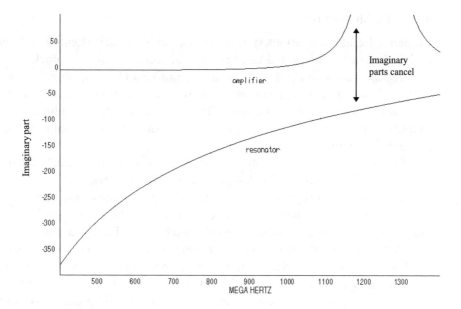

Figure 4.58   Plot of the imaginary parts of the impedances for the amplifier and the resonator.

Next, a transient analysis is done in SPICE for confirmation. Figure 4.59 shows the resulting (1-GHz) oscillations. This figure shows a 4.5-V output swing at an output frequency of approximately 1 GHz. At 0.5 ns, an initial impulse is given to the circuit to get the oscillation started, which models the effect of noise that would occur in silicon.

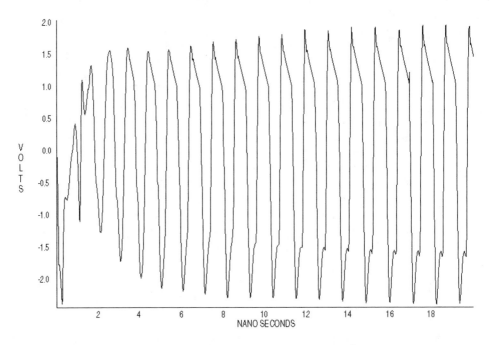

Figure 4.59   Startup transient response of the negative-resistance oscillator.

4.2.5   LC Multivibrators

Modern telecommunications systems, for example, cellular telephones, have caused an increase in RF integrated circuits. Several receiver building blocks can already be integrated on a single die. These include mixers [12, 13], low-noise amplifiers [14], and so forth. Frequency synthesizers used in modern telecommunications systems require very low phase noise. Therefore, in the design of high-performance frequency synthesizers using PLLs, the VCO becomes the key component in meeting phase-noise requirements.

The trend toward monolithic integration poses some major challenges. The toughest challenge lies in the design of a completely integrated, low-power, low-noise, wide-tuning-range VCO that achieves the low-phase-noise specifications of most mobile telecommunications systems. The very narrow channel spacings used in cellular telecommunications networks, such as the European Global System for Mobile Communications (GSM) system, make low phase-noise levels a requirement.

For example, one of the requirements specifies a single-sided spectral-noise power density of −100 dBc/Hz at 10 kHz from the carrier. At the moment, an LC-tuned oscillator with an external inductor achieves this phase-noise level. From the high-frequency requirements, this is the only type of oscillator that generates a pure enough signal to meet the spectral requirement.

Ring oscillators also operate at several gigahertz in silicon bipolar technologies [15, 16] or in the 900-MHz range in CMOS [17, 18]. But, in general, LC-tuned oscillators have much lower phase noise because of the bandpass filtering character-istic of the LC-tank that reduces the phase noise. Ring oscillators suffer from switching transients that introduce glitches into the power supply and have a Q of 1. Consequently, a ring oscillator has a worse phase noise than LC-tuned oscillators, which have low switching transients because of the LC tank resonance response and a Q greater than 10. The Q ratio of 10 to 1 gives the LC-tuned oscillator a 20-dB lower phase noise [19].The LC multivibrator has been the most popular choice of VCO for meeting GSM specifications. Consequently, we will analyze the characteristics of this VCO.

Figure 4.60 shows the basic circuit building block for the MOSFET LC oscillator. It has a cross-coupled pair of FETs (M_0, M_1) with an inductor load. The inductor resonates with the FET gate and drain junction capacitance to determine the oscillation frequency. The FET pair presents a negative resistance of $-1/g_m$ across the two drain terminals, which overcomes inductor loss. Saturation in the large-signal I-V characteristic of the FET pair limits the oscillation amplitude. Equation (4.18) computes the oscillation frequency:

$$\omega_o = 1/\sqrt{LC} \tag{4.18}$$

To oscillate, R must be greater than the abs$(1/g_m)$. So, if $1/g_m = 50\Omega$, then an R of 500Ω or greater would be necessary for oscillation. This assumes a 10-to-1 ratio padding to avoid losing oscillation over process, voltage, and temperature [20].

Let's mathematically describe the waveforms in the oscillator. Figure 4.61 shows the theoretical switching transient of waveforms in the oscillator and what they look like in the frequency domain. From doing Fourier series analysis, (4.19) computes the current at the fundamental frequency of oscillation:

$$I_1 = 2I_T/\pi \tag{4.19}$$

From Ohm's law, (4.20) computes the voltage across the load resistance at the fundamental frequency of oscillation:

$$V_1 = 2R \cdot 2I_T/\pi = 4RI_T/\pi \tag{4.20}$$

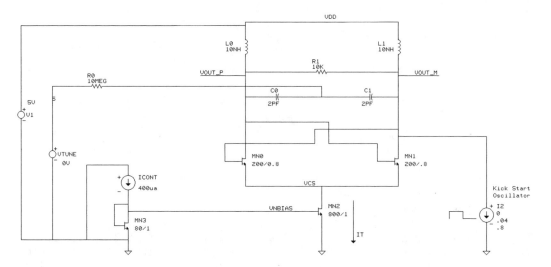

Figure 4.60 Schematic of LC multivibrator building block with SPICE test harness.

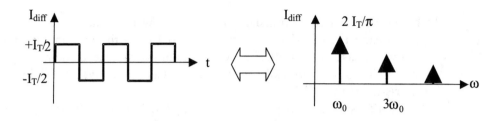

Figure 4.61 Time-domain and frequency-domain plots of the waveforms in the oscillator. (*From:* [20]. © 2001 Abidi. Reprinted with permission.)

Adding the dc bias to the output node gives (4.21) for the output voltage waveform [8]:

$$V_{out} = V_{dd} \pm 4RI_T/\pi \qquad (4.21)$$

Figure 4.62 shows the waveforms from the SPICE simulation of the oscillator. The top two waveforms show the current in the differential pair transistors. The current waveforms are opposite in phase. The bottom waveform shows the output of the VCO. The output shows a 1.4-V peak-to-peak voltage swing at a frequency of 1,050-MHz with a 5-V tune voltage. Setting the tune voltage to 0V produces an output frequency of 928 MHz. This computes a 24.4-MHz/V VCO gain [(1,050 MHz – 928 MHz)/5V].

Next, we look at the power-supply ripple rejection. Figure 4.63 shows the variation in period versus an injected sine wave on the supply with a frequency of

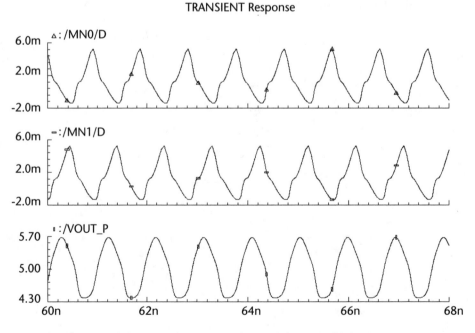

Figure 4.62 SPICE simulation time-domain waveforms of the LC multivibrator.

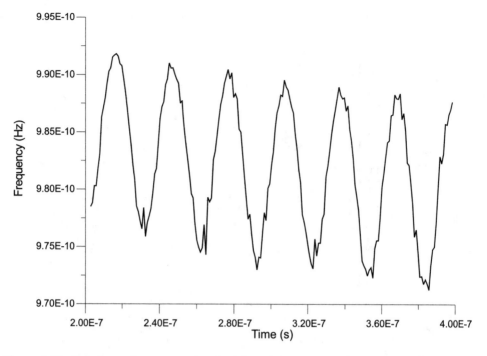

Figure 4.63   Period waveform versus time for injected sine wave on the power supply.

33 MHz and 1-V_{pp} amplitude. Noninteger harmonics of the output frequency are selected for injection to prevent injection locking of the VCO. The PLL is usually sensitive to frequencies between 1 and 20 MHz and especially around the 0-dB crossover frequency of the loop. With the application of the injected sine wave, Figure 4.63 shows that the output frequency varies from 1,008 to 1,025 MHz with 1-V_{pp} on the power supply. This computes to a 17 MHz/V [(1,025 − 1,008 MHz)/1V] power-supply sensitivity.

Next, we can compute the VCO phase-noise curve from simulation of the oscillator amplifier noise. Connecting the LC multivibrator as an amplifier allows measurement of the gain of the amplifier, the thermal noise floor, and the 3-dB corner-frequency location of the beginning of flicker noise. The example has a gain of 8 dB, a thermal noise floor of −167 dBV/Hz, and a flicker corner frequency of 2 MHz for 5V on the control line. The LC multivibrator has a Q of 10. The inductor Q is the main limitation on higher Q. Finally, the oscillator has an output frequency of 900 MHz. Substituting these values into (4.22) (Leeson's equation from Section 3.2.4) produces the VCO phase-noise curve in Figure 4.64:

$$\mathscr{L}(f) = \left(1 + \frac{f_l}{f}\right) \frac{10^{\frac{N_{amp}}{10}}}{\dfrac{V_{os}^2}{10^{\frac{G_n}{10}}}} \left[1 + \left(\frac{f_c}{2Q_l}\frac{1}{f}\right)^2\right] \tag{4.22}$$

From this equation phase-noise improvements can be determined. For example, lowering the gain of the delay element, increasing the gate width W, and increasing

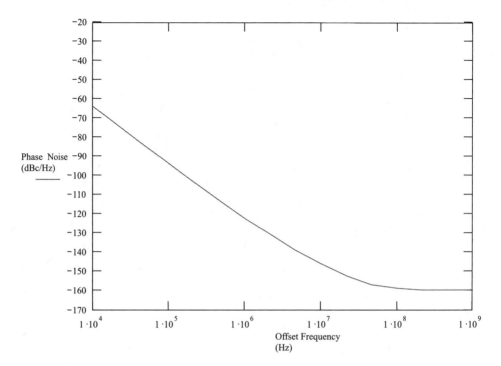

Figure 4.64 Phase-noise curve of the LC multivibrator from Leeson's equation.

the output voltage swing will reduce the far-from-the-carrier phase-noise levels. Increasing the gate length L will reduce the flicker corner frequency.

Figure 4.64 shows the phase-noise curve estimate using Leeson's equations from SPICE noise analysis of the oscillator amplifier. Figure 4.64 shows –135 dBc/Hz at 3-MHz offset frequency, which is close to the GSM specification of –139 dBc/Hz. The curve shows a 30-dB/dec slope with one break point at approximately 5 MHz. A break-point change from 30 to 20 dB/dec identifies the location of the flicker corner frequency. The curve slopes down until it hits the thermal noise floor at –160 dBc/Hz and about 80-MHz offset frequency.

To summarize, the last few sections showed several different types of oscillators that can be used for VCOs. Several key parameters for the VCOs are summarized in Table 4.2. Each VCO has advantages and disadvantages, but not one of them matches all the ideal requirements for a synthesizer. From the listing of these key parameters, it is easy to see that the selection of the VCO determines about 60% to 80% of the ideal requirements for a synthesizer. Consequently, we have to be very careful in selecting the best VCO to fit the application of interest.

4.3 Reference Oscillators

A frequency reference source is one of the most important elements used in frequency synthesis with a PLL. The reference source determines the stability and accuracy of the synthesized signal and contributes to phase noise (jitter) and spurious outputs associated with the signal.

Table 4.2   Summary of Key VCO Parameters

Type	Frequency Range (MHz)	Voltage Swing	Current Spikes (ma)	Linearity (%)	Gain	PSRR (MHz/V)	Far-Out Phase Noise (dBc/Hz)	Close-in Phase Noise at 1-MHz Offset (dBc/Hz)
Current starved single ring	3–110	5	1	30.9	20–150 MHz/V	150	–142	–97
Differential ring	200–500	0.2		2.2	840 MHz/V	90	–110	–95
Multivibrator	10–50	3	0.03	1.9	1.5 GHz/A	13.5	–132	–110
LC multivibrator	928–1,050	1.4			24 MHz/V	17	–160	–125

This section describes the basic design principles and operation of these devices. First, crystal oscillator specifications will be discussed. Then, desirable characteristics of the crystals themselves will be discussed. The design of the Pierce crystal oscillator with its design equations is shown. Equations for stability of the oscillator are presented and are followed by equations for the startup time. Then, the relationship between stability and startup time is shown.

A crystal resonator consists of a thin slab of quartz with electrodes plated on both sides. The piezoelectric effect links the mechanical and electrical properties of the resonator. A mechanical deflection of the quartz slab creates an electrical charge on the electrodes. Similarly, applying a voltage to the electrodes causes a deflection. Consequently, a sinusoidal electrode voltage produces a sinusoidal mechanical vibration in the crystal. Off resonance, the crystal response is small but increases greatly as the excitation frequency approaches the resonant frequency of the crystal. Crystal Q determines the magnitude of the increase. Actual displacement is miniscule—generally only a few atomic diameters. An external amplifier replaces energy dissipated by losses in the crystal in order to maintain oscillation [21].

A typical specification for a reference oscillator may consist of the following [22]:

- Center frequency;
- Long-term frequency stability (aging);
- Short-term frequency stability;
- Frequency stability with temperature and power-supply variations;
- Frequency accuracy;
- Warm-up time;
- Frequency tuning;
- Spurious outputs;
- Phase noise.

These specifications may have to be met by the PLL designer, or the designer will have to generate a specification for a vendor to build the oscillator.

The following discussion explains the less obvious specifications in more detail. *Long-term frequency stability* or *aging* specifies frequency drift over a 24-hour period. *Short-term frequency stability* specifies frequency drift over a 1-second period. A frequency accuracy specification with no restrictions assumes that the frequency is set at the time of shipment at room temperature, nominal power-supply voltage, and nominal load. The frequency accuracy decreases with longer times between applications of power to the circuit. Consequently, the accuracy of the oscillator when the user receives it is not the accuracy that was set at the factory. Determining this requirement is closely related to warm-up time and will be discussed next.

The response characteristic of the oven temperature-control circuit determines the *warm-up time* of a crystal oscillator placed in an oven. The selection of control-circuit parameters compromises between speed of response and circuit stability. An overdamped system has a stable slow response. An underdamped system has a fast response with ringing, which gives an unstable oven temperature and oscillator frequency until the ringing subsides. For fast warm-up time and good accuracy, the specification should include the accuracy to be achieved at a specific time after turn-on. After turn-on, limits on short-term frequency drift (ringing) thereafter should be specified. In addition, the length of the turn-off period that maintains the accuracy requirements should be specified. The oscillator oven usually has dc power applied at all times independently of the synthesizer operational status when the warm-up time, frequency accuracy, and turn-off period requirements cannot be simultaneously met.

Maintaining frequency accuracy over a long period may require periodic *frequency tuning* of the oscillator. Consequently, the resolution, total range, and type (manual or electronic) of the oscillator tuning mechanism should be specified.

From his own synthesis approach and the spurious-signal requirement for the synthesizer, a synthesizer designer should compute and specify the allowable level of *spurious output* signals from the reference oscillator. Familiarity with the approach used to derive the oscillator output frequency will help in making practical decisions about the spurious levels. Stringent requirements may make it impractical to filter all spurious signals out because of the limited space in the oven. Consequently, the designer may have to provide additional filtering of his own external to the oscillator. Finally, identifying the type of spurious signal, such as AM, FM, or non–harmonically related, may be helpful in designing the filter and help reduce cost.

Defining the phase-noise performance of the reference oscillator at offset frequencies of interest to the designer will help meet the phase-noise requirements of the overall synthesizer. If none of the standard models meets the requirements, narrowband crystal filter design can be used to do the additional filtering of the phase noise. Variation in phase noise for reference oscillators with equal performance and price has been found to vary by more than 6 dB from one manufacturer to the next [22]. The design technique for developing a low-phase-noise oscillator uses a very narrow-bandwidth crystal filter, followed by a low-noise amplifier.

Table 4.3 shows a list of reference oscillators from the least accurate and simplest circuit to the most accurate and most complicated circuit. The designer selects the accuracy required. Now let's list the other types of reference oscillators that are not shown in Table 4.3:

Table 4.3   Hierarchy of Reference Oscillators

Oscillator Type	Accuracy[a]	Typical Applications
Uncompensated crystal oscillator (XO)	1E–4 to 1E–5	Computer timing
Analog temperature-compensated crystal oscillator (TCXO)	1E–6	Frequency control in tactical radios
Microcomputer-compensated crystal oscillator (MCXO)	1E–7	Spread-spectrum (ECCM) system clocks
Oven-controlled crystal oscillator (OCXO)	1E–8	Navigation systems clock and frequency standard, MTI radar
Atomic frequency standard, rubidium-crystal oscillator (RbXO)	1E–9	Bistatic/multistatic radar, CCC satellite terminals

[a]Accuracy includes the effects of military environment and one year of aging.
[b]Sizes range from less than 1 in^3 to greater than 2,000 in^3.
[c]Costs range from less than $10 to more than $40,000.

- VCXO: Voltage-controlled crystal oscillator;
- TCVCXO: Temperature-compensated voltage-controlled crystal oscillator;
- OCVCXO: Oven-controlled voltage-controlled crystal oscillator.

Let's describe the reference-oscillator types. A simple XO does not provide a means for reducing the changing frequency-versus-temperature characteristics. In a VCXO, a voltage-controlled capacitor changes the phase around the loop, which changes the operating frequency of the oscillator. In a TCXO, the output signal from a thermister generates a voltage that is applied to a varactor in the crystal network. The reactance variations compensate for the frequency changes over the temperature of the crystal. This compensation provides up to twentyfold improvement over XOs. In an OCXO, a stable oven maintains the temperature of the crystal and other temperature-sensitive components. The oven temperature is set to where the crystal's frequency-versus-temperature curve has a zero slope. This compensation provides up to a thousandfold improvement over XOs. In an atomic frequency standard, a voltage-controlled crystal oscillator generates a spectrally pure 5-MHz output, which has its frequency stabilized by an atomic frequency discriminator. Using a narrow loop bandwidth preserves the low-noise spectrum of the VCXO and achieves exceptionally high degrees of accuracy and long-term stability, which closely approximates an atomic energy transition. This compensation provides a ten-thousandfold improvement over XOs.

Selection of a crystal requires studying key characteristics of their performance:

- Orientation and temperature coefficient:
  - AT versus BT cut;
  - SC cut;
- Overtones and spurious signals;
- Drive level;
- Aging.

Unlike amorphous solids, the physical properties of crystals depend upon the orientation in which they are measured. These materials are called *anisotropic*. For example, electrical conductivity can vary by a factor of 100 to 1, depending upon the direction of measurement. Temperature coefficient is the main concern for crystal orientation.

Two cuts (AT and BT) have a zero temperature coefficient over a temperature range near room temperature. AT-cut resonators follow a third-order curvature with a zero-crossing inflection point at 25C. Standard frequency versus temperature specifications are ±50 ppm from −20C to +70C. AT-cut crystals are used in fundamental (3.5 to 40 MHz) and overtone (40 to 70 MHz) modes [23].

Design specifications determine the plate thickness to be 66.4 mils/(output frequency in megahertz). Flexure and face shear coupling set the width-to-thickness design ratio. Process design limits of 0.02-inch thickness at low frequencies and 0.000166 inch thick at high frequency (40 MHz).

Furthermore, for overtone crystals, coupling restricts the upper frequency limit to 70 MHz. The harmonically and nonharmonically related spurious modes require tighter tolerances for repeatable device performance.

BT-cut resonator curves follow a parabolic curve with the maximum frequency nearly at room temperature. The frequency decreases at both high and lower temperature. Standard frequency versus temperature specifications are (150 ppm from −20C to +70C. BT-cut crystals are only used in the fundamental mode (30 to 45 MHz), are lower in cost than the AT-cut crystals, and are 1.521 times thicker.

Inverted mesa crystals use an etching technique to produce bilevel resonators. The thicker lapped portion is used for mounting support. The thinner etched portion contains the electrode. Fundamental crystals have frequency ranges from 60 to 120 MHz, and third overtones have a frequency range up to 350 MHz.

For ovenized oscillators, SC-cut quartz has improved performance over an AT-cut crystal. The SC quartz has improved frequency versus temperature coefficients (±5 ppm) and a higher turn point temperature (95C), and has less sensitivity to abrupt temperature changes.

Desirable crystal characteristics should include low-resistance crystals for better stability and reliability, but they are more costly than high-resistance parts. Next, load capacitance is specified as the amount of capacitance placed in parallel with the lead of a crystal that will cause the oscillator to operate. Stray capacitance around the board and the package capacitance contribute to the load. Typical values are 5–10 pF. Proper selection of C_1 and C_2 requires that these stray capacitances be considered for all calculations.

Crystal drive (given in milliwatts) represents the energy dissipated by losses in the crystal. Overdrive of the crystal causes excessive heating that changes the crystal temperature and frequency. Further increase in the drive increases the mechanical movement of the crystal. Ultimately, the crystal fractures. AT-cut crystal frequency rises slightly with drive level, and BT-cut frequency lowers slightly with drive level. Furthermore, AT-cut crystals have a tendency to mode-hop or break up over temperature at drive levels of approximately 2 mW. Typical AT-cut crystals have drive levels in the 10–100-mW region. In contrast, SC-cut crystals have drive levels as high as 8 mW without difficulty. This higher drive level allows far-from-the-carrier phase-noise improvement over AT-cut crystals. The increased drive level in SC-cut crystals improves the far-from-the-carrier phase noise by 10 dB.

Movement of mass to or from the crystal blank and the relaxation of stresses cause crystal aging (a slow, long-term frequency drift). Cleanliness improves the first problem. Extensive cleaning and handling procedures, a clean production environment, gold-plated housings, and low out-gassing bonding materials all contribute to low aging rates. Aging is also a function of blank mass. For this reason, a 10-MHz crystal has better aging than a 100-MHz crystal.

4.3.1   Oscillator Circuits, Stability, and Startup Time

The Pierce crystal oscillator is widely used among CMOS designers. Understanding the conditions for oscillation helps designers adjust their designs to maintain oscillations over voltage, process, and temperature variations. In addition, IC designers need to know when the output of the crystal oscillator from power up is high enough to start reliably running the software on the chip.

Figure 4.65 shows the typical Pierce crystal oscillator schematic in CMOS technology. Figure 4.66 shows the small-signal equivalent circuit for the oscillator.

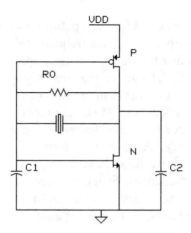

Figure 4.65 Pierce crystal oscillator schematic in CMOS technology. (*From:* [24]. © 2001 IEEE. Reprinted with permission.)

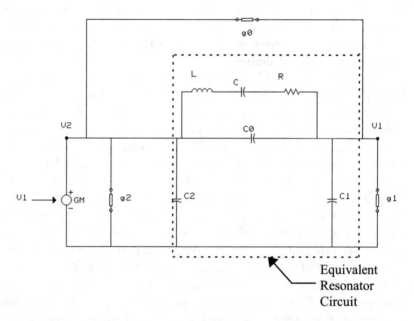

Figure 4.66 Small-signal equivalent circuit of oscillator. (*From:* [24]. © 2001 IEEE. Reprinted with permission.)

In the equivalent circuit, L and C are the motional components of the resonator, and R represents the losses. C_1 and C_2 are the capacitances necessary to form the oscillator. C_0 comprises the capacitances parallel to the resonator. The transconductance g_m in Figure 4.66 is the sum of the transconductances of the driver transistors p and n. The conductance g_0 represents the conductance of the feedback resistor R_0. The conductance g_1 represents the input conductance of the drive transistors. For a CMOS inverter, the input conductance is very low and can be neglected in the analysis. The conductance g_2 represents the output conductance of the drive transistors [24].

4.3.2   Equations for Oscillation

Equations are now presented that model the current as a function of time in the oscillator, determine the stability of the oscillation, and compute the startup time. First, equations for the current as a function of time are presented. Equation (4.23) shows the La Place transform of the current in the oscillator circuit:

$$i(s) = \frac{b_0 s^2 + b_1 s + b_2}{d_0 s^3 + d_1 s^2 + d_2 s + d_3} \qquad (4.23)$$

where

$$b_0 = 1$$

$$b_1 = \frac{R}{L} + s_1 + \frac{di(0)}{i(0)}$$

$$b_2 = \left[\frac{R}{L} + \frac{di(0)}{i(0)}\right] s_1$$

$$d_0 = 1$$

$$d_1 = \frac{R}{L} + s_1$$

$$d_2 = \frac{1}{LC_p} + \frac{R}{L} s_1$$

$$d_3 = \frac{s_1}{L}\left(\frac{1}{C_0} + \frac{1}{C}\right)$$

$$C_T = C_0 + \frac{C_1 C_2}{C_1 + C_2}$$

$$\frac{1}{C_p} = \frac{1}{C} + \frac{1}{C_T}$$

$$s_1 = \frac{g_m}{C_1 + C_2} \frac{C_0}{C_T}$$

The current as a function of time is described by the inverse transform of (4.23). The solution uses the method of Cardan [24]. Equation (4.24) shows the general form of the solution:

$$i(t) = i(0)\left\{k_1 e^{s_a t} + e^{\alpha t}[k_2 \sin(\beta t) + k_3 \cos(\beta t)]\right\} \qquad (4.24)$$

where

$$\beta = \frac{1}{\sqrt{LC_p}};$$

α = exponential growth coefficient.

The values of k_1, k_2, and k_3 depend on the initial conditions for current and the initial derivative of the current. Beta is the radian frequency of oscillation. The first term decays in time, and the exponential evolution of the second term is characterized by the coefficient α. If $\alpha \leq 0$, the circuit will not oscillate [21].

4.3.3 Stability of Oscillation

In this section, we will determine if the conditions for oscillation exist in the circuit. If the circuit oscillates, we will also determine how close the circuit is to not oscillating. An example will help in the explanation. Table 4.4 shows the values that a crystal manufacturer will give you. Most crystal manufacturers will not give inductance information. Consequently, (4.25) is used to make the calculations:

$$L = \frac{1}{\omega^2 C} \tag{4.25}$$

Table 4.5 shows the values that semiconductor foundries will give for an amplifier.

Negative-resistance analysis of an oscillator is an impedance method for computing the stability of an oscillation. Rearranging the small-signal equivalent circuit for the Pierce oscillator in Figure 4.66 produces the equivalent circuit for negative-resistance analysis in Figure 4.67. This figure shows two impedance groups. One impedance group, which we will designate as the resonance group, consists of only the crystal resonance components R, L, and C. All other components, including the amplifier, are in the second impedance group, which we will designate as the amplifier group.

The impedance of the amplifier group consists of a negative imaginary part indicating net capacitance and a negative real part. The negative real part, which is a synonym for the gain of the amplifier, indicates the potential for the circuit to oscillate. If the magnitude of the negative resistance at the frequency of resonance is greater than the crystal resistance, the net resistance will be negative. The net negative resistance causes an oscillation at the frequency where the inductance of the crystal cancels the capacitance of the surrounding circuitry. Equations (4.26) and (4.27) express these conditions [25]:

$$\text{Re } Z + \text{Re } Z_e = 0 \tag{4.26}$$

Table 4.4 Example of Crystal Parameters from a Vendor

R	C_o	C_1	C	f
75	3 pF	10 pF	8 fF	4 MHz

Table 4.5 Example of ASIC Amplifier Parameters from a Semiconductor Foundry

g_m	g_2	R_o	V_{dd}	V_t
10 μa/V	10E–6 mhos	1.0 MΩ	5V	0.75

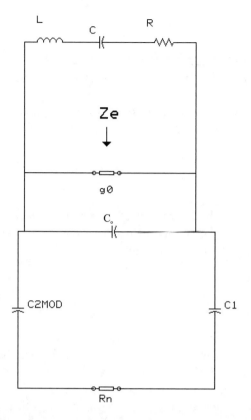

Figure 4.67 Equivalent circuit for Z (resonator group) and Z_e (amplifier group). (*From:* [21]. © 2001 IEEE. Reprinted with permission.)

$$\text{Im } Z + \text{Im } Z_e = 0 \tag{4.27}$$

Now equations are presented to compute the impedance of the two groups. First, some terms are defined. Equation (4.28) computes the modified C_2 capacitance in Z_e:

$$C_{2\,mod} = C_2 \frac{(\omega C_2)^2 + g_2^2}{(\omega C_2)^2 + g_m g_2 \dfrac{C_2}{C_1}} \tag{4.28}$$

Equation (4.29) computes the negative resistance:

$$R_n = \frac{g_m \dfrac{C_2}{C_1} - g_2}{(\omega C_2)^2 + g_2^2} - 1 \tag{4.29}$$

With $C_{2\,mod}$ and R_n defined, (4.30) computes the real part of the impedance Z_e presented to the resonator:

$$\text{Re } Z_e = \tag{4.30}$$

$$\frac{\omega^2 (C_1 C_{2mod})^2 R_n (1 + R_n g_0) + g_0 (C_1 + C_{2mod})^2}{[-\omega^2 C_0 C_1 C_{2mod} R_n + g_0 (C_1 + C_{2mod})]^2 + \omega^2 [C_0 (C_1 + C_{2mod}) + C_1 C_{2mod}(1 + R_n g_0)]^2}$$

With the real part of the amplifier group computed, (4.31) computes the inverse of the exponential coefficient for the growth of oscillation:

$$\alpha = \frac{R + |\text{Re } Z_e|}{2L} \tag{4.31}$$

The coefficient alpha characterizes the evolution of the oscillation current. Higher alphas mean faster growth and faster startup time. Equation (4.26) computes the time constant for the growth factor:

$$\tau = \frac{1}{\alpha} \tag{4.32}$$

Table 4.6 shows the calculated values for the example.

Figure 4.68 shows the variation of growth coefficient alpha with g_m in (4.24) and (4.25). This figure shows several characteristics about oscillators. First, the oscillator will not start for an alpha of zero. The figure shows a zero value for alpha at low g_m and high g_m. Consequently, too little gain or too much gain will prevent oscillation. In addition, the fastest startup time is for a g_m at the optimum

Table 4.6 Example Calculated Values for the Exponential Growth Coefficient

L	C_{2mod}	R_n	Re Z_e	α	τ
200 mH	9.9 pF	−160Ω	−37	280	4 ms

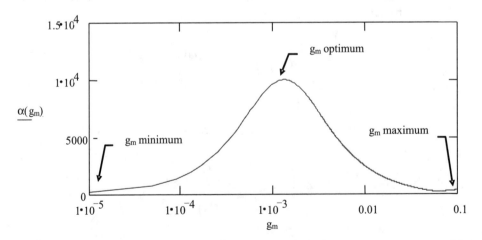

Figure 4.68 Variation of the exponential growth coefficient with the transconductance of the amplifier.

point. Finally, maximizing alpha and operating at the optimum g_m provides the greatest margin for maintaining a working oscillator over temperature and process [24].

For design purposes, it is easier to compute g_m minimum, maximum, and optimum rather than generating an alpha-versus-g_m curve. Equation (4.33) computes the total parallel capacitance to the resonator:

$$C_t = C_o + \frac{C_1 C_2}{C_1 + C_2} \tag{4.33}$$

Substituting (4.33) into (4.34) computes the minimum amplifier g_m for oscillation:

$$g_{m\_min} = \frac{(C_1 + C_2)^2}{C_1 C_2} (\omega C_t)^2 R \tag{4.34}$$

Equation (4.35) computes the maximum amplifier g_m for oscillation:

$$g_{m\_max} = \frac{1}{R} \frac{C_1 C_2}{C_o^2} \tag{4.35}$$

Equation (4.36) computes the capacitance (along with the inductance) that determines the resonant frequency of the circuit:

$$C_p = \frac{C C_t}{C + C_t} \tag{4.36}$$

Substituting (4.36) into (4.37) computes the optimum amplifier g_m for oscillation [24]:

$$g_{m\_opt} = \frac{C_1 + C_2}{\sqrt{L C_p}} \frac{C_t}{C_o} \tag{4.37}$$

Using (4.33) through (4.37) with the example oscillator produces the results in Table 4.7. In this example, the oscillator is close to not working because it is close to g_m minimum.

4.3.4 Startup Time

Startup time, as shown in Figure 4.69, consists of the amount of time for the amplifier to start $(t_1 + t_s)$ and the amount of time for the oscillations to grow to

Table 4.7 Example Calculation Results for the Minimum, Maximum, and Optimum Amplifier Transconductance

C_t	g_{m_min}	g_{m_max}	C_p	g_{m_opt}
8 pF	12 $\mu a/V$	148 ma/V	8 fF	1 ma/V

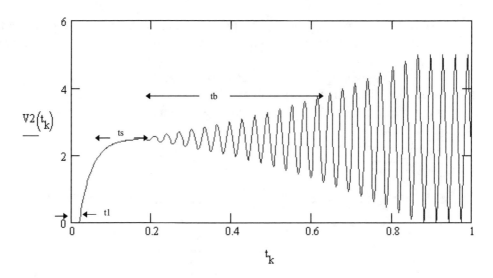

Figure 4.69   Illustration of the three characteristic time intervals for the startup of an oscillator.

half the supply voltage peak to peak (t_b). Without the amplifier in operation, there is no growth in oscillations. Consequently, equations are presented to compute the startup time for the amplifier. Equations are presented for the amplifier startup conditions of $V_1(0) = V_{dd}$ and $V_2(0) = 0$. Another possible startup condition is $V_1(0) = 0$ and $V_2(0) = 0$. The time response shows a transition of the output from V_{dd} to 0 that is not shown in the measured data. For convenience, (4.38) normalizes the voltages:

$$x_t = \frac{V_t}{V_{dd}} \tag{4.38}$$

On startup from $t = 0$, the capacitors C_o and C_1 begin to be discharged by R_o and have an RC time constant. From $t = 0$ to t_1, the n and p channel transistors do not conduct. After t_1, the n and p channel transistors conduct, and the p channel transistor is in saturation, while the n channel transistor is not in saturation. Equation (4.39) computes time t_1:

$$t_1 = R_o(C_o + C_1) \ln\left(\frac{1}{1 - x_t}\right) \tag{4.39}$$

At time t_s after t_1, the n channel transistor enters the saturation region, and the amplifier is operational. Equation (4.40) computes the time at which the n channel transistor enters saturation:

$$t_s = R_o C_1 \left[1 - \left(\frac{C_o}{C_1} - \frac{x_t}{1 - 2x_t}\right) \ln\left(\frac{x_t}{1 - x_t}\right)\right] \tag{4.40}$$

Summing the two previous equations, (4.41) computes the total amplifier turn-on time to operational status:

$$t_{amp} = t_1 + t_s \qquad (4.41)$$

Now, let's study the equations for the exponential growth factor of the oscillation. Equation (4.42) computes the time constant for the time period when the p and n channel transistors are not conducting:

$$t_o = R_o C_o \qquad (4.42)$$

Entering the results of (4.42) into (4.43) computes the initial current in the resonator when the n channel transistor enters the saturation region:

$$i_o = \frac{V_t C}{\sqrt{LC + t_o^2}} \qquad (4.43)$$

Equation (4.44) computes the time needed by the amplitude from initial conditions to reach the level of $V_{dd}/2$:

$$t_b = \tau \ln\left(\frac{V_{dd}\,\omega C_t}{i_o}\right) - \tau \ln(2) \qquad (4.44)$$

where

$\tau =$ time constant of the growth factor [(4.32)].

The level of $V_{dd}/2$ is chosen because most flip-flops are designed to begin toggling at this amplitude level. Equation (4.45) computes the total startup time due to the amplifier and the resonator [24]:

$$t_{tot} = t_{amp} + t_b \qquad (4.45)$$

Table 4.8 shows the calculated values for the example.

A second example is used to do calculations and compare the results with measured data. For the second example, a 32-MHz dielectric resonator is used with a 74 HCU04 amplifier. Table 4.9 shows the parameters provided by the resonator vendor, and Table 4.10 shows the parameters for the amplifier. Substituting the values into the oscillator equation produces time characteristics of t_1 equal to 1.1 μs, a t_s equal to 12 μs, and a t_b equal to 87 μs for a total response of

Table 4.8 Example Calculation Results for Startup Time

t_o	i_o	x_t	t_1	t_s	t_{amp}	t_b	t_{tot}
4.5 μs	1.3 nA	0.15	3.2 μs	17.2 μs	20.4 μs	46 ms	46 ms

Table 4.9 Example Two of Resonator Parameters from a Vendor

R	C_o	C_1	C	f
24.8 ohms	6 pF	12 pF	55 fF	32 MHz

Table 4.10 Example Two of Amplifier Parameters for a 74 HCU04

g_m	g_2	R_o	V_{dd}	V_t
20 ma/V	500E–6 mhos	1.0 MΩ	5V	0.75

100.1 μs. Figure 4.70 shows a startup time of approximately 90 μs, which is a good correlation.

One of the significances of this correlation is that the startup characteristics can be used to troubleshoot oscillators. For example, increasing startup time over a lot of devices shows a degradation in margin to oscillate. This characteristic can be monitored to make sure degraded devices do not get into the final product. Another example would be oscillators that fail to start up at all. It is difficult to troubleshoot circuits when there is nothing to measure. Consequently, a switch can be added for test purposes to increase the amplifier's gain significantly to force the oscillator to start, even though there might be something wrong with the crystal. Once the oscillator starts in this test mode, the measurement time for starting up can be measured, and the amount of degradation can be determined.

Next, let's look at the open-loop response of a crystal oscillator. The advantage of simulating this plot is that it can be verified in the laboratory. Figure 4.71 shows a open-loop plot of an example 4-MHz crystal oscillator. The 180° of the amplifier

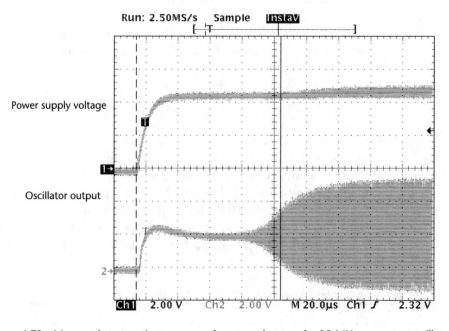

Figure 4.70 Measured startup-time response for example two of a 32-MHz resonator oscillator.

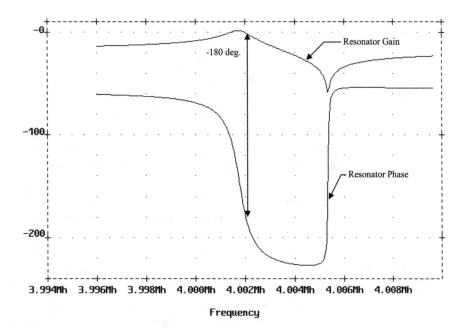

Figure 4.71   Open-loop ac plot of a 4-MHz crystal oscillator.

and the 180° indicated on the figure at 4.002 MHz combine to give 360° phase shift to meet the phase portion of the criterion to oscillate. Consequently, for a gain greater than 1, the circuit will oscillate at this point.

4.4 Summary

This chapter discussed detailed designs of programmable dividers, VCOs, and crystal oscillators. Design techniques of programmable dividers were discussed. Programmable dividers, pulse-swallowing counters, and fractional dividers were the type of dividers covered. Techniques were shown to prevent a PLL from hanging up on one of the power rails. A high-speed dual modulus prescaler with the pulse-swallowing technique isolates the high-frequency performance to a few D flip-flops. Optimizing the speed of these D flip-flops improves the performance of the whole chip.

This chapter showed that VCO performance accounts for most of a PLL's characteristics. An example current-starved ring oscillator was characterized to show the performance trade-offs. The ring was shown to have a very nonlinear tune-voltage-versus-frequency curve that restricted its frequency range and could cause moding. Then, the ring oscillator's phase-noise curve from simulation of the oscillator amplifier noise was computed. Next, the availability of quadrature outputs and 50% duty cycle were shown to make differential ring oscillators desirable for clock-recovery circuits, and these oscillators have many applications in telecommunications. However, it was shown that the voltage drop across the tail-current transistor makes the output have a low output voltage swing, which can cause edges to be skipped by the output differential-to-single-ended converter.

Multivibrators were shown to provide wide frequency range and high-speed operation because the effects of parasitic components were minimized; however, it was shown that a multioscillator condition can occur, and the current source can generate high switching current spikes that can cause interference with other circuits. Then, LC resonant oscillators were shown to rely on feedback control through a resonator to generate oscillations. From this feedback-control theory, a negative-resistance LC resonant oscillator was analyzed to show that a feedback path can exist from a single node connection. Understanding this concept is important in troubleshooting many applications where the oscillating feedback path is not clear. Next, the LC multivibrator was described as the most popular choice of VCO for meeting GSM specifications. Consequently, the characteristics of this VCO were analyzed, and it was shown to meet the low-phase-noise GSM requirements.

Crystal-oscillator design techniques were covered. This chapter showed that crystal reference oscillators are being incorporated into integrated circuits with the PLLs. The crystal reference oscillator is one of the more important elements used in frequency synthesis with a PLL because it determines the stability and accuracy of the synthesized signal and contributes to phase noise (jitter) and spurious outputs associated with the signal. This chapter showed the relationship of increasing startup time with degrading the margin to oscillate, which can be used to troubleshoot oscillators. The open-loop response simulation of a crystal oscillator was explored because it can easily be used to verify the performance of the oscillator in the laboratory. Techniques for maintaining oscillation and improving startup time were also discussed.

Questions

4.1 What limits the speed of a programmable divider?

4.2 Why should the input frequency be ramped to test frequency dividers?

4.3 What does the dual modulus prescaler need to be for a pulse-swallowing divider to have a programmable divider working up to 100 MHz, divide ratios of 2 to 16, a maximum locked output frequency of 1 GHz, and a maximum VCO frequency of 1.35 GHz?

4.4 What can you do to get divide-by-4, -5, -6, and -7 for a pulse-swallowing divider with the programmable divider of 2-to-16 divide ratios, pulse-swallowing counter divide ratios of 0 to 3, and a divide-by-4/5 dual modulus prescaler?

4.5 What is the divide ratio, output frequency, and loop natural frequency for a fractional-N loop with a divide-by-8/9 dual modulus prescaler, a 2-to-16 divide ratio programmable divider, an 8-bit accumulator for the fractional part, 33 for the word into the accumulator, a 1-MHz reference frequency, and a 100-MHz maximum frequency for the VCO?

4.6 What is the output frequency for a three-gate ring oscillator with a delay stage of 100 ps?

4.7 What is the output frequency for an eleven-gate ring oscillator with a delay stage of 100 ps?

4.8 What is the output frequency for a three-gate ring oscillator with a delay stage of 1,000 ps?

4.9 What is the output frequency for a four-gate differential ring oscillator with a delay stage of 800 ps?

4.10 What is the output frequency for a four-gate differential ring oscillator with a delay stage of 300 ps?

4.11 What capacitance is needed for a multivibrator at 160 MHz and 1,000 μa of current? Given $dv = 1.8$V and $dt = 1$ ns, what is the VCO gain in Hz/μa at 10 μa and 1,000 μa?

4.12 What capacitance is needed for a multivibrator at 900 MHz and 8,000 μa of current? Given $dv = 1.8$V and $dt = 0.4$ ns, what is the VCO gain in Hz/μa at 8,000 μa?

4.13 Design a Pierce LC feedback oscillator with a 300-MHz output frequency. Using (4.9) through (4.12), compute L, C_1, C_2, and C_3. Can this oscillator be fully integrated?

4.14 Design a Pierce LC feedback oscillator with a 10-nH monolithic inductor. Calculate the output frequency and components C_1, C_2, and C_3.

4.15 What is the startup time, g_m minimum, g_m maximum, and optimum g_m for a 20-MHz crystal oscillator, given an amplifier circuit with g_m of 3.74 ma/v, g_2 of 271.7 μmohs, V_{dd} of 5V, V_t of 0.82V, R_o of 300k, C_1 and C_2 of 10 pF, and a crystal with C_o of 4.7 pF, L of 3.9 mH, R of 7.9Ω, and C of 16.24 fF?

4.16 What is the startup time, g_m minimum, g_m maximum, and optimum g_m for a 32-MHz resonator oscillator, given an amplifier circuit with g_m of 20 ma/V, g_2 of 5 μmohs, V_{dd} of 5V, V_t of 0.75 V, R_o of 300k, C_1 and C_2 of 12 pF, and a crystal with C_o of 6 pF, L of 320 μH, R of 14Ω, and a C of 77.3 fF?

4.17 What is the startup time, g_m minimum, g_m maximum, and optimum g_m for a 32-kHz crystal oscillator, given an amplifier circuit with g_m of 5 μa/V, g_2 of 33.3 nmohs, V_{dd} of 4V, V_t of 0.8 V, R_o of 6,000k, C_1 and C_2 of 9 pF, and a crystal with C_o of 1.5 pF, L of 11.8 kH, R of 32 kΩ, and a C of 1.99 fF.

References

[1] Wiest, N. H. E., and K. Eshraghian, *Principles of CMOS VLSI Design*, Reading, MA: Addison-Wesley, 1993, pp. 319–321.

[2] "Making Programmable UHF Counters When None Are Available or . . . Pulse Swallowing Revisited," *Fairchild Journal of Semiconductor Progress*, Vol. 3, No. 4.

[3] Best, R. E., *Phase-Locked Loops: Design, Simulation, and Applications*, New York: McGraw-Hill, 1997.

[4] Stilwell, J., "A Flexible Fractional-N Frequency Synthesizer for Digital RF Communications," *RF Design*, February 1993, pp. 39–43.

[5] Kim, B., D. Helman, and P. Gray, "A 30-MHz Hybrid Analog/Digital Clock Recovery Circuit in 2-μm CMOS," *IEEE Journal of Solid State Circuits*, Vol. SC-25, December 1990, pp. 1385–1394.

[6]   Goldman, S., *Phase Noise Analysis in Radar Systems*, New York: Wiley Interscience, 1989, pp. 15–19.

[7]   Banu, M., "MOS Oscillators with Multi-Decade Tuning Range and Gigahertz Maximum Speed," *IEEE Journal of Solid State Circuits*, Vol. SC23, April 1988, pp. 474–479.

[8]   Razavi, B., *Monolithic Phase-Locked Loops and Clock Recovery Circuits*, New York: IEEE Press, 1996, pp. 17–21.

[9]   Lee, T., "Recent Developments in CMOS RF Integrated Circuits," Presentation at Texas Instruments, Dallas, TX, March 2000.

[10]   Soorapanth, T., et al., "Analysis and Optimization of Accumulation-Mode Varactor for RF ICs," *Symposium on VLSI Circuits Digest of Technical Papers*, 1998, pp. 32–33.

[11]   Pitalis, O., and T. Reeder, "Nonlinear CAE Software Optimizes Oscillator Design," *RF Design*, April 1990, pp. 29–31.

[12]   Abidi, A. A., "900-MHz Downconversion Mixer," *Proc. 1993 Euro. Solid-State Circuits Conf.*, Sevilla, Spain, September 1993, pp. 210–213.

[13]   Crols, J., and M. Styaert, "A Full CMOS 1.5-GHz Highly Linear Broad-Band Down-conversion Mixer," *Proc. 1994 Euro. Solid-State Circuits Conf.*, Ulm, Germany, September 1994, pp. 248–251.

[14]   Chang, J. Y.-C., and A. A. Abidi, "A 750-MHz RF Amplifier in 2-μm CMOS," *Tech. Dig. 1992 Symp. VLSI Circuits*, May 1993, pp. 111–112.

[15]   Razavi, B., and J. Sung, "A 6-GHz 60-mW BiCMOS Phase Locked Loop with 2-V Supply," *ISSCC Dig. Tech. Papers*, February 1994, pp. 114–115.

[16]   Pottbacker, A., and U. Langmann, "An 8-GHz Silicon Bipolar Clock Recovery and Data-Regenerator IC," *ISSCC Dig. Tech. Papers*, February 1994, pp. 116–117.

[17]   Thamsirianunt, M., and T. A. Kwasniewski, "A 1.2-μm CMOS Implementation of a Low-Power 900-MHz Mobile Radio Frequency Synthesizer," *Proc. IEEE 1994 Custom Integrated Circuits Conf.*, May 1994, pp. 383–386.

[18]   Min, J., et al., "An All-CMOS Architecture for a Low-Power Frequency-Hopped 900-MHz Spread Spectrum Transceiver," *Proc. IEEE 1994 Custom Integrated. Circuits Conf.*, May 1994, pp. 379–382.

[19]   Craninckx, J., and M. Steyaert, "Low-Noise Voltage-Controlled Oscillators Using Enhanced LC-Tanks," *IEEE Transactions on Circuits and Systems*, Vol. 42, No. 12, December 1995, pp. 794–804.

[20]   Rael, J., and A. Abidi, "How to Design an LC Oscillator First Time Right," *IEEE ISSC Conference*, February 8, 2001.

[21]   Long, B. R., "Quartz Crystals and Oscillators," *RF Expo 1989*, Santa Clara, CA, February 14–16.

[22]   Manassewitsch, V., *Frequency Synthesizers: Theory and Design*, 3rd ed., New York: John Wiley and Sons, 1987, pp. 522–531.

[23]   Veeser, M. A., "ASIC Oscillator Cells Reduce System Timing Costs," *EDN*, July 20, 1995, pp. 81–86.

[24]   Rusznyak, A., "Start-Up Time of CMOS Oscillators," *IEEE Transactions on Circuits and Systems*, March 1987, pp. 259–268.

[25]   Vittoz, E. A., M. G. R. Degrauwe, and S. Bitz, "High-Performance Crystal Oscillator Circuits: Theory and Application," *IEEE Journal of Solid State Circuits*, Vol. 23, No. 3, June 1988, pp. 774–782.

CHAPTER 5

Components, Part 2—Detectors and Other Circuits

In Chapter 4, we discussed dividers and oscillators. In this chapter we discuss detectors and other circuits that support PLLs. Detailed designs of phase detectors, lock detectors, and acquisition aids are studied.

Phase detectors are studied because understanding how phase detectors work is one of the major keys to understanding how PLLs work. Many systems use the lock detector to reset the system. This is a disastrous change to the operation of most systems. In a PLL, a reset can start the loop operating at a very low frequency (or with no output), which will then acquire lock at the normally much higher output operating frequency. Consequently, a small phase shift in the PLL that would marginally affect the system can cause a huge disruption in the operation of the system if a reset occurs. The key to lock detection is to alarm on behavior that shows that the PLL is broken. Quadrature phase detection, time-window edge comparison, tune-voltage window comparator, and cycle-slip detection are the lock-detection methods that are covered.

Open-loop sweep, closed-loop sweep, and discriminator-aided acquisitions are the methods that are covered. The phase/frequency detector uses discriminator-aided acquisition and is the most popular choice. Clock-recovery circuits cannot use phase/frequency detectors. Consequently, these circuits require an acquisition aid. Understanding the design details and trade-offs in these components is critical in designing monolithic PLLs.

A charge pump and operational amplifier provide a method to convert the differential up and down outputs of the phase/frequency detector to a single-ended positive and negative voltage source that charges and discharges capacitors in order to control the VCO in a PLL. These components also are used in synthesizing the loop compensation. For this function, the component that most closely models an ideal integrator is the best choice. Several different charge pump architectures are presented, and trade-offs between each are discussed. Then, operational amplifiers are presented. The evolution of operational amplifier architectures to the one that best fits a PLL is shown. Finally, trade-offs between charge pump and operational amplifier compensation designs are studied.

5.1 Phase Detectors

The phase detector in a PLL compares the reference-oscillator frequency with the VCO frequency and develops an error voltage for the loop to process. Consequently, phase detector components have a critical impact on the performance of a PLL. Tracking range, acquisition range, loop gain, and transient response depend on the characteristics of the phase detector. The major characteristics of a phase detector include the input phase-difference range, the response to a frequency difference, the sensitivity to input amplitude, and duty cycle.

This section begins by presenting the linear phase detector equations and a figure of merit for comparing phase detectors. Then, a variety of analog and digital devices for phase detection are presented, and the variety of different characteristics are studied. Some have a simple topology, others operate at high frequency, and still others have a very low content of undesirable signals in their output. Mixers, exclusive-OR gates, RS latches, flip-flops, and phase/frequency detectors are presented. They represent the major types that are used to detect phase error in integrated circuit PLLs.

5.1.1 Linear Model

Developing a linear model will help our discussion of phase detectors by giving us a standard for comparing the different types of phase detectors. Equation (5.1) establishes a linear model with an offset for a phase detector:

$$V_d = K_d \theta_e + V_{do} \tag{5.1}$$

where

K_d = phase detector gain (V/rad);

θ_e = phase error of the VCO output relative to the input signal;

V_{do} = offset voltage, or "free-running voltage."

This linear model breaks down for large enough θ_e (phase error). The values of θ_e for which the linear model is valid define the range of the phase detector. Range, offset, and gain characteristics distinguish the differences between the detectors. A large offset with respect to the gain of the detector ruins the effectiveness of the detector. Consequently, a figure-of-merit ratio will help compare the effectiveness of each detector circuit.

5.1.2 Phase Detector Figure of Merit

Is an offset V_{do} = 193 mV too large for a phase detector to be effective? The amount of output voltage per radian, which is the gain of the phase detector K_d, determines the usefulness of the phase detector. Therefore, the ratio of output voltage gain to offset voltage, K_d/V_{do}, indicates the size of the offset. We will call this the figure of merit, M, of the phase detector as shown by (5.2):

$$M = K_d/V_{do} \tag{5.2}$$

For example, a phase detector with a gain of 5 V/rad and an offset of 193 mV produces a figure of merit $M = (5 \text{ V/rad})/(193 \text{ mV}) = 26$. A phase detector should reasonably be expected to have $M \geq 15$, and M as high as 500 is possible with careful matching.

Range, offset, and gain characteristics distinguish the differences between the detectors. However, they all share the same characteristics as a multiplier. A multiplier has the characteristics of a phase detector by the trigonometric identity in (5.3):

$$\sin(A)\cos(B) = 0.5\sin(A - B) + 0.5\sin(A + B) \tag{5.3}$$

We already studied this equation in Section 2.2, but we will review it here and adjust it for our linear model of a phase detector. Applying the trigonometric identity for products of a trigonometric function to the multiplication of two signals (at frequencies RF and LO) produces (5.4):

$$V_1(t)\,V_2(t) = V_{p1}\,V_{p2}\,0.5\left[\cos(\omega_{rf}t - \omega_{lo}t + \theta_e) + \cos(\omega_{rf}t + \omega_{lo}t + \theta_e)\right] \tag{5.4}$$

Eliminating the high-frequency product with a lowpass filter yields (5.5):

$$V_1(t)\,V_2(t) = V_{p1}\,V_{p2}\,0.5\left[\cos(\omega_{rf}t - \omega_{lo}t + \theta_e)\right] \tag{5.5}$$

$$= V_{pbeat}\cos(\omega_{beat}t + \theta_e) \tag{5.6}$$

where

$\omega_{beat} = \omega_{rf} - \omega_{lo}$ for $\omega_{rf} > \omega_{lo}$;

$V_{pbeat} = V_{p1}\,V_{p2} \times 0.5 \times$ mixer losses;

V_{pbeat} = resulting voltage level after mixing (V).

The derivative of (5.6) calculates the incremental phase slope. For the mixer operating with a 0 frequency difference ($\omega_{beat} = 0$, which is dc) and 90° phase shift, the derivative of (5.6) produces (5.7) and (5.8):

$$K_d(\phi) = \frac{d}{d\phi}\left[V_{pbeat}\cos(\phi)\right] \tag{5.7}$$

$$K_d(\theta_e) = V_{pbeat}\sin(\theta_e) \tag{5.8}$$

Equation (5.8) shows the origin of the phase detector gain and that it relates to the multiplication of two signals.

For θ_e equals $\pi/2$ rad (quadrature) in (5.8), K_d equals V_{pbeat} (V/rad). For θ_e equals 0 rad in (5.8), K_d equals 0 V/rad. This shows that maximum phase sensitivity occurs for a 90° phase difference between the input signals, while a minimum phase sensitivity of 0 occurs for a phase difference of 0°. With the linear model established, let's look at the various phase detectors.

5.1.3 Balanced Mixer

First, we will look at a balanced mixer, sometimes called a double-balanced mixer, because it is very close to a four-quadrant analog multiplier. Furthermore, we will begin by looking at diode ring mixers that are used in PCB design because their operation has been well established. Figure 5.1 shows a diode ring mixer. This circuit consists entirely of passive components (baluns and diodes). Baluns and diodes usually compose the major components in a ring mixer for PCB designs. These components allow operation at extremely high frequencies (>18 GHz) and across wide frequency range.

5.1.3.1 Diodes

Only devices that have nonlinear current-voltage relationships or that change as a function of time (are time-variant), or both, cause mixing products. Multiplication of the two input signals yields one of the mixing products. A switch toggling between either a short or an open as a function of time makes a time-variant circuit. Mixers require very fast switching (at the LO frequency). In PCB designs, obtaining low-noise figure and fast switching speed make Schotkey-barrier diodes the usual choice in ring mixers. The nonlinear properties of diodes cause mixing products, so the time-variant and nonlinear properties of diodes help produce mixing products.

Simultaneously applying two or more signals across a nonlinear diode causes mixing to occur. This action produces single- and multiple-tone intermodulation products. Two voltages applied in series across a diode cause the current through it to contain the IF and higher-order intermodulation products of the two voltage inputs. A lowpass filter after the mixer must significantly reject these products to have a good phase detector.

5.1.3.2 Baluns

Baluns are the other major components in a ring mixer. Inductive transformers or a microwave coupling structure are used for baluns. The balun balances the diodes and interfaces them with the unbalanced system. In addition, the balun current, I,

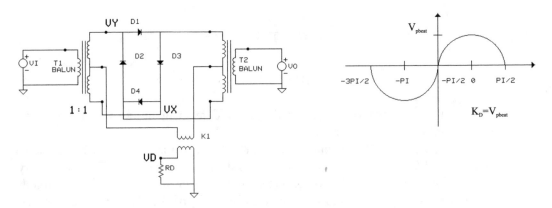

Figure 5.1 Diode ring mixer and transfer function of output voltage versus phase error.

matches system and diode impedance and provides interport isolation. The balun at the LO port provides the IF current return path to ground. Currents in the two balanced leads of a balun are 180° apart in phase and ±90° out of phase with respect to ground [1]. The amplitude and duty cycle of the input affect the phase detector gain. Consequently, these circuits should be used with large input signals.

The mixer circuit in Figure 5.1 operates with any waveform; however, using a square wave, like the V_o in Figure 5.2, eases understanding of the circuit's operation. The switching voltage V_o "turns on" either the bottom two diodes (D3, D4) or the top two diodes (D_1, D_2) depending on the polarity of V_0. A positive V_o makes the bottom two diodes (D_3, D_4) conduct. V_x equals the voltage at the midpoint of the secondary winding of transformer T_2, which is ground. Then, $V_y = V_i$, and $V_{davg} = 0.5V_y = 0.5\ V_i$. Similarly, a negative V_o makes the top two diodes (D_1, D_2) conduct. V_y becomes effectively ground. Then, $V_{davg} = -0.5V_i$. The bottom waveform in Figure 5.2 shows the resulting waveform of V_{davg}.

To determine the transfer function for the mixer, several SPICE runs must be done for several phase errors that cover the range of the detector (π). Figure 5.3 shows the average output voltage versus phase error for a mixer phase detector. Each curve is a SPICE run with a different phase-error difference input to the mixer. A lowpass filter is adjusted to filter out the harmonics of the 5-MHz input frequencies and not cause long simulation times. In this case, the simulation was adjusted so that the waveforms settle out in 2 μs with a 50-mV ripple. The remaining ripple can be minimized by averaging the waveform over the time period from 4 to 6 μs.

Figure 5.4 shows the transfer function plotted for mixer phase detector from the data shown in Figure 5.3. This figure shows the average detected voltage V_{davg} versus sweeping phase error. As θ_e increases with time, the average component V_{davg} varies sinusoidally. The maximum of the characteristic equals the phase detector gain:

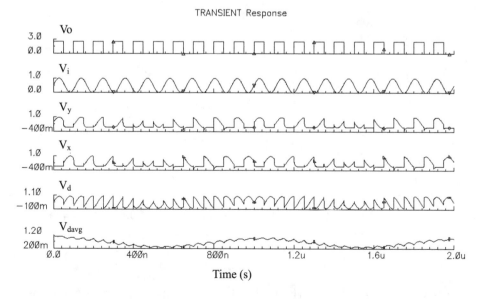

Figure 5.2   The switching waveforms in a diode ring mixer phase detector.

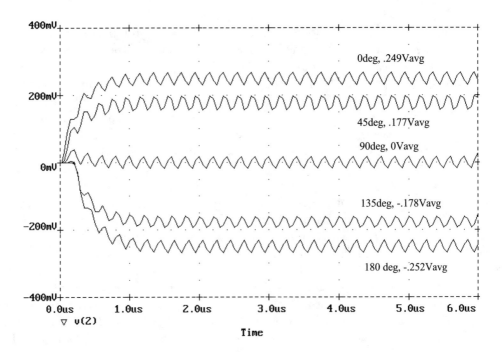

Figure 5.3   Average output voltage (lowpass filtered) versus phase error for a mixer phase detector.

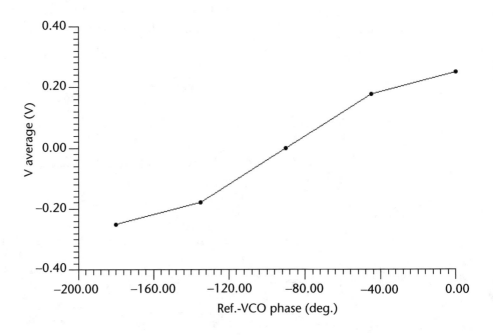

Figure 5.4   Transfer function for a mixer phase detector (average voltage versus phase-error difference in degrees).

$$V_{dm} = K_d = V_i / \pi \qquad (5.9)$$

This assumes that both transformers have primary turns equal to secondary turns. A high signal amplitude V_i keeps V_{dm} high and gives the best figure of merit. For the operation as described above, the signal level of V_i should not cause a diode pair to conduct. This corresponds to $V_i \leq 1.2V$ for equal transformer ratio. The fastest slope is at $-90°$ phase shift, and the slope is reduced for shifts approaching $-180°$ or $0°$. For the peak-to-peak voltage of 0.5V, we can calculate the slope at $-90°$. From (5.8) this computes to 0.25 V/rad because V peak equals V peak to peak divided by 2 (0.5/2).

A figure of merit $M \geq 400$ can be achieved with well-matched diodes in an integrated circuit that produce a V_{do} of less than 1 mV. If low offset voltage of the phase detector is achieved, then the active loop filter usually becomes a significant contributor. The active loop filter can have an offset voltage of 5 mV; however, high-performance opamps can be used that have an offset as low as 0.1 mV.

In Figure 5.5, two signals with different frequencies (5 MHz and 5.5 MHz) are connected into a mixer phase detector, followed by a lowpass filter. This connection produces a sinusoidal output waveform that shows the sinusoidal shape of the phase detector transfer function. This characteristic waveform will also be seen in other phase detectors and will explain some of the behavior that can occur. Figure 5.6 shows the CMOS transistor ring mixer that is equivalent to the diode ring-based mixer. This connection can easily be integrated while the diode ring cannot.

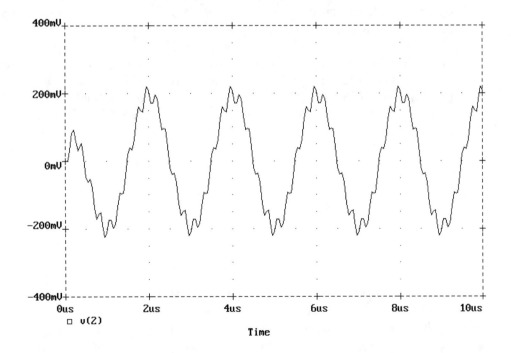

Figure 5.5 Waveform that results from two input frequencies into the mixer, showing the shape of the transfer function.

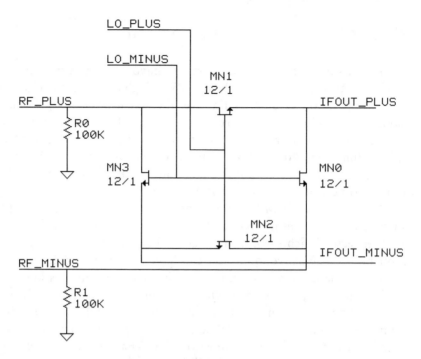

Figure 5.6   Schematic of CMOS ring mixer that can be used for a phase detector.

5.1.4   Gilbert Multiplier

The Gilbert multiplier circuit shown in Figure 5.7 also implements a four-quadrant multiplier. This circuit does not have a balun and can be more easily integrated into a monolithic circuit.

Figures 5.7 and 5.8 help describe the operation of the circuit. First, V_o splits the current I to the left as i_1 or to the right as i_2, according to the characteristic shown in Figure 5.8. V_i splits the current i_1 into currents i_3 and i_4, according to the characteristic shown in Figure 5.8. A similar characteristic holds for i_2. Similarly, V_i splits the current i_2 into currents i_5 and i_6. Combining the four currents produces (5.10):

$$V_{davg} = (i_4 + i_6)R_1 - (i_3 + i_5)R_2 \qquad (5.10)$$

$$= (i_4 + i_6 - i_3 - i_5)R$$

where

$$R_1 = R_2 = R.$$

In practical applications, R_1 and R_2 have different values, which causes a dc offset; we will discuss this effect later in (5.14) through (5.17).

Now let's study an equation for the average detected output voltage and its associated multiplier gain constant. Keeping V_i and V_o in the linear region of the characteristics in Figure 5.8 (amplitude less than 52 mV) produces (5.11) for the average detected output voltage V_d.

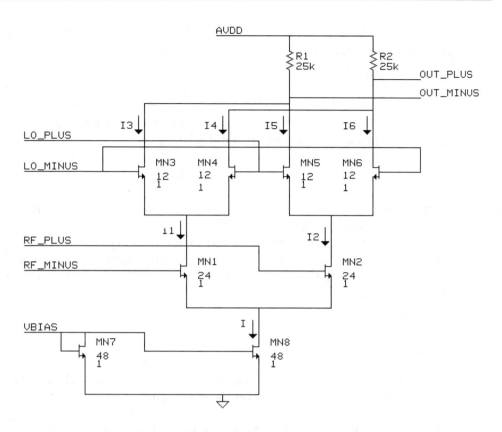

Figure 5.7   Schematic of CMOS Gilbert cell mixer.

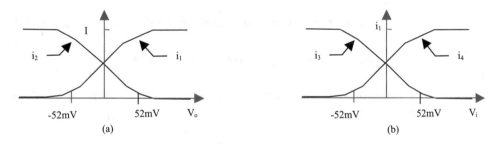

Figure 5.8   (a) Transfer function for the bottom difference amplifier and (b) the left-hand-side difference amplifier in Figure 5.7.

$$V_{davg} = K_m V_i V_o \qquad (5.11)$$

K_m is a constant in volts^{-1} associated with the multiplier. Studying Figure 5.8 gives us the calculation for the multiplier gain constant. If we assume all the tail current I goes to i_1 and none goes to i_2, i_5, and i_6, then from the transfer function on the left-hand side of Figure 5.8 ($i_1 = V_o I/52$ mV). Again, if we assume that all the current goes into i_4 and none goes into i_3, then from the transfer function on the right-hand side of Figure 5.8, $i_4 = V_i i_1/52$ mV. Substituting the previous equation into the last gives $i_4 = V_i V_o I/(52 \text{ mV})^2$. Multiplying by R and comparing the equation to (5.11) gives (5.12) for computing the multiplier gain constant:

$$K_m = RI/(52 \text{ mV})^2 \qquad (5.12)$$

Equation (5.13) computes the maximum voltage out of the phase detector, which gives us the phase detector gain for sinusoidal inputs:

$$K_d = K_m V_i V_o \qquad (5.13)$$

From (5.8) V_{dmax} equals K_d. Now, with the detector gain computed, let's look at the offset voltages that occur in the Gilbert mixer.

The mismatch of the transistors and of the resistors is the source of offset voltages. A mismatch in the transistors causes input offsets V_{io} of a few millivolts that add to the inputs V_i and V_o. Similarly, a mismatch between R_1 and R_2 causes the other offset to be added to the output. Equation (5.13) computes the effects of the mismatch:

$$V_{oo} = (R_1 - R_2)I/2 \qquad (5.14)$$

Next, let's compute the average voltage at the output of the detector. Equation (5.15) computes the total expression for the output:

$$V_{davg} = K_m(V_i + V_{io})(V_o + V_{io}) + V_{oo} \qquad (5.15)$$

$$= K_m V_i V_o + K_m\left(V_{io}V_i + V_{io}V_o + V_{io}^2\right) + V_{oo}$$

Taking the time average, as we did in going from (5.3) to (5.6), gives (5.16):

$$V_d = V_{dm}\sin(\theta_e) + K_m(V_{io})^2 + V_{oo} \qquad (5.16)$$

The dc terms in (5.16) can be combined as an effective offset voltage at the phase detector output to give (5.17):

$$V_{do} = K_m(V_{io})^2 + V_{oo} \qquad (5.17)$$

Figure 5.9 uses signal flow graphs to summarize these relationships for calculating dc offset voltage.

Example 5.1

The Gilbert multiplier circuit in Figure 5.7 has a current $I = 2$ mA and resistance $R = 5$ kΩ. R_1 and R_2 differ by 2%, and $V_{io} = 5$ mV. Find the dc offset V_{do}, the gain of the phase detector K_d, and the figure of merit.

From (5.12) and (5.13), $K_m = (10\text{V})/(52 \text{ mV})^2 = 3{,}698$, and $K_d = V_i V_o(3{,}698)$. K_d depends not just on the circuit but on the input levels. The maximum K_d

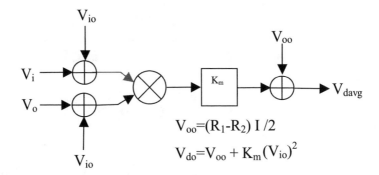

Figure 5.9 Signal flow graph of calculations for dc offset in Gilbert mixer. (*From:* [2]. © 1991 Prentice-Hall. Reprinted with permission.)

corresponds to $V_i = V_o = 52$ mV, which is the largest signal that (5.8) holds. Then, $K_d = (52 \text{ mV})^2 \, 3{,}698 = 10.0$ V/rad.

From (5.14), $V_{oo} = (0.02R)I/2 = (100\Omega) \, 2$ mA/2 = 100 mV. Also, $K_m(V_{io})^2 = (5 \text{ mV})^2 \, (3{,}698) = 93$ mV. Then, by (5.16), the total offset is $V_{do} = 92$ mV + 100 mV = 193 mV. Then, from (5.2), the figure of merit for the multiplier is $M = 10/0.193 = 51.8$.

5.1.5 Exclusive-OR Phase Detector

An exclusive-OR logic circuit has essentially the same characteristics as an overdriven multiplier circuit. Overdriving the multiplier saturates the output at either a positive value, corresponding to a logic "high," or a negative value, corresponding to a logic "low."

A voltage at V_i in Figure 5.1 causes current to flow through either the left or right leg of the diode bridge, depending on the polarity of V_i. As the signal at V_i changes polarity, the conducting leg switches to the other side. This switching action inverts the polarity of a smaller signal at V_o as it appears at the output V_d. A negative or positive V_i and V_o input makes the output V_d positive. One positive input, with the other input being negative, makes the output V_d negative. The truth table for a multiplier is shown in Table 5.1.

Comparing this logic with a truth table as shown in Table 5.2 shows the multiplier logically functions as an exclusive-NOR gate. Consequently, an overdriven multiplier is an exclusive-NOR, with "+" corresponding to a logic high (1) and "−" corresponding to a logic low (0).

Then, an exclusive-NOR gate can be used for phase detection. The transfer function will have a triangular shape. In addition, an exclusive-OR gate can be

Table 5.1 Multiplier Truth Table

V_i	V_o	V_{davg}
−	−	+
−	+	−
+	−	−
+	+	+

Table 5.2 Exclusive-NOR Gate
Truth Table

Input A	Input B	Output OUT
0	0	1
0	1	0
1	0	0
1	1	1

used. This gate just inverts the output logic. Consequently, a balanced output that is both positive and negative, where $V_{davg} = V_{bavg} - V_{aavg}$, uses an exclusive-OR output for V_{aavg} and its complement (exclusive-NOR) for V_{bavg}. Figure 5.10 shows a schematic of an exclusive-OR gate. This figure shows that it takes an inverter and four transistors to make an exclusive-OR gate. This is a minimum of transistor connections a (total of six transistors).

Figure 5.11 shows two square-wave inputs (A and B) to the exclusive-OR gate and the resulting output waveform at OUT. To determine the transfer function for the exclusive-OR phase detector, several SPICE runs must be done for several phase errors that cover the range of the detector (π). Figure 5.12 shows these multiple SPICE runs for 5-MHz reference and VCO frequency inputs. The figure contains lowpass-filtered output voltages versus phase error for an exclusive-OR phase detector. Each curve is a SPICE run with a different phase-error-difference input to the mixer. A lowpass filter is adjusted to filter out the harmonics and keep simulation times down. For this case, the output waveforms are settled at 2.5 μs. The remaining ripple can be minimized by averaging the waveform over the time period from 2.5 to 4 μs.

Figure 5.13 shows a plot of the average value of the output as a function of the phase difference between inputs. The phase error is plotted as the ratio of phase-error edge difference divided by the input source time period. This average value, which results from lowpass-filtering the gate output, detects phase error

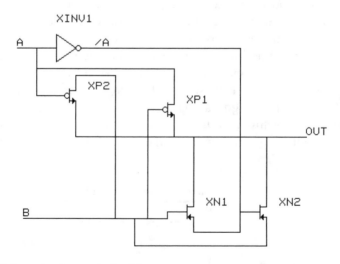

Figure 5.10 Schematic of an exclusive-OR gate.

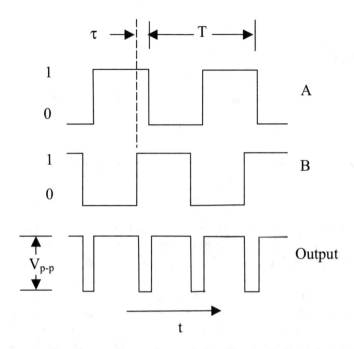

Figure 5.11 Input and output waveforms of an exclusive-OR gate that is used as a phase detector.

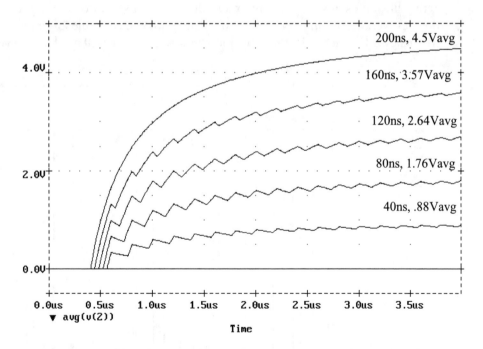

Figure 5.12 Lowpass-filtered output of an exclusive-OR gate versus the variation of input phase error.

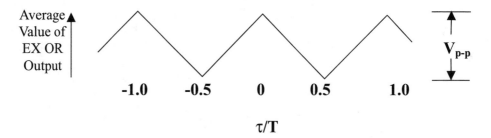

Figure 5.13 Transfer function of exclusive-OR phase detector showing average output voltage versus fractional phase error.

over a range of half a cycle before it starts repeating. For any input phase error, the ideal output does not contain any energy at the fundamental input frequency. At a 90° phase difference for the worst case, the second-harmonic has peak-to-peak amplitude 1.27 times the peak-to-peak range of the phase detector characteristic.

To generate the transfer function for a phase detector, we put in the same frequency 5 MHz with varying phase. This transfer function controls the behavior when the loop is close to lock. When the loop is unlocked, the phase detector has two different input frequencies. Figure 5.14 shows the PSPICE response of the detector with two different input frequencies (5 and 4 MHz). Figure 5.15 shows the unfiltered pulse-width modulation that comes out of the exclusive-OR gate. When this response is seen in other phase detectors, it shows the phase detector acting like an exclusive-OR gate or a multiplier with saturated inputs. A PLL uses the dc average of this waveform to acquire lock. As a rule of thumb, if this frequency difference is 10 times the loop bandwidth, lock will not be achieved, and

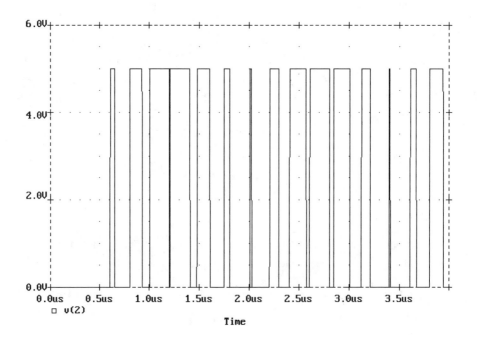

Figure 5.14 Exclusive-OR gate output for inputs with a frequency difference.

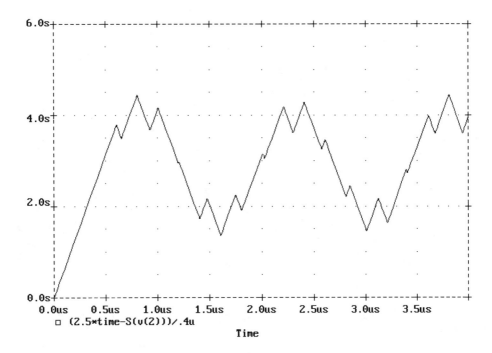

Figure 5.15   Lowpass-filtered (integrated) output of an exclusive-OR detector with a frequency difference at the input.

a frequency-aided acquisition circuit will have to be used. Between 2 and 10 times the bandwidth, the loop will lock in an increasing amount of time, depending on the frequency difference.

Lowpass-filtering the waveform in Figure 5.14 produces the waveform in Figure 5.15. The x-axis is simulation time, and the y-axis is average voltage. The average voltage was computed in PSPICE by integrating the output voltage $[V(2)]$ over the input period (0.2 μs) and offsetting the waveform (2.5 × time) to make the resulting waveform start at 0. The waveform in the figure shows the sawtooth transfer function and looks similar to the filtered output of the mixer.

So far, we have assumed that the square waves at the input to the detector have a symmetrical shape. Consequently, the input waveforms have a 50% duty cycle. Suppose that A has a duty cycle of 20%, and B has a duty cycle of 40%, as in Figure 5.16. A duty cycle not equal to 50% produces a nonzero, free-running voltage and reduces V_{dm} of the phase detector characteristic. If either, or both, inputs has other than 50% duty cycle, the exclusive-OR phase detector characteristic will have flat spots, as shown in Figure 5.16. The detailed harmonic content of the output of this type can be obtained as a function of duty cycle by Fourier analysis [2]. Using the digital exclusive-OR gate as a phase detector has the advantage of greater gain K_d, less offset V_{io}, and greater linear phase range.

5.1.6   RS Phase Detector (Two States)

An RS latch can be used as a phase detector and have similar characteristics to an exclusive-NOR gate. Figure 5.17 shows a schematic of the RS latch. Figure 5.18

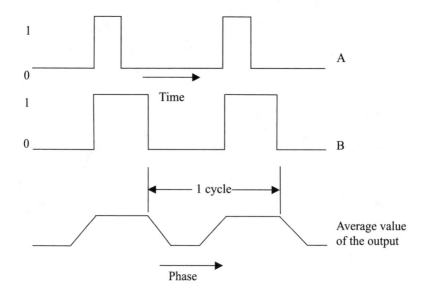

Figure 5.16   Exclusive-OR average output transfer function with non-50% duty-cycle inputs.

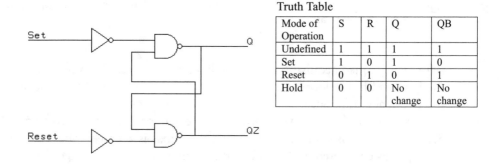

Figure 5.17   Schematic of an RS latch.

Truth Table

Mode of Operation	S	R	Q	QB
Undefined	1	1	1	1
Set	1	0	1	0
Reset	0	1	0	1
Hold	0	0	No change	No change

shows the expected waveforms for an RS latch used as a phase detector. The figure shows the input and output signals of the RS-latch phase detector. The input signals A and B consist of narrow pulses. They connect to the set and reset inputs of a set-reset latch. The average value of the Q output is proportional to the phase at A relative to B. This creates a sawtooth transfer function of voltage versus phase as shown in Figure 5.19. For either input with a finite width, the phase detector transfer function has a flat spot that corresponds to the width at a corner. For both inputs with finite widths, the flat spot, or the sum of two flat spots, will be as wide as the wider input.

Connecting one signal of arbitrary width to the clock input of an edge-triggered D flip-flop will give the same transfer function as the RS latch without causing the flat spot. Consequently, one edge of the waveform defines the switching time. The other input connects to the set or reset input. This signal still must be narrow to minimize the flat spot.

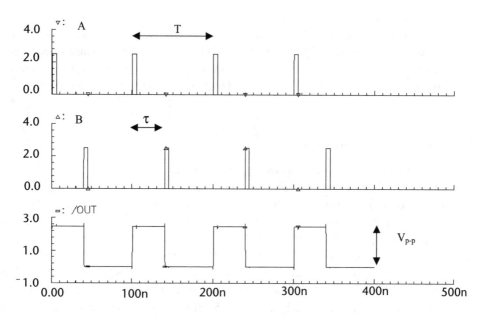

Figure 5.18   Input and output waveforms of an RS latch used as a phase detector.

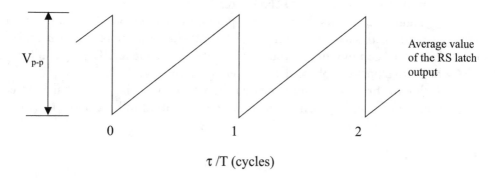

Figure 5.19   Transfer function for an RS latch used as a phase detector.

In the middle of the 360° linear phase range, the Fourier frequency output component at the fundamental frequency of the input signals has a peak-to-peak magnitude equal to 1.27 times the peak-to-peak voltage range of the output characteristic, which is V_{p-p} in Figure 5.18. The exclusive-OR phase detector has very little spectral energy at the fundamental frequency and has a lot more energy at twice the fundamental frequency. In some applications, this makes filtering out this spurious signal for an exclusive-OR gate easier.

Consequently, the RS latch has the disadvantage of higher spectral energy at the fundamental relative to the exclusive-OR; however, the RS latch has twice the phase-error range for a given gain as shown by comparing Figure 5.19 with Figure 5.13. The exclusive-OR gate and the RS latch are the building blocks for more complicated phase detectors. Understanding their behavior will help us understand the operation of more complicated phase detectors.

Figure 5.20 shows input and output waveforms of an RS latch used as a phase detector for unequal input frequencies (10 and 9.9 MHz). The bottom waveform shows the result of lowpass-filtering the output. This figure shows the pulse-width modulation in an RS flip flop (RSFF) that also occurs in an exclusive-OR gate for two input signals with different input frequencies (10 and 9.9 MHz). This waveform characteristic can be recognized in more complicated phase detectors.

5.1.7 Phase/Frequency Detector

This section studies the phase/frequency detector, which is one of the most commonly used detectors. As you will see, understanding the phase/frequency detector is key to understanding loop performance. First, we will study the gate logic. Next, we will cover the operating conditions. Finally, we will study the transfer function of the phase detector.

Figure 5.21 shows the PSPICE model for the phase/frequency detector, which is one of the most commonly used phase detectors [3, 4]. This model is based on logic gates in the device. The schematic shows that four RS latches make up the phase/frequency detector. FF1 and FF2 control the up and down pulses, and FF3 and FF4 control the reset signal for all the latches. The critical path is limited by just three gate delays, two from the cross-coupled two-input NAND gates and one from the four-input reset NAND gate.

Figure 5.22 shows the relationships of up and down outputs to input waveforms by plotting the average voltage of the individual outputs u and d and the u-d differential combination output versus phase or frequency. The logic in Figure 5.22 shows five possible phase detector relationships.

For the first relationship, the frequency of the VCO is greater than the frequency of the reference. This condition causes the up output to have a low logic level with

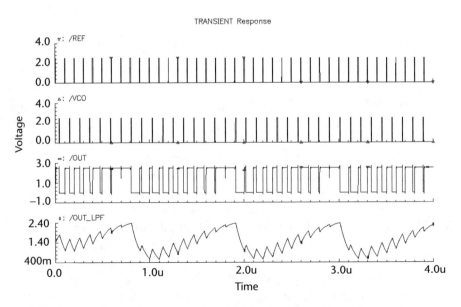

Figure 5.20 Input and output waveforms for RS latch with unequal input frequencies.

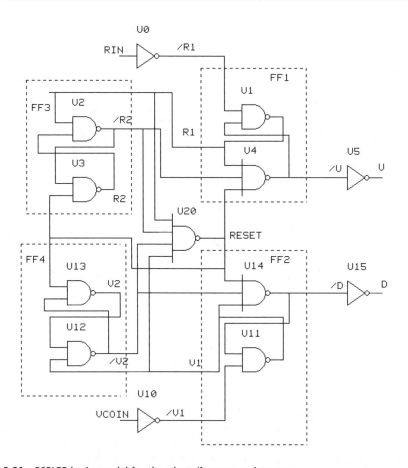

Figure 5.21 PSPICE logic model for the phase/frequency detector.

small positive-going pulses and the down output to have approximately a 50% duty-cycle (pulse-width-modulated) waveform. In this condition, decreasing the VCO frequency will lock the PLL.

For the second relationship, the frequency of the VCO is less than the frequency of the reference. This condition causes the down output to have a low logic level with small positive-going pulses and the up output to have approximately a 50% duty-cycle (pulse-width-modulated) waveform. In this condition, increasing the VCO frequency will lock the PLL.

For the third relationship, the frequency of the reference equals the frequency of the VCO, and the phase of the reference lags the phase of the VCO. This condition causes the up output to have a low logic level with small positive-going pulses and the down output to have positive-going pulses. In this condition, decreasing the VCO phase will lock the PLL.

For the fourth relationship, the frequency of the reference equals the frequency of the VCO, and the phase of the reference leads the phase of the VCO. This condition causes the down output to have a low logic level with small positive-going pulses and the up output to have positive-going pulses. In this condition, increasing the VCO phase will lock the PLL.

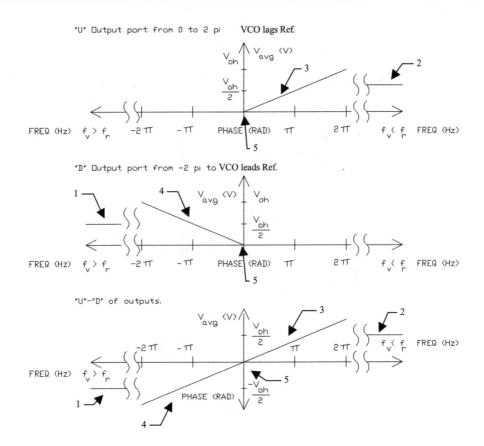

Figure 5.22 Relationships of up and down outputs to input waveforms.

In the third and fourth conditions, the phase difference between the reference and VCO input determines the duty cycle of the positive-going pulses at the down and up outputs. Theory states that a 180° phase difference produces a 50% duty cycle.

In the fifth condition, the frequency of the reference equals the frequency of the VCO, and the phase of the reference equals the phase of the VCO. This condition produces the lowest duty cycle and causes positive-going pulses out of the up and down outputs to occur simultaneously. This condition causes a phase distortion in the transfer function [5].

Let's use a SPICE model to see if we can generate the five relationships and to study the logic more closely. This can be accomplished by varying conditions of the two input sources into the phase/frequency detector. Let's use a reference frequency of 2.5 MHz for the phase detector test. The higher-frequency speeds up the simulation time and make it easier to see the narrow pulses on the plots. The CD that accompanies this book shows the hierarchical SPICE deck for the phase/frequency detector. Now that the SPICE circuit is defined, let's begin understanding the operation of phase/frequency detectors by studying the frequency discriminator function of the detector.

5.1.7.1 Frequency-Difference Response

Let's look at the operation of the phase/frequency detector when the VCO frequency is greater than the reference frequency. Figure 5.23 shows the response of the detector for a reference frequency of 40 MHz and a VCO frequency of 50 MHz, as shown in the first two waveforms at the top of the figure. The up output has narrow pulses and the down output has a pulse-width-modulated output, as shown by the next two waveforms. The loop-compensation processes the pulse-width-modulated down output to push the VCO frequency lower toward the lock condition. The next four waveforms in Figure 5.23 are the inputs to the four-input NAND gate that generates a reset for the latches in the phase detector. The reset signal is the last waveform in the figure. Next, the reset pulse, the up pulse, and the positive edge of the reference are in synch. When the VCO positive edge lags the reference edge by half a period, the reset pulse resets the RS latch (v_1) to get the correct phase detector pulse width.

Figure 5.24 shows the response of the detector to a reference frequency of 50 MHz and a VCO frequency of 40 MHz, as shown in the first two waveforms at the top of the figure. The reset pulse, the down pulse, and the positive edge of the VCO are in synch. When the VCO positive edge lags the reference edge by half a period, the reset pulse resets the RS latch (signal r_1) to get the correct phase detector pulse width. Next, the up output has pulse-width-modulated output, and the down output has narrow pulses, as shown in the next two waveforms. Loop-compensation processes the pulse-width-modulated up output to push the VCO to a higher frequency.

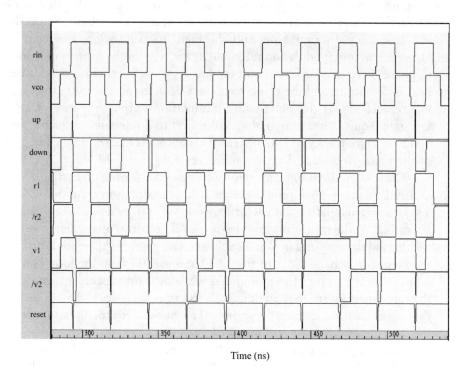

Time (ns)

Figure 5.23 VCO frequency greater than the reference frequency.

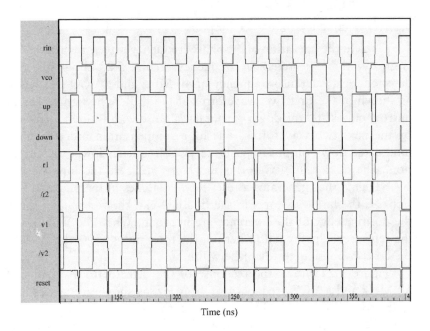

Time (ns)

Figure 5.24   VCO frequency less than the reference frequency.

Pulse-width modulation in the up and down outputs of the phase/frequency detector produces the frequency-discrimination mechanism; however, a potential hang up in the transition from frequency detector to phase detector can be caused by the frequency-discriminator mechanism. At high modulation rates, the detector is a frequency discriminator, and the high frequencies are filtered by the loop. However, at low modulation rates and at the edge of the transition from frequency to phase detector, one of the outputs has a slowly varying pulse-width-modulated beat-note output, and the phase frequency detector acts like an exclusive-OR gate. Figure 5.25 shows the pulse-width-modulation output response of an exclusive-OR gate (equivalent to an analog multiplier) to a reference frequency of 2.5 MHz and a VCO frequency of 1.825 MHz. This output is like the down and up responses of the phase/frequency detector in Figures 5.23 and 5.24.

The low-modulation-rate condition can occur for the phase/frequency detector in a PLL with a high damping factor. A low-modulation condition can arise from a slow approach path to lock with only one side (up or down) of the phase detector operating. At the transition from frequency detector to phase detector, it has pulse-width modulation similar to the exclusive-OR gate. If this modulation is less than the modulation bandwidth of the VCO and greater than the loop bandwidth, the loop will have to pull the modulated waveform into lock. This pull into lock is the same mechanism that analog PLLs use to achieve lock. This is a slow process that significantly increases lock time. The phase detector uses the dc average of the modulated signal to pull into lock.

An example measurement of a PLL shows the pull-into-lock mechanism that we have discussed. The PLL has a 270-MHz output frequency, 100-kHz reference frequency, 0.9 damping factor, and 5-kHz bandwidth. Figure 5.26 shows the zoomed-in transient response (output frequency on the y-axis and time on the

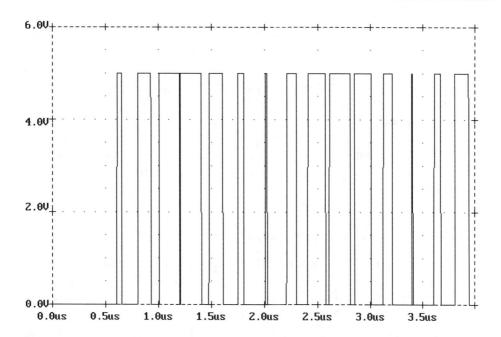

Figure 5.25   Output of exclusive-OR gate for different input frequencies.

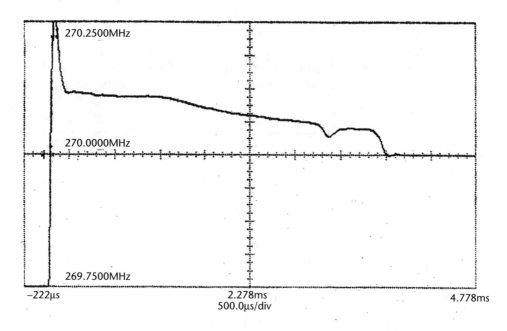

Figure 5.26   Zoomed-in measurement of transient response of a PLL with 0.9 damping factor that shows an analog pull-in mechanism.

x-axis) of the PLL, and it shows that the output frequency remains 50 to 120 kHz above the locked frequency for 3.5 ms before it locks at 270 MHz. The settling time of 3.5 ms is much greater than the 1/5 kHz (200 μs) one would expect for this loop.

Figure 5.27 shows the response of the example PLL with the damping factor changed to 0.5 and a loop bandwidth of 6 kHz. The loop now settles to 270 MHz in 500 μs and has a more classic PLL transient. This is a sevenfold improvement over the previous response. Consequently, this figure shows that the PLL transient response can get hung up in the transition from the frequency-discriminator to phase detector mode on one side of the phase detector and rely on an analog pull-in method to attain lock. Consequently, for fast transient response, a 0.5 damping factor is optimum. This completes the discussion of the frequency-difference response of the phase/frequency detector; we now move on to the phase-difference responses.

5.1.7.2 Phase-Difference Response

Let's study the operation of the phase/frequency detector when the VCO and reference frequency are equal but the phases are different. Figure 5.28 shows the phase/frequency detector output when the VCO phase leads the reference-oscillator phase, and their frequencies are equal to 50 MHz. The down output has a wide pulse width, and the up output has narrow pulses. The down output results from the logic AND of signals V_1 and $/V_2$. Also, the reset pulse, the up pulse, and the positive edge of the reference are in synch. The reset pulse resets the RS latch (v_1) to get the correct phase detector pulse width.

Figure 5.29 shows the phase/frequency detector output when the VCO phase lags the reference-oscillator phase. The down output has narrow pulses, and the up output has wide pulses. The up output results from the logic AND of signals R_1 and $/R_2$. Also, the reset pulse, the down pulse, and the positive edge of the

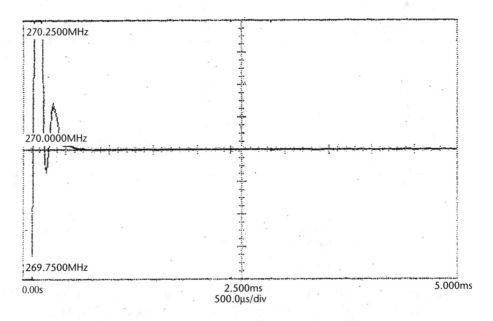

Figure 5.27   Zoomed-in measurement of transient response of example PLL that showed an analog pull-in mechanism with the damping factor changed to 0.5.

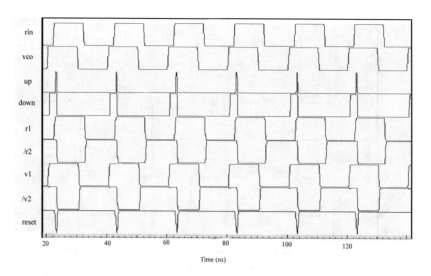

Figure 5.28 Phase/frequency detector output when VCO phase leads reference phase.

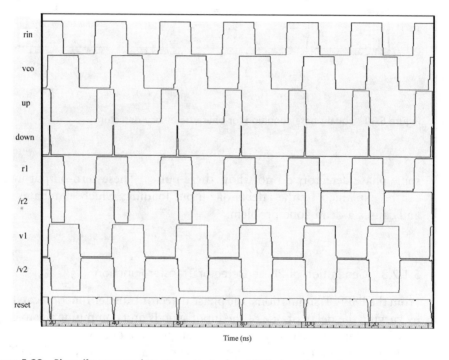

Figure 5.29 Phase/frequency detector output when VCO phase lags reference phase.

VCO are in synch. The reset pulse resets the RS latch (r_1) to get the correct phase detector pulse width.

Figure 5.30 shows that, for equal phase and frequency inputs, the up and down outputs produce narrow pulses. This results from the four-input logic AND of R_1, $/R_2$, V_1, and $/V_2$, which also generates the reset pulse. It is significant that the logic shows that output pulses should be present because measurements of

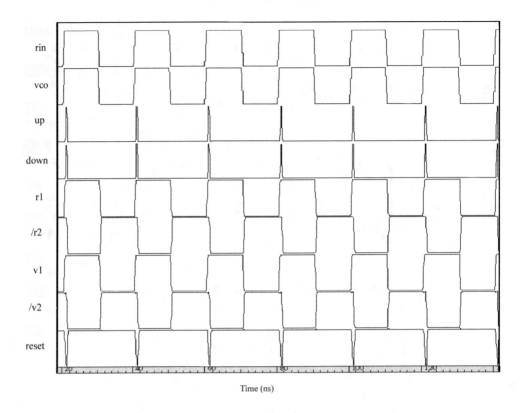

Time (ns)

Figure 5.30   Equal phase condition for phase/frequency detector.

some phase detectors do not show these pulses. These zero-output measurements can be explained by the effects of output loading, which eliminates these pulses and causes a dead-zone problem.

5.1.7.3   Generation of Phase Detector Transfer Function

From the PSPICE simulations, the phase detector transfer function can be computed by varying the delay of one of the input signals and computing the average voltage of the resulting waveforms. First, let's compute the output voltage versus the phase of the phase detector by varying the delay (phase) of the VCO input in SPICE.

Figures 5.31 and 5.32 show phase/frequency detector outputs overlayed on top of each other for the variation of the delay of the VCO with respect to the reference of –8 to 8 ns in 4-ns steps. Variation of delay is an equivalent phase shift. Figure 5.31 shows the up output. The y-axis is voltage and the x-axis is simulation time. The falling edge of the up signal depends on the positive edge of the VCO input. The positive edge of the up signal stays fixed with the positive edge of the reference input. For –8-ns, –4-ns, and 0-ns conditions, the up signal has minimum pulse widths of 1 ns that are overlayed on top of each other and centered at the simulation time of 26 ns. This figure shows that for a positive-

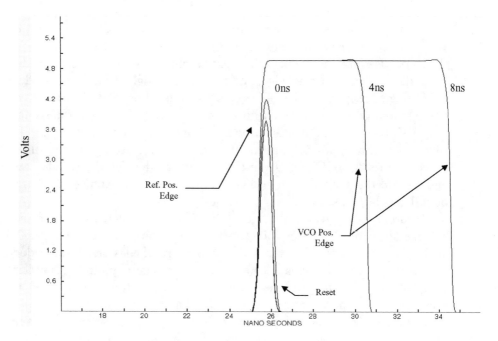

Figure 5.31   Up outputs versus delay in VCO phase.

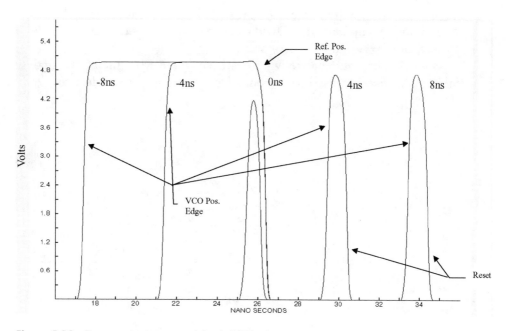

Figure 5.32   Down outputs versus delay in VCO phase.

sloped VCO tune curve, increasing the tune voltage moves the falling up edge from the far right- to the left-hand side of the page and toward the positive reference edge at the simulation time of 25 ns. The time period gets smaller so frequency increases.

Figure 5.32 shows the down output for the variation of the delay of the VCO with respect to the reference of −8 to 8 ns in 4-ns steps. The y-axis is voltage and the x-axis is simulation time. For negative phase errors, the positive edge of the VCO determines the rising edge, and the positive edge of the reference determines the falling edge. For positive phase errors of 0 ns, 4 ns, and 8 ns, the down signal has a minimum pulse of 1 ns with the positive edge following the rising edge of the VCO and centered at simulation times of 26 ns, 30 ns, and 34 ns. This figure shows that, for a positive-sloped VCO tune curve, decreasing the tune voltage moves the rising VCO edge from the far left- to the right-hand side of the page and toward the positive reference edge at the simulation time of 25 ns. The time period gets larger, so frequency decreases.

Figure 5.33 shows the average of the up output voltage waveform minus the average of the down output voltage waveform versus VCO delay. After 0.3 μs of an RC-type time response, the averages have reached a steady-state dc voltage. The average value of the last 100 ns of this voltage is plotted against the delay value of the VCO to produce a phase transfer function. Figure 5.34 plots the previous average dc voltage after 0.3 μs in Figure 5.33 and an additional ±18-ns delay points versus the delay of the VCO. This figure shows a linear transfer function going through zero phase error for the phase detector.

Calculations of the phase detector gain (K_d V/rad) are made from the slope of the phase detector transfer curve in Figure 5.34. The denominator of the gain is $2 \times \pi \times 2$ periods (one reference period = 20 ns) in the x-axis. The numerator is 5V minus −5V in the y-axis. Dividing the numerator by the denominator computes a gain of 0.79 V/rad.

Finally, let's summarize what has been found from the above analysis. First, there is always a pulse coming out of both outputs of the detector from a logic

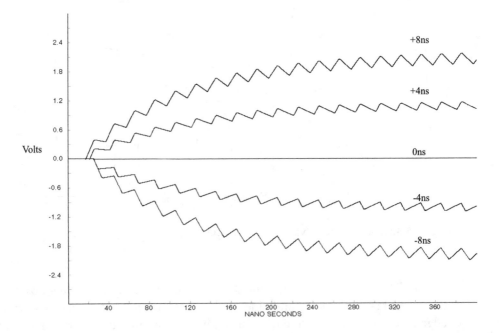

Figure 5.33   Average of the up minus the down waveform versus VCO delay.

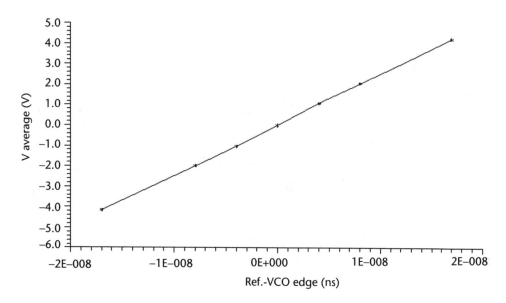

Figure 5.34 Phase transfer function for the phase/frequency detector.

viewpoint. This assumes no loading effects. There is no evidence of a dead zone because pulses are always present at the outputs, and Figure 5.34 shows a linear response around the 0° phase point where the dead zone should appear.

Phase Detector and Loop-Compensation Interface
Earlier we mentioned that loading effects could cause the minimum pulses to be eliminated. Consequently, let's look at loading effects on the phase/frequency detector to see if they cause pulses to be eliminated and a dead zone in the phase detector transfer function. The study of this interface consists of developing a PSPICE model, studying the stored charge on the loop-compensation capacitors, varying the delay between source inputs to vary phase relationship, varying capacitive loading, and computing the phase detector transfer function.

Let's look at the model for the phase/frequency detector and loop-compensation interface. To stay within the parameters of the evaluation version of PSPICE and to speed up calculations, an equivalent ECL output circuit is used and a 2.5-MHz reference frequency is used. The logic of the phase/frequency detector is replaced by voltage sources with a variable delay relationship. Figure 5.35 shows a PSPICE model schematic of the phase/frequency detector and loop-filter combination.

An active lowpass filter is used to model the effects of the loop compensation and also shows the circuitry used to measure the phase transfer function for the phase/frequency detector [6]. The included CD has a PSPICE netlist for studying the loading effects of the loop compensation on the outputs of the phase/frequency detector.

With this circuit, the mechanism that stores the phase error in the loop compensation can be studied, and in Section 5.1.7.4 the dead zone in the phase transfer function will be studied. First, Figure 5.36 shows charge being stored on each capacitor with the output waveform from the phase/frequency detector. Each pulse

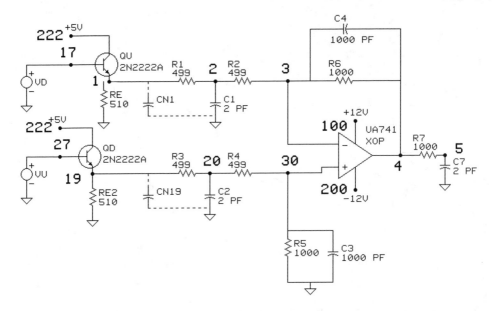

Figure 5.35 PSPICE model of phase/frequency detector and loop-compensation interface.

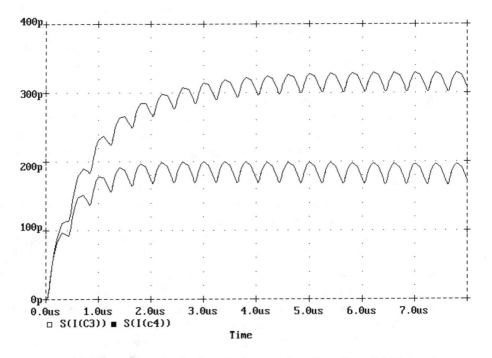

Figure 5.36 Charge stored on feedback and input capacitors in a differential filter.

out of the phase/frequency detector is integrated and stored on the capacitors with each phase comparison. Since the pulse width is a measure of the amount of error, the error is stored in the capacitors.

Figures 5.37 and 5.38 show four positive and three negative averaged-output voltage waveforms versus phase that are created for an ECL phase detector. The

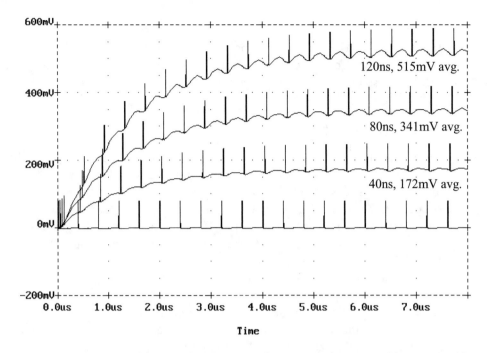

Figure 5.37   For positive delays, averaged output voltage versus phase created from an ECL phase detector.

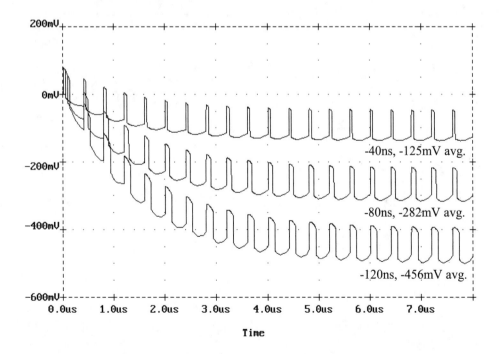

Figure 5.38   For negative delays, averaged output voltage versus phase created from an ECL phase detector.

average voltages of the last 2 ms are used to generate the phase detector transfer function curve. The even voltage spacing shows that a linear transfer function curve will be plotted. Consequently, we can use these plots to study the dead-zone effect on the phase detector transfer function curve.

5.1.7.4 Distortion Zone

The fifth condition of the phase/frequency detector, which has 0° phase difference between the inputs, causes a race condition in the IC. This race condition causes an undetermined state in the logic of the phase detector. In this condition, the output of the detector is sensitive to circuit-loading effects that can produce a nonlinearity in the phase-versus-output-voltage transfer function of the phase detector. This nonlinearity produces higher reference sidebands, loop instability, longer loop settling time, and higher phase noise.

The race condition depends on the rise and fall time and the propagation time of the IC. Consequently, the width of the nonlinear distortion zone depends on the rise and fall time and the propagation time of the IC. Rise and fall times and propagation times decrease with higher-speed digital logic families. Consequently, a high-speed phase detector (MC12040 ECL) has a smaller dead zone than a lower-speed phase detector (MC4044 TTL) [6].

Let's study the effects of loading on the phase detector transfer function by varying the output capacitance. Figure 5.39 shows output voltage variation with capacitive loading on the output level at a constant delay of 10 ns for a 400 ns

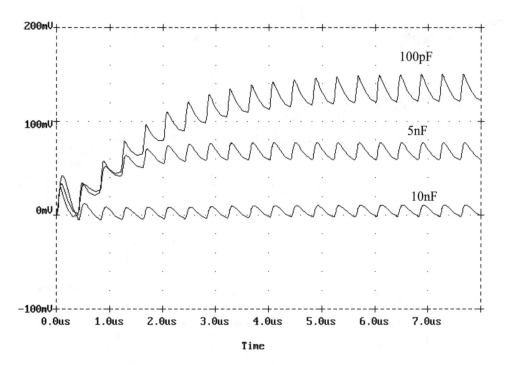

Figure 5.39 Variation of averaged phase detector output voltage for constant delay versus capacitive loading.

reference period. The reduction in output level with increasing load capacitance shows an enhancement of the dead-zone problem.

Next, the phase detector transfer function curve is calculated with 10-nF capacitive loading. Figure 5.40 shows the resulting curve, which shows the dead-zone effect on the phase detector transfer function with the output capacitively loaded.

Finally, there are some application and design considerations in using phase/frequency detectors. First, Figure 5.41 shows another form of the three-state phase/frequency detector, which uses D flip-flops [2]. The input reference R_{in} connects to the "clock" input of the top flip-flop. The VCO input connects to the "clock" input of the bottom flip-flop. If R_{in} precedes VCO_{in}, then the data input is logic one when the clock rises, the top flip-flop latches a one, and the output up is asserted. When the VCO edge occurs, the flip-flops are cleared. Similarly, if VCO_{in} precedes R_{in}, the flip-flop latches a one, and the output down is asserted. When the R_{in} edge occurs, the flip-flops are again cleared. The flip-flops are usually a master-slave dynamic topology. They have two transmission gates and two inverters to lower the device count and reduce silicon-area requirements. The conventional CMOS circuit uses an input pass-transistor structure, which gives a longer setup time than hold time, effectively introducing a time offset between R_{in} and VCO_{in}. In addition, the setup and hold timing for the reset signal can cause a dead zone or, in some cases, oscillation. Consequently, the RS flip-flop connection is preferred in most applications.

Next, the phase/frequency detector is an edge-sensitive device. Consequently, under noisy conditions (multiple clocks on rising edges of the input from the reference oscillator) can cause false pulse-width outputs that increase the error in the loop and can eventually cause loss of lock. Laboratory measurements have shown that at −20-dB spurious-to-signal ratio, the phase/frequency detector output produces enough false edges to cause the loop to lose lock. For an exclusive-OR

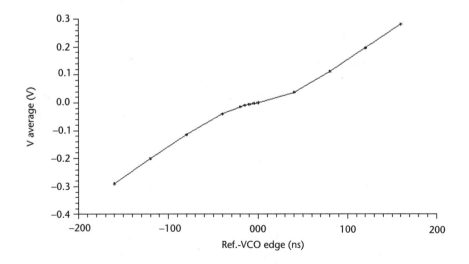

Figure 5.40   Dead-zone effect in phase detector transfer function from capacitive loading.

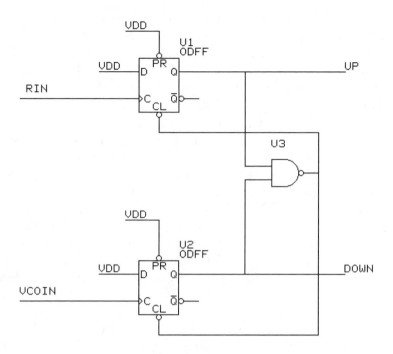

Figure 5.41 Three-state phase/frequency detector with D flip-flops.

phase detector (mixer or multiplier) that is not edge dependent, laboratory measure-ments have shown that a 0-dB spurious-to-signal ratio into these detectors produces enough error to cause the loop to lose lock. This is a 20-dB improvement. The draw back for the exclusive-OR detector is that frequency acquisition aids are needed initially to lock and maintain lock. Consequently, for most applications, the phase/frequency detector is preferred.

5.1.8 Conclusion

To summarize, modeling of the phase/frequency detector showed that the dead-zone problem is caused by capacitive loading of the output. Improved buffering of the output or increasing the driving capability of the gates can help alleviate this problem. In addition, simulations show that a pulse is always present at the output of the phase/frequency detector.

5.2 Lock Detection

Many systems use a lock-detection circuit to reset the system. This is a disastrous change to the operation of most systems. In a PLL, a reset can start the loop operating at a very low frequency (or with no output), which then acquires lock at the normally much higher output operating frequency. Consequently, a small phase shift in the PLL that would marginally affect the system can cause a huge disruption in the operation of the system if a reset occurs.

The key to lock detection is to alarm on behavior that shows the PLL is broken. A PLL can respond to a disturbance in a manner that makes it look like it is broken. Consequently, the best lock-detection schemes will allow some time for the loop to respond to a disturbance and recover. Otherwise, false alarms may occur. Quadrature phase detector, time-window edge comparison, tune-voltage window comparator, and cycle-slip detection are the most common methods for lock detection.

5.2.1   Quadrature Lock Detection

The quadrature phase detector technique as shown in Figure 5.42 uses two mixers (or analog multipliers) to compare the input reference with the in phase and quadrature phase of the fed-back VCO output. The in-phase comparison has a sin θ_e output relationship with phase error, and the quadrature comparison has a cos θ_e output relationship with phase error. The in-phase comparison is used to lock the loop because sin $\theta_e \approx 0$, and the quadrature comparison is used to detect lock because cos $\theta_e \approx 1$. The quadrature detector output is followed by a lowpass filter and a threshold comparator. A dc level above the threshold indicates that the circuit is locked. An unlocked loop produces a beat-note frequency at the output of the quadrature comparator that is lowpass-filtered to 0-V input to the threshold comparator. Without the proper filter, the lock-detection signal will flicker on and off because of noise, which will give false indications of lock or loss of lock. Too much filtering will cause an excessive delay in the lock or unlock signal. Consequently, the design of the output-smoothing filter is a vital part of a practical lock indicator. A compromise in the amount of smoothing is required. R. C. Tausworthe has performed a detailed analysis of the problem and has produced design curves [7, p. 88; 8, 9].

5.2.2   Tune-Voltage Window Comparator

A PLL can lose lock when the input frequency is to high, which pegs the tune voltage close to the supply rail or when the input frequency is too low, which pegs

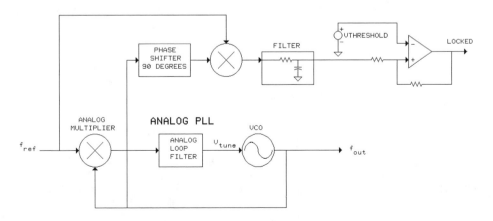

Figure 5.42   Block diagram of quadrature lock detection.

the tune voltage close to the ground supply rail. A tune-voltage window comparator, as shown in Figure 5.43, uses these two voltages to detect the unlocked condition. A tune-voltage window comparator monitors the tune line voltage with a high voltage threshold and a low voltage threshold. Crossing either the high or the low threshold causes an out-of-lock condition. Tune-voltage window comparison requires several operational amplifiers to achieve. Consequently, the size, power consumption, and threshold variation with process makes it impractical for integrated circuit design.

5.2.3 Time-Window Edge Comparison

The time-window edge comparison detector, as shown in Figure 5.44, uses an exclusive-OR gate to combine the up and down pulses out of the phase detector. Following the exclusive-OR gate with a lowpass filter rejects the narrow up and down pulses and takes the dc average of the wider up and down pulses that occur when the loop is unlocked. Monitoring the dc voltage at the output of the lowpass filter with a comparator sets a dc trip point when the loop is determined to be out of lock. Crossing the trip point to higher voltages sets the comparator to a high value (unlocked), and crossing the trip point to lower voltages sets the comparator to a low value (locked).

This detection scheme has several disadvantages. First, the time-window edge comparison (1 to 2 ns) will vary with process, which can cause false unlock conditions. The loop can have a constant phase offset that is out of the time window and will indicate a false unlock condition. A larger capacitor value is needed to increase the time window, which makes it consume area and renders it less integratable. An external disturbance can modulate the PLL so that the detector switches between the locked and unlocked condition. Even worse, the loop control logic may switch bandwidths when the loop is locked or unlocked. Switching bandwidths can disturb the loop enough that the time-window edge-comparison detector toggles, which can cause an oscillation condition to occur.

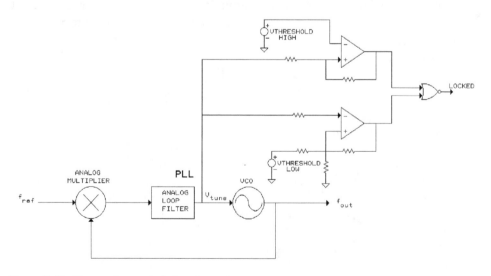

Figure 5.43 Tune-voltage window comparator.

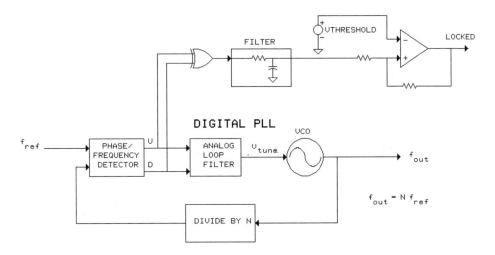

Figure 5.44 Time-window edge comparison.

In summary, the time-window-edge-comparison technique is too sensitive and gives an excessive number of unlocked conditions. Special care must be taken for this to be a successful implementation.

5.2.4 Cycle-Slip Detector

The cycle-slip detector detects frequency differences between its two input clocks that remain different for a time period greater than the inverse of the frequency difference. The detector generates a pulse when the phase/frequency detector output pulse width rolls over from its largest pulse width to its narrowest pulse width. Figure 5.45 shows the schematic for the cycle-slip detector [10]. The cycle-slip detector circuit consists of a phase/frequency detector, two cycle-slip detectors, and a final AND gate. The inputs to the circuit are the divided-down VCO waveform and the reference. The outputs of the phase/frequency detector are up pulses and down pulses. The outputs of the cycle-slip detector are active, low, up cycle slips and down cycle slips. A logical AND of the active, low, up cycle slips and down cycle slips produces an active, low cycle-slip detector output. This output is used to reset a counter and count reference clocks between cycle slips. For example, if 16 reference clocks are counted after a cycle slip, then the loop is set to the locked state by latching the carry out of the counter and disabling any further counts. Figure 5.46 gives an example connection for counting reference clocks after a cycle slip.

The cycle-slip detector does not detect loss of the reference clock or VCO feedback clock. The cycle-slip detector requires that both input clock signals be present. With one clock missing, the output of the cycle-slip detector stays a constant high, which indicates a locked condition to the following circuitry.

Figure 5.47 shows the response of the lock detector with a ramped VCO from below the lock condition to above the lock condition. The x-axis value is 0.1 ns, which makes the minor divisions equal to 5 ns. The VCO ramps from 250 to 400 MHz with the reference frequency set at 300 MHz. The left-hand side of

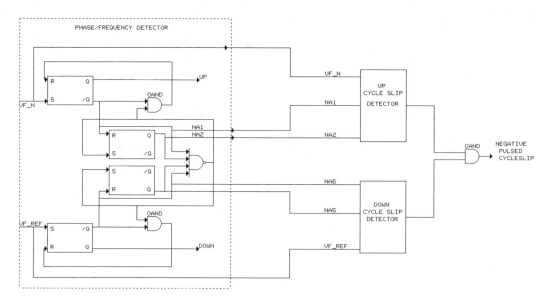

Figure 5.45   Schematic of cycle-slip detector.

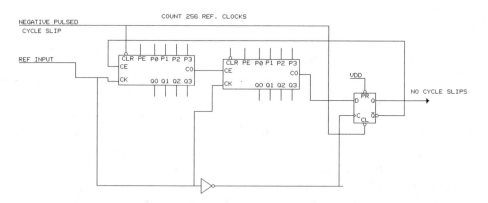

Figure 5.46   Schematic for cycle-slip counter to create lock-detection output signal.

Figure 5.44 shows that the up output of the phase/frequency detector is a pulse-width-modulated signal for a VCO frequency greater than the reference frequency. The down output of the phase/frequency detector has constant narrow pulses, which indicates no response. On the far left-hand side, the 50-MHz difference in input frequency causes a 50-MHz beat note in the pulse-width modulation output of the phase/frequency detector. Using the rising edge of the VCO as a sample clock to the reference waveform outlines a 50-MHz-period beat note. Using the rising edge to rising edge of the VCO waveform as a cycle bin and reflecting these bins on the pulse-width-modulated waveform output (VUP) of the phase/frequency shows a cycle bin where no edges occur. This is a cycle-slip up pulse condition. Sampling this nonresponse latches the cycle slip and causes one pulse per beat note.

The right-hand side of Figure 5.47 shows that the down output of the phase/frequency detector is a pulse-width-modulated signal for a VCO frequency greater

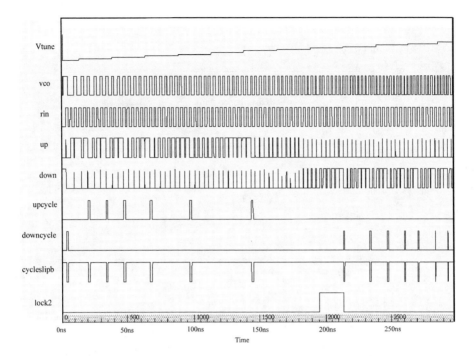

Figure 5.47   Response of the lock detector with a ramped VCO from below the lock condition to above the lock condition.

than the reference frequency. The up output of the phase/frequency detector has constant narrow pulses, which indicates no response. A cycle bin of the down output where no edges occur causes a cycle-slip down pulse to occur. The beat note on the high-frequency side does not represent the frequency difference because of aliasing in the sampling. In the middle of the figure, VCO and reference input frequencies are equal. The up and down cycles have no response. The bottom waveform shows the response of a simple 4-bit counter to generate a lock-detect signal. The wait time for the counter is set to 16. After 16 reference periods, the carry out of the counter goes high, and the counter is disabled until the next cycle slip occurs.

5.2.5   Cycle-Slip Detector Versus Time-Window Comparator

Another example compares the performances of a time-window comparator and a cycle-slip detector. The PLL has a 10-MHz input frequency and multiplies the input frequency by 24. The control system has a natural frequency of 300 kHz and a damping factor of 0.3.

Figure 5.48 shows a normal frequency and phase response of a PLL to a 1-MHz frequency step in the reference frequency. A time-window comparator set to ±10° would give an out-of-lock indication because the 10° threshold would have been crossed. In addition, the time-window comparator would toggle several times because the ±10° threshold would have been crossed several times. The cycle-slip detector would not alarm because 360° phase shift did not accumulate.

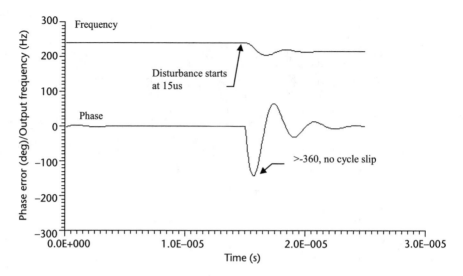

Figure 5.48 Frequency and phase-error response of a PLL to a 1-MHz frequency step in the reference frequency.

Figure 5.49 shows a normal frequency and phase response of a PLL to a 2.5-MHz frequency step in the reference frequency. Here, the cycle-slip detector is on the verge of alarm. Consequently, this figure shows that the cycle-slip detector alarms with frequency steps greater than 2.5 MHz. Figure 5.50 shows a normal frequency and phase response of a PLL to a 50-ns phase step. Here, the time-window comparator would again have multiple alarms, but the cycle-slip-based lock detector would not alarm.

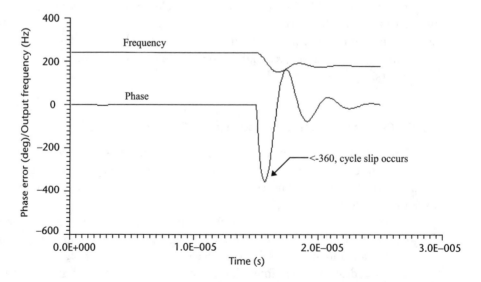

Figure 5.49 Frequency and phase-error response of a PLL to a 2.5-MHz frequency step in the reference frequency.

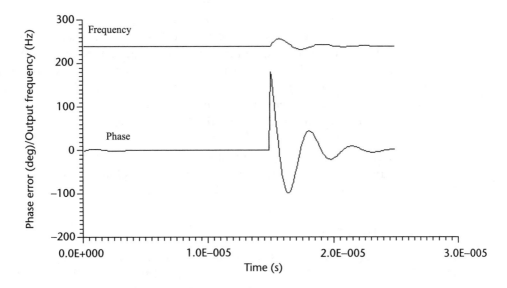

Figure 5.50 Frequency and phase-error response of a PLL to a 50-ns phase step.

Figure 5.51 shows the toggling of a window phase comparator versus solid cycle-slip lock detectors. A PLL with 16-MHz input reference frequency and 128-MHz output frequency shows how the lock detectors operate. The x-axis has the units of nanoseconds. The top waveform shows the tune voltage to the VCO, which shows that the loop locks after 2 μs. The second waveform shows the time-window lock detector. It starts toggling at 1.7 μs and continues to toggle. The bottom waveform shows the response of the cycle-slip detector with a state machine that waits 32 reference clock periods after no cycle slips occur. The cycle-slip detector detects lock at 5.3 μs and does not toggle after detection.

5.3 Acquisition Aids

If you cannot use a phase/frequency detector, then you will need an acquisition aid. One circumstance occurs for clock-recovery applications. In clock recovery,

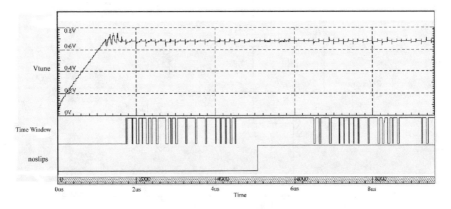

Figure 5.51 PLL transient response that compares a time-window and cycle-slip lock detectors.

multiple false locking conditions and missing clock edges occur. These conditions confuse the phase/frequency detector, and an exclusive-OR gate detector is generally used. Another circumstance occurs in frequency-generation applications. After the other loops are locked, the VCO in the translation loop is tuned to the lock condition. In only a narrow portion of the VCO tune range does the phase detector have an input. Otherwise, there is no input to the phase detector. Consequently, an acquisition aid must be used to tune the VCO until the phase detector gets a response.

In this section, we will study acquisition with an open-loop sweep and a closed-loop sweep and discriminator-aided acquisition. So, let's look at these acquisition strategies in more detail.

5.3.1 Open-Loop Sweep

As shown in Figure 5.52, a sweep can be applied to a type 2 loop in the simple manner of switching a current source in and out of the circuit. When the current source is connected, the voltage sweeps up, and when the current source is disconnected and there is no lock, the voltage sweeps down from the discharging of the capacitor. Otherwise, a separate sawtooth generator must be added to sweep the voltage directly into the VCO [7, pp. 79–87; 11, pp. 227–234; 12].

The phase detector has a maximum output of K_d rad. The output from the phase detector produces a current of K_d rad/R_1 into C in an active filter. This current causes a voltage ramp of slope K_d rad/$R_1 C$. The voltage ramp on the VCO tune line produces a frequency change with a slope that is described by (5.18):

$$K_d K_v /(R_1 C) = \omega_n^2 \qquad (5.18)$$

The frequency ramp in (5.18) is related to the natural frequency ω_n from servo mechanical definitions. Ramping the frequency faster will keep the loop unlocked. For a short time, the frequency can move faster due to the zero $R_2 C$ in the filter.

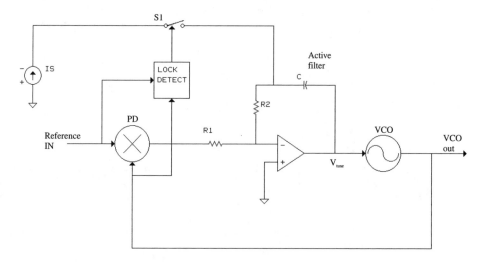

Figure 5.52 Open-loop sweep circuit for lock-acquisition aid.

A sudden change at the input to the active filter will be multiplied by R_2/R_1 and immediately appear at the filter output; however, the sustained ramp cannot move faster than (5.18). A lowpass pole from placing R across C reduces the current into C and the slope. In addition, a loop that cannot hold lock on an input signal will not be able to acquire lock on that input signal. Consequently, the maximum hold lock of ω_n^2 for a swept-input frequency applies to the acquisition limits. The 100% probability of acquiring lock with a swept input depends on the sweep rate and the amount of noise.

The sweep frequency circuit must be shut off once lock has been achieved. Shutting off the sweep circuit prevents loss of lock and reduces reference sidebands from loop-error corrections. In a closed loop, the tracking of the loop allows some leeway in time before turning off the sweeper; however, an open-loop sweep requires a quick detection of frequency agreement and a rapid shut off of the sweep.

5.3.2 Closed-Loop Sweep

One method for closed-loop sweep requires a positive-feedback circuit. Adding low-frequency positive feedback, as shown in Figure 5.53, causes the loop filter to oscillate, which sweeps the VCO frequency. When the loop locks, the large amount of negative feedback dominates the local positive feedback, which suppresses the oscillation and allows the loop to track normally. The feedback network can be a Wien bridge, and the limits of the sweep are set by the supply limits of the operational amplifier.

5.3.3 Discriminator Aided

Figure 5.54 shows a discriminator-aided frequency acquisition. This method has a phase detector loop and a frequency detector loop. With the loop out of lock, the frequency loop controls the VCO, and the phase loop has little effect. After locking, the phase loop has much greater dc gain than the frequency loop to

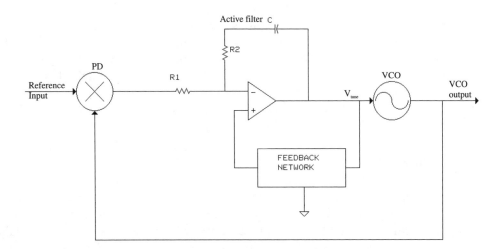

Figure 5.53 Closed-loop, low-frequency, positive-feedback sweep.

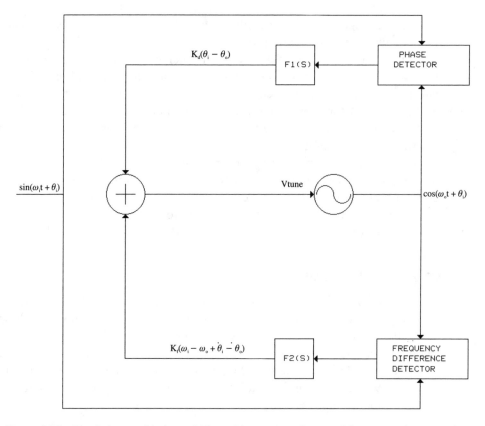

Figure 5.54 Discriminator-aided acquisition with separate phase and frequency detectors. (*From:* [7, pp. 79–87]. © 1979 Wiley Interscience. Reprinted with permission.)

dominate control of the VCO. Consequently, at lock, the discriminator can be disconnected. For a type 2 PLL, the transfer function for the frequency loop can be a type 1, which means function block $F_2(s)$ can be a simple integrator [7, pp. 79–87].

Figure 5.55 shows another discriminator-aided acquisition method. For this method, the frequency loop and phase loop share the same integrator. The response of the loop remains type 2, and the frequency path only adds to the damping factor. The frequency loop can remain connected if the relative contributions between the frequency and phase loops are adjusted to maintain the desired damping factor [7, pp. 79–87]

In a closed loop with frequency sweep activated, the coherent operations in a PLL allow locking to an input signal with low SNR. A discriminator does not distinguish between signal and noise. Consequently, for SNRs less than 6 dB, the sweep search may not successfully lock onto the input signal.

Figure 5.56 shows a frequency-difference detector (quadricorrelator). Equations in the figure describe the signal paths. The quadricorrelator mixes the input signal with an in-phase and quadrature output of the VCO to generate an in-phase and quadrature output of the frequency difference. Differentiating one signal path and mixing the two together produces a product that contains a dc component proportional to the frequency difference. Differentiating one signal path with a

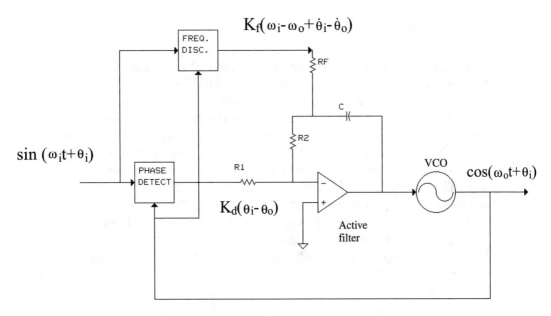

Figure 5.55 Discriminator-aided acquisition with frequency and phase loops sharing the same discriminator. (*From:* [7, pp. 79–87]. © 1979 Wiley Interscience. Reprinted with permission.)

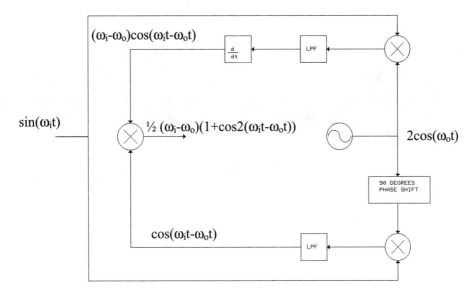

Figure 5.56 Quadricorrelator frequency-difference detector. (*From:* [7, pp. 79–87]. © 1979 Wiley Interscience. Reprinted with permission.)

highpass filter section disconnects the quadricorrelator for very small frequency differences, which allows the phase detector loop to take control of the VCO. The lowpass filters in the in-phase and quadrature signal paths suppress the intermodulation products out of the mixer and limit the operating range of the circuit. Using the in-phase signal path for the phase detector in the PLL reduces the complexity of the circuitry [7, pp. 79–87].

5.4   Charge Pumps

The charge pump in a PLL converts the differential up and down outputs of the phase/frequency detector to a single-ended, switched, positive and negative current source that charges and discharges a capacitor in order to control the VCO in a PLL. The number of its parameters makes the charge pump one of the most important circuits in a PLL (second to the VCO). The linearity of this circuit affects the stability of the loop, the amount of jitter, the amount of noise, the reference sideband levels, the phase tracking error, and the limits of the frequency operating range. The amount of peak charge pump current affects the amount of noise, the loop bandwidth, and the frequency slew rate of the PLL.

A charge pump consists of two switched-current sources driving a capacitive load as shown in Figure 5.57. The left part of Figure 5.57 shows the functional block diagram, and the right part of Figure 5.57 shows a simple implementation in CMOS. An up pulse out of the phase frequency detector turns a p channel MOSFET on to charge the capacitor in the positive voltage direction. A down pulse out of the phase/frequency detector turns an n channel MOSFET on to discharge the capacitor in the negative direction. For a pulse of width T, current I_1 deposits a charge equal to $I \times T$ on the capacitor C_p. As the phase error approaches zero, zero charge accumulates on the capacitor, and the voltage ideally remains constant [13].

Example 5.2

A phase/frequency detector drives a charge pump with a current source of 1 μa. Calculate the phase detector gain. If the charge pump has a figure of merit of 10, calculate the offset current for each current.

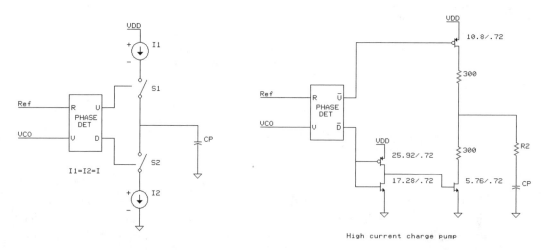

Figure 5.57   (Left) Functional diagram of phase/frequency detector with charge pump combination and (right) simple schematic of phase/frequency detector with a high-current charge pump combination.

The charge pump gain is the charge pump current divided by the range of the phase detector, which is 2π for the phase/frequency detector. Consequently, the charge pump gain is 1 μa/2π = 0.159 μa/rad. Dividing the charge pump gain 0.159 μa/rad by the phase detector figure of merit 10 computes the offset current at zero phase error to be 0.159 μa/rad/10 = 0.0159 μa.

Figure 5.58 shows a test fixture for charge pumps that helps generate the charge pump transfer function. This figure shows an ideal operational amplifier after the charge pump that processes the current out of the charge pump for testing purposes. The operational amplifier presents a high impedance load back to the charge pump output at the voltage that is set by VCPO. The high input impedance of the operational amplifier isolates the voltage source from the charge pump. Consequently, the output voltage to the charge pump can be varied from one supply rail to the next without loading down the charge pump. When switches S_1 or S_2 are closed, the current must go through the R_1 feedback resistor because of the negative feedback around the amplifier. Consequently, measuring the current in R_1 in a SPICE simulation measures the current sourced by the charge pump. Finally, the capacitor C_1 is used to attenuate high-frequency switching spikes that can occur. Figure 5.59 shows a low-current version of the charge pump where transistors are cascoded to make a better current source.

Figure 5.60 shows a differential charge pump, which is another variation. A differential charge pump can help minimize the distortion in the transfer function. In this circuit, the control from the phase detector turns on only pull-down currents. The control signals switch differential pairs $M_1 - M_2$ and $M_3 - M_4$. The pull-up currents Ip are always present. Consequently, if up and down controls are in the low state, a common-mode feedback circuit consisting of transistors $M_5 - M_9$ counteracts the pull-up currents to set a proper output common-mode level at

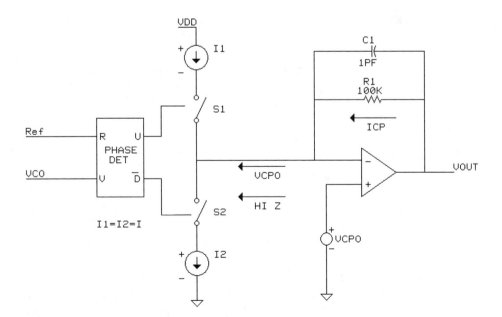

Figure 5.58 Circuit to process the output of a charge pump and generate the transfer function.

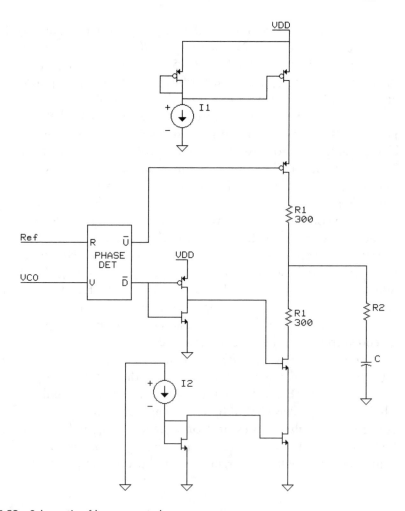

Figure 5.59 Schematic of low-current charge pump.

$V_{GS5} + V_{GS9}$. Each phase comparison momentarily disturbs the common-mode level. Consequently, care must be taken to avoid common-mode transients that can lead to differential settling components.

A dead zone can occur for a charge pump. If the output pulse widths of the phase/frequency detector are small, and there is not enough drive current or there is a high load capacitance, then the logic level will not be high enough to turn the charge pump on. This causes a low-gain time width around the zero phase point where a PLL locks. Zero phase detector gain makes the loop unlock and drift until enough error is generated to cause a correction. Consequently, the amount of jitter this causes is proportional to the time width of the dead-zone area.

Current offsets, transistor mismatches, and charge sharing can cause errors in the transfer characteristics of the charge pump. This can cause an offset and distortion in the transfer characteristics of the charge pump. An offset in the transfer characteristics of the charge pump produces a constant pulse width out of the charge pump when the loop is locked. This pulse width adds to the output jitter of the PLL.

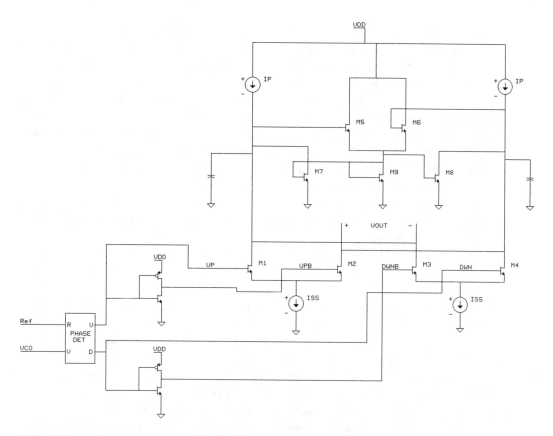

Figure 5.60 Schematic of a differential charge pump.

Using a charge pump has several advantages. Charge pumps have a simple implementation that takes up less area. In addition, this circuit has potential for high-frequency response. Finally, a charge pump is the preferred implementation for a loop filter outside an integrated circuit because it minimizes the number of external components (only needs one large capacitor, a smaller roll-off capacitor, and one resistor.

Using a charge pump also has some disadvantages. Any imbalance of the two transistors can cause a dead zone in the transfer function response. Charge sharing in the parasitic capacitors of the output transistors can be one of the major causes of this imbalance. Furthermore, to get narrow-bandwidth PLLs requires reducing the charge pump gain. At some level this causes a low SNR with these low currents, which adds noise to the PLL over the reference phase noise. Eventually, with a low enough SNR, the loop will not lock.

Next, an ideal charge pump with infinite gain requires an infinite output impedance. A practical charge pump has a finite output impedance, which significantly reduces the gain from the ideal charge pump case. Consequently, the finite output impedance of a practical charge pump gives a low gain when compared to operational amplifiers. This low gain reduces the PLL's tracking effectiveness and insensitivity to disturbances.

Next, as discussed previously, the imbalance of the charge pump transistor can cause an offset in the transfer function, which reduces the figure of merit. The high output impedance of a charge pump produces a significant slew-rate limit. If a large frequency step occurs, the loop is slew-rate limited by the charge pump current. This would be a disadvantage for frequency-hopping applications.

As discussed earlier, charge sharing is one of the major causes of imbalance in a charge pump. The charge pump circuit in Figure 5.61 shows one method of addressing this problem [14]. Similar to the other charge pump circuits, the up and down signals from a phase frequency detector switch current sources I_{up} and I_{dn} onto node $V_{control}$. The selected current source delivers a charge to move $V_{control}$ up or down. I_{up} and I_{dn} currents need to be equal to minimize offsets. Connecting a unity gain amplifier as shown in Figure 5.61 helps minimize these offsets. The amplifier accomplishes this balance by biasing nodes N_1 and N_2 when they are not switched to $V_{control}$. This bias suppresses any charge sharing from the parasitic capacitance on N_1 or N_2.

Figure 5.62 adds the connection of the integrating capacitor in the loop filter and the parasitic capacitors of the output transistors to the charge pump connection described in Figure 5.61. Figure 5.62 shows the charge-sharing mechanism in the charge pump attached to a loop filter. A pair of matched switched-current sources charges or discharges a dc current I_o into or out of a filter capacitor (a large n channel transistor), depending on the amount of phase error. The up and down switch-control signals add charge or remove charge for a fixed time, which pumps a fixed-size charge packet into or out of the capacitor on each cycle. The operational amplifier connected as a unity-gain buffer provides a low-impedance version of the capacitor voltage, and its output V_{CTRL} is sent to nodes N_1 or N_2.

The unavoidable parasitic capacitance C_p at the current-source output charge-shares with the filter capacitor. This charge sharing produces a charge-error term. Clamping each parasitic capacitor C_p of the current source by using the unity gain amplifier that is not charging the filter capacitor to the $V_{control}$ voltage prevents

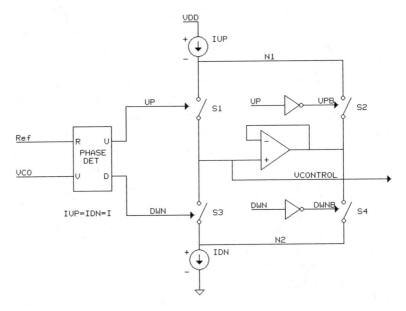

Figure 5.61 Differential charge pump with unity gain feedback to minimize charge sharing.

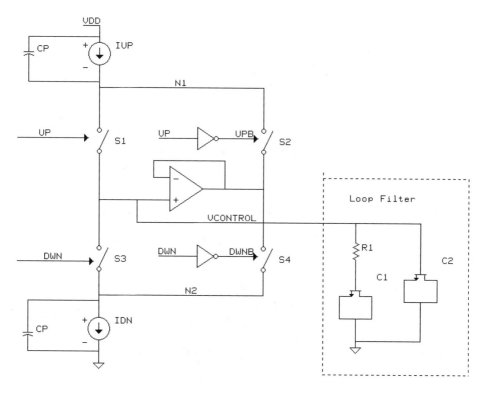

Figure 5.62   Loop-filter and parasitic capacitances that show charge-sharing effects in a differential charge pump with unity-gain feedback.

charge-sharing errors. Keeping the $V_{control}$ voltage on the parasitic capacitor C_p makes a very small ΔV when the parasitic capacitor later connects to the filter capacitor. Consequently, very little charge sharing can occur.

If one were going to use this approach, one might consider using the operational amplifier itself to convert differential to single-ended input. However, if the application requires a 1-μF external capacitor from the IC, a charge pump would be better because only one external capacitor and pad would be necessary, whereas a differential operational amplifier configuration would require two external capacitors and two pads. Many IC customers do not want this extra expense.

To show how the charge pump works, let's use an example. For this example, we will use the simple high-current charge pump circuit shown in Figure 5.57. It will be connected to a test fixture by the method shown in Figure 5.58 in order to get the transfer function.

Figure 5.63 shows the variation of charge pump transfer curve with output voltage. As the voltage approaches one rail or the other one of the output, transistors start to turn off, which causes an offset current. Furthermore, the gain of the charge pump detector varies with voltage and with negative and positive phase error. At the nominal condition (supply midpoint) of 2.5V, the negative phase-error slope equals 295 μa/rad, and the positive phase-error slope equals 390 μa/rad. At 3.5V the negative phase-error slope decreases to 199 μa/rad, and the positive phase error slightly increases to a slope of 429 μa/rad. At 1.0V the negative phase-error slope slightly increases to 356 μa/rad, and the positive phase-error slope decreases

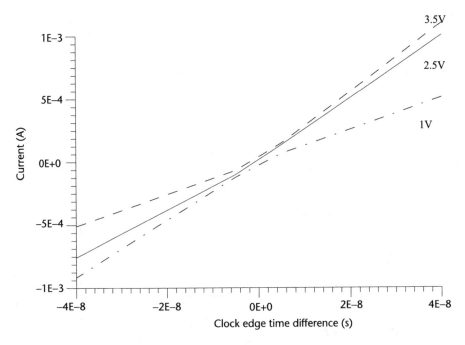

Figure 5.63 Variation of simple charge pump transfer function with output voltage.

to 199 μa/rad. Different gains about the zero-error point can cause oscillations in the loop, so that the loop never locks. A difference of less than 2-to-1 phase detector gain between positive and negative phase-error should be maintained to avoid oscillations.

Figure 5.64 shows the variation of charge pump slope with nominal, weak, and strong process conditions. The transfer function was generated by varying the positive-edge-to-positive-edge time difference between two 10-MHz input clocks with 50% duty cycles. For nominal conditions, the positive current phase-error slope is 390 μa/rad and the negative current phase-error slope is 295 μa/rad. For weak conditions, the positive current phase-error slope decreases to 219 μa/rad, and the negative current phase-error slope decreases to 127 μa/rad. For strong conditions, the positive current phase-error slope increases to 624 μa/rad, and the negative current phase-error slope increases to 644 μa/rad. This is a 490% worst-case variation in-phase detector gain over process, which is a large variation. Furthermore, a 72% variation in positive to negative phase-error slope borders on causing an unstable loop condition.

5.5 Design Considerations for Opamps in a PLL

An operational amplifier provides an alternative method to a charge pump. The operational amplifier in a PLL converts the differential up and down outputs of the phase/frequency detector to a single-ended positive and negative voltage source that charges and discharges capacitors in order to control the VCO in a PLL. The number of PLL parameters that the operational amplifier affects makes it one of the most important circuits in a PLL (second to the VCO).

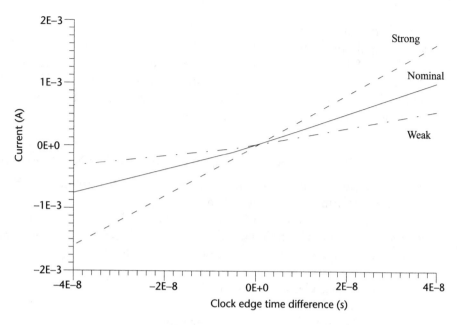

Figure 5.64 Variation of simple charge pump transfer function with process and temperature.

Listed in order of importance, the following operational parameters affect PLL performance:

- Rail-to-rail output voltage swing ≥ (wide PLL frequency range);
- High power-supply rejection ≥ (low jitter and low spurious);
- Low differential offset bias current ≥ (low-jitter and low reference sidebands);
- Wide input common-mode voltage range ≥ (wide PLL frequency range);
- Wide bandwidth ≥ (low noise and jitter);
- High dc gain ≥ (low tracking error and low noise);
- Low-noise figure ≥ (low noise, low jitter, and improved PLL stability);
- Output current loading ≥ (improved stability and low jitter);
- Fast slew rate ≥ (fast PLL switching time).

The linearity of this circuit affects the stability of the loop, the amount of jitter, the amount of noise, the reference sideband levels, the phase tracking error, and the limits of the frequency operating range.

Voltage offset of an operational amplifier is not a problem in PLLs because it operates like a switch capacitor circuit with correlated double-sampling auto-zeroing. With the PLL close to lock, there are small changes in-phase error at the rate of the reference-clock frequency. The amplifier offset is constant through the reference-clock period and its effect is held on the main integrator capacitor. The effect of the offset causes a onetime phase error (since it is constant) that is subtracted by the next phase-error sampling off the main integrator capacitor, which results in autozeroing of the offset of the amplifier.

We will begin the study of CMOS operational amplifiers for PLLs by studying the baseline two-stage amplifier and then comparing performance to this baseline.

Next, a folded-cascode operational amplifier is studied because of its high power-supply ripple rejection, which is crucial for PLLs. Then, the preferred AB-input, two-stage, folded-cascode operational amplifier is presented because it has high PSRR, low offset voltage, high output drive capability, rail-to-rail input operation, and rail-to-rail output operation. Rail-to-rail operation is the key to preventing the loop from hanging up at either power rail.

5.5.1   Architecture Selection, Comparison to Basic Two-Stage Opamp

For PLL design, we must select the operational amplifier architecture that best fits the list of performances mentioned earlier in this section. We will select and study a basic, two-stage, differential amplifier followed by common-source amplifier-type architectures. The following describes the advantages and disadvantages of some of the differential amplifiers to be considered:

- Folded cascodes and telescopics:
  - Advantages:
    - Good high-frequency power-supply rejection;
  - Disadvantages:
    - Reduced output swing;
    - Higher offset voltage;
    - Reduced voltage gain.
- Class B:
  - Advantages:
    - High output swing;
    - Good slew rate;
    - High gain.

The basic topology used for most CMOS opamps is shown in Figure 5.65. A differential input stage drives an active load, followed by a second gain common-source stage. An output stage may be added for driving heavy loads off-chip. This circuit configuration provides good common-mode range, output swing, voltage gain, and common mode rejection ratio (CMRR) in a simple circuit that can be compensated with a single capacitor.

5.5.2   Basic Opamp

In this section we will analyze the various performance parameters of the CMOS operational amplifier circuit. To do the analysis, we will define the transistor parameters for a 0.8-μm gate length with 5-V supply process and equations to help in the analysis and synthesis. Table 5.3 shows the transistor parameters. We will assume no loading of the first stage by the second stage because of the essentially infinite input resistance of MOS devices. This allows us to separate the two stages in order to find the voltage gain of the amplifier. Let's look at the gain of the first stage.

Using the small-signal equivalent circuit for the transistors allows the voltage gain to be calculated. Equation (5.19) computes the first-stage voltage gain:

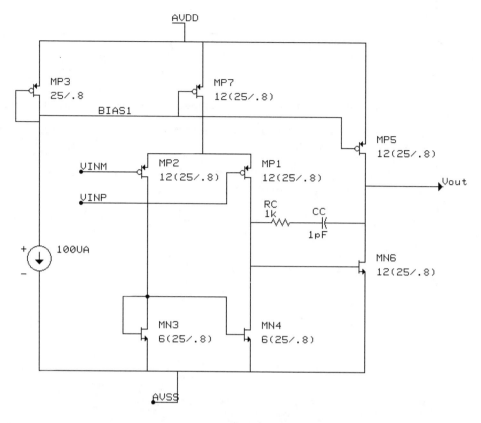

Figure 5.65 Basic two-stage operational amplifier topology.

Table 5.3 List of 0.8-μm Gate-Length Transistor Parameters Used in Studying Opamps

Transistor Parameter	Value
U_{ncox}	191 μA/V^2
U_{pcox}	76 μA/V^2
λ_n	0.33 μV
λ_p	0.14 μV

$$A_{v1} = \frac{g_{mp2}}{g_{dsp1} + g_{dsn4}} \tag{5.19}$$

where

g_{mp2} = transconductance of the input transistors;

g_{dsp1} = small-signal device-output conductance of the input transistor;

g_{dsn4} = small-signal device-output conductance of the active load transistor.

Similarly, (5.20) computes the second-stage common-source voltage gain:

$$A_{v2} = \frac{g_{mn6}}{g_{dsp5} + g_{dsn6}} \tag{5.20}$$

where

g_{mn6} = transconductance of the second-stage common source transistor;

g_{dsn6} = small-signal device-output conductance of the second-stage common source amplifier;

g_{dsp5} = small-signal device-output conductance of the active load for the second-stage common-source amplifier.

Consequently, combining gain equations gives (5.21) to compute the overall gain:

$$A_v = \frac{g_{mp2}}{g_{dsp1} + g_{dsn4}} \frac{g_{mn6}}{g_{dsp5} + g_{dsn6}} \tag{5.21}$$

The overall gain is thus related to the quantity $(g_m/g_{ds})^2$ as shown by (5.22) and (5.23).

$$\frac{g_m}{g_{ds}} = \frac{2L}{(V_{gs} - V_t)} \frac{1}{\dfrac{dX_d}{dV_{ds}}} \tag{5.22}$$

where

$V_{gs} - V_t$ = effective gate source voltage;

V_{ds} = drain to source voltage.

$$\frac{g_m}{g_{ds}} = \frac{2L}{(V_{gs} - V_t)} \frac{1}{\lambda} \tag{5.23}$$

where

λ = the output impedance constant (V^{-1}).

Thus, the overall voltage gain is a strong function of the bias point chosen for the devices and the channel length of the transistors used, and $(V_{gs} - V_t) = 0.2V$ for all devices.

Example 5.3

For example, consider a typical 0.8-μm gate length that has a channel-length modulation parameter of 0.33 μ/V for an n channel transistor and 0.14 μ/V for a p channel transistor, and set $(V_{gs} - V_t) = 0.2V$. Let's apply these values to the opamp circuit in Figure 5.65 and calculate the gain for an n channel and p channel transistor and the gain for the whole amplifier. For a typical 0.8-μm process,

λ has a value of 0.33 μ/V for an N-channel MOS transistor and 0.14 μ/V for a P-channel MOS transistor with 0.8-μm channel lengths.

Then (5.23) gives g_m/g_{ds} of $25 = 2 \times 0.8/(0.2 \times 0.14)$ for NMOS devices and 57 for PMOS devices. Equation (5.21) gives $A = -25 \times 57 = -1,385$.

Example 5.3 shows that either increasing the channel length of the devices used or reducing the bias current to reduce $(V_{gs} - V_t)$ increases the voltage gain. Unfortunately, both of these changes degrade the frequency response of the amplifier. Equation (5.24) computes the frequency bandwidth of the operational amplifier:

$$\omega_t = \frac{g_{mp2}}{C_c} \qquad (5.24)$$

where

$$g_{mp2} = 2I_{dp2}/V_{dsat} \qquad (5.25)$$

where

$V_{dsat} = (V_{gs} - V_t)$.

For an I_{d2} current of 55 μa, a V_{dsat} of 0.2, and a miller compensation capacitance C_c of 1 pF, substituting the g_{mp2} (550 μmhos) results from (5.25) into (5.24) computes an 87.5-MHz bandwidth. Next, from high-frequency analysis, the stability of the operational amplifier can be computed as shown by (5.26):

$$\theta_{pm} = \frac{\pi}{2} - \mathrm{atan}\left(\frac{\dfrac{g_{mp2}}{C_c}}{\dfrac{g_{mp5}}{C_L}}\right) - \mathrm{atan}\left(\frac{\dfrac{g_{mp2}}{C_c}}{\dfrac{g_{mp5}}{C_{gsp5}}}\right) \qquad (5.26)$$

where

C_L = the output load capacitance.

Substituting the previous values and a load capacitance of 1 pF, a g_{mp5} of 1.7 ma/V, and a C_{gsp5} of 141 fF into (5.26) computes a phase margin of 69°.

Next, the systematic offset voltage can be computed as shown by (5.27)

$$V_{oscm} = \frac{V_{icm}}{\mathrm{CMRR}} = \frac{V_{icm}}{2g_{mp2}\dfrac{1}{g_{dsp7}}g_{mn3}\dfrac{1}{g_{dsp2}}} \qquad (5.27)$$

where

V_{icm} = input common mode voltage.

Substituting an input common mode of 2.5V, g_{mn3} of 1.2 ma/V, g_{dsp2} of 19 μmhos, and a g_{dsp7} of 26 μmhos into (5.27) computes an offset of 0.93 mV. Furthermore, (5.27) shows that the CMRR is related to the offset voltage; it is given by (5.28):

$$\text{CMRR} = 2g_{mp2}\frac{1}{g_{dsp7}}g_{mn3}\frac{1}{g_{dsp2}} \qquad (5.28)$$

Substituting the previous values into (5.28) computes a CMRR of 2,672. Further degradation in offset can be expected due to the mismatch of the transistors in the process.

With the bandwidth computed, we now compute the slew rate of the amplifier, which depends on the miller capacitance and the loading capacitance of the operational amplifier. Equation (5.29) computes the slew rate for the operational amplifier assuming no capacitive loading:

$$SR = I_{p5}/C_c \qquad (5.29)$$

For an I_{p5} bias current of 110 μa and a C_c miller capacitance of 1 pF, (5.29) computes a 110 V/μs slew rate.

Finally, let's look at the PSRR of the operational amplifier. The PSRR to the power supply is a function of the input transistor transconductance, mismatch with the other input transistor transconductance, and the tail-current transistor output conductance. Equation (5.30) shows this relationship:

$$\text{PSRR} = \frac{g_{mp2}}{g_{dsp7}}\frac{2g_{mp2}}{\Delta g_{mp2}} \qquad (5.30)$$

where

Δg_{mp2} = mismatch ratio between input transistors.

Using a Δg_{mp2} of 0.01% and the previous values for the other variables computes a PSRR of 47 dB (ratio of 232).

Thus, we find fundamental trade-offs between frequency response, gain, and offset voltage in CMOS operational amplifier design. The above basic, two-stage amplifier with many variations has been widely used in a variety of applications; however, the evolution of several alternative circuits optimizes certain aspects of the performance of the amplifier. In the next section, we consider variations on the basic circuit and alternative architectures.

5.5.3 Folded Cascode

Many modern IC applications need CMOS opamps designed to drive only capacitive loads. Capacitive-only loads make it necessary to use a voltage buffer to obtain a low output impedance for the opamp. Opamp designs that realize this buffer have higher speeds and larger signal swings than those that must also drive resistive loads. Having only a single high-impedance node at the output of an opamp that drives only capacitive loads yields these improvements. All the other nodes in these opamps have relatively low impedance from the admittance seen at all the other nodes, being on the order of a transistor's transconductance. Having relatively low impedance on all the internal nodes maximizes the speed of the opamp. These low node impedances also result in reduced voltage signals at all nodes other than the output node; however, various transistors can have quite large current signals. With these opamps, the load capacitance compensates the amplifier. Increasing the load capacitance usually makes the opamp more stable, but it also makes it slower. The transconductance parameter (i.e., the ratio of the output current to the input voltage) distinguishes these opamps from the other modern opamps. Consequently, they are sometimes referred to as *operational transconductance amplifiers* or *transconductance opamps*.

Figure 5.66 shows an example of such an amplifier. This circuit is also known as the *folded cascode*. From CMOS amplifier theory, cascode configurations increase the voltage gain of CMOS transistor amplifier stages. A single common-source, common-gate stage gain can provide enough voltage gain for many applications. This opamp configuration also has a high output impedance.

The design converts a differential input to single-ended output. All the current mirrors in the circuit use wide-swing cascode current mirrors. The use of these mirrors results in high-output impedance for the mirrors (compared to simple current mirrors). This higher impedance maximizes the dc gain of the opamp. Even though a folded-cascode amplifier only has a single gain stage, it can have reasonable gain on the order of 700 to 3,000.

A normal cascode would have NMOS input devices M_{P1} and M_{P2} driving M_{N3} and M_{N4}. For ac signals in Figure 5.66, the NMOS devices M_{N3} and M_{N4} fold the cascode up to the PMOS input transistors M_{P1} and M_{P2}. This circuit is a

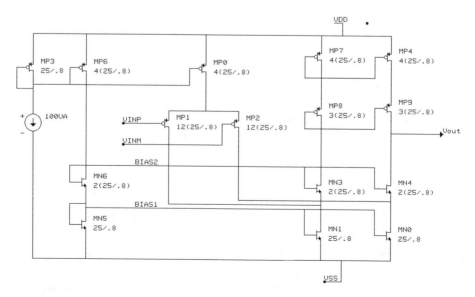

Figure 5.66   Schematic of a folded-cascode operational amplifier.

"folded-up" version of a standard cascode. Additionally, this circuit simultaneously performs a level-shifting function [15].

The folded-cascode opamp applies cascode transistors to the input differential pair, but by using transistors opposite in type from those used in the input stage. For example, the input differential pair has p channel transistors M_{P1} and M_{P2} in Figure 5.66, and the cascode transistors have n channel transistors M_{N3} and M_{N4}. This arrangement of opposite-type transistors allows the output of this single gain-stage amplifier to be taken at the same bias-voltage levels as the input signals [16].

Now, for the other transistors, the current mirror composed of M_{P4}, M_{P7}, M_{P8}, and M_{P9} converts differential output to single ended. The circuit in the figure uses a simplified bias network M_{P3}, M_{P8}, M_{N5}, and M_{N6} to simplify matters. The intent here is to show the basic architecture. In practical circuits, the bias for I_{DP0} might be replaced by a constant transconductance bias network.

The load capacitor C_L determines the compensation by setting the dominant pole. In applications where the load capacitance is very small, it is necessary to add additional compensation capacitance in parallel with the load to guarantee stability. If lead compensation is desired, a resistor can be placed in series with load capacitor C_L. In some applications, lead compensation may not be possible because the compensation capacitance is mostly supplied by the load capacitance; however, it is more often possible than many designers realize. For example, it is often possible to include a resistor in series with the load capacitance.

The input differential-pair transistors have bias currents equal to $I_{DP0}/2$. Making the currents in current sources M_{N1} and M_{N0} larger than $I_{DP0}/2$ provides the proper bias to the transistors. The bias current I_{DN3} of one of the n channel cascode transistors M_{N3} or M_{N4}, hence the transistors in the output-summing current mirror as well, equals the drain current of M_{N0} or $M_{N1 \text{ minus}} I_{DP0}/2$. Consequently, this bias relationship gives (5.31) and (5.32):

$$I_{DN3} = I_{DN4} = I_{DN0} - I_{DP0}/2 \tag{5.31}$$

$$I_{DN0} = I_{DN1} \tag{5.32}$$

An I_{dp0} of 110 μa and I_{DN0} of 135 μa give an 80-μa current through I_{DN3} and I_{DN4}. The ratio of $(W/L)_{N1}$ or $(W/L)_{N0}$ to bias transistor mirror M_{N5} establishes the drain current I_{DN0}. Since the bias current of one of the cascode transistors is derived by a current addition, for it to be accurately established, it is necessary that both I_{DN1} and I_{DP0} be derived from a single-bias network. In addition, any current mirrors used in deriving these currents should be composed of transistors realized as parallel combinations of unit-size transistors. This approach eliminates inaccuracies due to second-order effects caused by transistors having nonequal widths [16].

Let's look at the two paths of the gain from input to output. A differential ac input voltage V_{in} causes $g_{mp2} V_{in}/2$ ac current to flow in the input transistors. This current flows from transistor M_{P2} through transistor M_{N4} to the output. Transistor M_{P2} has a common-source amplifier connection, and transistor M_{N4} has a common-gate amplifier connection. From this connection, this circuit is also known as a common-source, common-gate amplifier. The current through transistor M_{P1} flows through transistor M_{N3} and is mirrored by transistors M_{P8} and M_{P7} through transistor M_{P4} and M_{P9} to the output.

Equation (5.33) computes the small-signal voltage gain of this circuit at low frequencies, which is a function of the input transistor transconductance and the output conductance contributions from M_{P9} and M_{N4}:

$$A_v = \frac{g_{mp2}}{g_{outp9} + g_{outn4}} \tag{5.33}$$

Equation (5.34) computes the output conductance contribution from M_{P9}:

$$g_{outp9} = \frac{g_{dsp9}}{1 + \dfrac{g_{mp9}}{g_{dsp4}}} \tag{5.34}$$

Equation (5.35) computes the output conductance contribution from M_{N4}:

$$g_{outn4} = \frac{g_{dsn9}}{1 + \dfrac{g_{mn4}}{g_{dsn0} + g_{dsp2}}} \tag{5.35}$$

Substituting (5.33) into (5.35), an g_{mp2} of 1.03 ma/V, a g_{dsp2} of 18.6 μmhos, a g_{mp9} of 0.93 ma/V, a g_{dsp9} of 13.9 μmhos, a g_{mn4} of 1.3 ma/V, a g_{dsn4} of 30.7 μmhos, a g_{dsp4} of 14.3 μmhos, and a g_{dsn0} of 32 μmhos into the equation computes a gain of 57 dB (ratio of 756).

Next, let's look at the stability of the amplifier. The load capacitance C_L performs the function of a compensation capacitor. This gives an important advantage to this circuit versus the previous circuit, which required the additional miller capacitance C_c to keep the amplifier from oscillating in a feedback loop connection. An important disadvantage of single-stage amplifiers is that the output voltage swing is degraded by the presence of the cascode transistors M_{N3}, M_{N4}, M_{P8}, and M_{P9}. A high-swing cascode configuration can be directly applied to the single-stage amplifier to improve the swing. The high swing cascode results in an available output voltage swing that extends to within $2(V_{GS} - V_t) = 2\,\Delta V_{GST}$ of each supply.

Analysis of the opamp begins with an expression for the gain. Equation (5.36) computes the gain bandwidth [17]:

$$\omega_t = \frac{g_{mp2}}{C_L + (C_{dbn4} + C_{dbp9})} \tag{5.36}$$

A capacitive load of 0.5 pF, M_{N4} drain capacitance of 26 fF, and M_{P9} drain capacitance of 55 fF substituted into (5.36) gives a 282-MHz unity-gain crossover frequency.

Next, stability is determined by evaluating the nondominant poles. We will use the three-pole approximation for the frequency response as shown by (5.37):

$$\theta_{pm} = 90 - \mathrm{atan}\left(\frac{\omega_t}{p_1}\right) - \mathrm{atan}\left(\frac{\omega_t}{p_2}\right) - \mathrm{atan}\left(\frac{\omega_t}{p_3}\right) \tag{5.37}$$

Pole 1 occurs at the junction of M_{N4} and M_{N0}. Equation (5.38) computes its value:

$$p_1 = \frac{g_{mn4} + g_{mbn4}}{C_{dbn4} + C_{dbn0}} \tag{5.38}$$

Pole 2 occurs at the junction of M_{P4} and M_{P7}. Equation (5.39) computes its value:

$$p_2 = \frac{g_{mp4}}{C_{gsp7} + C_{dbp8}} \tag{5.39}$$

Pole 3 occurs at the junction of M_{P9} and M_{P4}. Equation (5.40) computes its value:

$$p_3 = \frac{g_{mp9}}{C_{dbp9} + C_{dbp4}} \tag{5.40}$$

Other poles are less significant. A g_{mbn4} of 195 μa/V, a C_{dbn0} of 33 fF, a g_{mp4} of 1.01 ma/V, a C_{gsp7} of 53 fF, a C_{dbp8} of 68 fF, and a C_{dbp4} of 79 fF substituted into (5.37) computes a phase margin of 59°. For design purposes, we check that poles 2 and 3 are more than 10 times greater than ω_t so that they do not affect the stability of the amplifier and its crossover frequency point. Then, we adjust pole 1 to give us our phase margin.

Next, we analyze the slew rate. Equation (5.41) computes the slew rate of a folded cascode:

$$SR = \frac{I_{dp0}}{C_L + (C_{dbn4} + C_{dbp9})} \tag{5.41}$$

Using an I_{dp0} of 110 μa, a C_L of 0.5 pF, and the previous capacitor values computes a slew rate of 183 V/μs.

Figure 5.67 shows an AB-input, two-stage, folded-cascode operational amplifier for rail-to-rail operation. The folded-cascode state is followed with a common-source push-pull output stage for current boosting, higher gain, and output rail-to-rail swing. This circuit has gain from any common-mode input to the operational amplifier. This occurs because either the n channel or the p channel differential amplifier will be turned on at one of the rails. This is important in PLL designs to prevent the loop from hanging up on either rail. As long as there is gain at the rails, the integrator will operate and move the control voltage to the VCO back toward the locking condition. The equations developed in the earlier designs can also be applied to this design.

Table 5.4 compares the performance of operational amplifiers. This comparison shows the strengths and weaknesses of each design architecture. OP27 and 741 operational amplifiers are individual product ICs and are shown for reference. The AB-input, two-stage, CMOS operational amplifier is preferred for PLLs.

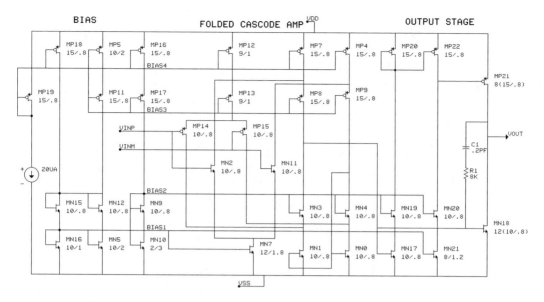

Figure 5.67   AB-input, two-stage, folded-cascode operational amplifier for rail-to-rail operation.

5.6   Differences Between Charge Pump and Operational Amplifier Compensation

One of the most important concepts to understand in PLL design is that synthesis of the compensation circuit only approximates an ideal integrator, which is needed to make an ideal type 2 PLL. A charge pump would have to have infinite output impedance, and an operational amplifier would have to have infinite gain in order for a compensation circuit to have an ideal integrator. Consequently, a decision has to be made as to which circuit makes a better integrator and what the difference in performance is.

Differences in performance show up in the following areas:

- Tracking a swept frequency;
- Noise suppression in phase-noise plots;
- Steady-state phase-error tracking.

5.6.1   Error Tracking of Charge Pump and Active Compensation

Active internal compensation allows better error tracking than charge pump compensation. An example design comparison between charge pump internal compensation and type 2 active compensation makes it easier to show the advantage. Table 5.5 lists the example design parameters. The 75-MHz reference frequency determines the limit for the loop bandwidth because of sampling effects. Consequently, a 1.5-MHz loop bandwidth was chosen because it is far enough away to minimize sampling effects.

Table 5.6 lists the charge pump solution and the type 2 active-compensation solution. For charge pump compensation, charge pump gain and C_1 are used to

Table 5.4 Comparison of Well-Known Opamp Performance Versus Various CMOS Architectures

Key Performance Parameters	OP27	LM741	Typical CMOS 3-μm Process	CMOS State of the Art	Basic Two-Stage CMOS	Folded-Cascode CMOS	AB-Input Two-Stage CMOS
Voltage gain	10^6	10^5	10^4	10^6	10^3	10^3	10^5
dc V_{os}	10 μV	2 mV	5 mV	20 μV	2 mV	50 mV	250 μV
Input bias current	10 na	300 na	—	—	0	0	0
Frequency of unity gain	8 MHz	1 MHz	10 MHz	500 MHz	150 MHz	300 MHz	180 MHz
Slew rate	2.8 V/μs	1 V/μs	10 V/μs	1,000 V/μs	44 V/μs	52 V/μs	83 V/μs
Power dissipation	90 mW	50 mW	1 mW	10 μW	1.65 mW	2.75 mW	3.6 mW
PSRR	1 μV/V	—	—	—	60 dB	60 dB	79 dB
Input noise density	3 nV/sqr Hz	20 nV/sqr Hz	—	—	9 nV/sqr Hz	10 nV/sqr Hz	22 nV/sqr Hz

Table 5.5   Example PLL Design
Parameters

Design Parameter	Value
Reference frequency	75 MHz
K_v	700E6 rad/s/V
N	1
Loop bandwidth	1.5 MHz
Damping factor	0.8

Table 5.6   Charge Pump and Type 2 Active-
Compensation Solutions to Example Design

Component	Charge Pump	Type 2 Active Compensation
K_d	1.3 mA/rad	0.4 V/rad
R_1	0	0.666 Megohms
R_2	22	10 Kohms
C_1	0.12 μF	20 pF
C_2	2,200 pF	0

adjust the bandwidth, and R_2 adjusts phase margin. For type 2 active-compensation phase detector gain, the ratio of R_2/R_1 to C_1 adjusts bandwidth, and R_2 adjusts phase margin.

Next, error tracking for active compensation will be compared with the charge pump compensation. Using a behavioral model and ac SPICE analysis (refer to Section 8.2), Figure 5.68 shows the resulting error transfer function for the charge

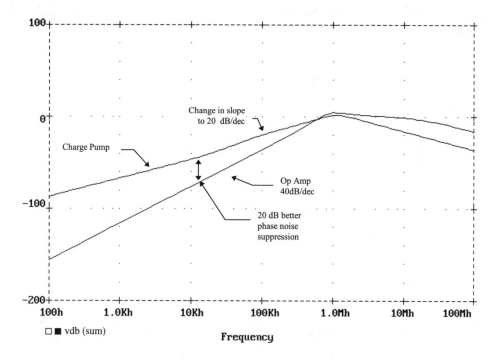

Figure 5.68   Error transfer function for the charge pump and active-based compensation.

pump and active compensation. This figure shows 10–20-dB lower error for active compensation. This is due to the lower gain for the charge pump when compared to an operational amplifier. A parallel 40Ω resistor with the current source was added to model the reduced gain in the charge pump. This accounts for the change in slope at 100 kHz. The difference in control-system response shows up in the amount of phase noise and error tracking. To demonstrate the error-tracking effect on loop performance, a simulation with a sweeping input frequency (triangle wave) will be used.

From doing transient analysis on the behavioral model, Figure 5.69 shows the results for charge pump compensation and active compensation with the swept-input frequency. The frequency of the triangle wave was increased until a significant difference in tracking occurred. Active compensation tracks the triangular modulation closer than charge pump compensation. Closer tracking of the input modulation means the loop will have higher immunity to disturbances. This is a relative measurement of whether the loop is locked with Post-it Note glue that a slight wind can blow off the lock or Super Glue that a strong wind cannot blow off the lock.

5.6.2 Phase-Noise Suppression

Suppression of close-in phase noise is another area where the difference between charge pump– and opamp-based compensation shows up. This occurs at the loop bandwidth of the PLL when the phase noise of the VCO is much greater than the multiplied phase noise of the reference crystal oscillator. This is the usual condition in integrated circuits because of the inherently noisy VCOs in integrated circuits.

As we did before, it is easier to make this phase-noise comparison by using an example. Consequently, we will use a comparison of two example PLLs. One has

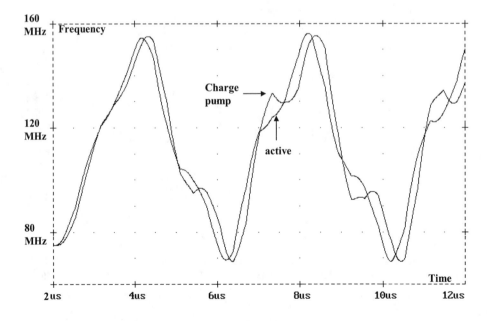

Figure 5.69 Charge pump and active-compensation response to a swept-input frequency.

a charge pump that multiplies 20 MHz by 4 to 80 MHz with a 1-MHz loop bandwidth. The other has an operational amplifier that multiplies 20 MHz by 4 to 80 MHz with a 1-MHz loop bandwidth. In this case, measured phase noise of the examples shows the comparison better than modeling the effects because of the difficulty of taking into account all the nonideal parameters. Figure 5.70 shows the measured comparison results.

This figure shows that the operational amplifier suppresses the phase noise of the VCO from its 1-MHz bandwidth value of −85 dBc/Hz until an offset frequency of 100-kHz value of −95 dBc/Hz. This 10-dB/dec reduction in noise confirms the presence of a type 2, −40-dB/dec slope suppression of the +30-dB/dec increase in noise from the VCO. Consequently, the operational amplifier has produced an ideal integrator over at least the 100-kHz to 1-MHz decade range. From 1 kHz to 100 kHz the loop follows the multiplied reference-oscillator phase noise. In the other example loop, the charge pump does not suppress the phase from its 500-kHz bandwidth value of −85 dBc/Hz to its 10-kHz frequency offset value of −85 dBc/Hz. This flat response shows that the charge pump does not act like an integrator that gives a type 2 PLL response to noise. Consequently, for low output phase noise close to the carrier, operational amplifier compensation is preferred.

5.6.3   Phase-Error Tracking for Changing Input Frequency

Figure 5.71 shows an oscilloscope picture of phase-error tracking for a PLL with an operational amplifier. The top waveform shows the output of the VCO, which multiplies the reference by 2. The bottom waveform shows the reference-oscillator input at 66 MHz. At 20-MHz input frequency, the edge of the VCO that is marked by the dotted line coincided with the y-axis. Slowly changing the input frequency

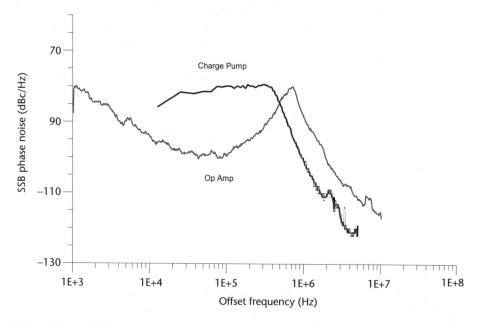

Figure 5.70   Phase-noise suppression comparison between charge pump– and opamp-based compensation.

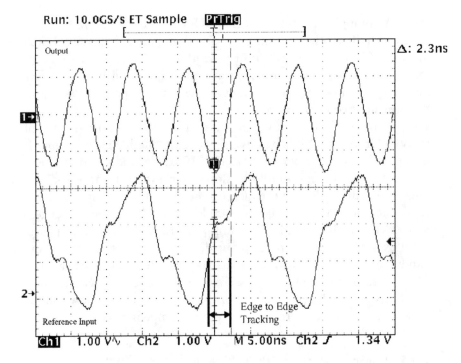

Figure 5.71 Oscilloscope picture of VCO output and reference input of a PLL example for phase tracking.

to its final value of 66 MHz showed the VCO output edge moving slowly to its final value of 2.3 ns from the y-axis. Consequently, we had a 50-ps tracking change per 1-MHz change in input frequency. No special phase matching was done to achieve this result. A repeat of this measurement with the charge pump PLL could not be done because the VCO edge moved so quickly across the oscilloscope screen that it was difficult to maintain the identity of each edge.

Another method for quantifying the effectiveness of the compensation is to do an ac SPICE analysis [18]. Figure 5.72 shows an ac SPICE simulation of a charge pump charging capacitors of 1, 10, and 100 pF. The plots show only a decade of integration above the 0-dB gain point.

Figure 5.73 shows an ac SPICE simulation of an operational amplifier with feedback capacitors of 10 pF and 100 pF. This plot shows 2.5 decades of integration above the 0-dB gain point. This increase should improve suppression of phase noise when compared to a charge pump integrator.

5.7 Summary

In the previous chapter we discussed dividers and oscillators. In this chapter, we discussed detectors and other circuit designs of the individual components in a PLL. Detailed designs of phase detectors, lock detectors, and acquisition aids were studied.

Phase detectors were studied because understanding how phase detectors work is one of the major keys to understanding how PLLs work. Next, lock detection

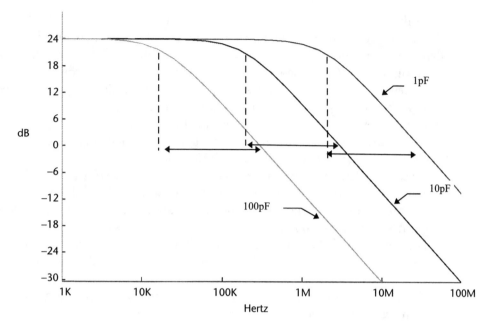

Figure 5.72   Integrator frequency range versus integrating capacitor for a charge pump.

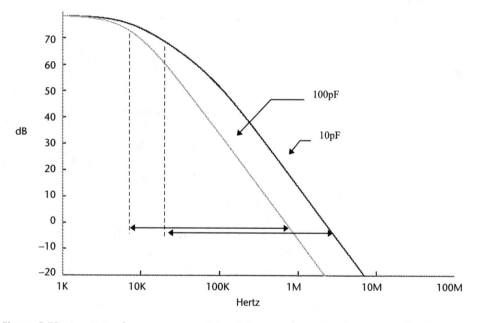

Figure 5.73   Integrator frequency range versus integrating capacitor for an operational amplifier.

was studied. Many systems use the lock detector to reset the system. This is a disastrous change to the operation of most systems. In a PLL, a reset can start the loop operating at a very low frequency (or with no output), which then acquires lock at the normally much higher output operating frequency. Consequently, a small phase shift in a PLL that would marginally affect the system can cause a huge disruption in the operation of the system if a reset occurs. The key to lock detection is to alarm on behavior that shows the PLL is broken. Quadrature phase detection, time-window edge comparison, tune-voltage window comparator, and cycle-slip detection are the lock-detection methods that were covered.

Then, acquisition techniques were studied. Open-loop sweep, closed-loop sweep, and discriminator-aided acquisitions were the methods covered. The phase/ frequency detector uses discriminator-aided acquisition and is the most popular choice. Clock-recovery circuits cannot use phase/frequency detectors. Consequently, these circuits require an acquisition aid.

Next, single-ended and differential charge pumps were studied. The number of parameters that the charge pump affects makes it one of the most important circuits in a PLL. The linearity of the circuit affects the stability of the loop, the amount of jitter, the amount of noise, the reference sideband levels, the phase tracking error, and the limits of the frequency range. The amount of peak charge pump current affects the amount of noise, the loop bandwidth, and the frequency slew rate. Next, basic and other opamp variations were studied as an alternative to charge pumps. An opamp affects the same parameters as the charge pump. Basic opamp design was covered. Folded-cascode and other variations were studied and compared in order to find the best configuration for a PLL. Finally, the trade-offs between operational amplifiers and charge pumps were studied in order to decide which circuit to use. Understanding the design details and trade-offs in these detector and supporting components is critical in designing monolithic PLLs.

Questions

5.1   A phase detector has a maximum voltage of 3.3V for a 2π phase error and 13.3 mV at zero phase error. Calculate the phase detector gain and the figure of merit M. Is this a reasonable phase detector?

5.2   A sinusoidal phase detector has 0.1V average output for a $0°$ phase error. What is the gain of the phase detector at $0°$ phase error? Is this a reasonable phase detector at $0°$ phase error?

5.3   A sinusoidal phase detector has a maximum output voltage of 0.6V and an offset voltage of 1 mV. Calculate the phase detector gain and the figure of merit M. Is this a reasonable phase detector?

5.4   A exclusive-OR phase detector has a maximum output voltage of 1.8V and an offset voltage of 11.1 mV. Calculate the phase detector gain and the figure of merit M. Is this a reasonable phase detector?

5.5   An RSFF phase detector has a maximum output voltage of 1.1V and an offset voltage of 21.1 mV. Calculate the phase detector gain and the figure of merit M. Is this a reasonable phase detector?

5.6   The Gilbert multiplier circuit in Figure 5.6 has a current $I = 2$ mA and resistance $R = 5$ kΩ. R_1 and R_2 differ by 2%, and $V_{io} = 5$ mV. Find V_{do}, K_d, and the figure of merit.

5.7   Run a simulation for an RSFF with a 10-MHz reference frequency and 50% duty cycle and one with a 9-MHz VCO frequency and 50% duty cycle. Draw the output waveform after it has been filtered. Describe any problems with this waveform.

5.8 How many regions of operation does a phase/frequency detector have? Describe the regions.

5.9 In a phase/frequency detector, what is the dead zone? What is one condition that can cause a dead zone? What can be done to the phase detector to minimize the dead zone? Can a type 2 PLL stay locked in the dead zone?

5.10 What happens to the transient response of a PLL with a phase/frequency detector and a damping factor of 3? What can be done to improve the transient response.

5.11 Why are there equal up-down pulses at $0°$ phase error in a phase/frequency detector? What would happen if we filtered the output pulses at $0°$ phase error so that there would be no output?

5.12 Your customer wants you to choose between a NAND gate and a D flip-flop phase/frequency detector that he plans to use over several process (0.8μ to 0.15μ gate lengths). Which one would you choose and why?

5.13 A PLL with 100-kHz loop bandwidth uses a quadrature lock detector and has a 3.3-V supply. The mixer has x gain. What bandwidth do you set the lowpass filter to in the detector and what voltage threshold would you use in the comparator?

5.14 A PLL with 10-kHz bandwidth uses a tune-voltage window comparator lock detector. What would you set the low-voltage and high-voltage thresholds to?

5.15 $K_d = 0.4$, $K_v = 1E9$, $R_1 = 300$ kΩ, and $C = 100$ pF. What is the maximum sweep rate and the natural frequency? For a 3-V supply, what is the fastest ramp time of the tune line that can be achieved?

5.16 $K_d = 0.16$, $K_v = 3E8$, $R_1 = 10$ kΩ, $R_2 = 1$ kΩ, and $C = 3300$ pF. What is the maximum sweep rate and the natural frequency? For a 5-V supply, what is the fastest ramp time of the tune line that can be achieved? What pulse width should you make the one shot?

5.17 What are the advantages and disadvantages of using a charge pump in a PLL?

5.18 Name and describe two variations of a charge pump?

5.19 What is the advantage of using an operational amplifier to feed back the voltage to the other leg in a differential charge pump?

5.20 A phase/frequency detector drives a charge pump with a programmable current source of 100 μa and 1 ma. Calculate the phase detector gains? If the charge pump has a figure of merit of 10, calculate the offset current for each current.

5.21 What opamp has the widest swing and largest PSRR?

5.22 Name a few operational amplifier configurations.

5.23 Name a few operational amplifier characteristics that pertain to PLLs.

5.24 Why do you use cascode transistors in a current mirror?

5.25 Would you use a charge pump or an operational amplifier to convert a differential signal to a single-ended signal?

5.26 Which has more flexibility in synthesizing compensation configurations, a charge pump or operational amplifier?

5.27 For which IC applications would charge pump compensation be a good choice?

5.28 What noise specification is significantly improved by using an operational amplifier?

References

[1] Henderson, B. C., "Predicting Intermodulation Suppression in Double-Balanced Mixers," *Watkins-Johnson Catalog*, 1985, pp. 720–724.

[2] Wolaver, D., *Phase-Locked Loop Circuit Design*, Englewood Cliffs, NJ: Prentice-Hall, 1991.

[3] "Phase-Frequency Detector MC4344-MC4044," MTTL Complex Functions, Motorola Semiconductor Products, Inc., Data Sheet, 1973.

[4] Rohde, U., *Digital Frequency Synthesizers Theory and Design*, Englewood Cliffs, NJ: Prentice-Hall, 1983.

[5] Goldman, S., *Phase Noise Analysis in Radar Systems*, New York: Wiley Interscience, 1989, pp. 95–97.

[6] Goldman, S. J., "Differential Circuit Minimizes Detector Distortion in PLLs," *Microwaves and RF*, May 1992, pp. 145–151.

[7] Gardner, F., *Phaselock Techniques*, New York: Wiley Interscience, 1979.

[8] Best, R., *Phase-Locked Loops: Design, Simulation, and Applications*, 3rd ed., New York: McGraw-Hill, 1997, p. 56.

[9] Tausworthe, R. C., "Design of Lock Detectors," *JPL SPS*, 37–53, Jet Propulsion Laboratory, Pasadena, CA, January 31, 1967, pp. 71–75.

[10] Goldman, S. J., "PLL Lock Detection Using a Cycle Slip Detector with Clock Presence Detection," U.S. Patent 6,466,058, granted October 15, 2002.

[11] Egan, W. F., *Phase-Lock Basics*, New York: Wiley Interscience, 1998.

[12] Black, M. F., "Design of Search Based PLL," *IEEE MTT*, Dallas Chapter Presentation, October 1993.

[13] Razavi, B., *Monolithic Phase-Locked Loops and Clock Recovery Circuits*, New York: IEEE Press, 1996, pp. 25–28.

[14] Young, I. A., "A PLL Clock Generator with 5 to 110 MHz of Lock Range for Microprocessors," *IEEE Journal of Solid State Circuits*, Vol. SC27, November 1992, pp. 1599–1607.

[15] Gray, P., and R. Meyer, *Analysis and Design of Analog Integrated Circuits*, New York: Wiley Interscience, 1993, pp. 460–466.

[16] Martin, K., and D. Johns, *Analog Integrated Circuit Design*, New York: Wiley Interscience, 1997, pp. 266–273.

[17] Laker, K., and W. Sansen, *Design of Analog Integrated Circuits and Systems*, New York: McGraw-Hill, 1994, pp. 587–591.

[18] Gray, P., and H. Khorramabadi, "High-Frequency CMOS Continuous Time Filters," *IEEE Journal of Solid State Circuits*, Vol. SC19, No. 6, December 1984, pp. 939–948.

CHAPTER 6

Loop-Compensation Synthesis Revisited

This chapter presents loop-compensation synthesis. The synthesis of loop-compensation components in the loop is application dependent. Consequently, several examples of the most popular compensation schemes are presented. This gives us an intuitive feel to extend these methods to designs that are not presented. These examples include the following design types:

- Passive;
- Active with maximum capacitor value;
- Sampling delay;
- Fast switching time;
- Minimum bandwidth;
- Maximum divide ratio;
- Optimum low phase-noise.

This chapter begins by ranking PLL requirements to help make the design trade-offs that optimize a PLL loop compensation for an application. Ideal design requirements for a synthesizer and a clock-recovery circuit are presented. Next, a nine-step cookbook method for doing compensation is presented to ease making decisions about the trade-offs in PLLs. For beginning- and intermediate-level PLL designers, it is best to start with the cookbook method. An example shows how to apply the cookbook method. Next, a simple example to illustrate the technique for active compensation and maximum capacitor value is presented. The biggest impact on IC design is the area of the capacitor in silicon. The limitations of the components in this low area PLL architecture are explored. The next example illustrates the technique for active compensation and compensating the design for sampling delay. Sampling delay is a major limitation to widening the loop bandwidth. At the limit of the sampling delay, the loop loses stability. A derivation of the new design equations that accounts for sampling delay is presented. Another example illustrates the technique for obtaining fast switching time. Misconceptions about the ideal damping factor are discussed, and equations are derived that give the desired result.

Next, an example is presented that determines the minimum bandwidth of a charge pump monolithic PLL with a current-starved ring VCO. Telecommunications, wireless, and pager technologies require narrow-bandwidth PLLs (10 to 100 Hz) because the lowest common denominator for the multiple frequencies required force a low reference frequency into the phase detector. The factors that affect the minimum bandwidth are then discussed. Also, the least-common-denominator requirement forces the multiplication ratios to higher and higher values. Knowing the maximum value can help us decide if we need to change to a multiple-loop architecture or a fractional-N architecture to meet this requirement. Consequently, another example shows the maximum divide ratio that can be achieved with a charge pump PLL. Equations are derived to meet this PLL synthesis goal.

Receivers need low-phase-noise PLLs to maintain high dynamic range and meet minimum signal-to-noise requirements. In designing a PLL, the general practice has been to select damping factor and loop bandwidth for switching time. This example presents a method for obtaining the optimum damping factor and bandwidth goals for low-phase-noise PLL design. Equations are derived for this purpose.

6.1 Ranking Requirements for PLLs

Optimization of PLL compensation requires an objective function. In the objective function, the requirements are weighted in order to reach a clear optimum solution. Otherwise, the optimizer would get into an infinite loop. One way of weighting the requirements is to rank them in order of importance.

Ranking PLL requirements helps make the design trade-offs to optimize the PLL architecture for the application. Let's look at the ideal design requirements for a synthesizer and a clock-recovery circuit. This will help identify the requirements and goals. Then, we will look at some specific examples to show how these trade-offs have been made for a specific application.

First, here are the ideal PLL requirements for a synthesizer so that we know in general which requirements to rank:

- Ideal PLL requirements for synthesizer:
  - Small frequency-step size;
  - Low noise and jitter:
    - Low close-in phase noise;
    - Low far-out phase noise;
    - Low noise ear;
    - Low reference sidebands;
    - Low spurious levels;
  - Fast time response;
  - Wide tuning range;
  - High stability;
  - High output frequency;
  - Fast power up;
  - High noise immunity;
  - Minimum tracking error:
    - No resets;
  - Small size;
  - Low power;
  - Low cost.

Next, here are the requirements for an ideal-clock-recovery PLL, which will have a significantly different ranking than a synthesizer list because of the difference in application and other requirements:

- Ideal requirements for clock-recovery PLL
  - Low additional jitter and noise to the input:
    - Low far-out phase noise (from VCO);
    - Low noise ear/peaking (control system);
    - High-order poles (jitter-tolerance improvement);
    - Low reference sidebands;
    - Low spurious levels;
    - Low multiplication factors to minimize accumulated jitter;
  - Fast time response for initial lock;
  - Zero phase startup;
  - Slow time response to changing data;
  - Wide frequency range;
  - High stability;
  - High output frequency;
  - Fast power up;
  - High noise immunity;
  - Minimum tracking error:
    - No resets;
    - Minimum reference-edge-to-output-edge skew;
    - Tracking of low-frequency input modulation;
    - Rejection of high-frequency input modulation;
    - Minimum sensitivity to missing input edges;
  - Small size;
  - Low power;
  - Low cost.

The clock-recovery list has the additional requirements of fast switching time, slow response to changing data, tight edge tracking, and minimum sensitivity to a missing edge. The requirements for minimum sensitivity to a missing edge and tight tracking error eliminate the commonly used phase/frequency detector in Figure 5.21 and lead to a whole range of phase detectors for solving these problems.

Let's look at an example of ranking to get a better feel for the thought process involved in making these decisions. For the first example, we will rank a PLL for a serializer-deserializer application that process 16 bits of data. A serializer-deserializer application takes parallel data at a slower speed (10 MHz) and ships it out serially at a much faster rate (16 bits times 10-MHz reference rate = 160 MHz) to reduce the amount of wiring between functional blocks in a system. A receiver at the other end receives the 16 bits of serial data at 160 MHz and converts it back to 16 bits of parallel data at 10 MHz and resynchronizes the parallel data.

Here are the rankings of requirements for an example serializer-deserializer PLL:

- Serializer/deserializer PLL design requirement rankings
  - Output-frequency dynamic range and input-frequency dynamic range;
  - Adequate (versus optimum) stability;
  - No resets for acquisition (all circuits work at high frequencies)
  - System-level timing (skew);
  - Maximum frequency output (tied heavily to system timing and spurious noise);
  - Noise—spurious;
  - Frequency resolution (the size of the dividers);
  - Acquisition time;
  - Power consumption;
  - Size;
  - Noise—Gaussian.

Once the rankings have been established, architecture decisions can be made about the design. In particular, decisions can be made about the phase detector, VCO, process, bipolar versus CMOS implementation, and external versus internal compensation.

First, select between an RS-latch-type phase/frequency detector (PFD) versus a D-type tristate PFD. The D-type PFD can have a serious dead-zone issue, which the RS-latch-type PFD does not have. The D-type has no advantages as compared to the RS-type, except perhaps a slightly smaller area and the fact that it has historically been used a great deal with charge pumps.

Next, select between active compensation versus charge pump compensation: Charge pumps don't have as much dc gain as an active filter with an opamp and thus do not have as good dc tracking capability. Also, dc offsets in charge pumps are harder to correct for than in active-filter compensation, which leads to phase misalignment at the PFD inputs since the loop has to correct for the offset. The two-pole, two-zero compensation network we have chosen cannot be easily incorporated into the charge pump design. This compensation network will allow a wider dynamic range of input frequencies and divide ratios than is possible with typical second-order PLL compensation. There is concern that the CMOS opamp used has a low 0-dB crossover frequency; thus, the output VCO control line will have high impedance to high frequencies. However, even though the opamp in the active-filter design is limited to a ~50-MHz crossover frequency, the active-filter feedback network will couple noise in at high frequencies. In the two-pole, two-zero active-filter design, high-frequency noise is coupled from the output back to the input of the opamp, where it is rejected by the opamp. In addition, the VCO will act as an integrator that does not react immediately to high-frequency noise that cannot be regulated out by the opamp. In contrast, the charge pump control voltage usually has a high-impedance node that will have trouble rejecting high-frequency noise. It is true that in the charge pump case the VCO will also tend to integrate the noise out on this line; however, the active filter has the advantage that the high-frequency noise is coupled back to its input, whereas, in the charge pump design, the control-line noise does not have a path for high-frequency rejection.

Next, select between a single-ended CMOS VCO versus a differential bipolar VCO. The single-ended CMOS inverter chain has the advantage of having about twice the useable frequency range as the differential bipolar. The bipolar differential VCO has the apparent advantage of common-mode noise rejection and low device noise; however, initial measurements indicate that the single-ended VCO with active-filter drive has as good noise rejection as the differential VCO driven by a charge pump.

Next, choose the technology. RFBiCMOS was chosen as our process technology for the serializer/deserializer project for several reasons. We need the flexibility that RFBiCMOS offers in terms of faster bipolar and faster CMOS than our predecessor linear process. RFBiCMOS has an isolated NMOS and a good bipolar for high ESD, which is a differentiating feature of our interface products. Smaller CMOS processes were evaluated, but rejected, because we need good bipolar for high ESD protection, which CMOS does not easily offer. Also, the high wafer cost of small metal pitches that accompany small CMOS processes is not a requirement for our designs. So, in short, RFBiCMOS offers maximum flexibility because it has comparatively fast CMOS and bipolar, an isolated NMOS, and high ESD protection, and it allows us to share cells easily between designers in order to meet tight time-to-market demands.

Next, choose the transistor type of CMOS versus bipolar technology within the system design. It is apparent that design improvements in terms of device-level $1/f$ noise and f_t speed can be made by using bipolar instead of CMOS technology. Since we have both available to us in RFBiCMOS, we would obviously try to use bipolar wherever possible for high-speed designs; however, due to system-level timing constraints of the project, we are predicting that very-high-speed data transfer will not be possible. Furthermore, the PLL design we have chosen is in CMOS, and layouts can be easily transferred to RFBiCMOS to meet tight schedule requirements. For the time being, we are planning on investigating future improvements of the design by replacing CMOS with bipolar circuits. This is especially the case in the PLL, whose performance in terms of jitter, speed, and overall noise immunity could probably be improved by the use of bipolar circuits. The most notable speed improvement could be made by using differential ECL divider chains.

Finally, select between external versus internal compensation. Internal compensation has the obvious advantage of user friendliness. It also has a hidden advantage in that a great deal of applications support is not needed. Disadvantages of internal compensation are large RC area and perhaps greater on-chip noise-coupling characteristics in the case of high resistor values, leading to a high impedance filter. Advantages of external compensation include the availability of large capacitor values, allowing small on-chip resistors for low on-chip noise coupling. In addition, the flexibility of setting PLL bandwidth for a given application for optimum performance is possible with external components. This improves ac performance of the PLL, such as acquisition time and optimum stability. Disadvantages of external compensation are the obvious user unfriendliness, necessary applications support, and potential for noise coupling from external components onto the VCO control line. The team has chosen internal compensation since acquisition time and optimum (versus adequate) stability are not high priorities for the serializer-deserializer PLL. This provides the chipset with maximum user friendliness and avoids potentially high application-support resource costs. It is accepted that acquisition times for the higher input frequencies may be compromised. The noise coupling trade-off is still a subject of debate.

Another example shows how the requirements and trade-offs change with application. Here is the ranking of requirements for a time base generator for a hard-disk drive read-channel operation:

1. Eight output phases;
2. Maximum frequency of 500 MHz;
3. Frequency range of 250 to 500 MHz;
4. BER (SNR of 40 dB);
5. Minimized wander (phase error) of less than 200 ps peak to peak;
6. Loop lock even in the absence of resets;
7. Correct division of M feedback divider to maximum VCO output frequency;
8. Schedule;
9. Cycle-cycle jitter of less than 0.25% of the VCO's output period;
10. Low noise ear;
11. Minimize noise after the phase detector;
12. Minimize $1/f$ noise;
13. Minimized nonharmonically related spurious signals (<-40 dBc);
14. Frequency resolution at the output of the VCO of 1-MHz frequency steps;
15. High power-supply rejection;
16. High damping factor;
17. Optimum phase-noise bandwidth (50 to 5,000 kHz);
18. M programmable feedback divider ratios of 1 to 511;
19. N programmable reference divider ratios of 1 to 255;
20. Area;
21. Power consumption.

This PLL generated the clock (time base) for a digital clock-recovery circuit (CRC) that recovered the clock out of the data from a hard disk drive.

The top requirement shows eight phases out of the VCO in order for the digital clock-recovery circuit to sample the input data evenly at 45° increments. This requirement makes the VCO a ring oscillator with an even number of gate delays in the ring, which makes it a differential ring. The second requirement shows a maximum output frequency of 500 MHz. The higher output frequency increases the throughput, which increases with the increasing amount of data capacity on a hard drive. Next, a wide output-frequency range gives higher flexibility in the hard-drive capacities for which the PLL will work. Next, low BER and minimization of wander mean a low-phase-noise design for the VCO, low multiplication factor, and low-noise loop-filter design. No reset requirements and a fast prescaler divider make the loop robust so that it can recover from either power-supply rail. Next, the schedule requirement reflects that any solution must be fast to market. After this the list reflects more low-noise parameters, PSRR, area, and power consumption, which will be difficult to meet after satisfying the first eight requirements.

6.2 Loop-Component Synthesis

There are many different approaches to synthesizing loop components. It is beyond the scope of the current discussion to list every possible approach. The goal here is to give enough examples for readers to understand how to extrapolate these techniques to their particular applications.

In the previous chapters, charge pump compensation was done as a simple introduction to compensation. In this section, we will show a nine-step cookbook method for doing compensation. For beginning- and intermediate-level PLL designers, it is best to start with the cookbook method. An example will show how to apply the cookbook method.

There are several architectures of loop compensation to select from. Figure 6.1 shows passive and charge pump loop-compensation configurations. Each has its advantages and disadvantages. Ideally, we would like to choose an architecture that provides the greatest amount of freedom in adjusting polynomial values.

The selection of architecture affects many of the loop's performance parameters. Requirements that loop filter can aid include

- Low noise and jitter:
  - Low noise ear;
  - Low reference sidebands;
- Fast time response;
- Wide tuning range without switching components;
- High stability;
- High noise immunity (outside loop bandwidth);
- Minimum tracking error;
- Small size;
- Low power.

Table 6.1 shows desired PLL characteristics that are affected by changing the design parameters for loop compensation. Blank entries in the table mean the effect is undetermined. First, from left to right, low noise ear means a high loop bandwidth. Low reference sideband means a low loop bandwidth. Fast time response means wide loop bandwidth. Wide tuning range means low loop bandwidth. High stability means low loop bandwidth. High noise immunity means high loop bandwidth. Minimum tracking error means high loop bandwidth. Small area means wide loop bandwidth. External components required means narrow loop bandwidth. The loop-bandwidth term also applies to natural frequency and 0-dB crossover point because there is an almost constant factor relationship between them.

Low noise ear means high damping factor. Fast time response means low damping factor. Wide tuning range means low damping factor. High stability means high damping factor. High noise immunity means high damping factor. Minimum tracking error means high damping factor.

Select an opamp active filter for low noise ear, low reference sidebands, fast time response, wide tuning range, high stability, high noise immunity, and minimum tracking error. Select a charge pump for high device bandwidth, small size, and low power and to use external components.

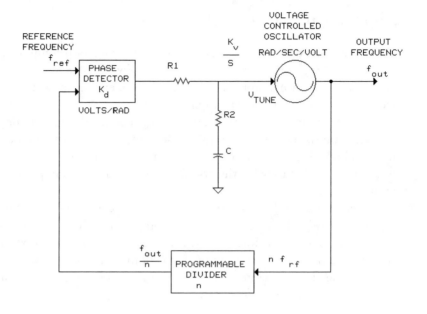

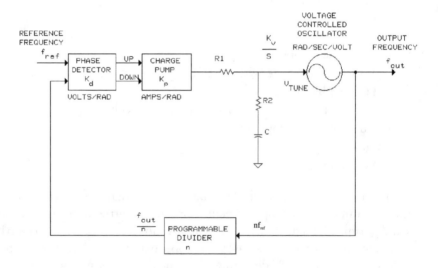

Figure 6.1 PLL interconnect for passive loop compensation and charge pump loop compensation.

Figure 6.2 shows the ideal stability geometry. For a 10-kHz 0-dB crossover point with 65° phase margin at the peak, the compensation puts a zero at 2 kHz (1/5 of 10 kHz) and a pole at 50 kHz (5 times 10 kHz). The zero adds phase margin up to 10 kHz, and the pole takes the phase margin away and helps filter out the phase noise and reference sidebands. A 65° phase margin peak at 10 kHz gives a 30° phase margin range from 1 to 100 kHz, which is enough to allow for a wide variation in divide ratios, temperature, process, and supply voltage.

For beginning- and intermediate-level PLL designer, it is best to start with the cookbook method. The following list shows a summary of a simple cookbook procedure to synthesize components [1, 2].

Table 6.1 Loop-Compensation Requirements Versus PLL Performance Characteristics

Design Parameter	Low Noise Ear	Low Reference Sideband	Fast Time Response	Wide Tuning Range	High Stability	High Noise Immunity	Minimum Tracking Error	High Device Bandwidth	Small Size	Low Power	External Components Required
Loop bandwidth	↑	↓	↑	↓	↓	↑	↑	↑		↓	
Damping factor	↑		↓	↓	↑	↑	↑				
Opamp/charge pump	Opamp	Opamp	Opamp	Opamp	Opamp	Opamp	Opamp	Charge pump	Charge pump	Charge pump	Charge pump

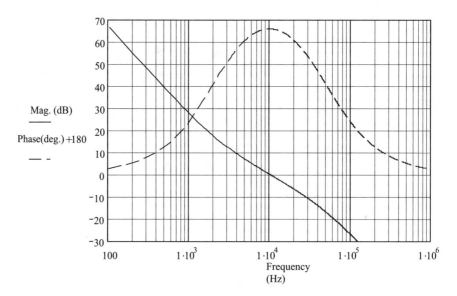

Figure 6.2 Ideal loop-component synthesis that is geometrically centered on the 10-kHz crossover frequency with a zero at 2 kHz and a pole at 50 kHz.

1. Determine output-frequency range.
2. Determine VCO gain.
3. Determine input-frequency range.
4. Determine divider range.
5. Determine largest capacitor value.
6. Determine maximum natural frequency.
7. Determine minimum damping factor.
8. Synthesize values.
9. Suppress reference sidebands (C_2).

The following procedure assumes a synthesizer application on an integrated circuit with all components on chip, and a phase frequency detector is used for a phase detector.

Step 1. In the first step, the output-frequency range needs to be identified. The output-frequency range and output frequency will determine which VCO to use. The performance of the VCO is the major performance circuit in a PLL. An RF frequency (>800 MHz) VCO would use an LC oscillator with a low VCO gain. A low frequency (<1 MHz) would use a multivibrator with a low VCO gain. A high-frequency and wide frequency range output (1 to 800 MHz) would use a ring oscillator with a high VCO gain. A single frequency output is easier to build than a wide range of frequency outputs. A low supply voltage with a high frequency and wide-frequency-range output gives the ring oscillator an even higher VCO gain. From a control-system point of view, we would like a low VCO gain.

Step 2. With the output-frequency range determined, and given the supply voltage and some margin, we can calculate an approximate VCO gain:

$$K_v = (\omega_{omax} - \omega_{omin})/(V_{tunemax} - V_{tunemin}) \qquad (6.1)$$

where

$$V_{tunemax} = V_{dd} - V_{margin}$$
$$V_{tunemin} = V_{ss} + V_{margin}$$

Step 3. In this step, the input-frequency range needs to be identified. A single input frequency is easier to build than a wide input-frequency range. A wide difference in input and output frequency makes synthesizing components more difficult to design. A widely varying input frequency, a widely varying output frequency, and a wide difference in input and output frequency is the most difficult to design.

Step 4. For this step, the feedback divider (or frequency-multiplication factor) needs to be identified. The ratio of the maximum output frequency to the minimum input frequency gives the maximum divide ratio, and the ratio of the minimum output frequency to the maximum input frequency gives the minimum divide ratio. A wide variation of divide ratios will effectively vary the divided-down gain of the VCO, which will change the natural frequency and the damping factor. For fixed VCO gain, phase detector gain, resistor and capacitor values, the increase in divide ratio will reduce the natural frequency and the damping factor.

Step 5. We need to select the largest capacitor value. For an integrated circuit, the loop capacitor takes up the most area. The area of the chip has a direct impact on the cost of the IC. A maximum of 100 pf for a capacitor is not an unusually large amount. For an off-chip capacitor, 1 μF is usually the maximum value.

Step 6. In this step, we determine the maximum natural frequency that will keep the loop stable. A natural frequency too close to the reference frequency will cause the loop to lose a significant amount of phase margin and eventually to lose lock because of sampling effects, which we will discuss in Section 6.4. A factor of one-tenth of the reference frequency is a good reference for the maximum natural frequency.

Step 7. We select the minimum damping factor. The minimum damping factor should at least take into account the variations in components with process to maintain stability. Comfortable stability margins are greater than 30° for phase margin in classical PCB designs, which corresponds to a 0.27 damping factor [1]. On ICs, damping factors can go as low as 0.1 and still maintain lock. Even though this low damping factor may be acceptable to the process, it may not be acceptable to the customer, who may want a minimum phase margin as high as 45° in order to see fewer than two significant overshoots.

Step 8. Synthesize R_1 and C_1 for the charge pump. We select a charge pump current (10 μa to 1 ma). Then, we use (2.27) to solve for R_2. Equation (2.26) solves for C_1, and this value determines if it is an acceptable value for the amount of area.

Step 9. Finally, we calculate the amount of sideband level. The difference between the sideband level calculated and the sideband level desired determines how much more suppression we need to meet specifications. Equation (6.2), from (3.212), computes the sideband level:

$$P_{sb} = \left\{ \left[\frac{2I_{pk} \sin\left(\dfrac{t_{pw} f_{ref} \, \pi}{2}\right)}{\pi} \right] R_2 \frac{K_v}{2\omega_{ref}} \right\}^2 \tag{6.2}$$

Next, we calculate the additional suppression of the sideband by, for example, adding a roll-off capacitor. Equation (6.3) computes the suppression:

$$\text{Suppression} = 20 \log(\omega_c / \omega_{ref}) \tag{6.3}$$

where

ω_c = radian frequency cut off of the roll-off filter (rad/s).

Summing the results of (6.2) and (6.3) gives the total sideband level. If the total sideband level does not meet requirements, then we will have to move the natural frequency to a lower frequency in step 6 and resynthesize the components or add filtering, as discussed in Section 3.6.2.6, to increase the suppression.

Example 6.1

Design a PLL with the cookbook method to increase 32 kHz to between 4 and 20 MHz with a process that has a power supply of 3.3V and a phase/frequency detector with a minimum output pulse width of 4 ns. The customer requires 0.9V of supply padding, a 1,000-pF maximum capacitor value, a damping factor of 0.5, and a roll-off capacitor with 1/10th of the value of the main capacitor.

The output-frequency range in step 1 was given to be 4 to 20 MHz. Substituting values into (6.1) in step 2 gives a VCO gain of 10.6 MHz/V. In step 3, the input-frequency range does not change. In step 4, the divider range is computed to be 125 to 625 by dividing the minimum and maximum output frequency by 32 kHz. In step 5, the customer has required the maximum capacitor size to be 1,000 pF. For extra margin in stability, the loop natural frequency is set to 1/40th of the clock frequency in step 6, which is 800 Hz. In step 7, the damping factor is set by the customer to 0.5. Rearranging (2.26) to solve for charge pump current computes 1.4 ma of charge pump current in step 8. Solving (2.27) gives an R_2 of 198Ω. In step 9, solving (6.2) gives a -44-dB reference sideband level without filtering, and setting the C_2 capacitor to 10 times less than C_1 gives -12 dB more suppression for a sideband level and a total reference sideband level of -56 dB.

Previously, we discussed the cookbook method of synthesizing the loop components. The cookbook approach is for beginning- and intermediate-level PLL designers. The resulting design will have average to good performance. To truly optimize PLL performance to the application requires more advanced techniques that manipulate the equations until the desired response is obtained. This manipulation may break the cookbook rules. To do this properly, we have to rank the requirements for the optimization carefully. Previously, we discussed the methodology of how to rank the requirements. Now, we will discuss the more advanced techniques for manipulating the equations to get more optimum design for a specific application. To do this, we will go through several examples that cover some common application requirements. These examples will show how the component values can drastically change, depending on how close you want to push the design to its optimum performance.

6.3   Active Compensation and Maximum Capacitor Value

The following is a simple example to illustrate the technique for active compensation and maximum capacitor value. The biggest impact on IC design is the area of the capacitor in silicon. In many cases this is the largest-area component in the design. First, the loop-compensation architecture is selected. In this case, an active-filter compensation is selected. Next, the limitations of the components in this PLL architecture are explored. The VCO gain and phase detector gain are usually given by the components chosen and cannot easily be adjusted. Consequently, the capacitor and resistor values are all that remain to synthesize the loop compensation.

Figure 6.3 show a differential active-filter compensation connection to a PLL. The biggest impact on IC design with the active filter is the area of the capacitor in silicon. A typical process may have an 8-fF/μm^2 integrated capacitor. This would require a 0.012-mm^2 area or a square 111×111 μm for a 100-pF capacitor. In many integrated PLLs, this could be as much as 10% to 20% of the total area for the whole PLL.

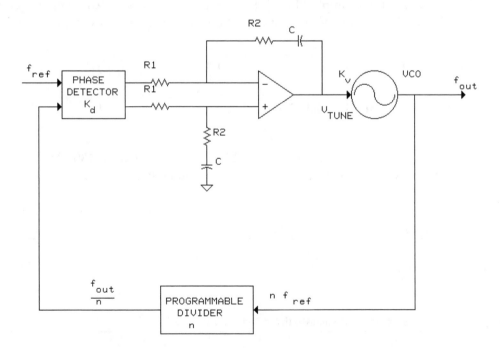

Figure 6.3   Block diagram of differential active-filter compensation.

External off-chip capacitors have a different size limitation. Capacitor values above 1 microfarad are electrolytic. Electrolytic capacitors are not recommended for filters. Therefore, a 1-microfarad capacitor or less will make the circuit component nonelectrolytic and insure excellent frequency-response characteristics.

For a selected C value, damping factor, and natural frequency, and given phase detector and VCO gain (K_d and K_v), rearranging (2.22) and (2.23) gives (6.4) and (6.5) to compute the final R_1 and R_2 component values:

$$R_1 = \frac{K_d K_v}{(\omega_n)^2 C n_{mf}} \tag{6.4}$$

$$R_2 = \frac{2\zeta}{\omega_n C} \tag{6.5}$$

There are many approaches to synthesizing the loop compensation. Applications mainly determine which approach is used. One approach minimizes capacitor values for integrated circuit design. Another approach minimizes the R_2 feedback resistor value (1 kΩ) in active compensation for low-noise applications. Other approaches include having one cycle of overshoot and one of undershoot to optimize fast switching time, optimizing phase noise by adjusting loop bandwidth for low-noise applications, and, finally, including the effects of sampling delay to optimize fast switching time.

A PLL design example requires a 1.9-MHz 0-dB crossover frequency, a damping factor of 0.99, and a reference frequency of 10 MHz. Table 6.2 shows the given values for the example PLL. Table 6.3 shows the calculated component values. From (2.24), (6.6) computes a 76° phase margin for a 0.99 damping factor.

$$\phi_m = \operatorname{atan}\left(2\zeta\sqrt{2\zeta^2 + \sqrt{4\zeta^4 + 1}}\right) \tag{6.6}$$

$$= \operatorname{atan}\left(2 \cdot 0.997\sqrt{2 \cdot 0.997^2 + \sqrt{4 \cdot 0.997^4 + 1}}\right)\frac{180}{\pi} = 76.27$$

From (2.25) and the zero term in (2.21), (6.7) computes a 480-kHz zero corner frequency for a 76° phase margin and no sampling-delay effects.

Table 6.2 Given Parameters for Example Sampling-Delay PLL

K_v	K_d	n_{mf}	f_{in}	f_{out}	f_x	ζ	C_{max}	Technology
6.2E9 rad/s/V	0.16 V/rad	14	10 MHz	140 MHz	1.9 MHz	0.99	20 pF	0.18-μm L_{eff}

Table 6.3 Calculated Values with No Sampling-Delay Effects

Phase Margin	f_x	R_2	f_z	G_{pll}	R_1	Filter Type
76°	1.94 MHz	16.75 kΩ	480.3 kHz	3.54E13, 135 dB	100 kΩ	Active

$$f_z = \frac{f_x}{2\zeta\sqrt{2\zeta^2 + \sqrt{4\zeta^4 + 1}}} \tag{6.7}$$

$$= \frac{1.94 \times 10^6}{2 \cdot 0.99\sqrt{2 \cdot 0.99^2 + \sqrt{4 \cdot 0.99^4 + 1}}} = 4.803 \times 10^5$$

Rearranging (2.25) to solve for natural frequency f_n computes a value of 947 kHz for a 1.9-MHz 0-dB crossover frequency and a 0.99 damping factor, as shown by substituting the natural-frequency and zero-frequency values into the zero term in (2.21). Substituting into (6.4) gives (6.8), which computes 100 kΩ for R_1:

$$\frac{0.16(6.2 \times 10^9)}{[2\pi(9.473 \times 10^5)]^2 (2 \times 10^{-11})14} = 1 \times 10^5 \tag{6.8}$$

Substituting into (6.5) gives (6.9), which computes 16.75 kΩ for R_2:

$$\frac{1}{2\pi(2 \times 10^{-11}) (4.751 \times 10^5)} = 1.675 \times 10^4 \tag{6.9}$$

Finally, Table 6.3 gives a summary of the computed values for the example. Notice, R_1 has a large value. Reducing the size of the feedback capacitor causes an increase in the value of R_1. A large resistor can also take up a large amount of silicon area. Consequently, the value of R_1 can also be another trade-off.

6.4 Sampling-Delay Synthesis

The following is an example to illustrate the technique for active compensation and compensating the design for sampling delay. Sampling delay is a major limitation to widening the loop bandwidth. At the limit of the sampling delay, the loop loses stability. Figure 6.4 shows a plot of phase error versus reference clock that shows the effects of sampling at the reference frequency. The y-axis has phase error in degrees, and x-axis has the number of reference periods. The dotted line is the envelope of the phase error, and the vertical lines are the samples at the reference clock with the height representing the value of the phase error. For sampling effects to occur, the loop bandwidth must be close to the reference frequency. There are seven samples from peak to peak of the damped oscillation of the phase error. Consequently, the natural frequency is about one-seventh of the reference frequency. This plot shows that with only seven samples, we are not getting an accurate measure of the phase-error envelope. The error in accurately sampling the phase error is the *sampling delay*. This error has a first-order approximation to a delay line of half a clock period. Later in the derivation, a factor will be derived to relate the 0-dB crossover frequency to the closed-loop, 3-dB bandwidth of the PLL.

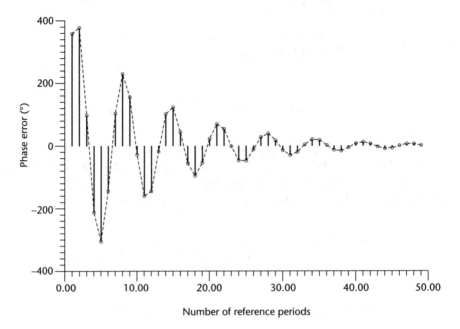

Figure 6.4 Reference-clock sampling of underdamped phase error from loop bandwidth set close to the reference-clock frequency.

Adding a delay term to the classic open-loop equation models [(2.16)] sampling-delay effects as shown by (6.10) [3, pp. 49–53].

$$G(s)H(s) = \frac{K_d K_v}{n_{mf} CR_1}\left(\frac{1}{s^2}\right)(sCR_2 + 1)\exp(-s\tau_{sd}) \qquad (6.10)$$

where

τ_{sd} = sampling delay (sec);

$\tau_{sd} = 1/f_{ref}$;

$\tau_{sd} = n_{mf} f_o$;

f_{ref} = reference frequency (Hz);

f_o = output frequency (Hz).

The last two terms of (6.10) contain phase variations versus frequency; however, the other terms have a constant phase-versus-frequency response. Therefore, for computing the phase response of $G(s)H(s)$, the other terms are ignored. Solving the last term in (6.10) for phase yields (6.11), which is used to plot the phase variations of the open-loop transfer function as a function of frequency [3, pp. 49–53]:

$$\phi(\omega) = \operatorname{atan}\left[\frac{\sin(-\omega\tau_{sd}) + \omega R_2 C \cos(-\omega\tau_{sd})}{\cos(-\omega\tau_{sd}) - \omega R_2 C \sin(-\omega\tau_{sd})}\right] \qquad (6.11)$$

where

$$\omega = 2\pi f;$$

f = frequency variable (Hz).

To find the frequency location of the zero and thereby insure the required loop stability, (6.11) has to be rearranged, and values have to be substituted into the variables. Damping factor is one of the values needed in (6.11) to find the frequency location of the zero. Damping factor expresses the stability requirement of the loop. Equation (6.12) converts the damping factor to a phase angle:

$$\phi_m = \text{atan}[2\zeta(\omega_x/\omega_n)] \tag{6.12}$$

where

ϕ_m = phase margin (rad);

ζ = damping factor.

Equation (6.13) computes the ratio of crossover frequency to natural frequency in terms of damping to make (6.12) only a function of damping factors.

$$\frac{\omega_x}{\omega_n} = \sqrt{2\zeta^2 + \sqrt{4\zeta^4 + 1}} \tag{6.13}$$

where

ω_x = angular frequency where the open-loop gain entry unity (rad/s);

ω_n = natural angular frequency (rad/s).

Substituting the 0-dB crossover frequency into the frequency variable in (6.11) computes a phase angle at the 0-dB crossover frequency. Now, the computed phase angle at the 0-dB crossover frequency in (6.11) is the phase margin. Solving for the zero frequency location produces (6.14):

$$f_z = \frac{-f_x \sin(\omega_x \tau_{sd}) \tan(\phi_m) + f_x \cos(\omega_x \tau_{sd})}{\tan(\phi_m) \cos(\omega_x \tau_{sd}) + \sin(\omega_x \tau_{sd})} \tag{6.14}$$

where

f_z = frequency location of the 3-dB point of the zero in the PLL (Hz);

f_x = frequency where $10 \log(|G(s)H(s)|)$, the open-loop gain, equals 0 dB (Hz).

For the servo control-system representation of a PLL, (6.15) defines the zero corner frequency for an active type 2 loop compensation [4]:

$$f_z = \frac{1}{2\pi R_2 C} \text{ for active compensation} \tag{6.15}$$

Equation (6.14) locates the zero frequency in terms of the 0-dB crossover frequency. To design the PLL for its closed-loop, 3-dB bandwidth requires the conversion formula in (6.16) [3, pp. 49–53]:

$$\omega_x = 2\pi f_{BW} \frac{\sqrt{2\zeta^2 + \sqrt{4\zeta^4 + 1}}}{\sqrt{1 + 2\zeta^2 + \sqrt{(1 + 2\zeta^2)^2 + 1}}} \tag{6.16}$$

where

f_{BW} = 3-dB bandwidth of the closed loop (Hz).

Equation (6.16) computes the 0-dB crossover frequency from the designer's desired closed-loop, 3-dB bandwidth, which would be specified for the PLL used as a modulator or demodulator.

6.4.1 Magnitude Response and Gain Constant of the Open-Loop-Gain Function

Knowing the frequency location of the zero and of the 0-dB crossover point allows more PLL parameters to be calculated. Equation (6.17) computes the magnitude response of (6.10) by using the frequency location of the zero and the gain constant of the loop:

$$|G(f)H(f)| = \frac{K_d K_v}{n_{mf} C R_1} \frac{1}{\omega^2} \sqrt{1 + \left(\frac{f}{f_z}\right)^2} \tag{6.17}$$

Substituting the gain constant into (6.17) gives (6.18):

$$|G(f)H(f)| = G_{pll} \frac{1}{\omega^2} \sqrt{1 + \left(\frac{f}{f_z}\right)^2} \tag{6.18}$$

where

G_{pll} = PLL gain constant.

Comparing (6.18) to servo terminology produces (6.19) for the gain constant:

$$G_{pll} = \frac{K_d K_v}{n_{mf} C R_1} \tag{6.19}$$

where

$$G_{pll} = (\omega_n)^2$$

Equation (6.18) computes the Bode plot of magnitude versus frequency for the open-loop transfer function. Next, the PLL gain constant in (6.18) is computed. The corner frequency of the zero is located by solving (6.14). The 0-dB crossover frequency is computed from (6.13). The gain constant is computed by setting (6.18) equal to unity at the 0-dB crossover frequency and solving for the gain constant as shown in (6.20):

$$G_{pll} = \frac{(2\pi f_x)^2}{\sqrt{1 + \left(\frac{f_x}{f_z}\right)^2}} \tag{6.20}$$

Equation (6.20) calculates the required gain constant as a function of the 0-dB crossover frequency and the zero-corner-frequency location. In addition (6.20) shows that the PLL parameters determine the value of the gain constant.

6.4.2 Solving for PLL Component Values

From the frequency location of the zero, the frequency location of the 0-dB crossover, and the determined value of the gain constant, the PLL parameters K_d, K_v, n_{mf}, C, R_1, and R_2 can be determined. The selection of the VCO and phase detector determines the K_v and K_d parameters. The ratio of the output frequency of the VCO to the reference frequency into the phase detector determines the frequency multiplication parameter n_{mf}. The PLL components C, R_1, and R_2 remain to be calculated. Equation (6.21) computes the capacitor parameter C by rearranging the frequency location of the zero formula in (6.15):

$$C = \frac{1}{2\pi R_2 f_z} \tag{6.21}$$

Substitution of (6.21) into (6.19) and solving for the R_2/R_1 resistance ratio produces (6.22):

$$\frac{R_2}{R_1} = \frac{n_{mf} G_{pll}}{K_d K_v 2\pi f_z} \tag{6.22}$$

Unfortunately, (6.21) and (6.22) have three unknown variables (C, R_1, and R_2). Therefore, engineering experience must be used to select one of the values of the components, then to solve for the other two component values. The values for the loop-filter components have limitations. On a printed circuit board, the capacitor values above 1 microfarad are electrolytic. Electrolytic capacitors are not recommended for filters. Therefore, a 1-microfarad capacitor or less will make the circuit component nonelectrolytic and insure excellent frequency-response characteristic.

Substituting the selected capacitance value into (6.21) and rearranging (6.21) computes the value for resistor R_2. Then, rearranging (6.22) computes resistor R_1.

The calculated resistance values should be within the range of resistances that insure proper operation of the phase detector and the operational amplifier. For example, a low value of R_1 should not make the phase detector exceed the output current limit, and a low value of R_2 should not make the operational amplifier exceed the output current limit. If any of the components R_1 and R_2 violate a specification limit, then recalculate until the specification limits are not violated. If repeated calculations do not yield a satisfactory solution, then the engineer should select another phase detector or operational amplifier that can accommodate the designed component values.

6.4.3 PLL Design with Sampling-Delay Examples

From the example in Section 6.3, we will use the component values in Table 6.2 without the effects of sampling delay and plot the open-loop response. We will then add the effects of sampling delay and see how much reduction in phase margin occurs. Next, we will compensate for this loss in phase margin by resynthesizing the loop-compensation values. Finally, we will look at the effects of changing the divide ratio.

Figure 6.5 shows a Bode plot of the open-loop transfer function without sampling-delay effects. Figure 6.5 shows a 475-kHz zero corner frequency, a 135-dB gain constant, a 1.9-MHz 0-dB crossover frequency, and a 76° phase margin, which were calculated in the previous example in Section 6.3. In the example without sampling effects, the reference frequency is a factor of 10,000 times greater than the loop bandwidth, so sampling-delay effects are minimized.

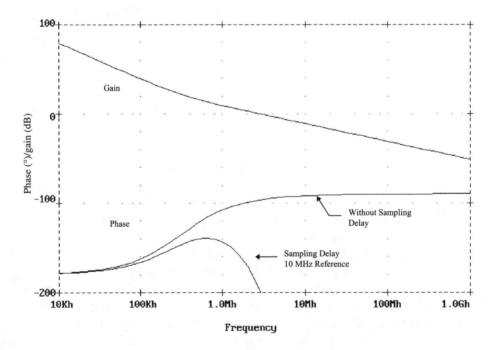

Figure 6.5 Open-loop-gain plots without sampling delay and with sampling delay.

Changing the reference frequency to 10 MHz brings the effects of sampling delay into the design and affects the component values in the PLL. Figure 6.5 shows the new Bode plot of the open-loop transfer function overlayed on top of the old one. Only the phase curve has changed, and the phase margin has been reduced to a small amount at the 1.9-MHz 0-dB crossover frequency. Substituting the 1.9-MHz 0-dB crossover frequency into (6.11) computes a phase margin of 6.3°, which is close to oscillation.

$$
\text{atan}\left\{\frac{\sin\left[\dfrac{-2\pi(1.945 \times 10^6)}{10 \times 10^6}\right] + \dfrac{1.945 \times 10^6}{4.751 \times 10^5}\cos\left[\dfrac{-2\pi(1.945 \times 10^6)}{10 \times 10^6}\right]}{\cos\left[\dfrac{-2\pi(1.945 \times 10^6)}{10 \times 10^6}\right] - \dfrac{1.945 \times 10^6}{4.751 \times 10^5}\sin\left[\dfrac{-2\pi(1.945 \times 10^6)}{10 \times 10^6}\right]}\right\}\frac{180}{\pi} = 6.253
$$

Recalculating the PLL design equations produces Table 6.4, which has new component values. By trial and error, the phase margin is changed in (6.14) until the zero corner frequency is a positive number, which is 19.5° phase margin and 17.56-kHz zero corner frequency. Then, the gain constant is multiplied by the ratio change in zero corner frequency (17.56 kHz/475 kHz = 0.037) to readjust the 0-dB crossover frequency back to 1.94 MHz. Then, (6.15) computes a 453-kΩ value for R_2.

$$
\frac{1}{2\pi(2 \times 10^{-11})(1.756 \times 10^4)} = 4.532 \times 10^5
$$

Finally, rearranging (6.22) computes a 681-kΩ value for R_1.

$$
\frac{4.533 \times 10^5}{\dfrac{14(5.199 \times 10^{12})}{0.16(6.2 \times 10^9)2\pi(1.756 \times 10^4)}} = 6.816 \times 10^5
$$

Figure 6.6 shows the new Bode plot of the open-loop transfer function overlayed on top of the old one. The Bode plot shows that the corner frequency of the zero has decreased to 17.56 kHz, and the gain constant has decreased to 121 dB to readjust the 0-dB crossover frequency back to 1.94 MHz. These values have changed because of the effects of sampling delay so that the 19.5° phase margin (0.17 damping factor) and the 1.94-MHz 0-dB crossover frequency can be maintained. The phase margin has been adjusted to 19.5° because it is the largest it can be without the zero corner frequency's going to zero or a negative value. Adjusting

Table 6.4 Calculated Values to Restore Some Stability with Sampling-Delay Effects

Phase Margin	f_x	R_2	f_z	Gpll	R_1	Filter Type
19.5°	1.94 MHz	453 kΩ	17.56 kHz	1.31E12, 121 dB	681 kΩ	Active

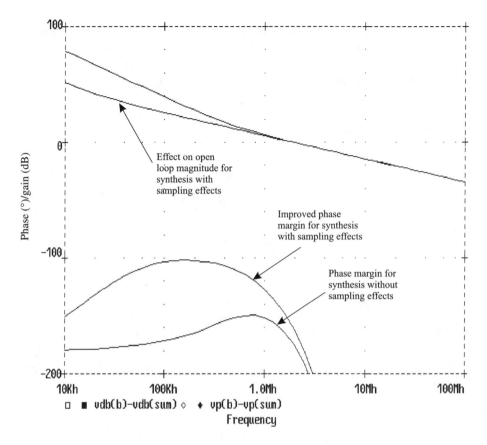

Figure 6.6 Bode plot of the open-loop gain with component values adjusted for sampling delay to get more stability.

the PLL parameters to maintain the 0-dB crossover frequency and phase margin indicates that the effects of sampling delay can be partially compensated.

Furthermore, setting the zero corner frequency to almost 0 Hz gives the maximum possible phase margin for the particular 0-dB crossover frequency and reference frequency because the full 90° improvement in phase margin is fully applied way in advance of the 0-dB crossover point. Consequently, a maximum phase margin of 20° at a five-to-one reference frequency to 0-dB crossover is not acceptable. At a 10-to-1 reference frequency to 0-dB crossover frequency, a maximum possible phase margin of 75° is possible, which is an acceptable amount and gives plenty of margin over process with the component values.

Now, we will look at the effects of changing the divide ratio. Since this occurs often in frequency-synthesizer design, it is important to understand the effects it has on the control system. Also, it shows another way that the phase margin can be improved with sampling effects.

Table 6.5 summarizes the new calculated values for changing the divide ratio. Taking the log of (6.19) computes the reduced gain constant 125.5 dB, and taking the square root of (6.19) and dividing by 2π gives a reduced natural frequency of 299 kHz.

Table 6.5 Summary of the New Calculated Values for Changing the Divide Ratio

Phase Margin	ζ	f_n	f_x	n_{mf}	f_z	G_{pll}	R_1	R_2	Filter Type
23°	0.315	299 kHz	0.33 MHz	140	0.475 MHz	3.54E12, 125.5 dB	100 kΩ	16.75 kΩ	Active

$$10 \log \left[\frac{0.16\,(6.2 \times 10^9)}{140\,(20 \times 10^{-12})(100 \times 10^3)} \right] = 125.494$$

$$\frac{\sqrt{\dfrac{0.16\,(6.2 \times 10^9)}{140\,(20 \times 10^{-12})(100 \times 10^3)}}}{2\pi} = 2.996 \times 10^5$$

Recalculating (6.11) and substituting into (6.12) computes the increased damping factor of 0.315. Equation (6.13) computes the reduced 0-dB crossover frequency of 0.33 MHz as shown by (6.23).

$$\frac{\sqrt{\dfrac{0.16\,(6.2 \times 10^9)}{140\,(20 \times 10^{-12})(100 \times 10^3)}}}{2\pi} \sqrt{2 \times 0.315^2 + \sqrt{4 \times 0.315^4 + 1}} = 3.306 \times 10^5$$

$$(6.23)$$

Substituting the new crossover frequency of 0.33 MHz into (6.11) computes the increased phase margin of 23° as shown by (6.24).

$$\phi = \operatorname{atan} \left[\frac{\sin\left(2\pi \dfrac{-f_x}{f_{ref}}\right) + \dfrac{f_x}{f_z} \cos\left(2\pi \dfrac{-f_x}{f_{ref}}\right)}{\cos\left(2\pi \dfrac{-f_x}{f_{ref}}\right) - \dfrac{f_x}{f_z} \sin\left(2\pi \dfrac{-f_x}{f_{ref}}\right)} \right]$$

$$= \operatorname{atan} \left\{ \frac{\sin\left[2\pi \dfrac{-(3.307 \times 10^5)}{1 \times 10^7}\right] + \dfrac{3.307 \times 10^5}{4.751 \times 10^5} \cos\left[2\pi \dfrac{-(3.307 \times 10^5)}{1 \times 10^7}\right]}{\cos\left[2\pi \dfrac{-(3.307 \times 10^5)}{1 \times 10^7}\right] - \dfrac{3.307 \times 10^5}{4.751 \times 10^5} \sin\left[2\pi \dfrac{-(3.307 \times 10^5)}{1 \times 10^7}\right]} \right\} \frac{180}{\pi}$$

$$= 22.935 \qquad (6.24)$$

Figure 6.7 shows the effect of changing divide ratio from 14 to 140 on the Bode plot of the open-loop gain. The 0-dB crossover point is lowered to 0.33 MHz, which puts it in a better position for phase margin. The phase plot is unchanged by changing the divide ratio.

The figure also shows the variation in the crossover frequency that has to be considered when doing a frequency synthesizer that covers a large frequency range.

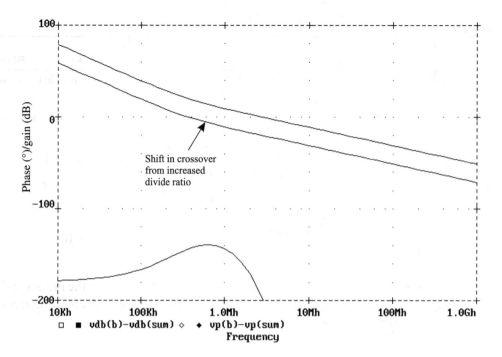

Figure 6.7   Bode plot of the open-loop gain for only changing divide ratio.

Furthermore, the worst-case conditions for a synthesizer are shown to be at the minimum divide ratio for high 0-dB crossover frequency and at the maximum divide ratio for low 0-dB crossover frequency. Centering this phase response is required for a wide-frequency-range synthesizer.

6.5   Fast Switching Time

This example illustrates the technique for active compensation, by compensating the design for sampling delay, and obtaining fast switching time. Fast switching time is a common design goal in many PLL references but is a less common customer requirement that will be addressed by this example. The example will achieve fast switching times without switching the loop bandwidths, which is another technique. Table 6.6 shows the given PLL parameters for the example PLL for fast switching response.

First, the classical transient-analysis equations are used by trial and error to find the damping factor that gives the fastest settling time to within 1% of the final value. Figure 6.8 shows an overlay of the results for 2 damping (dashed line) and 0.58 damping (solid line). The case with the damping factor of 2 settled to

Table 6.6   Given Parameters for Fast-Switching-Time Example PLL

K_v	K_d	n_{mf}	f_{in}	f_{out}	ζ	C	Process
1E9 rad/s/V	0.29 V/rad	4	12.5 MHz	50 MHz	0.58	100 pF	$0.35\text{-}\mu\text{m}\ L_{eff}$

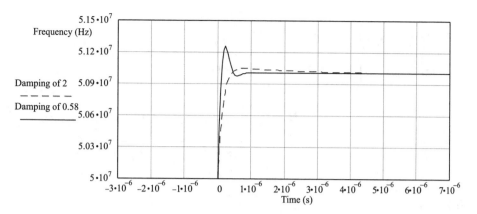

Figure 6.8   Overlay of settling-time response for damping factors of 2 and 0.58.

within 1% in 4.083 μs. The case with the damping factor of 0.58 settled to within 1% in 0.916 μs. This shows that the 0.58-damping-factor case settled 78% faster. From substituting (6.10) into (6.12), (6.25) computes the phase margin of 57.8° for a 0.58 damping factor:

$$\phi_{m\_goal} = \operatorname{atan}\left(2\zeta\sqrt{2\zeta^2 + \sqrt{4\zeta^4 + 1}}\,\right) \tag{6.25}$$

$$= \operatorname{atan}\left(2 \times 58\sqrt{2 \times 58^2 + \sqrt{4 \times 58^4 + 1}}\,\right)\frac{180}{\pi} = 57.828°$$

Setting zero corner frequency f_z in (6.11) to 0 makes the $R_2 C$ term a large number; this eliminates the first terms in the numerator and denominator since their maximum values can only be 1, which is small compared to the infinite large $R_2 C$ term. This simplification gives (6.26) for the maximum possible phase margin:

$$\phi_{m\_max} = \operatorname{atan}\left[\frac{\cos\left(-2\pi\dfrac{f_x}{f_{ref}}\right)}{-\sin\left(-2\pi\dfrac{f_x}{f_{ref}}\right)}\right]\frac{180}{\pi} \tag{6.26}$$

The reference frequency is usually a given requirement, which leaves 0-dB crossover frequency to be determined in (6.26). To get the fastest switching time, we want the 0-dB crossover frequency to be as high as possible. Because of the nonlinearity of (6.26), a simple optimization technique (Newton's method) must be used to get the maximum 0-dB crossover frequency of 1.1 MHz.

With the maximum crossover frequency and phase margin, the location of the zero corner frequency can be computed by substituting $1/f_{ref}$ for τ_{sd} in (6.14) to produce (6.27):

$$f_z = \frac{-f_x \sin\left(2\pi\dfrac{f_x}{f_{ref}}\right)\tan(\phi_m) + f_x \cos\left(2\pi\dfrac{f_x}{f_{ref}}\right)}{\tan(\phi_m)\cos\left(2\pi\dfrac{f_x}{f_{ref}}\right) + \sin\left(2\pi\dfrac{f_x}{f_{ref}}\right)}$$

$$= \frac{-(1.118\times10^6)\sin\left(2\pi\dfrac{1.118\times10^6}{1.25\times10^7}\right)\tan(0.942) + 1.118\times10^6\cos\left(2\pi\dfrac{1.118\times10^6}{1.25\times10^7}\right)}{\tan(0.942)\cos\left(2\pi\dfrac{1.118\times10^6}{1.25\times10^7}\right) + \sin\left(2\pi\dfrac{1.118\times10^6}{1.25\times10^7}\right)}$$

$$= 7.483\times10^4 \tag{6.27}$$

From the zero corner frequency, (6.28) computes component value R_2:

$$R_2 = \frac{1}{2\pi C f_z} = \frac{1}{2\pi(1\times10^{-10})(7.426\times10^4)} = 2.143\times10^4 \tag{6.28}$$

Gain constant at 0-dB crossover frequency with frequency variable set to 0-dB crossover frequency substituted into (6.20) gives (6.29) to compute the gain constant:

$$G_{pll} = \frac{(2\pi f_x)^2}{\sqrt{1+\left(\dfrac{f_x}{f_z}\right)^2}} = \frac{[2\pi(1.118\times10^6)]^2}{\sqrt{1+\left(\dfrac{1.118\times10^6}{7.426\times10^4}\right)^2}} = 3.27\times10^{12} \tag{6.29}$$

From the gain constant and proper substitution into (6.22) and solving for R_1, the component value R_1 can be computed by (6.30):

$$R_1 = \frac{K_d K_v 2\pi f_z R_2}{n_{mf} G_{pll}} = \frac{0.525(1,000\times10^6)\,2\pi(7.426\times10^4)\,R_2}{4(3.271\times10^{12})} = 4.013\times10^5 \tag{6.30}$$

Table 6.7 summarizes the calculated results for the fast-switching-time example PLL and completes the design example.

6.6 Minimum Bandwidth of a PLL

This example determines the minimum bandwidth of a charge pump monolithic PLL with a current-starved ring VCO. Several applications in telecommunications,

Table 6.7 Calculated Values for Fast-Switching-Time Example PLL

Phase Margin	f_{x_max}	R_2	f_z	G_{pll}	R_1	Filter Type
58°	1.1 MHz	21.4 kΩ	74.3 kHz	3.27E12, 125.1 dB	360 kΩ	Active

wireless, and pager technologies require narrow-bandwidth PLLs (10 to 100 Hz). Consequently, this example computes the minimum PLL bandwidth (theoretical limit) allowed by the phase noise of the VCO and the minimum bandwidth (component limit) for a monolithic PLL.

Several factors determine the minimum bandwidth. The phase noise of the VCO is one factor because the PLL will lose lock with fast modulation. The size of the loop-filter capacitor is another factor because the largest external ceramic capacitor is 1 μF. Electrolytic and tantalum capacitors leak currents that can cause a PLL to become unstable. The charge pump and phase detector gain combination and VCO gain are other factors because lowering the gain reduces the bandwidth of the PLL. The SNR generated by the loop to the input of the VCO is the final limit because a low SNR will cause the loop to unlock.

6.6.1 VCO Phase-Noise Limit

Computing the VCO phase-noise limit begins by measuring the phase noise of the VCO. Figure 6.9 shows extrapolated measured phase noise of the VCO for the example PLL with a current-starved ring VCO. Measured data points were taken between 100 kHz and 20 MHz, and the rest of the curve was extrapolated based on the slope's not changing and a white noise floor of −155 dBc/Hz.

Next, equations from Section 3.2.6 are used to model the phase noise [3, pp. 24–25]. From (3.65), (6.31) models phase noise:

$$\mathscr{L}(f) = \text{if}\left\{f(i) < f_{int},\ 10 \log\left[af(i)^{b}\right],\ \mathscr{L}_{f_{int}}\right\} \tag{6.31}$$

where

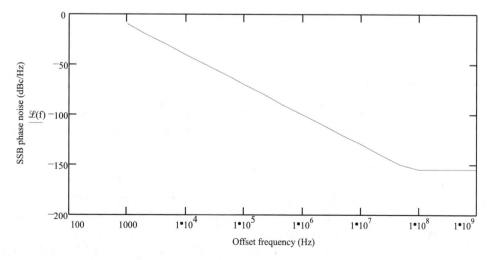

Figure 6.9 Measured phase noise of the VCO for the example PLL.

$\mathscr{L}(f)$ = phase noise (dBc/Hz);

$f(i)$ = array of frequency points in log steps (Hz);

f_{int} = frequency of intercept with white phase noise (Hz);

$\mathscr{L}_{f_{int}}$ = white phase-noise level (dBc/Hz).

Equation (6.32) describes phase noise with an exponential power slope in (6.31):

$$P_{slope} = 10\,\log[af(i)^b] \qquad\qquad (6.32)$$

where

a = multiplication factor;

b = noise-power-slope exponent.

The variable b is known from the slope of the phase-noise curve (in this case, −3, 30 dB/dec). The multiplication factor is solved by selecting a point ($\mathscr{L}_{fa} = -100$ dBc/Hz, $f_a = 10$ MHz) on the phase-noise curve and calculating a with (6.33):

$$a = 10^{\dfrac{\mathscr{L}_{f_a} - 10[b\,\log(f_a)]}{10}} \qquad\qquad (6.33)$$

Equation (6.33) computes the multiplication factor a to equal 1E8. For the final parameter in the phase noise, (6.34) is used to compute the frequency of intercept:

$$f_{int} = \left[\dfrac{10^{\left(\dfrac{\mathscr{L}_{f_{int}}}{10}\right)}}{a}\right]^{1/b} \qquad\qquad (6.34)$$

Equation (6.34) computes a frequency of intercept with white noise of 68.13 MHz.

Integrating the phase noise computes the amount of residual phase error from the VCO that the phase/frequency detector will have to operate with without losing lock. The maximum phase error allowed is π for the phase/frequency detector. Consequently, if the residual phase error from the VCO equals this value, the loop cannot lock because there is no headroom for the phase detector to adjust out the error in the PLL. Using Newton's method for the starting frequency, the phase-noise curve is segmented, integrated, and summed from the starting frequency to infinity until the phase error equals π. For instance, integrating from the first segment from the frequency of intercept to infinity computes the integrated phase error as shown by (6.35):

$$\Delta\phi_{\text{floor}} = \int\limits_{f_{int}}^{10^9} 10^{\left(\frac{\mathscr{L}_{f_{int}}}{10}\right)} df \qquad (6.35)$$

$$\Delta\phi_{\text{floor}} = 2.947 \times 10^{-7} \text{ rad}$$

Studying the phase-error contribution from each frequency band helps us find the main noise contribution. Table 6.8 shows the calculated results from each integration segment.

Summing column 3 in Table 6.8 for a 4–20-kHz starting frequency band computes 3.125 rad of phase noise. Consequently, the absolute minimum bandwidth limit is 4 kHz. Headroom from the minimum bandwidth is required for practical circuits. Therefore, the minimum bandwidth should be set on the order of ten times the absolute minimum bandwidth, or in this case 40 kHz. This assumes the feedback divider ratio is 1.

6.6.2 Component Limits of Standard APLL

To determine the component limits, we will use an example 1.3-ma charge pump PLL. The example PLL has a 66 MHz/V VCO gain, 0.03-μF loop-filter capacitor, and 22Ω compensation resistor. We will analyze the 3-dB bandwidths for a 1.3-ma charge pump PLL, a 1.3-ma charge pump PLL with a 1-μF loop-filter capacitor, and a 1.3-ma charge pump PLL with charge pump gain divided by 10 and a 1-μF loop-filter capacitor. Figure 6.10 shows a schematic of the PSPICE circuit for analog behavioral analysis of PLLs.

Figure 6.11 shows the closed-loop response for the 1.3-ma charge pump PLL, the 1.3-ma charge pump PLL with a 1-μF loop-filter capacitor, and the 1.3-ma charge pump PLL with charge pump gain divided by 10 and a 1-μF loop-filter capacitor. The capacitor C_1 and the charge pump gain, the G current source, in Figure 6.10 are changed to produce the closed-loop gain responses in Figure 6.11.

From Figure 6.11, the 1.3-ma charge pump PLL has a 3-dB bandwidth of 2 MHz. Increasing the loop-filter capacitor, C_1, to 1 μF reduces the bandwidth. The PLL with a 1-μF loop-filter capacitor has a 3-dB bandwidth of 400 kHz. Decreasing the phase detector and charge pump gain combination or the VCO gain by 10 further minimizes the bandwidth close to the targeted 40-kHz bandwidth minimum allowed by analysis of the VCO phase noise. For the analysis, all the change in gain is lumped into the phase detector and charge pump gain combination. The PLL with charge pump gain divided by 10 has a 3-dB bandwidth of 70 kHz.

Table 6.8 Calculated Phase Error for Each Integration Segment

Starting Frequency (Hz)	Ending Frequency (Hz)	Integrated Phase Error (rad)
68E6	1E9	0.294E–6
1E6	68E6	76.3E–6
100E3	1E6	5E–3
20E3	100E3	120E–3
4E3	20E3	3.0

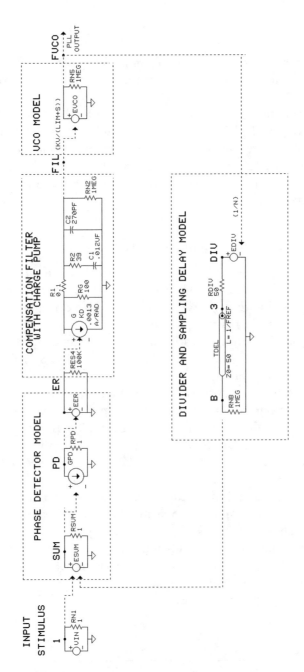

Figure 6.10 PSPICE schematic for analog behavioral analysis of PLLs.

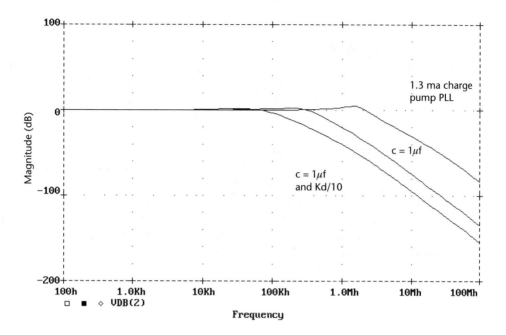

Figure 6.11 Closed-loop gain of the 1.3-ma charge pump PLL with modified parameters.

Finally, other limitations on bandwidth are considered. First, a low SNR at the input to the VCO will cause the loop to unlock. This puts a limit on reducing the phase detector and charge pump gain combination because a lower gain reduces the SNR into the VCO. Loop noise measurements will be needed to determine this limit. Next, a small frequency tuning range for the VCO will cause the loop to lose lock. This puts a limit on reducing the gain of the VCO because the resulting narrower tuning frequency range might not cover the specified frequency range over process corners.

In summary, the 1.3-ma charge pump PLL has a minimum bandwidth of 400 kHz with a 1-μF compensation capacitor. The 1.3-ma charge pump PLL would have to be changed (phase detector gain reduced by a factor of 10) to reduce the bandwidth to 40 kHz. The minimum PLL bandwidth with the VCO in the 1.3-ma charge pump PLL is 4 kHz. Consequently, bandwidths less than 4 kHz require that the VCO phase noise be reduced or that a higher feedback divide ratio be used. For example, a divide ratio of 100 would reduce the minimum bandwidth limit from the VCO to 40 Hz.

6.7 Maximum-Divide-Ratio Example for Loop-Component Synthesis

Frequency synthesis for a large number of frequencies occurs in cell phones for all the various formats that have to be met. Frequency resolution and the least common denominator among the formats forces the multiplication ratios to higher and higher values. Knowing the maximum value can help us decide if we need to change to a multiple loop architecture or fractional-N architecture to meet this requirement.

Consequently, finding the maximum divide ratio with a charge pump PLL gives us another PLL synthesis technique.

The following example makes it easier to illustrate the synthesis technique for finding the maximum divide ratio. First, the example loop-compensation architecture is selected. In this case, the charge pump compensation shown in Figure 6.12 is selected because the application is for an integrated circuit with external loop compensation.

Next, the limitations of the components in this PLL architecture are explored. The VCO gain, the charge pump gain, and the output frequency in Table 6.9 are given by the components chosen or by specifications. Consequently, they cannot be easily adjusted or have a narrow adjustment range.

Next, Table 6.10 shows the control-system specifications requirements to give the desired response for stability and switching speed. The natural frequency will be computed in (6.37) and will be at least a factor of 20 less than the reference

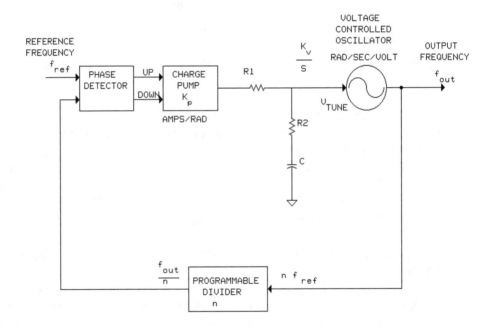

Figure 6.12   Charge pump compensation PLL.

Table 6.9   Component Values for Charge Pump Example

F_{out} (Hz)	K_v (rad/s/V)	K_p (A/rad)
622E6	4,000E6	0.0034

Table 6.10   Desired Control-System Parameter

ζ
0.9

frequency to avoid sampling-delay effects. Higher bandwidths can be achieved but not with the single-loop approach used in this example. In the example, we set the capacitor to the highest nonelectrolytic capacitor value of 1 μF because that will give us the lowest possible loop bandwidth, which will maximize the value of N.

Consequently, the divide ratio N and R_2 resistor values are all that remain to synthesize the loop compensation. With the loop capacitor, damping factor, and output frequency selected, and given K_p and K_v, equations for N_{max} and R_2 can be derived. First, (6.36) computes the relationship of the output frequency to the input frequency for a reference frequency 20 times the natural frequency:

$$N_{max} = \frac{f_o}{20 f_n} \tag{6.36}$$

Rearranging (2.26) and solving for the natural radian frequency produces (6.37):

$$\omega_n = \sqrt{\frac{K_v K_p}{2\pi CN}} \tag{6.37}$$

Substituting (6.37) into (6.36) produces (6.38) for computing the maximum multiplication factor, assuming the output frequency, VCO gain, and phase detector gain are fixed, and the maximum allowable capacitor value is used:

$$N_{max} = \left[\frac{f_o}{20 \dfrac{\sqrt{\dfrac{K_v K_p}{2\pi C}}}{2\pi}} \right]^2 \tag{6.38}$$

Equation (2.27) solves for R_2 as shown by (6.39):

$$R_2 = 2 \frac{2\zeta \omega_n \pi N}{K_v K_p} \tag{6.39}$$

Equations (6.38) and (6.39) compute the final N and R_2 component values. Table 6.11 shows the calculated values. An additional capacitor (C_2) across the VCO tune line to ground is used for additional filtering of reference sidebands, and its value is usually set to one-tenth the value of C.

The above calculations assume ideal phase noise for the reference oscillator and the VCO. The amount of phase noise added to the reference oscillator is

Table 6.11   Calculated Component Values for Maximum Frequency Multiplication for Charge Pump Example

N	R_2	C	f_n	f_{ref}
17,000	162	0.1E–6	1.7 kHz	35.2 kHz

20 log of 17,000, which equals 85 dB. This amount of multiplication to the input could make the loop unstable or give it a lot of jitter. For instance, to keep the jitter reasonable, we set a −20 dBc/Hz maximum limit (wideband FM begins at >−20 dBc/Hz). Consequently, the input reference oscillator must have a phase-noise level less than −105 dBc/Hz. In addition, the divided-down integration of the VCO must be less then 3.14 rad for an offset frequency of 170 Hz (one-tenth of the natural frequency) in order to have a stable loop. Finally, the input reference-oscillator input must be protected from coupled-in signals (approximately <−105 dBc) by using hysterises and low-noise-circuit layout techniques because any FM or jitter on the input will be multiplied by 17,000 (85 dB) and can cause the loop to become unstable.

Measured data was taken to confirm the feasibility of the above calculations. A PLL test chip with 0.35-μm gate length process was used to make measurements for the maximum multiplication factor. The PLL used a charge pump phase detector and a current-starved ring VCO. The external compensation values were 121Ω for R_2, 0 for R_1, 0.2 μF for C_1 and 0.005 μF for C_2. The input was multiplied by 512, and the output was divided by 8 because of the high frequencies involved. Figure 6.13 shows the hold-in range of the PLL. The y-axis is the ratio of output frequency over input frequency with major divisions of 5. Deviation from divide-by-64 means the loop is unlocked. The reference frequency is swept from 10 kHz to 2 MHz and then from 2 MHz to 10 kHz, as shown on the x-axis. The figure shows the loop is locked from a 400-kHz to 1.5-MHz reference input frequency, which produces a 204.8–768-MHz output frequency.

Figure 6.14 shows the phase noise of the divide-by-8 output. This shows a PLL bandwidth around 30 kHz. The peak phase noise is −73 dBc/Hz at the divide-by-8 output but −54 dBc/Hz at the output of the PLL (add 18 dB). The ratio of

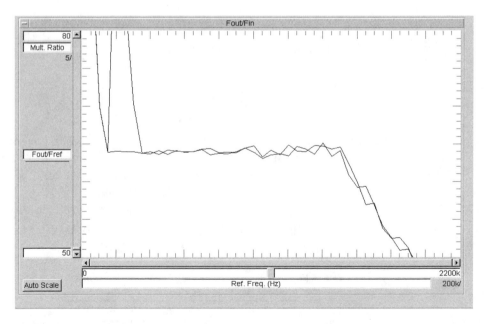

Figure 6.13   Hold-in range of the maximum-divide-ratio test chip PLL05.

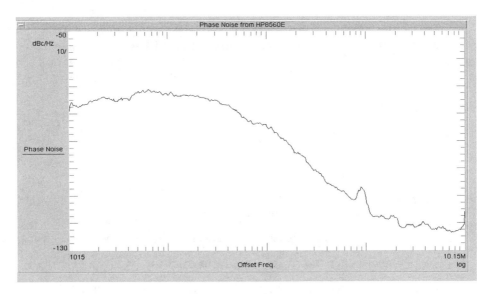

Figure 6.14   Phase noise of the maximum-divide-ratio test chip PLL05.

17,000 to 512 is 33, and 20 log of 33 is 30 dB. For a multiplication factor of 17,000, the phase noise at the output would be −24 dBc/Hz, which would be close to the wideband FM region and close to creating an unstable loop. Extrapolating from the above laboratory data, multiplication factors of approximately 3,000 and above can be risky.

6.8   Optimum PLL Design for Low-Phase-Noise Performance

Electronic warfare receivers, test instruments, communications systems, cell phones, and radar systems often require low-phase-noise performance from PLLs. Receivers need low-phase-noise PLLs to prevent large signals from degrading small-signal detection levels. Test instruments need low-phase-noise PLLs to make measurements over a large dynamic range and to improve the accuracy of measurements. Radar systems need low-phase-noise PLLs to detect small signals from ground clutter or chaff. In designing a PLL, the general practice has been to select damping factor and loop bandwidth for switching time [5]. This section shows that switching-time design goals are different from low-phase-noise-level design goals. In designing a PLL for switching-time requirements, references recommend using a 0.7 damping factor for minimum settling time and a bandwidth greater than the inverse of the switching time. This section will determine the optimum damping factor and bandwidth goals for low-phase-noise PLL design.

6.8.1   PLL Phase-Noise Equations

To design low-phase-noise PLLs requires equations that mathematically describe phase noise in PLLs. Making these equations a function of the damping-factor and loop-bandwidth design goals will ease the study of optimum phase-noise design goals. From Section 3.2.9, we will review the phase noise in PLL equations. Equation (6.40) computes phase noise at the output of the PLL with control-systems variables [3]:

$$\mathscr{L}_{pll}(f_m) = \mathscr{L}_{ref}(f_m)\left[\frac{1}{1 + G(f_m)H(f_m)}\right]^2 + \mathscr{L}_{vco}(f_m)\left[\frac{G(f_m)}{1 + G(f_m)H(f_m)}\right]^2$$

$$(6.40)$$

where

f_m = offset frequency from the carrier (Hz);

$\mathscr{L}(f_m)_{ref}$ = single-sideband phase-noise ratio from the reference oscillator (rad/Hz);

$\mathscr{L}(f_m)_{vco}$ = single-sideband phase-noise ratio from the VCO (rad/Hz);

$\mathscr{L}(f_m)_{pll}$ = single-sideband phase-noise ratio at the output of the PLL (rad/Hz);

$G(f_m)$ = forward transfer function;

$H(f_m)$ = feedback transfer function;

$G(f_m)H(f_m)$ = open-loop transfer function.

Converting (6.40) to a function of damping factor and natural frequency will produce an equation of damping factor and loop bandwidth because (6.41) relates loop bandwidth to natural frequency [3]:

$$f_n = \frac{B_{pll}}{\sqrt{1 + 2\zeta^2 + \sqrt{(1 + 2\zeta^2)^2 + 1}}}$$

$$(6.41)$$

where

B_{pll} = 3-dB bandwidth of the PLL (Hz);

ζ = damping factor ratio of the PLL;

f_n = natural frequency of the PLL (Hz).

To convert (6.40) to a function of damping factor and natural frequency requires an equation for the open-loop transfer function and the forward transfer function in terms of damping factor and natural frequency. Equation (6.42) computes the open-loop transfer function in terms of damping factor and natural frequency [3]:

$$|1 + G(f_m)H(f_m)| = \sqrt{\left[1 - \left(\frac{f_n}{f_m}\right)^2\right]^2 + \left(\frac{2\zeta f_n}{f_m}\right)^2}$$

$$(6.42)$$

Equation (6.43) computes the forward transfer function in terms of damping factor and natural frequency [3]:

$$|G(f_m)| = n_{mf} \frac{(f_n)^2}{(f_m)^2} \sqrt{1 + \left(\frac{2\zeta f_m}{f_n}\right)^2} \qquad (6.43)$$

where

n_{mf} = multiplication factor of PLL.

Combining (6.40) and (6.43) mathematically describes the output phase noise of a PLL in terms of damping factor and PLL bandwidth. To complete the model of output phase noise in a PLL requires an equation for modeling the phase noise of the oscillators in the PLL. Equations 6.44 through 6.45 model phase noise for the oscillators in the PLL [3]:

$$\mathcal{L}(f_m) = a f_m^b \qquad (6.44)$$

where

$$a = \exp_{10}\left[\frac{\mathcal{L}(f_a) - 10b \log(f_a)}{10}\right] \qquad (6.45)$$

$$b = \frac{\mathcal{L}(f_a) - \mathcal{L}(f_b)}{10[\log(f_a) - \log(f_b)]} \qquad (6.46)$$

and where

f_a = start frequency of slope (Hz);

f_b = stop frequency of slope (Hz).

Equations (6.40) through (6.46) model the output phase noise of a PLL. Consequently, these equations allow damping-factor effects and loop-bandwidth effects on the output phase noise of a PLL to be studied.

6.8.2 Damping-Factor Effect

Studying an example PLL shows the effect of damping factor on phase noise. For the example, a 30-MHz crystal oscillator and 1-GHz VCO are combined to form a PLL with a multiplication factor of 33 [3]. Table 6.12 shows the PLL parameters that describe the example. Table 6.13 shows the phase-noise characteristics for the crystal oscillator and VCO.

Table 6.12   Parameters That Describe the Optimum Phase-Noise Example PLL

K_v	K_d	n_{mf}	f_{in}	f_{out}	ζ	C	Process Gate Length
6.28E9 rad/s/V	0.286 V/rad	33	30 MHz	990 MHz	2	100 pF	0.18 μm

Table 6.13 Phase-Noise Characteristics for Optimum Phase-Noise Example PLL

L_{vco}	f_a	VCO Slope	L_{ref}	Reference Slope	L_{add}
−30 dBc/Hz	1 kHz	−2	−132 dBc/Hz	0	3.1

We begin our study by plotting the optimum phase-noise plot for the example PLL and then deviating from the optimum to show the effects of damping factor. Figure 6.15 shows the desired optimum phase-noise performance for the PLL combination of the 30-MHz crystal oscillator and the 1-GHz VCO. The figure shows a short dashed line for the VCO phase noise, a long dashed line for the reference-oscillator phase noise, and a solid line for the optimum combination of phase noise in the example PLL.

Varying the damping factor in Figure 6.16 shows the peaking effect that damping factor has on phase noise. In Figure 6.16, a damping factor less than 1 starts to add a significant amount of phase noise. Consequently, where practical, a damping-factor range of 1 to 1.5 minimizes the amount of additive phase noise; however, a damping factor of 1 to 1.5 degrades optimum switching time to a slower overdamped response. With the range of damping factors determined, the optimum bandwidth remains to be determined.

6.8.3 PLL Bandwidth Effect

Again, the example PLL can be used to study the effects of bandwidth on the output phase noise of a PLL. Figures 6.17 and 6.18 show the effect of varying the bandwidth on the output phase noise of the PLL. The bandwidths below optimum in Figure 6.17 follow the far-from-the-carrier phase noise of the VCO; however,

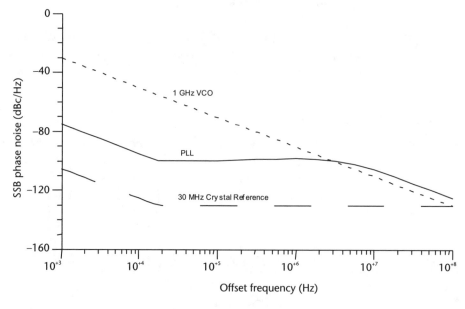

Figure 6.15 Phase noise of reference oscillator, VCO, and PLL.

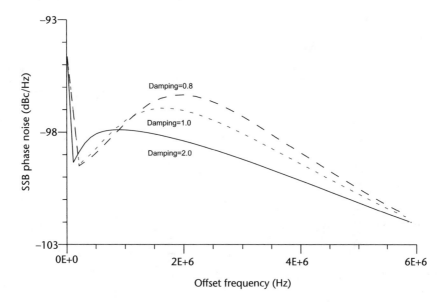

Figure 6.16   Phase noise versus overdamped conditions.

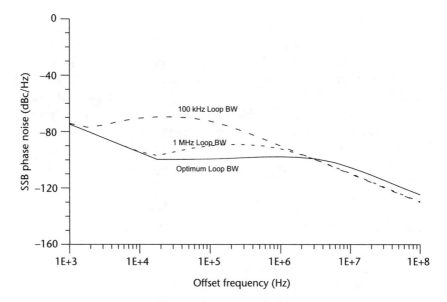

Figure 6.17   Phase noise for the bandwidths less than the optimum bandwidth.

these bandwidths add phase noise to the multiplied close-to-the-carrier phase noise of the reference oscillator.

The bandwidths above optimum in Figure 6.18 follow the multiplied close-to-the-carrier phase noise of the reference oscillator; however, these bandwidths add phase noise to the far-from-the-carrier phase noise by shifting the phase-noise curve of the VCO. From these figures, any switching-time performance requirement faster than the inverse of the optimum loop bandwidth will have to be sacrificed in order to obtain optimum phase-noise performance.

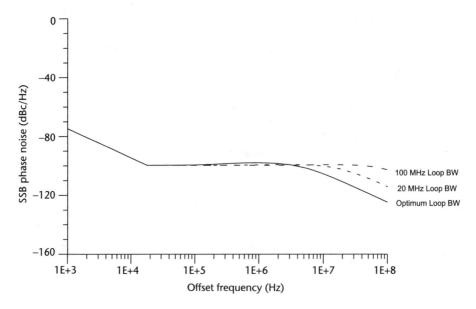

Figure 6.18 Phase noise for bandwidths greater than the optimum bandwidth.

Until now, the general practice has been to select the loop bandwidth to be the phase-noise-response intersection of the multiplied reference oscillator and the VCO (between 2 and 8 MHz in the example); however, Figures 6.17 and 6.18 show a different criterion to achieve optimum loop bandwidth. At optimum bandwidth, the PLL has output phase noise that follows the multiplied close-to-the-carrier phase noise of the reference oscillator. In addition, the optimum loop bandwidth adds a phase-noise contribution from the multiplied phase noise of the reference oscillator equal to the far-from-the-carrier phase noise of the VCO. This last bit of information will be used to compute the optimum bandwidth for the PLL.

6.8.4 Equations to Compute Optimum PLL Bandwidth

Equalizing the phase-noise-contribution characteristics for the optimum bandwidth and making assumptions about the phase noise of the VCO and reference oscillator makes calculating the optimum bandwidth possible. A 20-dB/dec roll-off for the phase noise of the VCO and a constant phase noise for the multiplied reference oscillator are assumed to compute the optimum bandwidth of the PLL. Substituting a 20-dB/dec phase-noise roll-off into (6.46) produces (6.47) for modeling the phase noise of the VCO:

$$\mathcal{L}(f_m)_{vco} = a_{vco} f_m^{-2} \qquad (6.47)$$

where

$\mathcal{L}(f_m)_{vco}$ = phase-noise response of VCO (rad/Hz);

a_{vco} = VCO power slope.

A constant phase noise in (6.44) produces (6.48) for modeling the phase noise of the reference oscillator:

$$\mathcal{L}(f_m)_{ref} = \text{constant} \tag{6.48}$$

where

$\mathcal{L}(f_m)_{ref}$ = constant phase-noise level of reference oscillator (rad/Hz).

From the previous section, optimum bandwidth produces equal phase-noise contributions from the reference oscillator and the VCO for phase noise far from the bandwidth of the PLL. Consequently, (6.49) mathematically describes this optimum PLL bandwidth condition:

$$\mathcal{L}(f_m)_{pll} = \mathcal{L}(f_m)_{vco}\mathcal{L}(f_m)_{add} \text{ for } f_m \gg B_{pll} \tag{6.49}$$

where

$\mathcal{L}(f_m)_{pll}$ = output phase-noise level of PLL (rad/Hz);

$\mathcal{L}(f_m)_{add}$ = amount of phase noise added to VCO phase noise far from the bandwidth of the PLL.

Studying the output phase noise of the PLL for frequencies greater than the bandwidth of the PLL allows several mathematical simplifications to be made. Equation (6.40) can be simplified for high frequencies to (6.50) because the numerator of (6.40) reduces to a magnitude of 1:

$$\frac{1}{1 + G(f_m)H(f_m)} = 1 \tag{6.50}$$

Then, substituting (6.49) and (6.50) into (6.40) reduces (6.40) to (6.51):

$$\mathcal{L}(f_m)_{vco}\mathcal{L}(f_m)_{add} = \mathcal{L}(f_m)_{ref}G^2(f_m) + \mathcal{L}(f_m)_{vco} \tag{6.51}$$

Equation (6.43) can also be simplified for frequencies much greater than the bandwidth of the PLL. For high frequencies, (6.43) reduces to (6.52):

$$G(f_m) = \frac{n_{mf}2\zeta f_n}{f_m} \tag{6.52}$$

Now, a few more algebraic steps with (6.47), (6.51), and (6.52) will produce an equation for optimum bandwidth. Substituting (6.52) into (6.51) produces (6.53):

$$\mathcal{L}_{vco}(f_m)[\mathcal{L}_{add}(f_m) - 1] = \mathcal{L}_{ref}(f_m)\left(\frac{n_{mf}2\zeta f_n}{f_m}\right)^2 \tag{6.53}$$

Substituting (6.47) into (6.53) produces (6.54):

$$a_{vco}(f_m)^{-2}[\mathscr{L}_{add}(f_m) - 1] = \mathscr{L}_{ref}(f_m)\left(\frac{n_{mf}2\zeta f_n}{f_m}\right)^2 \qquad (6.54)$$

Canceling the frequency variable and solving for the natural frequency produces (6.55):

$$f_n = \sqrt{\frac{a_{vco}[\mathscr{L}_{add}(f_m) - 1]}{\mathscr{L}_{ref}(f_m)(n_{mf}2\zeta f_n)^2}} \qquad (6.55)$$

Equation (6.55) computes the optimum natural frequency, which has a relatively simple conversion to loop bandwidth.

Now, however, the VCO slope constant and additive phase-noise constant must be determined to compute the optimum natural frequency in (6.55). From (6.44) and (6.47), (6.56) computes the VCO constant in (6.55) from one data point on the phase-noise curve for the VCO:

$$a_{vco} = \exp_{10}\left[\frac{\mathscr{L}(f_a)_{vco} - 10 \times -2 \times \log(f_a)}{10}\right] \qquad (6.56)$$

where

$$\mathscr{L}(f_a)_{vco} = \text{phase noise at (dBc/Hz)};$$

$$f_a = \text{frequency of data point on VCO phase-noise curve (Hz).}$$

Next, (6.57) computes the amount of additive phase noise in (6.55) by assuming equal reference-oscillator phase noise and VCO phase noise:

$$\mathscr{L}_{add}(f_m) = \frac{1}{10^{-3/20}} = 1.4125 \text{ for 3 dB} \qquad (6.57)$$

The 3-dB amount of additive phase noise provides a good compromise between close-to-the-carrier phase-noise performance and far-from-the-carrier performance; however, the amount of additive phase noise can be adjusted to lower values to improve far-from-the-carrier phase noise at the expense of higher close-to-the-carrier phase noise. Conversely, the amount of additive phase noise can be adjusted to higher values to improve close-to-the-carrier phase noise at the expense of higher far-from-the-carrier phase noise.

Now, all the equations are present to solve for the optimum loop bandwidth. First, (6.55) through (6.57) solve for the optimum natural frequency. Then, (6.41) converts optimum natural frequency to optimum 3-dB PLL bandwidth. These equations are new and provide a more accurate solution than those generally used in the past.

For instance, (6.41) and (6.55) through (6.57) compute a 5.2-MHz, 3-dB bandwidth and a 1.25-MHz natural frequency for the example PLL with a 1.5

damping factor. Computing this optimum bandwidth is more accurate than graphically estimating between 2 and 8 MHz, which has been the practice in the past for determining the loop bandwidth for optimum phase-noise performance.

Equations (6.58) through (6.60), derived from (6.13) and from (2.22) and (2.23), are used to finish the loop-filter synthesis and compute the 0-dB crossover frequency for the example. Table 6.14 shows the calculated results.

$$f_x = \sqrt{2(\zeta)^2 + \sqrt{4(\zeta)^4 + 1}} f_n \tag{6.58}$$

$$R_1 = \frac{K_d K_v}{(\omega_n)^2 C n_{mf}} \tag{6.59}$$

$$R_2 = \frac{2\zeta}{\omega_n C} \tag{6.60}$$

To summarize, low-phase-noise damping-factor and loop-bandwidth design goals have been shown to be different from switching-time design goals. For low-phase-noise design goals, a damping factor of 1 to 1.5 has been graphically shown to minimize additive phase noise in a PLL. For loop bandwidth, equations have been derived that compute the optimum loop bandwidth. Now, engineers can compute the optimum bandwidth instead of graphically estimating the optimum bandwidth.

6.9 Summary

This chapter presented loop-compensation synthesis. An optimizer's dependence on an objective function that ranks requirements in order to reach a clear optimum solution was shown. The ideal PLL requirements for a synthesizer and an ideal-clock-recovery PLL were presented. Consequently, synthesis of loop-compensation components in the loop is application dependent with many different strategies for loop compensation. Then, requirements that loop-filter architecture design can aid were presented. Several examples of the most popular compensation schemes were presented.

These examples included a cookbook method, active compensation, maximum capacitance design, compensation taking into account sampling delay, maximum loop bandwidth for fast switching time, maximum divide ratio, and optimum phase-noise design. For beginning- and intermediate-level PLL designers, it is best to start with the cookbook method. Active filter compensation with maximum capacitor value showed that we can reduce capacitor value to save silicon area by increasing R_1 to keep the same loop bandwidth.

Table 6.14 Calculated Values for Active Filter

a_{vco}	f_n	f_{3db}	f_x	R_1	R_2
1E3	1.25 MHz	5.3 MHz	5.0 MHz	8.8K	5.1K

For sampling effects to occur, the loop bandwidth must be close to the reference frequency. Sampling effects occur with undersampling that gives an inaccurate measure of the phase-error envelope. The sampling-delay example showed how to design around the effects of sampling delay. The error from sampling delay has a first-order approximation to a delay line of half a clock period.

The changing-divide-ratio example showed that increasing the divide ratio decreases the loop bandwidth, and changing the divide ratio does not change the phase of the transfer function; however, changing the divide ratio does change the phase margin. Next, the fast-switching-time example showed that the maximum loop bandwidth equals the reference frequency divided by 10. The fastest 1% settling time occurs with a damping factor of 0.58, a low divide ratio to keep the bandwidth wide, and a high reference frequency to reduce sampling-delay effects.

The minimum-bandwidth example showed that the phase noise of the VCO limits the minimum bandwidth because the PLL will lose lock with fast modulation. The charge pump and phase detector gain combination and VCO gain are other factors because lowering the gain reduces the bandwidth of the PLL. The VCO integrated-phase-noise effect can be reduced by using a higher feedback divide ratio. Finally, the SNR generated by the loop to the input of the VCO limits the minimum bandwidth because a low SNR will cause the loop to unlock.

The maximum-divide-ratio example showed that maximum divide ratio was determined by the ratio of the output frequency to the input frequency divided by 20, which is the widest stable natural frequency. The amount of phase noise added to the reference oscillator by a large feedback divide ratio can make the loop unstable or give it a lot of jitter. Also, the input reference-oscillator input must be protected from coupled-in signals because the large multiplication factor can cause the loop to become unstable. Next, the optimum low-phase-noise bandwidth example showed that low-phase-noise damping-factor and loop-bandwidth design goals are different from switching-time design goals. A damping factor of 1 to 1.5 was shown to minimize additive phase noise in a PLL. In addition, the optimum low-phase-noise loop bandwidth occurred at the phase-noise-response intersection of the multiplied reference oscillator and the VCO. Finally, these examples gave us an intuitive feeling for extending these methods to designs that are not presented.

Questions

6.1 Why is it that power consumption, jitter, and area cannot be the top requirement in the rankings?

6.2 A PLL design requires a 15-Hz, 3-dB, closed-loop bandwidth, a damping factor of 0.27, a multiplication factor of 64, and a reference frequency of 100 kHz. Given a K_v of 100,000 rad/s/V and a K_d of 0.4 V/rad, synthesize the loop-compensation values for an active filter without the effects of sampling delay, and do a Bode plot of the open-loop gain. Then, change the reference frequency to 100 Hz, resynthesize the loop-compensation values for an active filter with the effects of sampling delay, and do a Bode plot of the open-loop gain.

6.3 A PLL design requires a 6.75-MHz, 0-dB crossover frequency, a damping factor of 0.85, a multiplication factor of 10, a third-order pole at 31 times the crossover frequency, and a reference frequency of 125 MHz. Given a K_v of 1.6 GHz/V, an I_{rep} of 200 μa, a 0.5 main replication current multiplication factor (m_1), and one secondary replication current multiplication factor (m_2), synthesize the loop-compensation values for the filter in Figure 6.19 (C_{p1}, g_m, C_{p2}) without the effects of sampling delay, and do a Bode plot of the open-loop gain.

6.4 Why can we not lower the charge pump current to an extremely low value to achieve a small bandwidth?

6.5 Why can we not have external capacitor values greater than 1 μF?

6.6 Why can we not have an internal capacitor value greater than 100 pF?

6.7 Given a PLL with a VCO gain of 2 GHz/V, charge pump current of 100 μA, a maximum on-chip capacitor value of 100 pF, an output frequency of 960 MHz, and a damping factor of 0.6, compute the maximum multiplication factor, the minimum reference frequency, the natural frequency, and the value of R_2.

6.8 Given a PLL with a VCO gain of 50 MHz/V, a charge pump current of 100 μA, a maximum on-chip capacitor value of 100 pF, an output frequency of 12.8 MHz, and a damping factor of 0.6, compute the maximum multiplication factor, the minimum reference frequency, the natural frequency, and the value of R_2.

6.9 Given a PLL with a VCO gain of 50 MHz/V, a charge pump current of 1 μA, a maximum on-chip capacitor value of 100 pF, an output frequency of 48 MHz, and a damping factor of 0.99, compute the maximum multiplication factor, the minimum reference frequency, the natural frequency, and the value of R_2. Is this circuit practical?

6.10 Using graphical means, describe in words the location of the optimum phase-noise bandwidth.

6.11 What are the equations if the VCO slope is −30 dB/dec instead of −20 dB/dec?

6.12 Given a PLL with a VCO gain of 1 GHz/V, a damping factor of 2, a multiplication factor of 33, a flat reference phase noise of −131 dBc/Hz, a VCO phase noise of −90 dBc/Hz at 1-MHz offset frequency, and a 20-dB/dec slope, calculate the optimum bandwidth for a 3.162-ratio

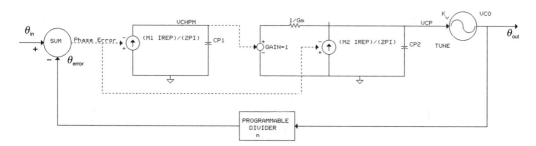

Figure 6.19 Loop compensation block diagram for Question 6.3.

phase noise added to VCO phase noise, and synthesize the loop-compensation values for a charge pump with a 3.4-ma maximum current.

6.13 Given a PLL with a VCO gain of 1 GHz/V, a damping factor of 2, a multiplication factor of 512, a flat reference phase noise of -131 dBc/Hz, a VCO phase noise of -120 dBc/Hz at 10-MHz offset frequency, and a 30-dB/dec slope, calculate the optimum bandwidth for a 3.162-ratio phase noise added to the VCO phase noise and synthesize the loop-compensation values for an active filter with a 1.8-V supply and a 100-pF maximum capacitor value.

References

[1]   Best, R., *Phase-Locked Loops: Design, Simulation, and Applications*, 3rd ed., New York: McGraw-Hill, 1997, pp. 66–70.

[2]   "Phase-Frequency Detector MC4344-MC4044," MTTL Complex Functions, Motorola Semiconductor Products, Inc., Data Sheet, 1973.

[3]   Goldman, S. J., *Phase Noise Analysis in Radar Systems*, New York: Wiley Interscience, 1989.

[4]   Egan, W. F., *Phase-Lock Basics*, New York: Wiley Interscience, 1998, p. 48.

[5]   Manassewitsch, V., *Frequency Synthesizers: Theory and Design*, 2nd ed., New York: Wiley Interscience, 1980.

CHAPTER 7

Test and Measurement

This chapter discusses test and measurements of PLLs. Measurements of PLLs are critical in verifying the performance of the PLL against its designed performance. Many of the tests can only be done on the bench because the length of the test time (greater than tens of milliseconds) makes it impractical to test on the tester in production. Understanding how a PLL is tested can help the designer incorporate testing aids into the PLL.

Initial measurements of a PLL are very important because measurements of an unlocked loop or a very noisy loop give erroneous data and may stop any further measurements. Consequently, measurements of hold-in range and spurious signals are presented to give some minimal stability and accurate noise information. Next, reference frequency-step-response measurements are discussed to verify damping and bandwidth design goals. The PLL bandwidth can also be accurately measured with a network analyzer. Several phase-noise measurements are presented. An explanation of the advantages and disadvantages of each method are presented. Jitter and spurious-signal measurements are also discussed. Understanding these frequency and time-domain issues will help in designing lower-noise loops.

In many applications, PLLs are embedded in systems on a chip. Loop performance depends on the floor planning done on these chips. A less sensitive PLL helps ease floor planning constraints. Consequently, measurements of coupled signals into the PLL are presented. Noise-injection tests on the reference clock and power supply are made to measure the performance of PLLs under low signal-to-noise conditions.

In monolithic PLLs, power is often cycled between off and on to save power. Consequently, measurements of how fast the PLL takes to acquire lock from a power-on condition are presented. Many ICs with PLLs have a crystal oscillator on board. Measurements of the open-loop response of the oscillator are presented to help set the output frequency and ensure that the circuit oscillates. An in-depth study of desirable equipment features is presented to make economically efficient decisions and allow other equipment to be substituted in case of obsolescence. Finally, this chapter presents troubleshooting techniques for PLLs. Testing and troubleshooting PLLs in integrated circuits presents unique problems. In order to determine which tests need to be run on a particular design, the reader must understand how to troubleshoot various problems that occur in a PLL.

Test measurements are made to measure the loop parameters. A list of the measurements and the loop parameters helps keep track of the results. The following lists the measurements to be presented:

- Spurious signals, hold-in range, and lock range;
- Phase noise;
- Frequency-switching time;
- Jitter;
- Closed-loop bandwidth;
- Oscillator open-loop test;
- Noise immunity to injected signals on supply;
- Noise immunity to injected signals on the reference input;
- Power-on switching.

The following lists the loop parameters that can be obtained from the measurements:

- Bandwidth;
- Phase margin;
- Jitter;
- Noise immunity/sensitivity;
- Power-on switching time;
- Hold-in range;
- Lock range;
- Spurious-signal levels;
- Noise levels;
- Stability of oscillation.

7.1 Hold-In Range, Lock Range, and Spurious Signals

Initial measurements of a PLL are very important because measurements of an unlocked loop or a very noisy loop give erroneous data. Consequently, measuring hold-in range and spurious signals gives some minimal stability and accurate noise information. High spurious signals in a PLL cause jitter in the output signal. A narrow hold-in range can mean some of the PLL parts are not working or signify an unlocked loop.

Figure 7.1 shows the test setup to measure the hold-in and lock range of a PLL. A low-noise synthesizer (HP8662A) that has no or very low (<−60 dBc) spurious signals of its own is connected to the trigger input of a pulse generator. The pulse generator (HP8082), which has a fast output rise time (<2 ns), converts the sine wave output of the synthesizer to a square wave input to the PLL under test. An oscilloscope (TDS784C), which can easily adjust to changing input voltage levels and frequency, measures the output frequency of the PLL as the reference input frequency is swept. A computer with a GPIB interface and software program automates the test.

An example procedure shows how the hold-in and lock ranges are measured. First, set the input frequency to the center of the operating range and lock the loop. Then, press the automatic waveform find function button on the oscilloscope to set the horizontal and vertical scales of the scope. Then, reduce the input

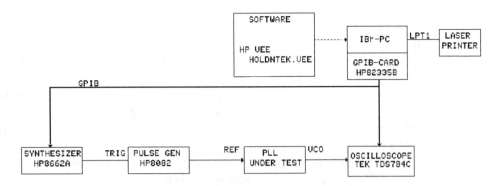

Figure 7.1   Test setup to measure lock and hold-in range.

frequency until the loop loses lock. Then, increase the input frequency to the loop until the loop locks. This is the lower limit of the lock range. Continue to increase the input frequency until the loop loses lock. This is the upper hold-in range. Do the reverse direction (decrease input frequency) for limit of the upper lock-in range. Continue to reduce the frequency until the loop loses lock. This is the lower limit of the hold-in range, which completes the range test. Figure 7.2 shows an example plot to determine the hold-in and lock range of a PLL. This figure shows a plot of the ratio of output frequency divided by input frequency. A constant ratio of 1 shows that the loop is locked. This figure shows limits of 20-MHz for the lower hold-in and lock ranges and 210-MHz for the upper hold-in and lock ranges. This PLL uses a phase/frequency detector, which usually makes the limits for the hold-in and lock ranges identical.

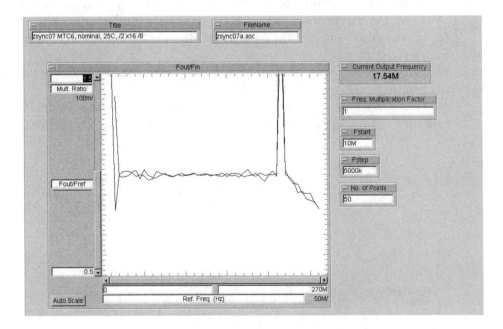

Figure 7.2   Example measurement of a PLL's hold-in and lock ranges.

Figure 7.3 shows the test setup to measure the spurious signals of a PLL. A low-noise synthesizer (HP8662A) that has no or very low (<–60 dBc) spurious signals of its own is connected to the trigger input of a pulse generator. The pulse generator (HP8082), which has a fast output rise time (<2 ns), converts the sine wave output of the synthesizer to a square wave input to the PLL under test. A spectrum analyzer (HP8560E), which also has low spurious signals (<–60 dBc), measures the spectrum of the PLL output. A computer with a GPIB interface and software program automates the test.

An example procedure shows how spurious signals are measured. After the initial setup, the procedure is to center the waveform on the spectrum analyzer and adjust the peak waveform level to the top reference line. Minimize the attenuation selection in the analyzer. Select a 10-MHz span to start. Select a video bandwidth that smoothes the curve and a resolution bandwidth rounded off to the nearest tens decimal place (i.e., 100 kHz, 10 KHz, 1 KHz, and so on). Then, take a plot of the waveform. Adjust the span to 1 MHz, 100 KHz, and lower until the waveform is unpronounced at the center or the span is 100 Hz. Make a plot at each span. Figure 7.4 shows an example spurious-signal-measurement plot of a PLL. This figure shows a 120-MHz center frequency and upper and lower reference sidebands equally spaced at 25 MHz away from the center frequency. The upper sideband is –30 dBc, and the lower sideband is –35 dBc. This picture also has a nonharmonically- and nonreference-related sideband at 12.5 MHz above the main output frequency and has an amplitude of –48 dBc.

7.2 Switching Time

Switching-time measurements are made by switching the input reference frequencies (frequency transition must be less than 100 μs to have a valid measurement) and observing the output-frequency time response on a modulation-domain analyzer [1, 2]. An added benefit of the modulation-domain analyzer is that it can measure the response directly out of the PLL, which is essential in monolithic PLLs. In nonmonolithic PLLs, the voltage-control line on the VCO can be monitored for the time response.

Figure 7.5 shows the test setup to measure the switching time. The pulse generator controls the frequency hopping amount and rate of the reference genera-

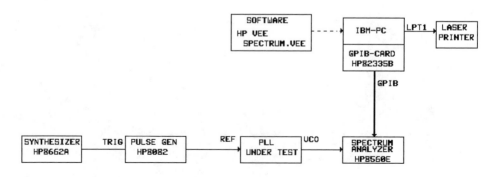

Figure 7.3 Test setup to measure spurious signals.

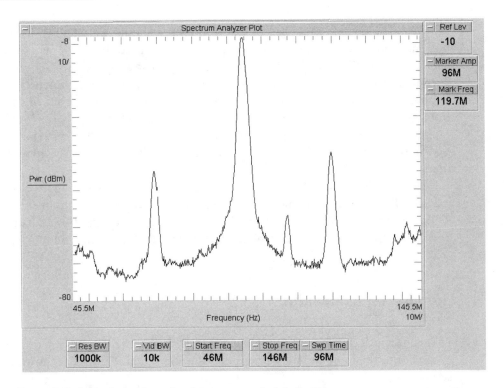

Figure 7.4   Example spurious-signal-measurement plot of a PLL.

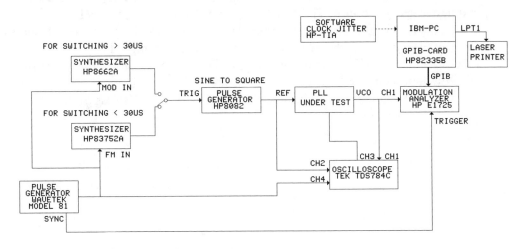

Figure 7.5   Frequency-switching-time test setup.

tor. The switching time of the test equipment must be faster than the switching time of the loop for an accurate measurement. The oscilloscope monitors the waveform out of the pulse generator. The modulation-domain analyzer instantaneously measures the output frequency versus time.

An example procedure on an example PLL that multiplies 27 to 81 MHz shows how switching time is measured. After the initial setup, the pulse generator should be set to a low voltage swing (100 mv peak to peak) and a repetition rate of

10 kHz. The center frequency of the synthesizer should be set to 27 MHz and 0 dBm. Connect the output of the synthesizer to the modulation-domain analyzer. The amplitude of the pulse generator is adjusted for a 100-kHz variation of the output frequency of the synthesizer. A 100-kHz variation was chosen because it is small enough to give a linear response and large enough to be out of the noise floor of the modulation-domain analyzer. Now, connect the synthesizer output to the reference input of the PLL and measure the output of the PLL with the modulation-domain analyzer. Figure 7.6 shows an example plot from the modulation-domain analyzer.

Bandwidth and stability are verified with the time response. Figure 7.6 shows a 2 μs 10% to 90% rise time and settling response with no overshoot. From type 2 PLL equations, a 2 μs 10% to 90% rise time computes a 200-kHz bandwidth and a damping factor greater than 2 (phase margin greater than 86°).

7.3   Closed-Loop Bandwidth

PLL bandwidth can be accurately measured with a network analyzer by injecting a signal into the loop and measuring the resulting signal. Figure 7.7 shows the test setup.

An example procedure shows how the closed-loop bandwidth is measured with a network analyzer. After the initial connection, sweeping the network analyzer output frequency produces a measured response. This is a closed-loop PLL response. Measuring the 3-dB point measures the bandwidth of the loop. Figure 7.8 shows an example plot from the network analyzer for an example PLL. Figure 7.8(a) shows the closed-loop response of the PLL. Figure 7.8(b) shows the variation of the 3-dB bandwidth with the output frequency of the loop. In many cases, this test may not be feasible for monolithic PLLs because there is no access to the injection and measurement nodes.

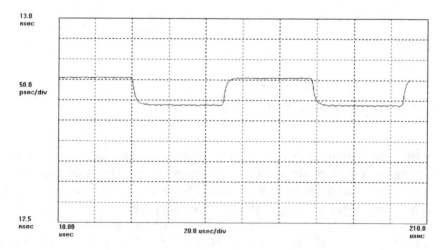

Figure 7.6   Switching-time measurement of example PLL (27-MHz input and 81-MHz output).

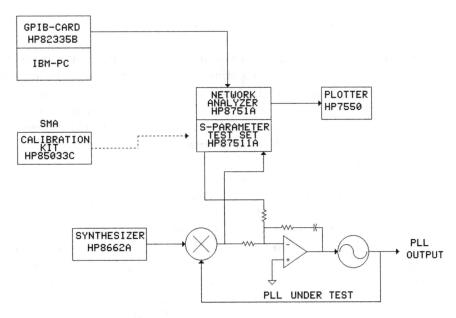

Figure 7.7   Test setup to measure closed-loop bandwidth by injecting a test signal.

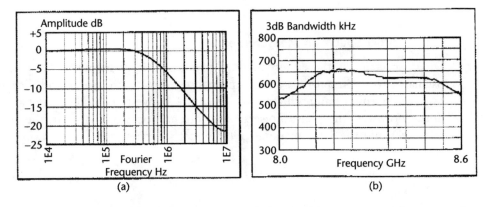

Figure 7.8   (a) Closed-loop response and (b) 3-dB bandwidth versus output frequency using a network analyzer.

7.4   Measurement of Phase Noise

Accurate measurement of phase noise from an oscillator and of additive phase noise from a component produces accurate phase-noise-analysis results. In the references, several phase-noise-measuring techniques and several pieces of phase-noise-measuring equipment are used. A variety of oscillator phase-noise curves are used to show which technique or equipment accurately measures the phase noise of that oscillator. A 10-MHz crystal oscillator, a 7-GHz dielectric resonator oscillator (DRO), and a 10-GHz yttrium iron garnet (YIG) oscillator represent a wide spread of phase-noise curves. Most phase-noise measurements of oscillators will occur between these ranges of phase-noise curves. Comparing phase-noise-measurement-system sensitivities to these curves identifies which measurement technique will measure which oscillator type. The methods of direct-spectrum measurement and carrier-suppression measurement are presented, and curves for the noise floor of each measurement technique are calculated. A phase-noise measurement technique is selected or modified by comparing the phase-noise measurement noise floor of each technique with the predicted noise curve of the oscillator. Now, these current phase-noise measurement methods, phase-noise measurement equipment, and phase-noise analysis methods allow sources to be evaluated to determine if they pass electronic system specifications before operating the sources and systems in the field.

Figure 7.9 shows various oscillator phase-noise curves that range from low stability to high stability. The 10-MHz crystal oscillator represents a very low-noise curve, and the 10-GHz voltage-controlled YIG oscillator represents a noisy curve. These ranges of phase-noise curves represent phase-noise levels that are encountered in phase-noise measurements, and they indicate the sensitivity of each phase-noise measurement method. The method and equipment chosen to measure phase noise depends on the phase-noise curve of the oscillator and the amount of time and money allotted for the measurement. In the references, several phase-noise measurement techniques are used, but the most common methods for measuring phase noise use the direct-spectrum method and the carrier-suppression method [3].

7.4.1 Direct-Spectrum Measurements

Direct-spectrum is the easiest method of measuring phase noise. Connecting an oscillator to the input of a spectrum analyzer directly measures and displays the power spectrum of an oscillator, which can be converted to single-sideband phase noise; however, the phase noise of the internal local oscillator in the spectrum analyzer limits the dynamic range of the direct-spectrum measurement. In addition,

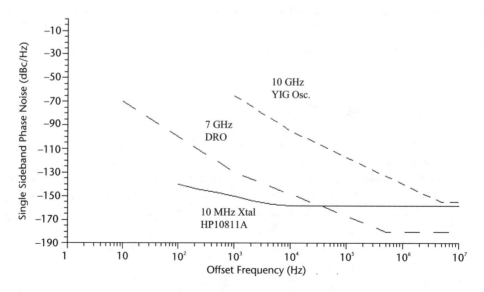

Figure 7.9 Low-stability to high-stability phase-noise plots of oscillators.

errors may arise due to the level of amplitude noise because a spectrum analyzer does not distinguish amplitude noise from phase noise. Equation (7.1) converts these direct-spectrum-analyzer power readings to a single-sideband phase-noise measurement [4]:

$$\mathscr{L}(f_m) = P_m(f_m) - P_c - 10 \log(B_{sa}) + P_{la} - P_{sbc} \qquad (7.1)$$

where

P_c = power level at the carrier frequency (dBm);

P_{la} = +2.5-dB error correction for logarithmic-amplifier characteristics in spectrum analyzer;

P_{sbc} = −6-dB correction factor to convert a carrier-to-sideband measurement to a single-sideband phase-noise measurement;

$P_m(f_m)$ = power level measured at the offset frequency from the carrier frequency (dBm);

B_{sa} = resolution bandwidth of the spectrum analyzer (Hz).

Equation (7.1) contains two correction factors. The 2.5-dB error-correction factor adjusts the measurement for the characteristics of the log amplifier in the spectrum analyzer. The −6-dB correction factor adjusts the measurement for converting a carrier-to-sideband measurement to a single-sideband phase-noise measurement. Section 5.1.2 explains the −6-dB factor for converting a carrier-to-sideband measurement to a single-sideband phase-noise measurement.

The direct-spectrum-measurement noise floor determines the sensitivity of this measurement method. Equation (7.2) calculates the expected noise floor for a direct-spectrum measurement from the specified noise floor of the spectrum analyzer:

$$N_{fm}(f_m) = -P_c + N_{sa}(f_m) - 10 \log(B_{sa}) + P_{la} - P_{sbc} \qquad (7.2)$$

where

$N_{fm}(f_m)$ = noise floor of measurement (dBc/Hz);

$N_{sa}(f_m)$ = spectrum analyzer noise floor (in dBc/Hz for direct measurements, but dBm/Hz for the other measurements).

Figure 7.10 presents the direct-spectrum noise floors of some commonly used spectrum analyzers. From a comparison of Figures 7.9 and 7.10, the direct-spectrum method limits measurement of oscillator phase noise far from the carrier and low phase-noise levels close to the carrier. Programming (7.1) and (7.2) into a computer, using the direct-spectrum noise floor of an HP8566 spectrum analyzer, and modeling the phase-noise curve of the YIG oscillator in Figure 7.9 computes the phase-noise level of the YIG and the noise floor of the direct-spectrum measurement. Figure 7.11 plots these results. Figure 7.9 shows the YIG oscillator to be the noisiest

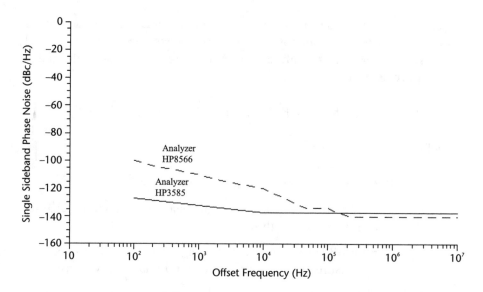

Figure 7.10   Direct-spectrum measurement noise floor for two spectrum analyzers.

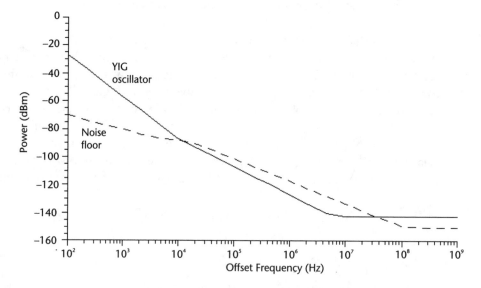

Figure 7.11   Comparison of the phase-noise response of a 10-GHz YIG oscillator to the direct-spectrum-measurement noise floor of a spectrum analyzer.

source in Figure 7.9. Yet, Figure 7.11 shows that the direct-spectrum measurement method has trouble measuring the phase noise of the YIG at frequency offsets greater than 10 kHz. The direct-spectrum-analyzer method adequately measures phase noise close to the carrier of noisy oscillators or microwave oscillators, but the direct-spectrum-analyzer method does not have enough sensitivity to measure low-noise oscillators [3].

An example procedure shows how phase noise is measured with the phase-noise test set up in Figure 7.12 for spectrum analyzers. This configuration measures

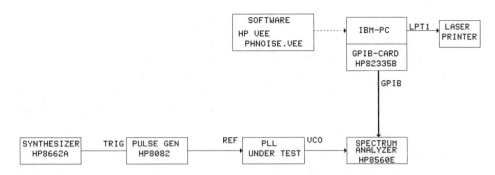

Figure 7.12   Test setup to measure phase noise with a spectrum analyzer.

phase noise greater than –130 dBc/Hz. Begin the procedure for making this measurement by pressing the user-module button on the spectrum analyzer. Then, select the phase-noise test from the menu. Then, follow the menu in the program to set up and measure phase noise. Figure 7.13 shows an example plot from the HP8560E for an 80-MHz PLL. After measuring the PLL phase noise, measure the phase noise of the reference source. After adjusting for the multiplication factor [20 log 10 (n_{mf})], this measurement gives the noise-floor limits for the measurement. Comparison with measured data helps determine the accuracy of the measurement.

From the measured phase-noise data, the PLL characteristics of bandwidth, stability, and jitter performance can be determined. The approximate bandwidth can be measured by finding an abrupt change in slope of the phase noise to a flat or reversed slope of the phase noise. In Figure 7.13, for example, the abrupt slope change occurs at 100 kHz. The approximate stability can be measured by observing the noise ear at the bandwidth of the PLL. A noise ear greater than –20 dBc down

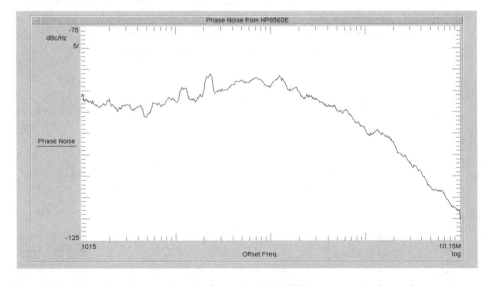

Figure 7.13   Example phase-noise plot from the HP8560E Spectrum Analyzer phase-noise user module.

from the carrier indicates a potentially unstable loop. In Figure 7.13, the noise ear is down −90 dBc, which is a stable loop.

Finally, the time jitter of the PLL can be calculated from the phase-noise curve. Equation (3.89) calculates time jitter from phase noise. The equation converts phase noise to jitter variance by integrating over the measurement bandwidth. These calculations help verify jitter measurements that were made by other methods.

7.4.2 Carrier-Suppression Measurements

The sensitivity of a phase-noise measurement will be improved by a method that suppresses the carrier. Consequently, several phase-noise measurement methods have been devised by engineers to suppress the carrier. Figure 7.14 shows a test setup for one method of carrier suppression.

This method requires two oscillators with equal phase-noise levels. One of the oscillators is voltage tunable, and the other is fixed. The voltage-tunable oscillator allows a narrow-bandwidth PLL to be used to keep the mixer inputs in phase quadrature. The lowpass filter eliminates the high-frequency mixing products after the mixer. Using the mixer as a phase detector requires that the inputs be in phase quadrature. The test setup must have enough isolation between the two oscillators to prevent them from injection locking. A phase shifter in one of the signal paths adjusts the dc voltage at the IF port output of the mixer. The dc voltage is adjusted to 0V to maintain phase quadrature between the RF and LO ports of the mixer. With the input signals at phase quadrature, the mixer converts phase fluctuations to voltage fluctuations (phase detector). A constant phase at the input of the mixer converts to a dc voltage; therefore, the IF port of the mixer must be dc coupled [4–13].

7.4.3 Mixer as a Phase Detector in a Measurement System

Carrier-suppression measurements commonly use mixers as phase detectors. Consequently, understanding the performance of a mixer as a phase detector will allow equations to be derived to convert carrier-suppression measurements to phase noise. Section 5.1.2 derives (7.3) for the mixing of two signals:

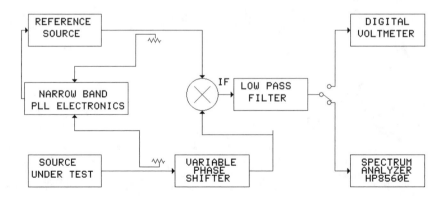

Figure 7.14   Carrier-suppression phase-noise measurement configuration for oscillators with equal phase-noise levels.

$$V_1(t)\, V_2(t) = V_{pbeat} \cos(\omega_{beat}t + \phi) \tag{7.3}$$

where

$\omega_{beat} = \omega_{RF} - \omega_{LO}$ for $\omega_{RF} > \omega_{LO}$;

$V_{pbeat} = V_{p1}\, V_{p2} \times 0.5 \times$ mixer losses;

V_{pbeat} = resulting voltage level after mixing (V).

The derivative of (7.3) calculates the incremental phase slope. For the mixer operating with a 0-beat frequency ($\omega_{beat} = 0$, which is dc) and 90° phase shift, the derivative of (7.3) produces (7.4) and (7.5):

$$V_{ps}(\phi) = \frac{d}{d\phi}[V_{pbeat} \cos(\phi)] \tag{7.4}$$

where

$V_{ps}(\phi)$ = phase slope (V).

$$V_{ps}(\phi) = V_{pbeat} \sin(\phi) \tag{7.5}$$

For $\phi = \pi/2$ rad (quadrature) in (7.5), $V_{ps} = V_{pbeat}$ (V/rad). For $\phi = 0$ rad in (7.5), $V_{ps} = 0$ V/rad. This shows that maximum phase sensitivity occurs for a 90° phase difference between the input signals while a minimum phase sensitivity of 0 occurs for a phase difference of 0°. With a 0-Hz beat frequency in (7.3), adjusting the phase shifter in Figure 7.14 for 90° phase difference produces 0V at the IF port of the mixer and gives maximum phase sensitivity for a measurement. Adjusting the phase shifter in Figure 7.14 for 0° phase difference produces a maximum voltage and gives minimum phase sensitivity for a measurement. A 0° phase difference measures amplitude noise because of the zero sensitivity to phase noise at this operating point.

Equations (7.3) and (7.5) will be used to derive equations for converting a voltage measurement to a phase measurement using a mixer as a phase detector. Today, most of the test equipment measures power level in milliwatts; however, to convert a voltage measurement to a phase measurement requires measuring the phase slope in volts per radian. Equation (7.6) converts the measured phase slope (V/rad) to a power level in milliwatts:

$$P_{ps} = \frac{(V_{pbeat})^2}{2\dfrac{R_L}{P_{mW}}} \tag{7.6}$$

where

R_L = 50Ω load resistance;

P_{mW} = 1,000 mW/W to convert watts to milliwatts;

P_{ps} = power level in a 50Ω load of the phase slope (mW/rad^2);

P_{ps} = measured power level of the beat note.

Equation (7.6) will be used to convert measured phase-noise power levels (mW/Hz) to phase-noise levels (dBc/Hz). One way of computing the power level of the phase slope is measuring the maximum dc level at the IF port by adjusting the phase shifter in Figure 7.14 for zero phase difference; however, another method must be used to measure the power level of the phase slope when adjusting the phase shifter is difficult.

Disconnecting the voltage to the voltage tune port on the VCO also produces a sinusoidal beat frequency. A power meter or a spectrum analyzer measures the power level of sinusoidal beat frequency into 50Ω that is referenced to 1 mW. Equation (7.6) relates the measured beat-frequency power level to the phase slope. This relationship will be used to derive an equation that converts noise-power measurements to phase-noise measurements.

Equation (7.6) makes it possible to derive an equation to convert measured noise-power levels to phase-noise levels using a mixer as a phase detector. To begin with, a relationship between modulation index, measured voltage, and measured power must be presented. Equation (7.7) converts measured RMS voltage to RMS phase fluctuations, which is the RMS value of the modulation index:

$$\eta_{rms}(f_m) = V_{\phi rms}(f_m)/V_{pbeat} \tag{7.7}$$

where

$V_{\phi rms}(f_m)$ = measured RMS voltage at the output of the phase detector;

$\eta_{rms}(f_m)$ = RMS value of the modulation index;

$\eta_{rms}(f_m)$ = phase fluctuations.

Single-sideband phase noise is related to the square of the RMS value of the modulation index. Equation (7.8) converts measured voltage fluctuations to the square of the RMS value of the modulation index, which is phase fluctuation:

$$[\eta_{rms}(f_m)]^2 = [V_{rms}(f_m)]^2/(V_{pbeat})^2 \tag{7.8}$$

Since power-level measurements are in dBm, dividing the numerator and denominator in (7.8) by the load resistance and watts-to-milliwatts ratio produces (7.9) for the modulation index in milliwatts:

$$[\eta_{rms}(f_m)]^2 = \frac{[V_{\phi rms}(f_m)]^2/(R_L/P_{mW})}{(V_{pbeat})^2/(R_L/P_{mW})} \tag{7.9}$$

$$= P_{if}(f_m)/(2P_{ps})$$

where

$P_{if}(f_m)$ = measured power level (mW);

$P_{if}(f_m) = [V_{\phi rms}(f_m)]^2/(R_L/P_{mW})$

Equation (7.9) converts the square of the RMS value of the modulation index in (7.8) to a function of measured power levels and the measured phase-slope power level. Dividing the modulation index and the right-hand side of (7.9) produces (7.10):

$$[\eta_{rms}(f_m)]^2/2 = P_{if}(f_m)/[2(2P_{ps})] \tag{7.10}$$

Equation (7.11) converts (7.10) for the square of the RMS value of the modulation index divided by 2 to a single-sideband phase noise and computes single-sideband phase noise from noise measurements using a mixer:

$$\mathcal{L}(f_m) = 10 \log\{[\eta_{rms}(f_m)]^2/2\} \tag{7.11}$$

$$= P_m(f_m) - P_{psdb} - P_{sbc}$$

where

P_{sbc} = 6-dB conversion factor for converting measured power to single-sideband phase noise in a mixer;

$P_m(f_m)$ = measured power (dBm);

$P_m(f_m) = 10 \log[P_{if}(f_m)]$;

P_{psdb} = power of the phase slope (dBm);

$P_{psdb} = P_c - L_{mc}$;

$P_{psdb} = 10 \log(P_{ps})$;

L_{mc} = mixer's conversion loss (dB).

Equation (7.11) shows the effect of a mixer on phase-noise measurements because a value of −6 dB is added to convert measured sideband power levels −(P_{psdb} − P_{ifdb}) in dBc/Hz to single-sideband phase noise. Now various carrier-suppression measurement methods can be analyzed.

7.4.4 Carrier-Suppression Measurement Model

The first measurement method modeled measures phase noise with a narrow-bandwidth PLL, and the measurements are made outside the bandwidth of the PLL, so PLL effects can be ignored. Equation (7.12) converts the spectrum analyzer's measured noise-power levels from the carrier-suppression technique in Figure 7.14 to single-sideband phase-noise measurements:

$$\mathcal{L}(f_m) = -(P_c - L_{mc}) + [P_m(f_m) - 10 \log(B_{sa}) + P_{la}] - P_{sbc} - P_{eq} \quad (7.12)$$

where

P_{eq} = −3-dB factor for equal phase-noise contribution of each source.

Equation (7.12) has three correction factors. The −6-dB correction factor, which is also in (7.11), converts measured power to phase noise. The −3-dB factor accounts for the equal phase-noise contributions from each oscillator to the total measured phase noise. Finally, the 2.5-dB correction factor accounts for the error from the logarithmic amplifier in measuring noise. Several other correction factors, which are mentioned in the references, correct for the phase detector characteristics of the mixer. These factors correct for distortions from large input signal levels, phase-curve shifts from residual dc voltages, and phase shifts from quadrature by the two input sources.

Another variation of the carrier-suppression method measures phase noise when the narrowband PLL does not phase lock because of a noisy source. Lack of phase lock makes a wide PLL bandwidth necessary. Unfortunately, phase-noise measurements inside the loop bandwidth are not valid because of the noise improvement inside the loop bandwidth of the PLL; however, adding the inverted PLL error function to (7.12) compensates for the phase-noise improvement that occurs inside the bandwidth of the loop. Equation (7.13) converts measured data inside the loop bandwidth to single-sideband phase noise by adding the inverted PLL error function to (7.12):

$$L(f_m) = -(P_c - L_{mc}) + \{P_m(f_m) - 10 \log(B_{sa}) + P_{la} + 10 \log[1 + G(s)H(s)]\} \\ - P_{sbc} - P_{eq} \quad (7.13)$$

where

$G(s)$ = PLL gain function;

$H(s)$ = PLL feedback function.

Computing the smallest phase-noise level that this measurement method can measure determines the oscillator types that this measurement method can measure. Equation (7.14) calculates the smallest phase-noise level the carrier-suppression measurement system can measure:

$$P_{ss}(f_m) = -(P_c - L_{mc}) + [N_{sa}(f_m) - 10 \log(B_{sa}) + P_{la}] - P_{sbc} - P_{eq} \quad (7.14)$$

where

$P_{ss}(f_m)$ = smallest phase-noise level the system can measure (dBc/Hz).

In addition, (7.14) shows that the sensitivity of the spectrum analyzer determines the smallest phase-noise level that the measurement system can measure.

Inserting a low-noise buffer amplifier between the spectrum analyzer and the output of the phase detector improves the dynamic range of the carrier-suppression method; however, this method requires characterizing the gain and noise figure of the buffer amplifier to compute single-sideband phase noise and the noise floor of the measurement system. Subtracting the gain of the buffer amplifier from (7.13) compensates the measured data for the gain of the buffer amplifier. Equation (7.15) converts measured data to single-sideband phase noise for the carrier-suppression method with a buffer amplifier:

$$L(f_m) = -(P_c - L_{mc}) + P_m(f_m) - 10 \log(B_{sa}) + P_{la} + 10 \log[1 + G(s)H(s)]$$
$$- P_{sbc} - G_v - P_{eq} \tag{7.15}$$

where

G_v = video-amplifier gain (dB).

Computing the noise floor of the measurement system with a buffer amplifier allows the smallest measured phase-noise level to be computed. Equation (7.16) computes the measurement system's noise floor for the carrier-suppression method with a buffer amplifier:

$$N_{mf}(f_m) = 10 \log \left[10^{\frac{N_{sa} - 10 \log(B_{sa}) + P_{la}}{10}} + kT_0 \left(10^{\frac{G_v + N_v}{10}} \right) \right] \tag{7.16}$$

where

$N_{mf}(f_m)$ = measurement noise floor (dBm/Hz);

k = Boltzmann's constant (mW/K · Hz);

$k = 1.38 \times 10^{-20}$;

T_0 = temperature (K)

$T_0 = 290$ at room temperature;

$10 \log(kT_0) = -174$ (dBm/Hz);

N_v = video-amplifier noise figure (dB).

The noise floor of the measurement system in (7.16) does not depend on the noise floor of the spectrum analyzer if the amplifier gain plus noise figure is greater than the noise floor of the spectrum analyzer. Equation (7.17) computes the smallest phase-noise level that the carrier-suppression measurement system with an additional buffer amplifier can measure:

$$P_{ss}(f_m) = -(P_c - L_{mc}) + N_{mf}(f_m) - P_{sbc} - P_{eq} \tag{7.17}$$

The insertion of the buffer amplifier into the measurement system adds more error factors because the gain and noise figure of the amplifier varies with frequency;

therefore, a substantial amount of time must be spent in characterizing the response of the amplifier.

7.4.5 Generating a Calibration Signal

The previous sections helped select a phase-noise measurement method. Next, confidence in the accuracy of the measurement and confidence in the calibration of the measurement system must be reinforced. A calibration signal helps establish this confidence. In addition, injecting a calibration signal at the input of the measurement system and measuring the response of the system can correct all the errors without computing each error component in the system. To meet the above goals, the calibration signal must produce a known sideband-to-carrier ratio at the input of the measurement system. Inserting a directional coupler between the source under test and the measurement system allows a small calibration signal to be injected. Adding a small signal at an offset frequency from one of the sources injects a known sideband-to-carrier ratio into the measurement system as shown in Figure 7.15. A spectrum analyzer measures the resulting carrier-to-sideband ratio. From narrowband FM theory and AM theory, the signal produced in Figure 7.15 must contain equal phase and amplitude modulation.

Knowing the input carrier-to-sideband ratio allows the response of the system to be calibrated. Comparing the measured carrier-to-sideband ratio of the measurement system with the input carrier-to-sideband ratio measured on the spectrum analyzer calibrates the measurement system. The calibration factor must be measured at each measurement offset frequency to ensure accurate results.

For comparison to the measured calibration factor, (7.18) computes the theoretical calibration factor that will adjust the measured results to single-sideband phase-noise measurements:

$$P_{cal} = P_{sbm} - P_{sbin} - P_{ampm} \qquad (7.18)$$

where

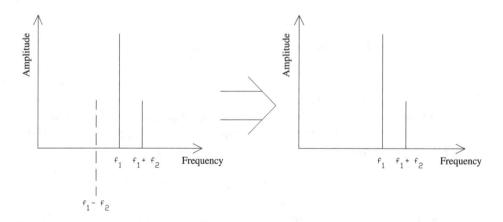

Figure 7.15 Spectral plot showing equal AM and PM components for single-sideband test signal.

P_{ampm} = a 3-dB factor that accounts for equal AM and PM;

P_{cal} = calibration factor (dB);

P_{sbm} = sideband-to-carrier measurement (dB);

P_{in} = sideband-to-carrier input (dB).

The P_{ampm} factor in (7.18) accounts for the equal AM and PM sideband levels. Equation (7.19) computed single-sideband phase noise using the calibration factor:

$$L(f_m) = P_m(f_m) + P_{cal} + P_{la} - P_{eq} - P_{sbc} - 10 \log(B_{sa}) \qquad (7.19)$$

Equation (7.19) has four correction factors. The 2.5-dB correction factor compensates the measurement for the spectrum analyzer's logarithmic-amplifier noise-response characteristics, the −3-dB correction factor compensates the measurement for equal oscillator noise contribution, the bandwidth factor normalizes the measurement to a 1-Hz bandwidth, and the −6-dB factor converts the measurement to single-sideband phase noise [6].

7.4.6 Phase-Noise Measurement Equipment

Several spectrum analyzers are available today that can be used in a custom phase-noise measurement system. This section provides a method for comparing spectrum analyzers but is not intended as an endorsement of one vendor over another. In addition, the accuracy of the equipment characteristics measured or obtained from data sheets is not guaranteed. Please contact the vendors to obtain up-to-date characteristics. Comparing analyzer sensitivities is one way to select the optimum analyzer. Figure 7.16 shows the sensitivities for some of these instruments. The HP3585 spectrum analyzer can be used for low-cost phase-noise measurements using the carrier-suppression method. The HP8566 spectrum analyzer is a higher-priced option that provides ability to measure phase noise using the direct-spectrum

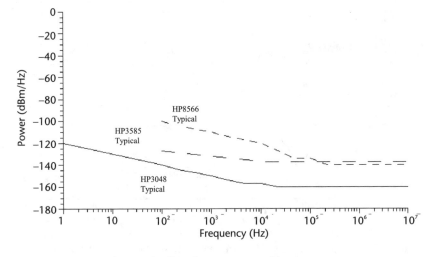

Figure 7.16 Spectrum-analyzer noise floor for a suppressed carrier.

method and the carrier-suppression method. Using the HP8566 or the HP3585 requires that the user make a custom phase-noise measurement system. The user must be careful not to induce noise into the custom phase-noise measurement system. Furthermore, obtaining believable and repeatable measurements is very time-consuming.

The HP3048 phase-noise measurement system lets an engineer avoid the problems involved in making a custom phase-noise measurement. The HP3048 is a more expensive option than making a custom phase-noise measurement system; however, the HP3048 contains a sensitive analyzer, a low-noise buffer amplifier, a phase-detecting mixer, and PLL circuitry. The PLL circuitry produces a control voltage for tuning VCOs in the carrier-suppression method. More importantly, the HP3048 phase-noise measurement system contains software that calibrates the system for the errors factors that have been discussed in this chapter; consequently, this equipment requires a minimal amount of time and knowledge to obtain reliable measurements, and it has high sensitivity and flexibility to do the special interconnections for specific applications.

7.4.7 Phase-Noise Measurements with the HP3048

Test vendors have equipment that uses the carrier-suppression measurement method but they require external low-phase-noise reference sources. Accurate phase noise, bandwidth, and stability measurements must be made with a clean signal generator for a reference. Figure 7.17 shows the phase-noise plot of several low-noise synthesizers, which shows a wide variation in phase noise between them. Pulse generator sources were not measured because, in general, their phase noise is more than the plots shown. In most cases, the choice of synthesizer determines the noise-floor of the measurement. Figure 7.18 shows the phase-noise test setup that measured the resulting spectrum [1, 3] in Figure 7.17.

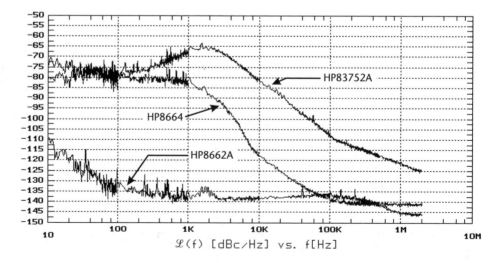

Figure 7.17 Phase-noise plot of several low-noise synthesizers to show the wide variation in phase-noise sources.

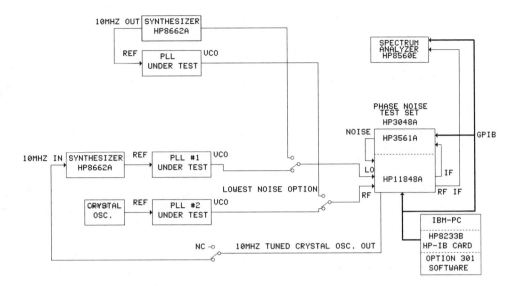

Figure 7.18   Phase-noise test setup.

The lower-phase-noise test configuration measures the phase noise of two identical devices under test with a phase-noise test set (HP3048). This configuration measures phase noise greater than −165 dBc/Hz. This tests low-noise PLLs and crystal oscillators.

The upper-phase-noise test configuration measures phase noise in comparison with a reference HP8662A synthesizer. This configuration measures phase noise greater than −140 dBc/Hz. This is adequate for a variety of PLLs. This tests PLLs with noise 30 dB above the HP3048 test's noise floor; however, it is easier to make this measurement. The reference generator provides a low-noise source so that the phase-noise contribution of the PLL can be accurately measured.

An example procedure shows how phase noise is measured by selecting the lower-phase-noise test in Figure 7.18 for a phase-noise test set. First, follow the in-structions from the HP3048 manual to take a measurement of the PLL and the reference generator. Figure 7.19 shows an example plot from the HP3048 for a 400-MHz PLL. After measuring the PLL phase noise, measure the phase noise of the reference source. After adjusting for the multiplication factor $[20 \log 10(N)]$, the measurement of the reference generator plus the multiplication factor $[20 \log 10(N)]$ gives the noise-floor limits for the measurement. Comparison with measured data helps determine the accuracy of the measurement.

7.4.8   Variations of the Carrier-Suppression Technique

Several other variations of the carrier-suppression technique are used for more specific measurements. Reexamining Figure 7.14, we see that the phase-noise test set configuration will measure only the reference source's phase noise if the voltage-controlled source's phase noise is lower than the reference source's phase noise. Using an HP8662 as a voltage-controlled source in Figure 7.14, we can make phase-noise measurements on untunable oscillators with operating frequencies

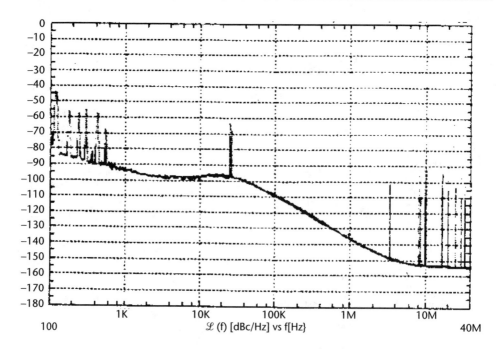

Figure 7.19 An example plot from the HP3048 for a 400-MHz PLL.

from 5 to 1,280 MHz. For higher-frequency sources, the HP11729A/B down-converts the source's operating frequency to the HP8662's operating frequencies, allowing the HP8662 to be phase-locked to the source under test. Down-converting with the HP11729A/B extends the phase-noise measurement range of sources to 18 GHz. The downconversion method makes phase-noise measurements of fixed frequency sources easier and makes phase-noise measurements on a wide variety of oscillators possible; however, using the downconverter degrades the noise floor of the measurement system. Figure 7.20 shows the noise floor of this downconversion method for a 10-GHz source.

Figure 7.21 shows another method of carrier suppression that uses a delay-line discriminator to measure the phase noise of an oscillator. The delay-line-discriminator method does not require a second source that could degrade the noise floor of the measurement system; however, the delay in the delay-line-discriminator method degrades the noise floor of the spectrum analyzer. The delay degrades the analyzer noise floors in Figure 7.16 by 20 dB/dec from a frequency offset of one half the inverse of the time delay in the discriminator.

Figure 7.22 shows the degraded phase-noise floor for the delay-line-discriminator methods that use the HP3048 or HP3585 analyzers. In general, the delay-line-discriminator method makes adequate measurements for YIG and ring oscillators at RF and microwave frequencies. Figure 7.23 shows the results of a computer model that simulates the phase-noise measurement of a YIG oscillator using the delay-line-discriminator method [4, 7, 8, 10, 14–16].

Finding a method of measuring the effects of various components on phase noise makes state-of-the-art phase-noise performance in systems achievable. Figure 7.24 shows a test configuration that measures a component's additive phase noise,

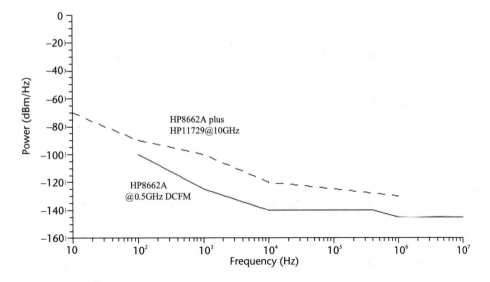

Figure 7.20   Carrier-suppression phase-noise measurement floors for a PLL using a voltage-controllable HP8662A frequency synthesizer and an additional HP11729 downconverter.

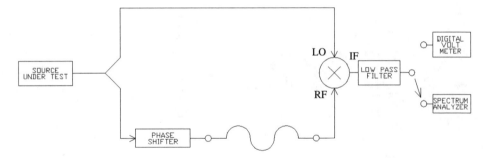

Figure 7.21   Carrier-suppression phase-noise measurement for one source. This configuration uses a delay-line discriminator.

which is another variation of the carrier-suppression method. The additive phase-noise measurement method requires two major conditions to obtain the best additive phase-noise sensitivity. First, equal time delay between the power splitter and mixer ports must be achieved to cancel the phase noise of the source. Second, the source must be very stable so that its phase noise is not measured.

The additive phase-noise measurement method makes measuring a wide variety of components possible. Variations of the additive phase-noise measurement configuration measure the additive phase noise of amplifiers, vibrating devices, and frequency translation devices. Inserting an amplifier in one path to the mixer measures the additive phase noise of the amplifier. Vibrating devices in one path measures the additive phase-noise effects of mechanical vibrations. Inserting frequency translation devices (digital dividers, multipliers, and mixers, as shown in Figure 7.25) in both paths measures the additive phase-noise effects due to translations in frequency [4, 8, 17].

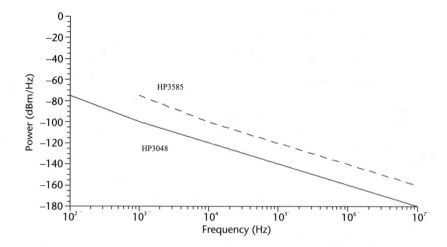

Figure 7.22   The degradation of the phase-noise measurement noise floor from using a delay.

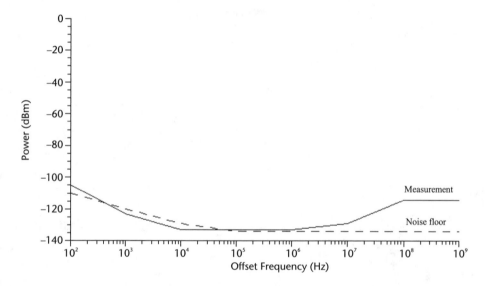

Figure 7.23   Theoretical measurement of a 10-GHz YIG oscillator using a delay-line discriminator.

In summary, selecting the best method and equipment produces the most accurate phase-noise measurements. This section helps an engineer select the best method and equipment for making accurate phase-noise measurements. In the past, phase-noise measurements required several pieces of equipment and some custom circuits. Today, the HP3048 measurement system eliminates the need for custom circuits by integrating the equipment and custom circuitry into one system. With the included software, the HP3048 makes accurate phase-noise measurements easy and provides several measurement configurations. In this chapter, equations were developed that model various measurement configurations. From the modeled measurement configurations, the feasibility of measuring the phase noise of a particular source was studied. Making phase-noise measurements on complex

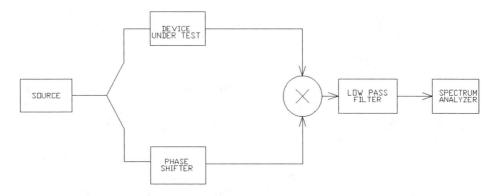

Figure 7.24 Additive phase-noise measurement configuration for components.

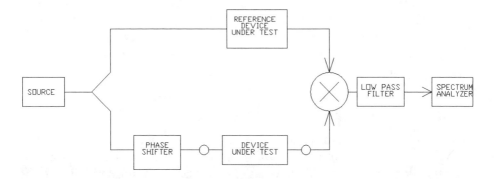

Figure 7.25 Additive phase-noise measurement configuration for frequency-translatable components.

sources and systems allows engineers to determine the feasibility of passing specifications with these sources. In addition, making additive phase-noise measurements of devices allows engineers to reduce a device's effect on the phase noise of a complex system or oscillator. Obtaining accurate phase-noise data and additive phase noise makes the phase-noise analysis accurate.

7.5 Testing for Jitter

In this section, we will study the equipment for measuring jitter. For each piece of equipment the advantages and disadvantages will be presented. Example benchmark measurements will be shown to help engineers understand which equipment best suits the measurement they require. Next, example isolation measurements in silicon are presented to show the importance of isolation in making low-jitter ICs, and, finally, example test procedures are presented to make it clear how the measurements are to be made and what plots should be made.

Available equipment for testing jitter in PLLs are oscilloscopes, time-interval analyzers (TIAs) (modulation-domain analyzer), spectrum analyzers, and clean high-frequency sources for a reference. Main issues for accurate jitter analysis in these instruments include maximum useable frequency, input bandwidth, jitter noise floor, smallest one-shot resolution versus linearity, throughput, and frequency and time correlation.

Figure 7.26 shows the relationship between measurements of an FM waveform with a modulation-domain analyzer (frequency versus time), oscilloscope (voltage versus time), and spectrum analyzer (voltage versus frequency). This picture helps in understanding the portion of the waveform that is being measured and its relationship to the other domains.

After making jitter measurements, we would like to correlate the measurements between each of the instruments. Correlation issues can make this a difficult task. Consequently, let's list issues that can uncorrelate measurements. Correlation issues applicable to all instruments are probing and test fixturing. Correlation issues that depend on particular instruments are equipment resolution and accuracy versus range and other specifications, instrument noise floor, record length (ability to resolve low-frequency modulation), asynchronous versus synchronous measurements, instrument calibration techniques, operation beyond equipment limitations, instrument setting, and changes in the external environment (e.g., temperature, supply noise, line noise).

7.5.1 Oscilloscope Jitter Measurements

Figure 7.27 shows the relationship between an ideal clock, FM clock, and modulating function. The top waveform shows the ideal clock. The next waveform shows the clock with the FM modulation. The last waveform shows the modulation

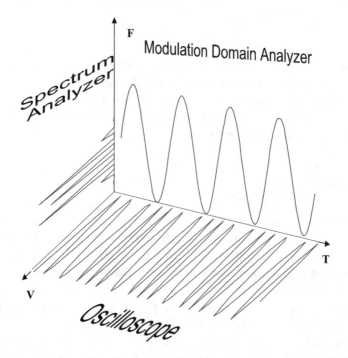

Figure 7.26 Measurement relationships of an FM waveform between a modulation-domain analyzer, oscilloscope, and spectrum analyzer.

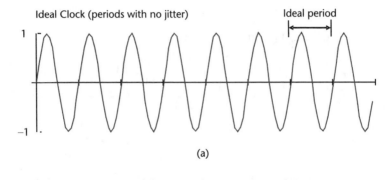

(a)

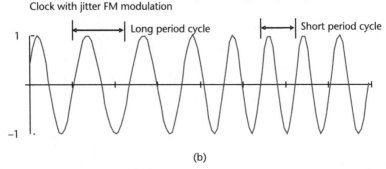

(b)

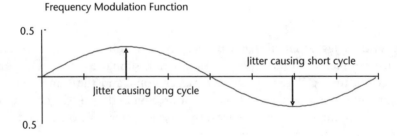

(c)

Figure 7.27 Relationship between (a) ideal clock, (b) modulation domain of an FM waveform, and (c) time domain of an FM waveform. (*From:* [18]. © 1997 Wavecrest. Reprinted with permission.)

waveform. The peak of the modulation waveform gives the longest time period and the minimum gives the shortest period. The short-period cycle can cause a design margin error, to which DSP and microprocessor ICs are most sensitive. An oscilloscope can be used to measure the FM waveform in the time domain.

Figure 7.28 shows the difference in measuring one period and multiple periods. Measuring multiple periods captures the period change of multiple periods (six in the figure). This is equivalent to dividing the clock by six and measuring the edge variation. This is not the desired measurement. Measuring the variation of one period is the desired measurement.

Figure 7.29 shows digital-sampling oscilloscope modes [2]. The real-time sampling method captures an entire waveform with a single trigger event. The sequential, repetitive sampling method acquires a new sample at a certain time interval

1 clock period oscilloscope measurement with jitter FM modulation

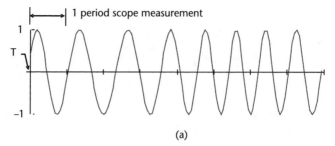

(a)

6 clock periods oscilloscope measurement with jitter FM modulation

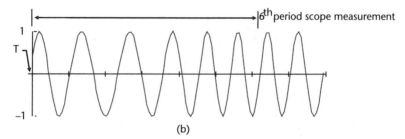

(b)

Figure 7.28 Oscilloscope measurement of (a) one period and (b) multiple periods of jitter. (*From:* [18]. © 1997 Wavecrest. Reprinted with permission.)

after the trigger. The instrument increases this time delay by a fixed amount after each sample. Random, repetitive sampling is similar to sequential sampling except that the time difference between trigger point and the sample point is random. Sampling is done constantly and is not precipitated by a trigger event [1]. The real-time sampling mode must be used to measure jitter. The other modes can give false measurements.

Key performance specifications for an oscilloscope measuring jitter are vertical resolution and accuracy, linearity, reconstruction error, record length, range error, time-based resolution and accuracy, noise floor, trigger threshold jitter, time-based delay error, crystal stability, and update rate. Advantages in measuring jitter with an oscilloscope are that the whole waveform can be captured, low jitter can be measured on some instruments, and it has a wide input-frequency range. Disadvantages in measuring jitter with an oscilloscope are that low-frequency modulation is difficult to measure, modulation domain cannot be measured, it has low sensitivity with fast Fourier transform (FFT) of waveform, the setup can vary from person to person, accuracy depends on the internal delay generator, spurious signals cannot be easily identified, there can be confusion between minimum and maximum time jitter (1 period) and time-interval jitter (multiple periods), and it can have a narrow input-amplitude dynamic range

7.5.2   TIA and Spectrum Analyzer Jitter Measurements

The operation of a TIA begins when the sampling logic triggers a sample. The contents of a counter and interpolator are read into the next free memory location.

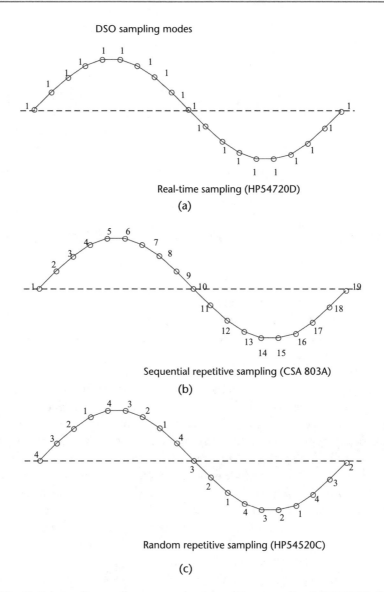

DSO sampling modes

Real-time sampling (HP54720D)

(a)

Sequential repetitive sampling (CSA 803A)

(b)

Random repetitive sampling (HP54520C)

(c)

Figure 7.29   Digital-sampling oscilloscope modes: (a) real-time sampling (HP54720D), (b) sequential repetitive sampling (CSA 803A), and (c) random repetitive sampling (HP54520C). (*From:* [18]. © 1997 Wavecrest. Reprinted with permission.)

The interpolator is then reset for the next sample. Samples are taken until a specified number is reached, and the processor then reads the memory for calculations and display. Unlike a counter, the display is often graphic based.

Figure 7.30 shows TIA sampling [2] of an FM waveform. Only the zero-crossing time stamps are measured. Consequently, more data is collected than is done by an oscilloscope because the oscilloscope must store the whole waveform to find the zero crossing. Advantages in measuring jitter with a modulation-domain analyzer are that the modulation domain is directly measured, the frequency domain can be calculated, low-frequency modulation can be measured easily, a large number of data points can be stored, and spurious signals can be identified. Disadvantages

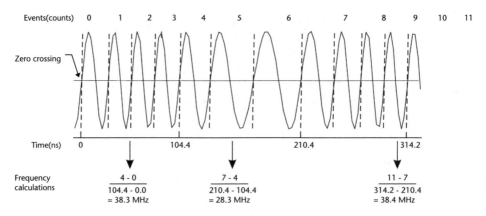

Figure 7.30 TIA sampling of an FM waveform. (*From:* [18]. © 1997 Wavecrest. Reprinted with permission.)

in measuring jitter with a modulation-domain analyzer are that the noise floor is greater than the Wavecrest DTS 2070/2075 and the Tektronix oscilloscope, the time domain cannot be directly measured, and it has a narrow input-amplitude dynamic range.

A spectrum analyzer measures the frequency-domain content of a waveform. Jitter shows up as FM sidebands on a spectrum analyzer. The offset frequency from the carrier and the level below the carrier level determines the amount of jitter produced. Advantages in measuring jitter with a spectrum analyzer are that the frequency domain is directly measured, spurious signals are easily identified without an FFT, it has a wide input-frequency range, it has a low noise floor and it has a wide input-amplitude dynamic range. Disadvantages in measuring jitter with a spectrum analyzer are that the time domain cannot be measured, the modulation domain cannot be measured, jitter must be calculated, AM and FM sidebands are not easily identifiable, it cannot trigger on an event, it has blind spots in sweep (noncontinuous) where it can miss an event, and it has a low input resistance of 50Ω, which requires a high current drive.

The Wavecrest jitter analyzer measures jitter with the combination of several precision comparators and software algorithms in the time domain. Several postprocessing software routines help analyze the content of the jitter. Advantages in measuring jitter with the Wavecrest DTS are that it has a wide input-frequency range, a fast throughput, and a low-jitter noise floor. Disadvantages in measuring jitter with the Wavecrest DTS are that it has a narrow input-amplitude dynamic range (>100 mV) and a low input resistance of 50Ω, which requires a high current drive, it has nonconsecutive periods (>25 μs between periods), and it cannot measure time step response.

7.5.3 Minimum Noise-Floor Measurements of TIA, Oscilloscope, and Digital Time Scope

Several measurements were used to provide the noise-floor accuracy of the measuring equipment. The following section provides a method for comparing jitter noise floors in equipment, but it is not intended as an endorsement of one vendor over another. In addition, the accuracy of the equipment characteristics measured or obtained from data sheets is not guaranteed. Please contact the vendors to obtain up-to-date and accurate characteristics. A comparison of measured data with calibration numbers gives a relative-confidence factor for the measured data. Table 7.1 shows jitter measurements using the Tektronix scope. A 10-mV window on the Tektronix scope was used to make the measurements. The first measurement measures jitter with no modulation. The second shows 30 kHz of FM modulation was needed to double the standard deviation of jitter with no modulation. The next measurement shows the amount of high-frequency FM sideband level that was needed to double the amount of jitter.

Table 7.2 shows jitter measurements using the modulation-domain analyzer. The first measurement shows jitter of the internal noise of the modulation-domain analyzer because the 10-MHz signal from the HP8662 was measured referenced to the external standard from the HP8662. The second measurement measures jitter with no modulation. Notice that subtracting the internal noise (TIA) measurement produces numbers close to the Tektronix measurements (50 − 34 = 16 ps standard deviation, and 440 − 200 = 240 ps peak to peak). The next shows the 30 kHz of FM modulation that was used in the previous table. Subtracting the internal-noise measurement produces numbers close to the Tektronix measurements (65 − 34 = 31 ps standard deviation, and 440 − 200 = 240 ps peak to peak). The last measurement shows the amount of high-frequency FM sideband level that was needed to double the amount of jitter.

Table 7.3 shows jitter measurements using a digital time scope. An HP8082 pulse generator was used to convert the sine wave output of the HP8662 to a square wave, which provides the best waveform to the scope. The first measurement

Table 7.1 Jitter Calibration for the Tektronix Oscilloscope

Source Description	Modulation	Jitter (Standard Deviation)	Jitter (Peak to Peak)
HP8662 into HP8161 @33.33 MHz	None	13 ps	84 ps
HP8662 into HP8161 @33.33 MHz	1 kHz, 30 kHz deviation FM	26 ps	182 ps
HP8662 into HP8161 @33.33 MHz	1.33-MHz injected sideband, −35 dBc	32 ps	216 ps

Table 7.2 Jitter Calibration for the Modulation-Domain Analyzer

Source Description	Modulation	Jitter (Standard Deviation)	Jitter (Peak to Peak)
HP8662 @10 MHz and TIA external reference	None	34 ps	200 ps
HP8662 into HP8161 @33.33 MHz	None	50 ps	440 ps
HP8662 into HP8161 @33.33 MHz	1 kHz, 30 kHz deviation FM	65 ps	440 ps
HP8662 into HP8161 @33.33 MHz	1.33-MHz injected sideband, −26 dBc	100 ps	630 ps

Table 7.3 Jitter Calibration of the Wavecrest Digital Time Scope

Source Description	Modulation	Jitter (Standard Deviation)	Jitter (Peak to Peak)
HP8662 @10 MHz into HP8082	None	4 ps	38 ps
HP8662 into HP8082 @33.33 MHz	None	4.6 ps	40 ps
HP8662 into HP8082 @33.33 MHz	1.33-MHz injected sideband, −53.7 dBc	9.2 ps	66 ps
HP8662 into HP8082 @200 MHz	None	4.7 ps	34 ps

shows jitter of the internal noise because the 10-MHz signal from the HP8662 was measured referenced to the external standard from the HP8662. The second measurement measures jitter with no modulation. Both measurements show the lowest noise floor. The next measurement shows the amount of high-frequency FM sideband level that was needed to double the amount of jitter. This measurement shows that the digital time scope has a noise floor 18-dB lower than the oscilloscope and 28-dB lower than the modulation-domain analyzer. The last measurement shows that a low amount of jitter is obtained even with a higher input frequency.

Consequently, the comparison of this data shows jitter measurements with less than 400-ps peak-to-peak accuracy require using the Tektronix scope or the Wavecrest DTS. For jitter measurements less than 100 ps peak to peak, we need to use the Wavecrest Digital Time Scope. For jitter less than 40 ps peak to peak, phase noise must be measured and converted to jitter. The modulation analyzer can measure deterministic noise with accuracy because averaging can be used, and the modulation-domain analyzer limit with averaging is 1 ps. Frequency step responses, power-supply injection jitter, and power-up response can also be accurately measured with the modulation-domain analyzer because they are deterministic measurements.

7.5.4 Isolation Measurements Between PLLs in Silicon

Figure 7.31 shows the output of one PLL at 48 MHz while another PLL on the silicon chip is operating at 63 MHz. Both I/O buffers for the PLLs are operating simultaneously. The two PLLs are located at opposite corners of the die. From the figure, the isolation is 43 dB. This is an acceptable level.

Figure 7.32 shows the effect of increasing the frequency to 142.5 MHz and bypassing one PLL and increasing it to 157.5 MHz and bypassing the other PLL. This shows the isolation between the I/O output buffers. From the figure, the isolation decreases to 20 dB at 142.5 MHz. Consequently, isolation between these two PLLs depends on the isolation between the I/O buffers. A sideband 20 dB down and 15 MHz away from the 150-MHz output signal would cause 200 to 400 ps of additional jitter. The current trend to higher output frequencies and more PLLs on a chip requires that we should study various isolation techniques in Section 3.6.3 in order to meet our customers' demands for lower jitter.

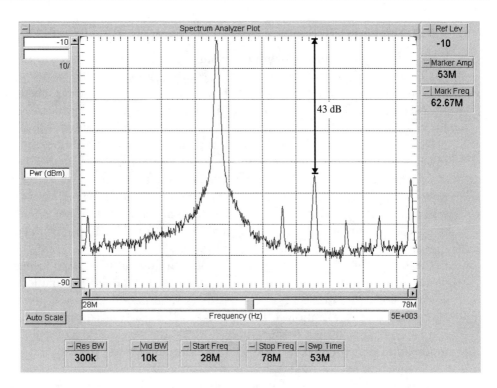

Figure 7.31 Output of one PLL at 48 MHz with another PLL operating at 63 MHz.

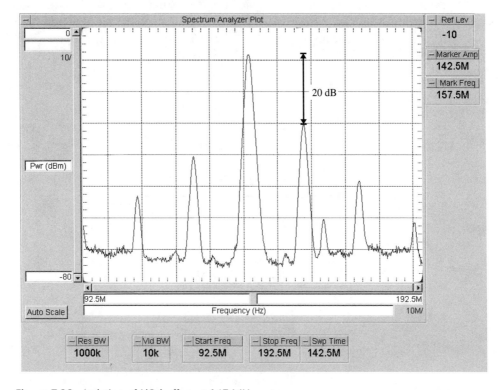

Figure 7.32 Isolation of I/O buffers at 147 MHz.

7.5.5   Time-Jitter Test Setups

Time jitter can be measured with the modulation-domain analyzer [3], the oscillo-scope, and the Wavecrest time-domain system. Figure 7.33 shows the test setup. Although the Wavecrest DTS is not shown in the figure, it can be substituted for the oscilloscope. The modulation-domain analyzer has a 200-ps peak-to-peak noise floor. For lower jitter measurements the Tektronix oscilloscope or the Wavecrest DTS is used. The upper switch selects the modulation-domain analyzer, and the lower switch selects the oscilloscope or Wavecrest DTS.

An example procedure on an example PLL that multiplies 27 to 81 MHz shows how jitter is measured with the modulation-domain analyzer. After making the initial connections, select the input menu and select the appropriate impedance for the measurement probe (50Ω for cable, 1 $M\Omega$, or 10 $M\Omega$). Then, adjust the threshold to the 50% point of the output waveform from the PLL. Leave the hysterises at 50%. Then, follow the modulation-domain manual to measure jitter versus time. Figure 7.34 shows an example plot from the modulation-domain analyzer for an example PLL. The y-axis is time period and the x-axis is time. The peak-to-peak variation in time period is 293 ps. Consequently, the data has a quantized appearance because it is close to the single-shot noise floor of the modulation-domain analyzer and its minimum quantization value.

Figure 7.35(a) shows jitter versus time interval. This measurement is made by varying the number of skipped output clock edges. This is equivalent to inserting a programmable divider between the PLL output and the modulation-domain-analyzer input and varying the divide ratio. Figure 7.35(a) shows the amount of jitter versus the time interval between the edges that were not skipped. Converting to divide ratios, the x-axis varies from divide-by-5 at 60 ns to divide-by-3,125 at 40 μs. The solid line is the peak-to-peak jitter, and the dashed line is the RMS jitter. At a time interval of 300 ns, the RMS jitter flattens out, which is approximately the

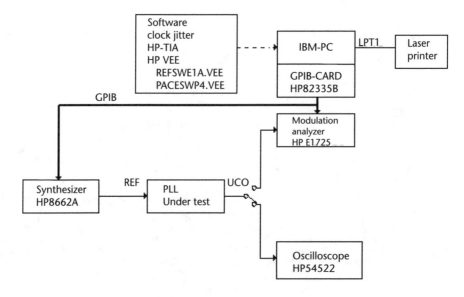

Figure 7.33   Test setup to measure jitter with an oscilloscope or the modulation analyzer.

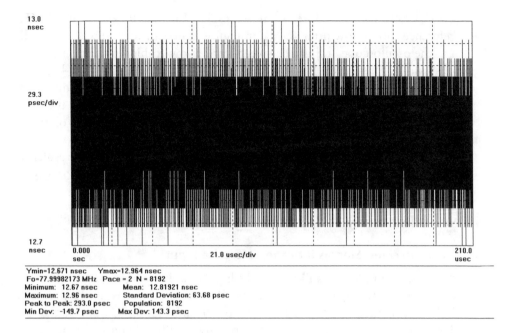

Ymin=12.671 nsec   Ymax=12.964 nsec
Fo=77.99982173 MHz   Pace = 2   N = 8192
Minimum: 12.67 nsec   Mean: 12.81921 nsec
Maximum: 12.96 nsec   Standard Deviation: 63.68 psec
Peak to Peak: 293.0 psec   Population: 8192
Min Dev: -149.7 psec   Max Dev: 143.3 psec

Figure 7.34   Example plot from the modulation-domain analyzer for an example PLL.

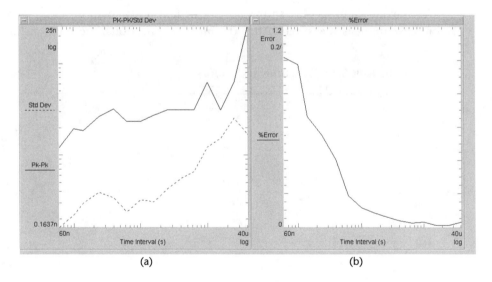

Figure 7.35   (a) Jitter and (b) +/− percentage error versus time interval.

loop bandwidth of the example PLL. At a time interval of 2 μs, the RMS value starts increasing because the TIA loses accuracy at this point.

Figure 7.35(b) shows the +/− percentage error of the period versus time interval. The PLL has 1% peak jitter at 60 ns and drops to 0.025% peak jitter at 4 μs. This decreasing characteristic shows that the noise is not uniformly distributed and not Gaussian. These plots allow users to determine if the PLL will meet jitter specifications at the divided-down clock frequency that they are using.

An example procedure on an example PLL that locks to an 54-MHz input reference shows how jitter shown in Figure 7.36 is measured with an oscilloscope. On the oscilloscope, adjust the sweep until a low-to-high transition of the clock is observable. Then, with the delayed sweep, highlight the next low-to-high transition. Increase the vertical and horizontal scales. In the histogram mode, adjust a measurement box around the transition to 10 mV vertically and 20 times the trace width horizontally. Then, with infinite persistence, measure the width of the jitter by selecting the scope's measurement for standard deviation, peak to peak, and the number of hits in the box. The figure shows the measurement results and was printed using the option on the scope to save the PCX file to a floppy disk.

7.6   Noise Immunity to Injected Signals

7.6.1   Injected Signals into the Reference Input

Signals can couple into a PLL from the high input impedance of the reference crystal oscillator. Noise-injection tests on the reference-clock measure the performance of PLLs under these low signal-to-noise conditions. Figure 7.37 shows the test setup. Two sources are combined with a resistive splitter. The combined signals are connected to the reference-frequency input of the PLL. The output period variation of the PLL is measured by the modulation-domain analyzer.

An example procedure shows how noise immunity to an injected signal at the reference input is measured. After the initial connections, the relationship of the

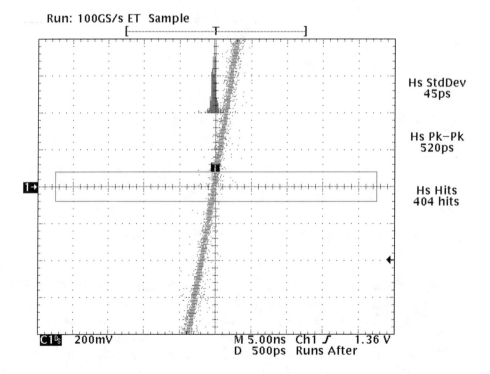

Figure 7.36   Jitter measurement from the Tektronix oscilloscope.

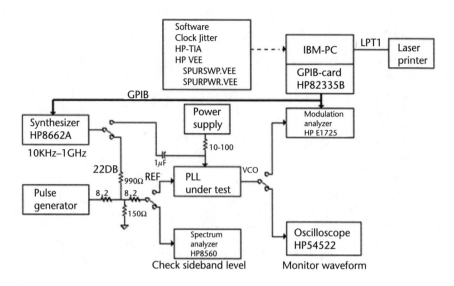

Figure 7.37   Noise-immunity test setup for injecting spurious signals at the reference input and on the power supply.

sideband (51 MHz, 1 MHz off the carrier) to the carrier signal (50 MHz in this example) is adjusted by connecting the resistive divider output to a spectrum analyzer and setting, in this example, the sideband level to –10 dBc. Then, the resistive divider output is connected to the reference input of the PLL under test, and the output of the PLL is connected to the modulation-domain analyzer. The HP8662A injected spurious frequency is logarithmically swept, and the period and jitter of the output of the PLL are measured by the modulation-domain analyzer. An HPVEE program (SPURSWP.vee) was written to control the instrumentation and make the measurements.

Figure 7.38 shows the measured jitter versus frequency of the injected spurious signal for a 50-MHz PLL. Figure 7.38(a) shows the standard deviation and the peak-to-peak variation of period (jitter) versus the frequency of the injected signal. Figure 7.38(b) shows minimum, mean, and maximum period versus frequency of the injected signal. Both figures show a significant increase in jitter at 400 kHz. However, with –10 dBc coupled to the input, this figure shows good noise immunity because it did not lose lock. Further tests for high noise immunity can be made by varying the spurious level at the 400-kHz sensitive frequency spot.

7.6.2   Injected Signals on Supply

Signals can couple into the PLL from the supply lines. Noise-injection tests on the power-supply line measure the performance of PLLs under these low signal-to-noise conditions. Figure 7.37 shows the test setup to make the measurement. The synthesizer injects a spurious signal through a 1-μF capacitor to the supply line. The decoupling capacitors on the PLL should be disconnected or kept to a minimum, and the resistance to the power supply should be raised in order to maximize the output swing on the supply line.

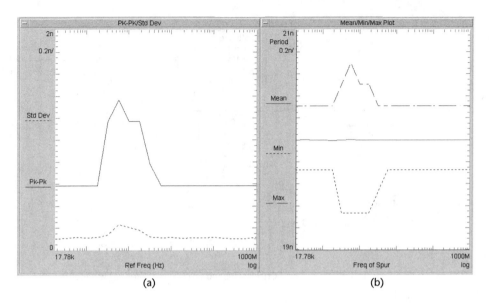

Figure 7.38 Measurement of jitter versus frequency of injected spurious signal coupled into the input: (a) standard deviation and peak-to-peak jitter, and (b) mean, minimum, and maximum period.

An example procedure shows how noise immunity to a signal injected into the supply is measured. After the initial connections, the HP8662A injected spurious frequency is logarithmically swept, and the period and jitter of the output of the PLL are measured by the modulation-domain analyzer. An HPVEE program (SPURSWP.vee) was written to control the instrumentation and make the measurements. The amplitude of the synthesizer is adjusted until the response shows distinguishing features without losing lock. If the amplitude is too low, the response will be a flat line. If the response is too high, the mean value on the right side in Figure 7.39 will significantly change, which shows the loop is losing lock.

Figure 7.39 shows the results. The figure on the left shows the standard deviation and the peak-to-peak variation of period (jitter) versus the frequency of the injected signal. The figure on the right shows minimum, mean, and maximum period versus frequency of the injected signal. The injected-signal frequency is logarithmically swept from 10 kHz to 1,000 MHz to find the most sensitive frequency spot. In this example, 31 MHz was the most sensitive spot. Further tests for high noise immunity can be made by varying the spurious level at the 31-MHz sensitive frequency spot, or the level of the injected signal can be measured on an oscilloscope, and the period can be measured on the modulation-domain analyzer to measure the supply sensitivity (MHz/V) at the worst condition.

7.7 Power-On Switching Time

In monolithic PLLs, power is cycled between off and on to save power. Consequently, IC designers need to know how fast the PLL takes to acquire lock from a power-on condition. Figure 7.40 shows the test setup to make the measurement.

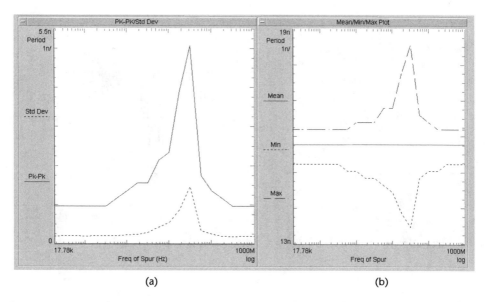

(a)   (b)

Figure 7.39 Example measurement of jitter and period versus frequency of injected signal: (a) standard deviation and peak-to-peak jitter; and (b) mean, minimum, and maximum period versus injected supply signal frequency.

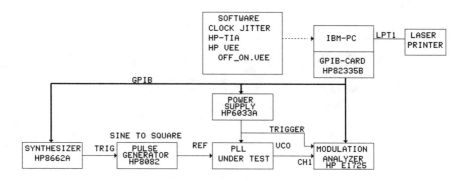

Figure 7.40 Test setup to measure switching time with power on.

The modulation-domain analyzer measures the output clock frequency by waiting for a trigger generated when the supply powers on.

Figure 7.41 shows the resulting measurement for power-on of an example PLL with an output frequency of 81 MHz. This measurement shows that it takes over 4 ms for the output frequency to settle. The results of this measurement depend on the decoupling capacitors on the test fixture. Consequently, they should be kept to a minimum to get an accurate measure of the power-on time for the PLL.

7.8 Oscillator Open-Loop Test

Many ICs with PLLs have a crystal oscillator on board. Measuring the open-loop response of the oscillator helps set the output frequency and ensure that the circuit oscillates. Figure 7.42 shows the test setup.

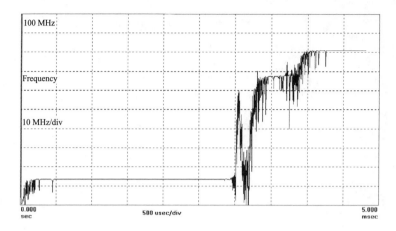

Figure 7.41 Example measurement of switching time with power on.

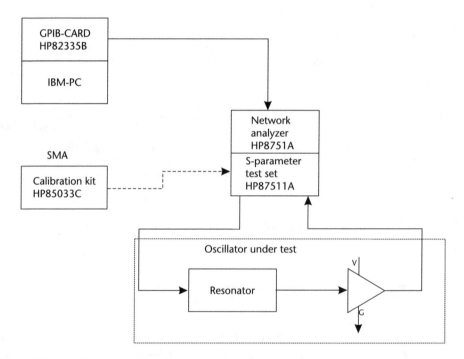

Figure 7.42 Oscillator open-loop test setup using a network analyzer.

An example procedure shows how the open loop of an oscillator is measured with a network analyzer. After the initial setup, sweep the frequency, and measure magnitude and phase. Find the 0° point that has gain greater than 0 dB, and the frequency of oscillation is found. The amount of magnitude above 0 dB, and the amount of magnitude above 0 dB for frequencies about the 0° point, determine the potential for the circuit to continue to oscillate over process and environmental conditions. Figure 7.43 shows an example plot from the network analyzer for an example oscillator.

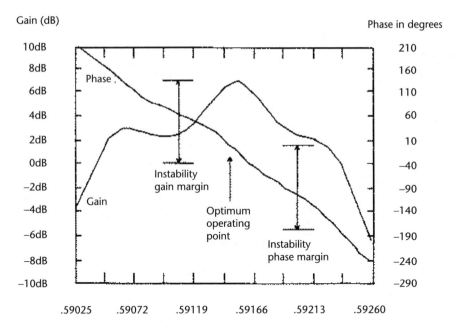

Gain (dB)

Phase in degrees

Figure 7.43 Example measurement of an open-loop oscillator.

7.9 Test Equipment

Table 7.4 shows a list of desired test equipment for making measurements on PLL parameters. The test equipment model numbers and estimated prices are given for performance and cost references. Performance of equivalent test equipment can be compared to the specification sheets for the equipment listed in the table to ease buying equipment or using existing equipment decisions. The cost of all the equipment in Table 7.2 adds up to over $300,000, which can put a significant dent in anyone's capital expenditure for the year. The following section provides a method for comparing equipment to be used in measuring PLLs, but is not intended as an endorsement of one vendor over another. In addition, the accuracy of the equipment characteristics and the estimated cost obtained from data sheets is not guaranteed. Please contact the vendors to obtain up-to-date and accurate information.

Figure 7.44 shows a picture of a PLL test bench. The PLL under test is on a blue ESD mat in the middle of the picture. The test fixture easily allows PLL integrated circuits to be tested in the PCB. Space is left around the PCB to allow the temperature thermal stream to be dropped down for temperature testing. On the first level on the left are the power supplies, multimeter, and temperature thermal stream. The middle has ESD protection fan and power supplies for a FET probe. The right of the picture has a low-noise signal generator, pulse generator to convert sine to square wave, and a jitter analyzer. The top level on the left has a phase-noise test set, a modulation-domain analyzer, a low-frequency spectrum analyzer for the phase-noise test set, and a GPIB programmable power supply. On the right side is a high-frequency spectrum analyzer and a high-frequency oscilloscope. This completes the description of the equipment in the picture.

Table 7.4   Test Equipment for Evaluating PLLs

Functional Description	Test Equipment Model # or Equivalent	Measured Parameter or Function	Estimated Cost (Thousands)
Oscilloscope	HP54522 or TEK784A (preferred)	Jitter, duty cycle, time response	$30
Spectrum analyzer, phase-noise utility	HP8560E, HP85671A	Phase noise, spurious, bandwidth, stability	$26
Fast-switching generator, 70-dB step attenuator, high stability, N connector	HP83752A opt1E1, opt1E5, opt1ED	Time response 10% to 90% 100 ns, stability, phase noise, jitter, bandwidth	$26
Modulation-domain analyzer (time-interval analyzer) (jitter analyzer)	HPE1725 Option 001, Option 243, delete IBM-PC, jitter-analysis software Or Wavecrest DTS-2075 with software	Jitter 400-ps peak-to-peak accuracy, time response Jitter, 3-ps RMS noise floor	$30, $54
Phase-noise analyzer—DOS software	HP3048A opt 301	Phase noise, stability, bandwidth, jitter	$35
Synthesized generator, low noise	HP8662A Option 001 of HP3048	Time response 4-μs, stability, phase noise, jitter, bandwidth	$42
PC with IEE488 card	PC with HP82335A card	Hardware controller for making measurements	$3, $0.3
Equipment controller software	HPVEE	Software controller for making measurements	$1
Pulse generator	HP8116A	Control of frequency step response	$5
Pulse generator, sine to square wave	HP8082 or HP8130 or HP8131	Sine-to-square-wave conversion	$15
Network analyzer with s-parameter test-set calibration kit	HP8751A and HP87511A, HP85033C, DOS floppy-disk option	Oscillator open-loop response and line impedances	$26, $4

An in-depth study of desirable equipment features is necessary to make an economically efficient decision and to allow other equipment to be substituted in case of obsolescence. For an oscilloscope, some of the key parameters are high accuracy and low jitter of the delay generator, high sampling speed, and wide bandwidth. Other features also make testing easier. Printing to a floppy disk in a TIF format makes taking data easier. Table 7.5 shows these key parameters for the TDS784A oscilloscope in Table 7.4.

For a spectrum analyzer, some of the key parameters are low noise floor, wide frequency range, high 1-dB compression point, and low levels of internally generated spurious signals. Other features also make testing easier. A zoom feature to zoom in and out from the center frequency is very desirable. Also, converting spectral-noise power to phase noise is a nice feature. Table 7.6 shows these key parameters for the HP560E spectrum analyzer in Table 7.4.

For the synthesized signal source, some of the key parameters are wide frequency range, low phase noise, high output power levels, low nonharmonically related sidebands, wide FM deviation, and fast 10% to 90% frequency step response.

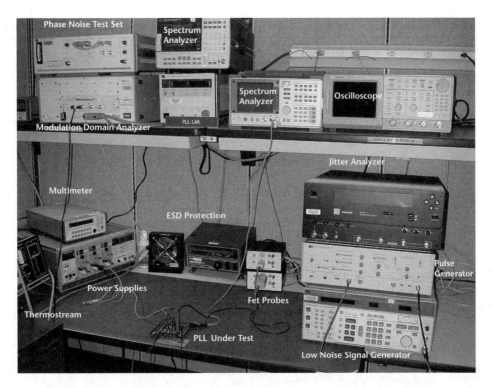

Figure 7.44 PLL laboratory with all the equipment to characterize a PLL on the bench.

Table 7.5 Example Values for Key Parameters for the TDS784A Oscilloscope

Delay Generator Accuracy	Maximum Sampling Rate	Bandwidth	Delay Generator Jitter	Maximum Record Length	Jitter Measurement Time
37 ps	4 GS/s	1 GHz	7 ps	500k	~2 min/100 hits

Table 7.6 Example Values for Key Parameters for the HP8560E Spectrum Analyzer

Frequency Range	Noise Floor	1-dB Compression	Internally Generated Spurious Signals
30 Hz–2.9 GHz	–120 dBm@10 kHz –140 dBm@1 MHz	>–5 dBm	<–80 dBc with –10-dBm input

Other features also make testing easier. A minimum of transient disturbances at the rollover frequencies (e.g., 1 MHz) and a minimum of rollover frequencies help in the testing. The rollover frequencies occur when selection between one of the PLLs in the synthesizer is switched. This switching can cause amplitude and frequency disturbances that can be mistaken for transients of the PLL under test. Table 7.7 shows the key parameters for the HP8662 low-noise synthesizer signal source in Table 7.2.

For the pulse generator that converts the sine to a square wave, the key parameters are highest output frequency and fastest rise and fall time. In addition, the conversion of the sine to a square wave should not increase the phase noise by

Table 7.7   Example Values for Key Parameters for the HP8662 Low-Noise Synthesizer Signal Source

Output-Frequency Range	Phase Noise, Typical (dBc/Hz)	Output Power Range (dBm)	Nonharmonically Related Spurs (dBc)	Frequency Step Response	FM Deviation
10 kHz–1,280 MHz	−126@100 −139@100 kHz	−139–+13	−90 dBc	4 μs	200 kHz

more than 6 dB. A few other features are also needed. An external gate is needed that triggers on the rising edge of the clock. A variable adjustment of this trigger level is also convenient. A control button that selects a 50% duty cycle and does not require readjustment when the frequency changes is also desirable. Finally, output amplitude and offset that can easily be adjusted for the logic levels of the devices under test and that also do not need to be readjusted for a frequency change are also desirable. Table 7.8 shows the key parameters for the HP8082 pulse generator in Table 7.4.

For the pulse generator that is used in the frequency-step-response test and the injected signal on the reference and injected signals on the supply measurements, the key parameters are highest output frequency and fastest rise and fall time. The functions of sine, square, and pulse waveforms should be available. Jitter should be low enough that it can be substituted for the reference frequency and not significantly affect the measurement. A control button that selects a 50% duty cycle and does not require readjustment when the frequency changes is also desirable. Finally, output amplitude and offset that can be easily adjusted for the logic levels of the devices under test and that also do not need to be readjusted for a frequency change are also desirable. Table 7.9 shows these key parameters for the HP8116A function generator in Table 7.4.

For measurements of jitter equipment, some of the key parameters are noise floor, input dynamic range, maximum input frequency, speed of a measurement (number of periods measured per minute), and minimum jitter measurement. Table 7.10 shows these key parameters for the HPE1740 jitter analyzer in Table 7.4.

For measurement of phase-noise equipment, some of the key parameters are input-frequency range, offset-frequency range, and phase-noise floor. Table 7.11 shows these key parameters for the HP3048 phase-noise test set.

For a network analyzer that measures the open-loop transfer function of a crystal oscillator or the open-loop transfer function of a PLL, the key parameters

Table 7.8   Example Values for Key Parameters for the HP8082 Sine-to-Square-Wave Pulse Generator

Maximum Frequency	Rise/Fall Time	Period Jitter	External Trigger Level
250 MHz	1 ns	<0.1% + 50 ps	>200 mV$_{\text{p-p}}$

Table 7.9   Example Values for Key Parameters for the HP8116A Function Generator

Maximum Frequency	Rise/Fall Time	Functions
50 MHz	7 ns	Sine, square, and pulse

Table 7.10   Example Values for Key Parameters for the E1740 Jitter Analyzer

Noise Floor	Input Range	Max Sampling Rate	Memory Size
400 ps peak-to-peak, no averaging	150 MHz (18 GHz with converter)	80 MHz	512k

Table 7.11   Example Values for Key Parameters for the HP3048 Phase-Noise Test Set

Input-Frequency Range	Offset-Frequency Range	Phase-Noise Floor
5 MHz—18 GHz	0.01 Hz–40 MHz	−170 dBc/Hz at 10 kHz–40 MHz −130 dBc/Hz at 1 Hz

are input-frequency range, input level range, input impedance, output power level, level accuracy, crosstalk, and the directivity of the couplers. Table 7.12 shows these key parameters for an HPE5100 network analyzer.

This concludes the in-depth study of desirable equipment features. The lists of desirable features in the equipment will help make an economically efficient decision and allow other equipment to be substituted in case of obsolescence.

7.10   Troubleshooting PLLs

PLLs can be troubleshot at several different levels:

- System;
- PCB;
- Integrated circuit:
  - Tests on the bench;
  - Tests on a tester;
- Simulation.

At the system level, the VCO, loop filter, phase detector, and feedback divider are individual modules. The modules can be individually tested to specifications so that, at system integration, there is a high degree of confidence that the system will work.

At the printed circuit board level, the VCO, loop filter, phase detector, and feedback divider are individual integrated circuits. Test points are made available on the printed circuit board, and switches can be set so that the PLL can be tested. At this point, it gets a little more challenging, but easy access to test points makes troubleshooting the PLL still pretty easy.

Table 7.12   Example Values for Key Parameters for the HP E5100 Network Analyzer

Frequency Range	Input Level	Input Impedance	Output Power Level	Level Accuracy	Crosstalk	Directivity
10 kHz to 300 MHz	0 to −130 dBm	50Ω/1 MΩ	−50 to +15 dBm	±0.5 dB	<−100 dB	>33 dB

The bigger challenge occurs for troubleshooting integrated circuits at the bench level, and troubleshooting is still more challenging at the tester level.

7.10.1 Integrated Circuit

Troubleshooting monolithic PLLs presents unique problems because of the limited access to test points. At a minimum, the output of the VCO, the divided-down output of the VCO, and the reference input port are available. Additionally, the clock inputs to the phase detector are usually available, and there might be a lock-detect signal.

The following bench test equipment is recommended:

- Oscilloscope;
- Spectrum analyzer;
- Low-noise synthesizer;
- Pulse generator;
- Time-interval analyzer.

7.10.2 Functional Check

First, turn power on and check for the output of the VCO. The VCO should be running at its highest frequency with no input. If there is no output, recheck the switch settings, then measure the power-supply current. Low power-supply current indicates an extreme failure. If all the settings still look correct, then input the reference-frequency clock at high frequency to move the VCO to its highest output frequency.

If the VCO output toggles, use an oscilloscope to check that the peak-to-peak swing of the VCO is close to the supply rails and that the duty cycle is between 30% and 70%. Next, use a spectrum analyzer to check that the spurious-signal levels are less than −30 dBc and that a solid spectral line appears at the VCO frequency. Significant FM modulation of the VCO can indicate multimode oscillation problems or significant coupling of signals into the VCO.

If the VCO output has a reasonable duty cycle, no significant FM modulation, and no large spurious signals, then the feedback divider can be tested. Using an oscilloscope that is triggered to the output of the VCO, observe the output of the feedback divider as the divide-ratio control bits are changed. By observing the change in period with the change in control bits, you can check the divider operation.

With operation of the feedback divider checked, the phase detector can be checked. Set the feedback divide ratio to a nominal value, and connect a clock to the reference input at a frequency much higher than the maximum frequency for the device and view the output of the VCO or the output of the feedback divider on a spectrum analyzer. The connection of the reference clock should change the output frequency of the VCO to its minimum value. If this occurs, then the phase/frequency detector is functional.

If the phase/frequency detector is functional, then change the reference frequency to a nominal value for the PLL to lock. If the loop locks, then the loop-filter connection is checked, and the hard failures have been checked. Cycle the power supply several times to verify that the loop maintains lock. Inconsistency on power up indicates a PLL problem.

Increase the reference frequency until the loop loses lock. Check that the lock-detector output changes states to the out-of-lock condition. Then, decrease the frequency until the loop relocks. Check that the lock-detector output changes state to the lock condition. The loss-of-lock frequency and the relock frequencies should be close together. Decrease the input frequency until the loop loses lock at a low frequency. Check that the lock-detector output changes state to the out-of-lock condition. Increase the input frequency until the loop relocks. Check that the lock detector output changes state to the lock condition. Again, the loss-of-lock frequency and the relock frequencies should be close together. This test also verifies that the lock-detection circuit works. If the loop passes these tests, then the loop is functional.

7.10.3   Requirement Compliance Checks

For bench testing, manually ramp the input frequency up and down and check for loss of lock. Slowly ramp the input frequency and watch that the reference-clock edge and the feedback clock edge track within the expected requirements.

The first test should be to check if the hold-in range and lock-in range match analysis. A low high-frequency lock-in range can indicate that the I/O buffers do not work at this high a frequency. Checking the divided-down output will verify if the loop is still locked at the high frequency.

Match frequency step response with the circuit analysis. A 10% to 90% response within (30% should be expected. Loss of lock with a step response is a bad indication. This can indicate low gain, high gain, narrow bandwidth, low phase margin, coupled signal into the VCO, or disconnected components. A large amount of ringing on the step response indicates low phase margin. This can indicate poor analysis, poor transistor models, or parasitics that have not been taken into account.

Spurious signals greater than −40 dBc can significantly effect jitter. For high spurious signals, power-down any signals that toggle, especially I/O buffers, and check if the spurious signals remain. If the spurious signals are gone, then the signals are coupling into the PLL, and by process of elimination, the offending function can be found. If the spurious signals are still there with every other circuit powered down, then the signals are from the PLL. Disconnect the reference clock, and check the output of the VCO at its highest output frequency unlocked. If the spurious signals go away, the signals are generated in the PLL. Connect the reference clock at a high frequency, and see if the spurious signals go away at the VCO's low operating frequency. If the spurious signals go away, the signals are generated in the PLL. Connect the reference clock at the nominal value for the loop to lock and vary the input frequency while observing the VCO output on the spectrum analyzer. If the spurious signals move with the VCO and keep an offset consistent with the reference-clock frequency input to the phase detector, then the spurious signals are reference sidebands. If the signals are not consistent with the reference-clock frequency, then the VCO is multimoding, or some other signals are coupling into the VCO from the PLL.

In order to test jitter, spurious signals must be eliminated in the previous step. With low spurious signals, high jitter numbers can result from problems with the VCO (poor transistor noise figure or many flaws in the semiconductor, causing a high flicker corner frequency).

Inject a sinusoid on the supply voltage to check the sensitivity of the PLL to the power supply. Sweep the frequency from 10 kHz to 1 GHz and measure the peak-to-peak voltage on the supply and the peak-to-peak frequency change on the output of the VCO. If there is no change in the peak-to-peak frequency, increase the sinusoidal amplitude until a clear peak-to-peak frequency can be identified.

7.10.4 Simulation

Troubleshooting at the simulation level is much easier than at the integrated-circuit level because the designer has access to all the nodes in the circuit. First, run individual simulations on each major block (VCO, dividers, phase detector, and charge pump) to make sure they function properly. From the individual cell, check of the VCO; it should be clear how to start the oscillation in the VCO. The best VCO designs require no startup pulse. Check the behavioral model for a PLL to make sure the loop compensation provides stability.

Next, check that the VCO starts up with power up. You may have to wait several microseconds for the oscillator to work. If the VCO does not work, a startup pulse may have to be applied to the oscillator. This can be done with a voltage or current source. Compare the reference-oscillator edge with the divided-down VCO edge at the input to the phase detector. The loop is locked when these edges line up to within approximately 1 ns of each other. Run the simulation several microseconds after lock to make sure the loop does not lose lock at a later time. Run another simulation that checks the loop locks from the highest VCO frequency condition. This can be done by initially setting the reference clock to a very high frequency. Then, after the tune voltage has reached the supply level, the reference clock can be switched to its nominal operating frequency.

Power-supply-injection testing and testing by adding series inductors on the supply give methods to test the sensitivity of the PLL to the power supply. For the power-supply-injection test, a sinusoidal modulation is applied to the power supply with a 0.1 V_{pp} amplitude and a frequency between 1 and 10 MHz. The frequency of the modulation must be a prime number to the output frequency of the PLL in order to avoid injection locking. A frequency of 3.33 MHz is a good example. In addition, the modulation frequency should be higher than the loop bandwidth because inside the loop bandwidth, the control system will reject the modulation. After the PLL is locked, the modulation should be applied, and the output frequency of the PLL versus time should be measured. The peak-to-peak frequency modulation of the output over the peak-to-peak amplitude of the supply modulation gives the sensitivity of the PLL. The units can be megahertz per volt of supply or percentage output-frequency change over percentage power-supply change.

For the inductor test, a series 10–20-nH inductor is added in series with the ground-supply line and the power-supply line. This simulates extremely poor power-supply routing on the silicon, which can happen with systems on a chip.

The peak current spikes of the PLL will then modulate the supply line. This will cause a PLL with high sensitivity to the power-supply line to lose lock.

In summary, this chapter discussed several measurement methods. Initial measurements of a PLL were shown to be very important because measurements of an unlocked loop or a very noisy loop give erroneous data and may stop any further measurements. Consequently, measurements of hold-in range and spurious signals were presented to give some minimal stability and accurate noise information. Next, reference frequency-step-response measurements were discussed. Results from step-response measurements verify damping and bandwidth design goals. Measurement of the PLL bandwidth that can be accurately measured with a network analyzer by injecting a signal into the loop and measuring the resulting signal were presented. Making these measurements insures the accuracy of the next set of measurements.

Several phase-noise measurements were presented. An explanation of the advantages and disadvantages of each method and several plots of phase-noise sources and measuring equipment can aid the reader in selecting the appropriate phase-noise method and test equipment. Jitter and spurious-signal measurements were also discussed. Understanding these issues will help design lower-noise loops.

Noise-injection tests on the reference clock were presented to measure the performance of PLLs under low signal-to-noise conditions. Noise-injection tests on the power-supply line were also presented to measure the performance of PLLs under low signal-to-noise conditions. Measurements of how fast the PLL takes to acquire lock from a power-on condition were presented. Next, measurements of the open-loop response of the oscillator were presented to help set the output frequency and ensure that the circuit oscillates. An in-depth study of desirable equipment features was presented to help make economically efficient decisions and allow other equipment to be substituted in case of obsolescence. Finally, this chapter presented troubleshooting techniques for PLLs in order to help determine which test needs to be run on a particular design.

Questions

7.1 Suppose a phase-noise measurement system has a buffer amplifier with 40-dB gain and 10-dB noise figure, a high-level mixer with 8-dB conversion loss, and an analyzer with a −140-dBm/Hz noise floor. Calculate the system's phase-noise floor for an oscillator with a +10-dBm output power level.

7.2 Suppose a phase-noise measurement system has a buffer amplifier with 40-dB gain and 10-dB noise figure, a high-level mixer with 8-dB conversion loss, and an analyzer with a −110-dBm/Hz noise floor. Calculate the system's phase-noise floor for an oscillator with +0-dBm output power level.

7.3 Suppose a phase-noise measurement system has a buffer amplifier with 40-dB gain and 10-dB noise figure, a high-level mixer with 8-dB conversion loss and +8-dBm input saturation level, and an analyzer with a −140-dBm/Hz noise floor. Calculate the system's phase-noise floor for an oscillator with +15-dBm output power level.

7.4 Suppose a phase-noise measurement system has a −130-dBm/Hz spectrum analyzer noise floor. Calculate the required buffer-amplifier gain that amplifies the thermal noise level to the spectrum analyzer's noise floor when the buffer amplifier has a 5-dB noise figure.

7.5 Calculate the gain in Question 7.4 when the buffer amplifier has a 10-dB noise figure.

7.6 Adjusting the phase shifter in Figure 7.14 through 360° range, the digital voltmeter reads a maximum of +1V and a minimum of −1V. Calculate the phase slope in volts per radian. Also, calculate the phase slope's power level in dBm.

7.7 An amplifier has a 8-dB noise figure, 33-dB gain, and 2-W output saturation level. Calculate the additive phase-noise relative to the carrier at the amplifier's output in a 1-Hz bandwidth for an input signal level of −5 dBm.

7.8 After varying the phase shifter in the carrier-suppression test set, the minimum dc voltage on a multimeter reads −0.5V. Calculate the power level of the beat frequency.

7.9 After making the measurement in Question 7.8, you discover a dc bias of +0.1V. Calculate the change in the phase slope.

7.10 A calibration signal is injected into a carrier-suppression measurement system. The calibration signal is adjusted to measure −40 dB below the carrier on a spectrum analyzer. The measurement system measures a −60-dBm power level. Calculate the power level measured for a signal 0 dB below the carrier.

7.11 A calibration signal is injected into a carrier-suppression measurement system. The calibration signal is adjusted to measure −40 dB below the carrier on a spectrum analyzer. The measurement system measures a −20-dBm power level. Calculate the power level measured for a signal 0 dB below the carrier. If the 1-dB input compression point of the mixer is +8 dBm and the conversion loss is 8 dB, explain why the output power level of the measurement system is larger than the input calibration signal level.

7.12 Prove that a single-sideband signal has equal AM and PM magnitude components.

7.13 A measurement is made with a phase detector phase slope of 0.5 V/rad, and a sine wave is assumed to be the shape of the beat-frequency waveform. Later, the waveform on the oscilloscope shows a triangular beat-frequency shape. Calculate the actual phase slope using the first five harmonics of the Fourier series of a triangular wave.

7.14 A measurement is made with a phase slope of 0.1 V/rad, and a sine wave is assumed to be the shape of the beat-frequency waveform. Later, a spectrum analyzer measures a third harmonic 30 dB, a fifth harmonic 35 dB, and a seventh harmonic 40 dB below the beat carrier frequency. Calculate the actual phase slope using the first, third, fifth, and seventh harmonics of a Fourier series.

7.15 In Questions 7.4 and 7.5, the buffer-amplifier gains were calculated. Calculate the input signal level to the amplifier that is equal to the equivalent thermal noise input level for both amplifiers. Which amplifier should be used for the measurement system to have the best sensitivity?

7.16 Suppose a phase-noise measurement system has a −130-dBm spectrum analyzer noise floor. Calculate the required buffer-amplifier gain that amplifies the thermal noise level to the spectrum analyzer's noise floor if the buffer amplifier has a 44-dB noise figure. What are the advantages of using this amplifier?

References

[1] Best, R. E., *Phase-Locked-Loops: Design, Simulation, and Applications*, New York: McGraw-Hill, 1997, pp. 265–279.

[2] Coombs, C. F., *Electronic Instrument Handbook*, New York: McGraw-Hill, 1995, pp. 19.4–19.7, 22.7–22.18.

[3] Goldman, S. J., *Phase Noise Analysis in Radar Systems Using Personal Computers*, New York: Wiley Interscience, 1989, pp. 91–113.

[4] Temple, R., "Choosing a Phase Noise Measurement Technique—Concepts and Implementation," unpublished Hewlett-Packard RF and Microwave Measurement Symposium, Dallas, TX, February 1983.

[5] Scherer, D., "The Art of Phase Noise Measurement," *HP RF and Microwave Symposium and Exhibition*, Dallas, TX, May 1983.

[6] Fisher, M., "An Overview of Modern Techniques for Measuring Spectral Purity," unpublished Hewlett-Packard RF and Microwave Symposium and Exhibition, Dallas, TX, January 1979.

[7] "Spectrum Analysis . . . Noise Measurements," Hewlett Packard Application Note 150-4, April 1974.

[8] Gibbs, J., "Computer-Aided Phase Noise Measurements," unpublished Hewlett Packard Phase Noise Measurement, a Seminar on Modern Techniques, May 1982.

[9] Lance, A. L., et al., "Automating Phase Noise Measurements in the Frequency Domain," *Proc. 31st Annu. Symp. Frequency Contr.*, Atlantic City, NJ, June 1–3, 1977, pp. 347–358.

[10] Ashley, J. R., T. A. Barley, and G. J. Rast Jr., "The Measurement of Noise in Microwave Transmitters," *IEEE Transactions on Microwave Theory and Techniques*, Vol. MTT-25, No. 4, April 1977, pp. 294–318.

[11] Leeson, D. B., and G. F. Johnson, "Short-Term Stability for a Doppler Radar: Requirements, Measurements, and Techniques," *Proceedings of the IEEE*, Vol. 54, No. 2, February 1966, pp. 244–248.

[12] Manassewitsch, V., *Frequency Synthesizers: Theory and Design*, 2nd ed., New York: John Wiley and Sons, 1980, pp. 128–147.

[13] Rohde, U. L., *Digital PLL Frequency Synthesizers: Theory and Design*, Englewood Cliffs, NJ: Prentice-Hall, 1983, pp. 98–106.

[14] Tykulsky, A., "Spectral Measurements of Oscillators," *Proceedings of the IEEE*, Vol. 54, No. 2, February 1966, p. 306.

[15] Ondria, J. G., "A Microwave System for Measurements of AM and FM Noise Spectra," *IEEE Transactions on Microwave Theory and Techniques*, Vol. MTT-16, No. 9, September 1968, pp. 767–781.

[16] Grauling, C. H., Jr., and D. J. Healy, "Instrumentation for Measurement of the Short-Term Frequency Stability of Microwave Sources," *Proceedings of the IEEE*, Vol. 54, No. 2, February 1966, pp. 249–257.

[17] Sann, K. H., "The Measurement of Near-Carrier Noise in Microwave Amplifiers," *IEEE Transactions on Microwave Theory and Techniques*, Vol. MTT-16, No. 9, September 1968, pp. 761–766.

[18] "Jitter Analysis 101, A Foundation for Jitter Measurements," Wavecrest Seminar, Dallas, TX, May 1997.

CHAPTER 8

Simulation

This chapter covers simulation techniques for PLLs. The types of PLL simulators vary from SPICE transistor-level solutions, to behavioral-model solutions, to equation-based solutions. Transistor-level simulations are presented to show the many conditions that can be tested using this method. The characteristics of supply-voltage sensitivity responses, pulse feed-through levels, slew rate, powering up, powering down, and loop recovery from supply-rail conditions are unique conditions that are shown.

Behavioral-model simulations are presented to show that significant improvement in computing time with a minimum loss of accuracy can be achieved. The advantages and disadvantages of the model are discussed, and an example simulation is done to show its performance. Finally, an example of numerical errors is shown to give some insight into the types of errors that can occur with a behavioral model. Difference-equation-based simulations are presented not only to show increased computing speed but to show unique characteristics of digital PLL simulations. The speed of the behavioral and difference-equation models allow many iterations of what-if questions to be studied. The choice of simulation depends on the speed and accuracy of the results. For instance, accurate SPICE simulations can take several weeks to run when looking at acquisition times. This chapter will help the reader make an informed choice between methods.

8.1 Transistor Level

In this section, we will study transistor-level simulations. An example PLL will be used to show transistor-level simulation characteristics that could not be studied using behavioral models or equation-based simulators. Power-on from both supply rails, small frequency step response, large frequency step response, supply-voltage step response, sinusoidal-signal-injected-on-the-power-supply response, and power-down response simulations are presented.

There are several choices between transistor-level simulators. SPICE and Powermill can be used. Powermill characterizes transistors to a look-up table to speed up SPICE simulations. These simulators give a good look at the functionality of the PLL. For more accurate simulations, SPICE must be used; however, Powermill can be used to get node voltages in the steady-state condition, and these voltages can be used as the initial conditions for SPICE to improve accuracy.

An example PLL will help show the capabilities of transistor-level simulation. Figure 8.1 shows a block diagram of a transistor-level SPICE simulation circuit.

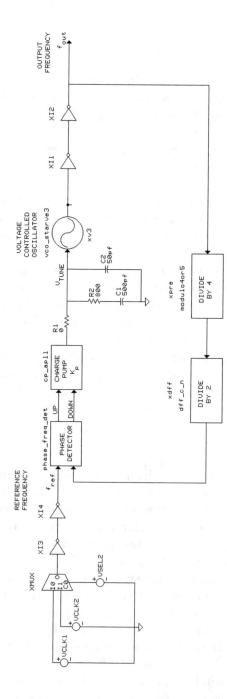

Figure 8.1 Block diagram of transistor-level SPICE simulation circuit.

This is the example PLL circuit that will be used to demonstrate the transistor-level SPICE simulation. The PLL consists of a phase/frequency detector, a simple charge pump, RC-loop compensation, a current-starved ring oscillator, and simple D flip-flop dividers. Two pulse generators are muxed at the input to do frequency-step tests.

Table 8.1 lists the design parameters for the example transistor-level PLL. Table 8.2 lists the calculated and rounded loop-compensation values. A transistor-level simulation of a PLL produces the most accurate results; however, the simulation time can take from several days to weeks. A simulation using Powermill or ADM speeds up SPICE by 10 to 100 times and only sacrifices some accuracy. Behavioral modeling with clock edges significantly speeds up simulation time also by 10 to 100 times but sacrifices even more accuracy. Improvements in speed of over 1,000 times occur if the classical equations are solved, but even more accuracy is lost. Each simulation has its advantages and disadvantages. Because of the long simulation times of SPICE, the faster techniques should be used to determine functionality, and SPICE simulations should only be used for the few cases where the faster simulators are inaccurate or suspicious responses must be verified. The following simulations show the capabilities of the SPICE simulators.

Figure 8.2 shows a power-on acquisition at 4 MHz from the supply rail with a feedback divide ratio of 8. The x-axis scale has the units 0.1 ns with the top scale showing the zoomed-in time. The circuit powers up in approximately 2 μs. The first waveform in the figure shows the reference input clock, and the next waveform is the divided-down input clock from the VCO to the phase detector. The third waveform is the up output out of the phase detector, and the next waveform is the down output out of the phase detector. The fifth waveform is the output clock from the VCO, and the next waveform is the output of the divide-by-4 frequency divider. The final waveform is the control voltage to the VCO, which is the filtered and compensated output of the charge pump.

Table 8.1 List of Design Parameters for Example Transistor-Level Simulation

Design Parameter	Value
Reference frequency	4–16 MHz
K_v	600E6 rad/s/V
N	8
Loop bandwidth	2 MHz
Damping factor	0.6

Table 8.2 Calculated Compensation Values for Transistor-Level SPICE Simulation

Component	Charge Pump
K_d	1.3 mA/rad
R_1	0
R_2	800Ω
C_1	500 pF
C_2	50 pF

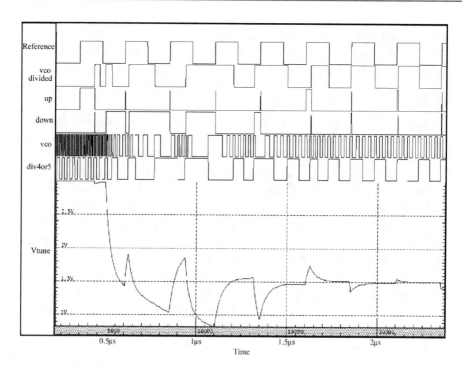

Figure 8.2 Powermill simulation showing power-on acquisition from the supply rail.

The figure shows the power-on acquisition from the supply-voltage rail of the PLL to an output frequency of 32 MHz. The detail of the simulation can be seen in this figure. The changing of the up and down pulse widths of the phase detector output can be seen. The modulation of the VCO and divider frequencies from the pulse widths of the phase detector can be seen. Consequently, these details show why transistor-level simulation is accurate.

In Figure 8.3 Powermill simulation shows power-on acquisition from 0V. The input frequency is 4 MHz, which produces a 32-MHz output for a divide ratio of 8. The power-on acquisitions have been from a 3-V supply. This simulation shows it takes 3 μs to power up. This is 50% slower than powering up from the supply

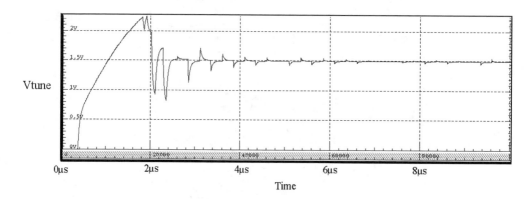

Figure 8.3 Power-on acquisition from 0V using Powermill simulation.

rail. Simulations of acquisition from both power supplies are necessary to make sure the loop does not get lost. Other simulation methods would not be able to check this characteristic.

Figure 8.4 shows the control-voltage response to a small frequency step at 5 μs from 4 to 8 MHz with a feedback divide ratio of 8. This produces an output frequency that steps from 32 to 64 MHz. The 10% to 90% rise time is approximately 200 ns, and there is a 25% overshoot on the step up. From (3.142), we can compute natural frequency and stability by fitting the simulated data to the equation. This results in a computed natural frequency of 850 kHz and a damping factor of 0.6. Equation (2.25) computes a 1.2-MHz 0-dB crossover frequency, and (2.24) computes a 59° phase margin. This figure shows that the step up and step down in frequency response is not symmetrical. Other simulation methods (behavioral and equation based) would not show this detail. In addition, other simulation methods would not show the feed-through of the pulses from the charge pump that modulate the VCO and cause reference sidebands.

The voltage-control waveform in Figure 8.5 shows power-on acquisition followed by a large frequency step at 12 μs from 4 to 16 MHz with the feedback divide ratio set to 8. A large frequency step shows any slew-rate limitations that the circuit might have. As this case shows, the 10% to 90% point has a 3-μs rise time, which is much larger than the 200-ns rise time for the small-frequency-step case. Other simulation methods (behavioral or equation based) would not show the asymmetry between rise time and fall time or the slew-rate limitation in this example.

Figure 8.6 shows power-on acquisition with 20 nH of inductance on the ground and supply leads and a feedback divide ratio of 8. At 7.5 μs, the input frequency changes from 4 to 8 MHz, which adds more difficulty to the test. This simulation identifies nonlinear and high current draws from the supply. In addition, integrated circuits use bond wires to the package, test sockets, and evaluation boards with circuit board traces that produce a series inductance (5 to 10 nH). The inductor

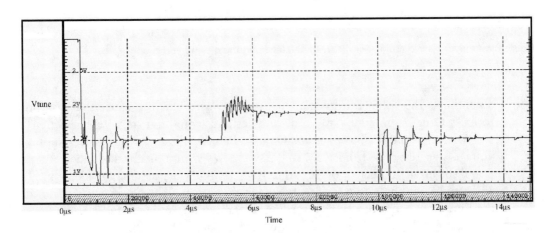

Figure 8.4 Control-voltage response to a small input-frequency step of 4 to 8 MHz with a feedback divide ratio of 8.

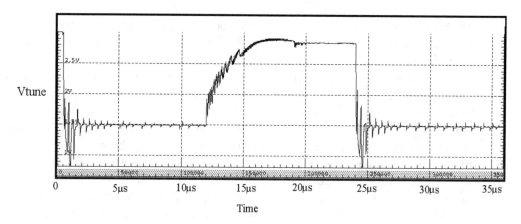

Figure 8.5 Power-on acquisition followed by a large frequency step.

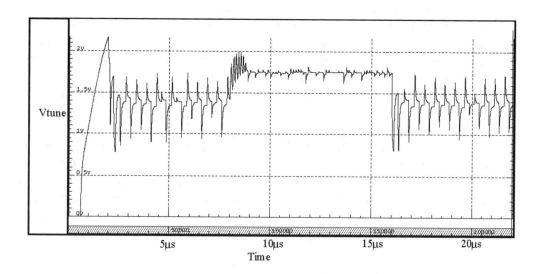

Figure 8.6 Power-on acquisition with 20-nH inductance on the ground and supply leads.

accentuates the voltage drop that this would cause. This can cause the loop not to lock. Other simulations use ideal supplies and would not show this sensitivity.

After the initial power-on acquisition, Figure 8.7 shows the power-supply-voltage step response from 4.5V to 5.5V at 6 μs with a feedback divide ratio of 8 and an input frequency of 10 MHz. Other simulators assume an ideal supply and would not show this sensitivity to supply voltage. These transients occur all the time since more functions are being integrated on an integrated circuit.

Figure 8.8 shows the output-frequency response of the PLL when a 0.5 V_{pp} sine wave at 37 MHz is injected into the power supply of the PLL. The input frequency to the PLL is 12 MHz, and the feedback divider is at 8. The sine wave

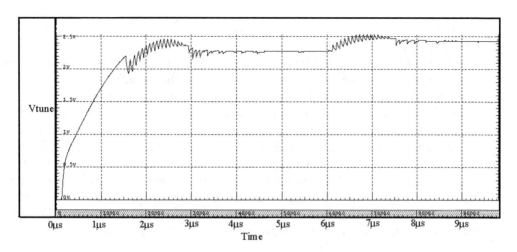

Figure 8.7 Power-supply-voltage step response from 4.5V to 5.5V.

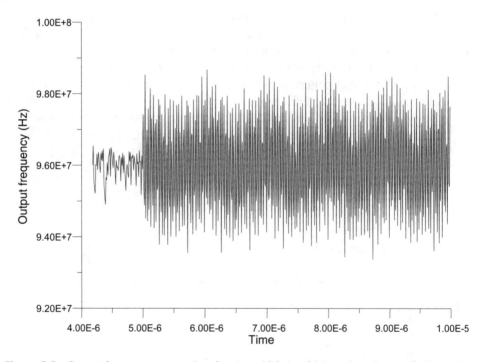

Figure 8.8 Output frequency versus time for sinusoidal signal injected on the supply line at 5 μs.

is injected after 5 μs. A 5.3-MHz peak-to-peak modulation of the output frequency occurs. Consequently, this shows a 10.6-MHz/V sensitivity to the supply line. A voltage regulator on the supply line would reduce the sensitivity. Sinusoidal signals are the most common disturbance on the supply line and can cause the worst-case disturbance to a loop. In general, the worst-case frequency is a little bit greater than the loop bandwidth of the PLL. Furthermore, prime frequencies with the output frequency need to be used because the PLL can be injection locked and look like there is no disturbance at all. Unusually low sensitivity should be checked by doing a couple of different injection frequencies to identify the injection-locked condition.

Figure 8.9 shows the control-voltage response for powering down at 8 μs and back up at 12 μs. The control voltage is held by the capacitor in the loop filter until power is returned. Other simulation methods (e.g., behavioral) would not show the capacitor holding the control voltage when the power was turned off.

Clearly, the above simulations show the many conditions that can be tested using transistor-level simulation. Supply-voltage sensitivity responses, pulse feed-through levels, slew rate, powering up, powering down, and loop recovery from supply-rail conditions are unique conditions tested by transistor-level SPICE simulations. The biggest disadvantage of transistor-level simulation is the amount of computer time that it can take to get a result. Depending on the circuitry, simulator, and computer, some of these simulations can take several weeks.

8.2 Behavioral Modeling of PLL with PSPICE

The goal of the behavioral model is to significantly improve computing time with a minimum loss of accuracy. Consequently, this section presents a behavioral model that achieves these goals. The advantages and disadvantages of the model are discussed, and an example simulation is done to show its performance. Finally, an example of numerical errors is shown to give some insight into the types of errors that can occur with a behavioral model.

Most of the components in a PLL can be analyzed on PSPICE with a behavioral model. These components can be classified under the major categories of control systems, phase detectors, oscillators, and frequency translators:

- Control systems:
  - Loop compensation filters:
    - Passive;
    - Active;
  - Notch filters;

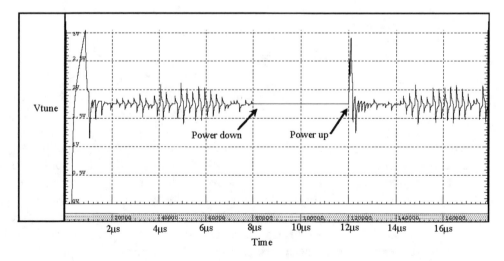

Figure 8.9 Control-voltage response for powering down at 8 μs and back up at 12 μs.

- Phase detectors:
  - Mixers;
  - Phase/frequency detectors;
  - Sampling phase detectors;
- Oscillators:
  - Reference oscillators;
  - VCOs;
- Frequency translators:
  - Mixers;
  - Multipliers;
  - Dividers.

Let's look at a simplified model of the overall PLL control system to try to get execution improvement. The behavioral model to analyze the control systems of a PLL consists of several key components:

- Simplified phase detector;
- Compensation model;
- Simplified VCO;
- Simplified divider and sampling delay.

The simplified phase detector, VCO, and divider are used to eliminate clock edges. Eliminating edges converts the PLL into a voltage regulator where megahertz are mapped to megavolts. This conversion significantly improves computing time with a minimum loss of accuracy. Computing time is improved because clock rising and falling edges of 100 ps require time-sample steps of less than 1 ps for the simulation. Consequently, to simulate a response of 1 ms requires a minimum of 1 ms/1 ps (1E9) iterations. Eliminating the clock edges increases the step size to 100 ns, which reduces the number of iterations to 1 ms/100 ns (1E4). In addition, eliminating the clock edges allows SPICE ac analysis to be performed, which computes Bode plots for open-loop, closed-loop, and error transfer functions. Figure 8.10 shows a graphical depiction of a PSPICE model for a PLL.

This model has several advantages and disadvantages. First, closed and open loops can be analyzed in one run. Next, the evaluation version of PSPICE (low-cost software) can be used to do the analysis. Next, the hierarchal subcircuit connection allows flexibility in modeling a variety of PLLs. Finally, eliminating clock edges by mapping frequency to voltage significantly decreases execution time of the program for transient analysis. This model also has some disadvantages. First, frequency is not directly displayed. The simplicity of the models can detract from the accuracy of the time-domain response.

Figure 8.10 shows the circuit detail for the phase detector model, the VCO model, and the divider-and-sampling-delay model. The phase detector model has a voltage-controlled voltage source, which is used in a polynomial mode to provide a comparison feedback path of the output with the input and with unity gain. Then, a voltage-controlled current source senses the output voltage of the summation loop and is used to multiply the resulting error comparison by the phase detector gain

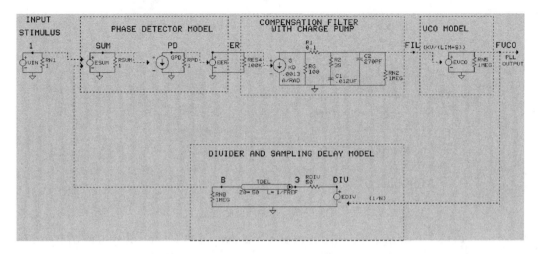

Figure 8.10   Schematic of behavioral model for PSPICE.

constant. A final voltage-controlled voltage source senses the voltage of the phase detector gain circuit and drives the input of the loop-compensation filter.

The VCO model uses a voltage-controlled voltage source to sense the output voltage of the compensation filter circuit. The La Place mode of the voltage-controlled voltage source is used to model the VCO gain constant and integral transfer function. The LIM variable in the La Place transform function prevents the numerical integration from blowing up. The inclusion of the LIM variable puts a pole at a very low frequency to model the ideal integrator.

The frequency divider is modeled by a voltage-controlled voltage source that senses the output voltage of the VCO circuit. The output is then multiplied by the divide ratio used in the La Place mode of the voltage-controlled voltage source. The divider voltage source provides the source for a transmission line that models sampling delay. The length of the transmission line models the delay-line effect of sampling. Finally, the output of the delay line is fed back to the input from the voltage sense of the voltage-controlled voltage source in the phase detector model.

The compensation model uses a charge pump. The charge pump is modeled as an ideal current source that drives a passive RC network. The variable ZOTA puts a finite driving impedance to the current source to model the charge pump more accurately. Otherwise, the charge pump would have infinite gain.

8.2.1   Example Behavioral Model of 270-MHz PLL

Let's use an example 270-MHz PLL to illustrate the modeling and analysis capabilities of a behavioral model. Table 8.3 lists the parameters for the example PLL. Obtaining a switching speed of less than 50 μs with 100-kHz frequency steps is the main goal of the example. The accompanying CD has the PSPICE listing for the control systems. The error tolerances had to be increased by a factor of 10,000 in order to get convergence. Parameter variables are used for easy direct substitution into the model.

Let's study the example PLL by calculating the open-loop-gain response and time response. Figure 8.11 shows the open-loop gain of the example 270-MHz

Table 8.3   Parameters for Example Behavioral Model 270-MHz PLL

Design Parameter	Value
Output frequency (f_o)	270 MHz
Phase detector gain (K_d)	0.4 V/rad
VCO gain (K_v)	81.6E6 rad/s/V
Reference frequency (f_{ref})	1,000 kHz
Switching time (t_{sw})	<50 μs
Closed-loop bandwidth (BW)	40 kHz
Divide ratio (n_{mf})	270

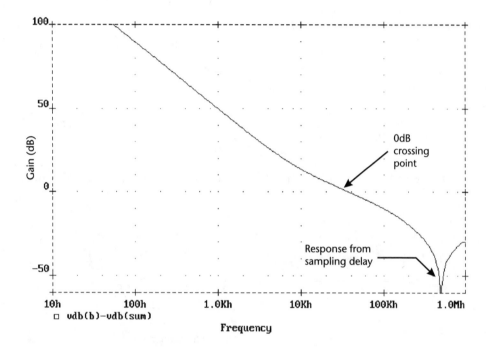

Figure 8.11   Open-loop-gain plot of the example behavioral model 270-MHz PLL.

PLL. The y-axis is the magnitude in decibels, and the x-axis is frequency in hertz. This PSPICE calculation shows a 40-kHz 0-dB crossing point and a zero response break at 8 kHz.

Figure 8.12 shows the open-loop phase response of the example 270-MHz PLL. The y-axis is the phase margin in degrees, and the x-axis is frequency in hertz. This PSPICE calculation shows a 65° phase margin at the 40-kHz 0-dB crossing point and a resonance at 500 kHz because of the sampling-delay effect.

Figure 8.13 shows the closed-loop and error responses. The y-axis shows magnitude in decibels, and the x-axis shows frequency in hertz. The closed-loop response shows the frequency multiplication of 270 in decibels (48.6 dB). The error response shows a 20-dB/dec response from 40 to 8 kHz and goes to a 40-dB/dec response at control frequencies less than 8 kHz.

Figure 8.14 shows the PSPICE calculated time response for an input-frequency step from 1 to 1.1 MHz. Consequently, the output frequency steps from 270 to

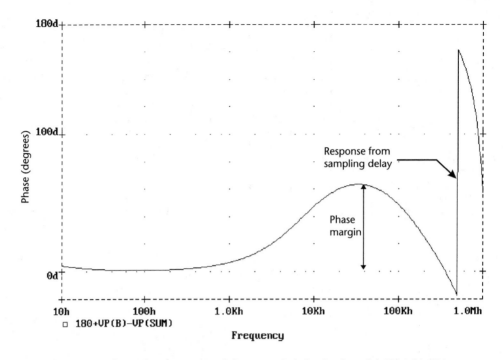

Figure 8.12   Open-loop phase response of the example behavioral model 270-MHz PLL.

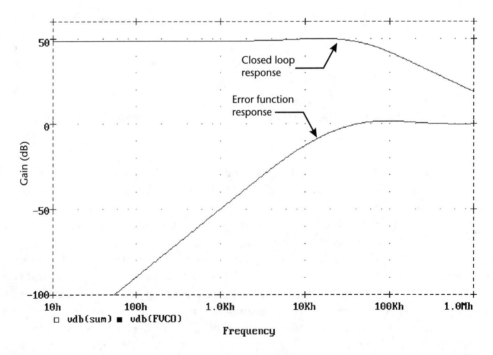

Figure 8.13   Closed-loop and error-function responses of the example behavioral model 270-MHz PLL.

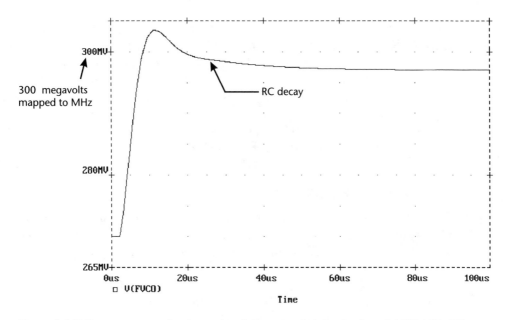

Figure 8.14   Frequency-step time response of the example behavioral model 270-MHz PLL.

297 MHz. Notice, the y-axis voltage at the output represents frequency in this model. Consequently, 270 MHz represents 270 MV in this figure. The 10% to 90% rise time is approximately 8 μs. The overshoot of the response is 2.4% (304 MHz/297 MHz). Because the clock edges are not used in this model, the execution time is in seconds.

Figure 8.15 shows the classical calculation of time response from (3.142). Figure 8.15 shows a 10% to 90% rise time of 8 μs. Consequently, the rise times are in close agreement. The figure also shows an overshoot of 1.7% (302/297).

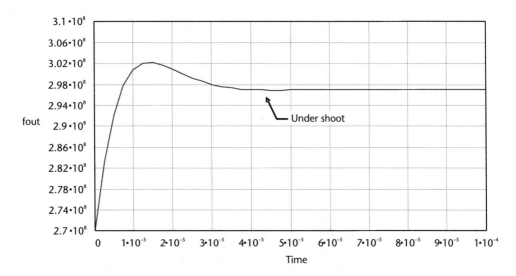

Figure 8.15   Classical frequency step response of the example 270-MHz PLL.

The overshoot shows a discrepancy between the classical settling-time response and the PSPICE settling-time response. In addition, the PSPICE response shows an RC time constant droop in the settling response, and the classical response shows a slight undershoot.

Other behavioral models (e.g., SIMULINK) use clock edges [1], and execution time significantly increases. It remains to be seen if this time penalty improves the accuracy of the response. In addition, frequency-domain responses of the loop transfer functions cannot be directly calculated.

8.2.2 Model for Error Tracking

A behavioral model of error tracking can show that active internal compensation allows better error tracking than charge pump compensation. An example design comparison between charge pump internal compensation and type 2 active compensation shows the model.

Table 8.4 lists the example design parameters. The 480-MHz reference frequency determines the limit for the loop bandwidth because of sampling effects. Consequently, a 1.5-MHz loop bandwidth was chosen because it is far enough away to minimize sampling effects.

Table 8.5 lists the charge pump solution and the type 2 active-compensation solution. For charge pump compensation, charge pump gain and C_1 are used to adjust the bandwidth, and R_2 adjusts phase margin. For type 2 active-compensation phase detector gain, the ratio of R_2/R_1 and C_1 adjusts bandwidth, and R_2 adjusts phase margin.

Next, error tracking for active compensation will be compared with the charge pump compensation. Figure 8.16 shows ac SPICE simulation of the model to compute the error transfer function for the charge pump and active compensation. This figure shows a 10–20-dB lower error for active compensation. This is due to the lower gain for the charge pump when compared to an operational amplifier.

Table 8.4 Example PLL Design Parameters

Design Parameter	Value
Reference frequency	480 MHz
K_v	5,000E6 rad/s/V
N	1
Loop bandwidth	1.5 MHz
Damping factor	0.8

Table 8.5 Charge Pump and Type 2 Active-Compensation Solutions to Example Design

Component	Charge Pump	Type 2 Active Compensation
K_d	1.0 mA/rad	0.4 V/rad
R_1	0	1.2 Meg
R_2	10	10k
C_1	0.01 μF	60 pF
C_2	1,000 pF	0

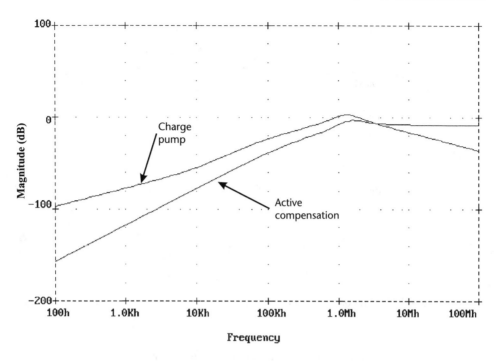

Figure 8.16 Error transfer function for the charge pump and active compensation.

A parallel 100Ω resistor with the current source was added to model the reduced gain in the charge pump. This accounts for the change in slope at 10 kHz. The difference in control-system response shows up in the amount of phase-noise and error tracking. To demonstrate further the error-tracking effect on loop performance, a transient SPICE simulation with a sweeping input frequency (triangle wave) will be used.

The frequency of the triangle wave in the model was increased until a significant difference in tracking occurred. Figure 8.17 shows the results for charge pump compensation and active compensation with the swept-input frequency. Active compensation tracks the triangular modulation more closely than charge pump compensation. The higher tracking error of the charge pump shows up as a delay in following the waveform, as shown in the figure. Closer tracking of the input modulation means the loop will have higher noise immunity. A transistor-level swept-frequency transient SPICE simulation would take much longer to simulate and have less accuracy than this behavioral model.

8.2.3 Identifying Numerical Errors

Numerical errors can occur in any simulation. Recognizing the error is important to obtaining reliable data. In PLLs, numerical errors occur because the phase detector takes the difference of two large input numbers. The result of the calculation can be so small that it is less than the numerical resolution of the computer. An example shows the numerical error that can be generated when doing behavioral analysis. Table 8.6 lists the parameters for the example PLL.

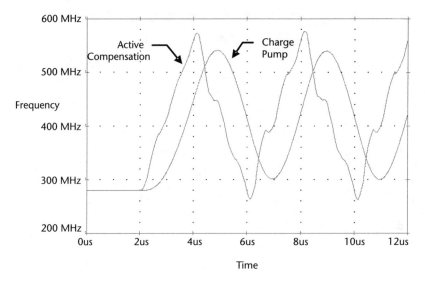

Figure 8.17   Charge pump and active-compensation response to a swept-input frequency.

Table 8.6   Parameters for
Numerical-Error Example PLL

Parameter	Value
K_v	31.4E6 rad/s/V
K_d	0.4 V/rad
Filter type	Passive
R_1	795Ω
R_2	100Ω
C	100 pF

Figure 8.18 shows a 210-kHz overshoot and a 20-μs settling time. The settling response does not have a classic second-order differential response as shown by the ragged waveform after the first undershoot. Figure 8.19 shows that the classical response has 100-kHz overshoot and settles to a steady-state solution in 2 μs. A comparison shows the behavioral model has a significant difference in overshoot and settling-time response.

Further study into the simulation model shows an unusual tune-voltage waveform. Figure 8.20 shows a 140-V spike on the tune line in the behavioral model. A 140-V spike is unrealistic for a circuit with a 5-V supply. Consequently, this shows that numerical error caused the large spike to occur.

Let's summarize the advantages and disadvantages of this model. The PLL model has the advantages of multiple responses in one run so that we can analyze open-loop gain, closed-loop gain, error response, and notch-filter responses. It has the advantages of fast execution, gives us a time response, and runs on the low-cost evaluation version of PSPICE. The PLL model has the disadvantages of settling-time response inaccuracies; also it maps frequency into voltages, the simple phase detector model will not show loss of lock response, and the phase detector model will not show cycle slips in the time domain. Weighing the advantages and disadvan-

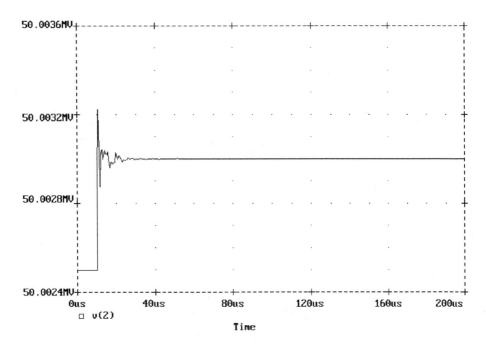

Figure 8.18 Frequency step response of numerical-error example PLL.

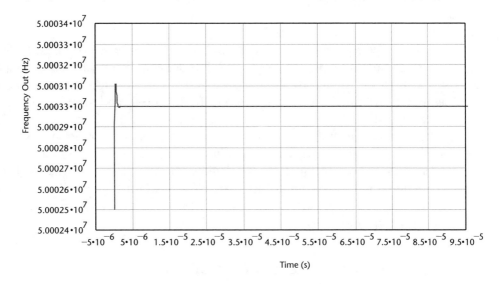

Figure 8.19 Classical frequency step response of numerical-error example PLL.

tages, this model is a useful design tool. The control-systems analysis done by the behavioral model is the preferred method for adjusting and centering the compensation circuit.

8.3 Difference-Equation Modeling of PLLs

Modeling PLLs with equations gives fast computing speed and, with careful modeling, accuracy is not sacrificed. Differential, La Place transforms, Z-transforms,

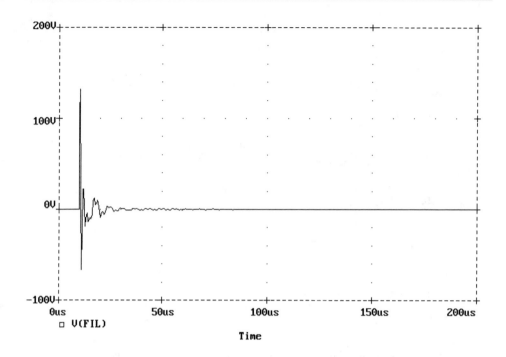

Figure 8.20 Control-voltage response to a frequency step for the numerical-error example PLL.

and difference equations are used to model the performance of PLLs. Difference equations model PLL performance in the time domain. Besides increased computing speed, difference equations allow noise effects, frequency step response, frequency ramp, quantization limits, nonlinear transfer functions, and acquisition responses to be studied.

The inclusion of PLL parameters allows a computer program model to be developed that closely models the response of measured PLL performance [2]. The speed of the difference-equation model allows many iterations of what-if questions to be studied. It is still recommended that SPICE runs be made to verify performance; however, the difference equations should be used to minimize the number of SPICE runs. Consequently, they are time savers.

8.3.1 Review of Difference-Equation Derivation

First, the derivation of the difference equations for a charge pump PLL will be reviewed [3]. Understanding the derivation helps apply the difference equations to other applications. Figure 8.21 shows the equivalent circuit for a charge pump PLL that will be used in the derivation. From Figure 8.21, (8.1) through (8.8) are valid for the entire first cycle, and (8.1) through (8.3) are valid from narrowband FM theory:

$$\theta_i(t) = \theta_i(0) + \omega_i t \tag{8.1}$$

$$\omega_o(t) = \Omega_o + K_v V_c(t) \tag{8.2}$$

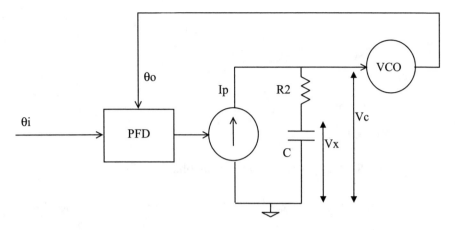

Figure 8.21 Equivalent circuit of charge pump PLL.

$$\theta_o(t) = \theta_o(0) + \Omega_o t + K_v \int_0^t V_c(\tau)\, d\tau \tag{8.3}$$

$$I_p = I_{pk}\, \text{sgn}[\theta_e(0)] \quad 0 < t < T_{pk} \tag{8.4}$$

$$\theta_e = \theta_i - \theta_o \tag{8.5}$$

$$T_{pk} \approx |\theta_e|/\omega_i \tag{8.6}$$

$$V_c(t) = I_p R_2 + V_x \tag{8.7}$$

$$V_x(t) = \frac{1}{C} \int_0^t I_p(\tau)\, d\tau \tag{8.8}$$

where

T_{pk} = pulse width out of the phase detector after comparing reference and VCO rising edges (sec);

K_v = VCO gain normalized to divide ratio of 1 (rad/s/V).

The VCO gain includes the divide ratio ($K_v = K_v/N_{div}$). This normalizes the analysis, and unnormalize it becomes a matter of multiplying any result that includes K_v by $N_{div} \times K_v$.

Ordinary linear network analysis produces (8.9) and (8.10):

$$\theta_o(T_{pk}) = \theta_o(0) + \Omega_o T_{pk} + K_v\left(V_{xo} T_{pk} + I_p R_2 T_{pk} + \frac{I_p T_{pk}^2}{2C}\right) \tag{8.9}$$

and

$$V_x(T_{pk}) = V_{xo} + \frac{I_p T_{pk}}{C} \qquad (8.10)$$

where

$$V_{xo} = V_x(0);$$

$$V_{xpk} = V_x(T_{pk}).$$

Equation (8.11) gives the VCO output phase at T_2:

$$\theta_o(T_2) = \theta_o(T_{pk}) + \Omega_o(T_2 - T_{pk}) + K_v(T_2 - T_{pk})V_{xp} \qquad (8.11)$$

Substituting from (8.9) into (8.11) produces (8.12):

$$\theta_o(T_2) = \theta_o(0) + \Omega_o T_2 + K_v\left(I_p R_2 T_{pk} - \frac{I_p T_{pk}^2}{2C} + V_{xo}T_2 + \frac{I_p T_{pk} T_2}{C}\right) \qquad (8.12)$$

Substituting $I_p \times T_{pk} = I_p \times \theta_e/\omega_i$ and $T_2 = 2\pi/\omega_i$ into (8.12) produces (8.13):

$$\theta_o(T_2) = \theta_o(0) + \frac{\Omega_o 2\pi}{\omega_i} + \frac{K_v V_{xo} 2\pi}{\omega_i} + \frac{K_v I_p \theta_e}{\omega_i}\left(R_2 - \frac{\theta_e}{2C\omega_i} + \frac{2\pi}{C\omega_i}\right) \qquad (8.13)$$

Dropping the $\theta_e/(2C\omega_i)$ term by assuming small θ_e, $\Delta\Omega = \omega_i - \Omega_o$ and $\theta_e = \theta_i - \theta_o$, produces (8.14):

$$\theta_e(T_2) = \theta_i(0) + \omega_i T_2 - \theta_o(0) - \frac{\Omega_o 2\pi}{\omega_i} - \frac{K_v V_{xo} 2\pi}{\omega_i} + \frac{K_v I_p \theta_e(0)}{\omega_i}\left(R_2 + \frac{2\pi}{C\omega_i}\right) \qquad (8.14)$$

Simplifying (8.14) and substituting into (8.10) produces (8.15) and (8.16):

$$\theta_e(T_2) = \theta_e(0) + \frac{\Delta\Omega 2\pi}{\omega_i} - \frac{K_v V_{xo} 2\pi}{\omega_i} - \frac{K_v I_p \theta_e(0)}{\omega_i}\left(R_2 + \frac{2\pi}{C\omega_i}\right) \qquad (8.15)$$

$$V_x(T_2) = V_{xo} + \frac{I_p \theta_e(0)}{C\omega_i} \qquad (8.16)$$

Equations (8.15) and (8.16) complete the review of the derivation [3, 4].

8.3.2 Extending the Difference Equations for Computer Simulation

Now, let's extend the equations to array index notation, which will make the equations closer to a programmable format. Converting (8.16) to array index format produces (8.17) and (8.18):

$$\theta_e(n+1) = \theta_e(n) + \frac{\Delta\Omega\, 2\pi}{\omega_i} - \frac{K_v I_p\, \theta_e(n)}{\omega_i}\left(R_2 + \frac{2\pi}{C\omega_i}\right) - \frac{K_v 2\pi}{\omega_i} V_x(n)$$

(8.17)

$$V_x(n+1) = V_x(n) + \frac{I_p\, \theta_e(n)}{C\omega_i}$$

(8.18)

Combining (8.17) and (8.18) produces (8.19):

$$\theta_e(n+1) = \theta_e(n) + \underbrace{\frac{\Delta\Omega\, 2\pi}{\omega_i}}_{\substack{\text{Previous}\\\text{Error}}} - \underbrace{\frac{K_v I_p\, \theta_e(n)}{\omega_i}}_{\substack{\text{VCO Offset}\\\text{Frequency}}}\underbrace{\left(R_2 + \frac{2\pi}{C\omega_i}\right)}_{\substack{\text{Proportional}\\\text{Part}}} - \underbrace{\frac{K_v 2\pi}{\omega_i}\left(\sum_{m=0}^{n+1}\frac{I_p}{C\omega_i}\theta_e(m)\right)}_{\substack{\text{Integral}\\\text{Part}}}$$

(8.19)

In (8.19), the first term to the right of the equal sign is the previous error. The second term is the offset frequency for the VCO (output frequency with 0V on the control line) minus the reference frequency. The third term is a proportional response to the phase error. The fourth term is an integral response to the phase error, which is a summation.

8.3.3   Example PLL

An example PLL is presented to show the application of the difference equation and the unique nonlinear conditions that can be solved. Figure 8.22 shows a block diagram of the example PLL. This loop contains an integral and proportional filter that generates currents proportional to the phase error. The output currents of the filters are summed together and are connected to the input of a current-controlled oscillator (CCO). The output of the CCO is at 16.8 MHz in the locked condition and is divided down by 512 to 32 kHz. A phase detector compares the divided-down CCO output with the reference input clock at 32 kHz and generates a phase-error pulse into the loop filter. At power up the loop is in a wide-loop-bandwidth mode (approximately 4 kHz) for a few milliseconds and switches to a narrow-loop-bandwidth (approximately 300 Hz) mode after lock is achieved.

Redoing the derivation of the difference equation for the example PLL produces (8.20), which is the linear difference equation:

$$\theta_e(n+1) = \theta_e(n) + \frac{2\pi(\omega_i - \omega_b)}{\omega_i} - K_{cco}\,\theta_e(n)\frac{K_p}{f_i} - \frac{2\pi K_{cco}}{\omega_i}\sum_{m=0}^{n+1}\frac{K_i}{f_i}\theta_e(m)$$

(8.20)

where

K_p = proportional gain (A/rad);

K_i = integral gain (A/rad-s);

K_{cco} = gain of CCO (rad/s/A).

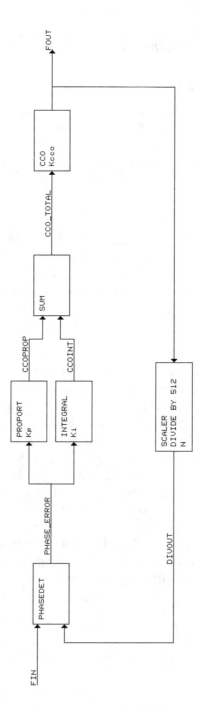

Figure 8.22 Block diagram of example PLL.

Let's study the proportional and integral filters to understand and determine the effects of the proportional and integral gain on the performance of the loop.

Figure 8.23 shows waveforms for the ideal integral filter and the example PLL integral filter. The integral filter does coarse adjustments to the output frequency of the loop by controlling a digital-to-analog converter (DAC) that generates current to change the frequency of the CCO. The top two waveforms show a typical phase relationship between the reference input oscillator and the divided-down CCO. Ideally, the amplitude out of the phase detector would be proportional to the phase error, as shown by the third waveform; however, the phase detector output, as shown in the fourth waveform, produces a pulse that has a pulse width proportional to phase error. In the fifth waveform, an ideal integral filter would integrate the ideal phase detector output for the full input period, which produces a ramped current output to $(TK_i\theta_e)$ for the ideal integral filter; however, the example PLL integrates only for the pulse width (T_{pk}) out of the phase detector and holds the value until the next period (T). The output current is determined by the number of clock pulses inside the phase detector output pulse times LSB size of the current DAC [5].

For example, given a feedback divide ratio of 512 and using the CCO clock divided by 4 to count the pulse size t_p produces 128 clock pulses in 2π of the input period. Multiplying the clock count by a 1-μa LSB size gives 128 μa of current in 2π. To equate the actual circuit with the ideal and normalize to 2π, the 128 μa is divided by 2π and the input period T (1/32 kHz = 30.5 μs) to produce the integral gain (slope) of 0.65 A/rad-s. Multiplying the integral gain by the input period and the phase error in radians produces the output current of the integral filter for the next period.

Figure 8.24 shows waveforms for the ideal proportional filter and the example PLL proportional filter. The proportional filter does fine adjustments to the loop's output frequency. The top four waveforms are the same as shown in Figure 8.23.

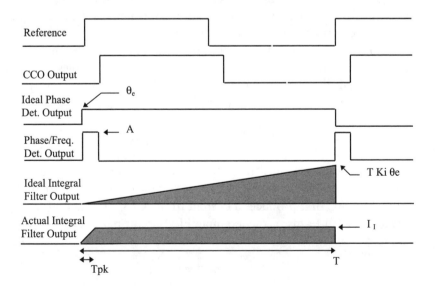

Figure 8.23   Integral gain waveforms.

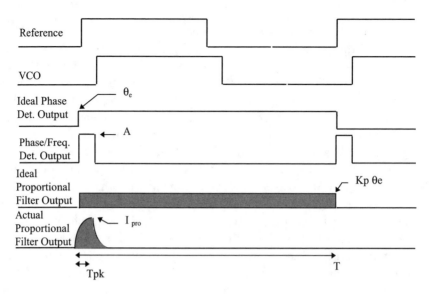

Figure 8.24 Proportional gain waveforms.

The last waveform shows the current output of the proportional filter for the example PLL, which is RC filtered to minimize the jitter of the CCO. The RC filter reduces jitter by reducing the peak for small phase errors and spreading the change in phase over a longer reference time period. The area of the waveform computes the phase adjustment to the loop $(K_p\,\theta_e T \approx I_{pro}T_{pk})$. Since $T_{pk} = \theta_e/\omega_i$ and $2\pi\,1/T = \omega_i$, then, by substitution, K_p equals $(I_{pro}/(2\pi$ in μa/rad)$. K_p also equals the phase detector slope after the charge pump. The average current of the waveform computes the frequency change $(I_{pro}T_{pk}/T)$. For example, a peak phase detector output current of 100 μa, a reference frequency of 32 kHz, and a phase detector output pulse width of 0.5 μs produces a proportional gain of 15.9 μa/rad, a phase change of 5E–11 rad, and an average output current of 1.6 μa [5].

Let's make the example PLL equation equivalent to the charge pump equation. Then, we can relate the example solutions back to the more familiar charge pump equation. Comparing the new (8.20) with the charge pump equation [(8.19)] produces equations that can relate charge pump variables to the integral and proportional-filter parameters in the example PLL. Comparing equation produces (8.21) for the integral gain in terms of charge pump parameters:

$$K_I = I_p/(2\pi C) \tag{8.21}$$

The integral gain measures the phase error out of the phase detector with a clock divided down from the output of the CCO. The number of clocks counted control a DAC that produces a current to the CCO. Equation (8.22) computes the integral gain in terms of the example PLL circuit parameters:

$$K_i = \frac{I_{step}N_{pdclk}}{2\pi\left(\dfrac{1}{f_i}\right)} \tag{8.22}$$

where

I_{step} = LSB DAC current (A);

N_{pdclk} = number of integral-filter clocks in one input period.

To make the example PLL parameters equivalent to the charge pump PLL parameters, set $I_p = I_{step} \times N_{pdclk}$ (e.g., $128 \times 0.5 \ \mu a = 64 \ \mu a$) and $C = 1/f_i$ (e.g., $1/32 \ kHz = 32 \ \mu F$).

The proportional gain produces an RC-shaped current pulse from the output pulse of the phase detector. Figure 8.25 shows the current waveform that is produced from a 500-ns pulse out of the phase detector. The average current of the waveform is also shown in the figure. The average current is the change in frequency that the waveform produces at the output of the CCO.

Comparing equations produces (8.23) for the proportional gain in terms of charge pump parameters with $C = 1/f_i$:

$$K_p = \frac{I_p}{2\pi}(R_2 + 1) \tag{8.23}$$

Equation (8.24) computes the proportional gain in terms of the example PLL parameters:

$$K_p = \frac{I_{pro}}{2\pi} \tag{8.24}$$

where

I_{pro} = maximum current out of proportional-filter charge pump (A).

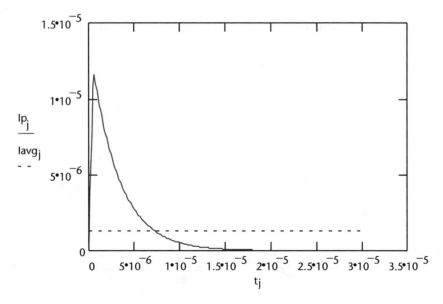

Figure 8.25 Proportional output current and average proportional output current from a detected 500-ns phase-error output.

Setting $R_2 = 1$ and $I_{pro} = I_p/2$ produces (8.25) for the proportional gain in terms of the charge pump PLL:

$$K_p = \frac{I_p}{2\pi} (2) \tag{8.25}$$

Equation (8.26) produces a ratio of integral and proportional gain current so that one I_p variable can be used:

$$K_{ip} = \frac{I_{step} N_{pdclk}}{I_{pro}} \tag{8.26}$$

For example, an integral gain current of 128-μa proportional gain current of 20 μa produces a ratio of 6.4 (128 μa/20 μa). From (8.26), (8.22) can now be rewritten in terms of I_{pro} as shown in (8.27):

$$K_i = \frac{I_{pro} K_{ip}}{2\pi\left(\frac{1}{f_i}\right)} \tag{8.27}$$

Equation (8.28) shows the example PLL difference equations in terms of the equivalent charge pump PLL term with $C = 1/f_i$, $I_p = I_{pro}/2$, and $R_2 = 1$:

$$\theta_e(n+1) = \theta_e(n) + \frac{2\pi\Delta\Omega}{\omega_i} - K_{cco}\theta_e(n)\frac{I_p}{\omega_i}(2) - \frac{2\pi K_{cco}}{\omega_i}\sum_{m=0}^{n+1}\theta_e(n)\frac{I_p K_{ip}}{\omega_i C} \tag{8.28}$$

Next, the example PLL integral and proportional parameters are related to the servo terminology for a transfer function with natural frequency and damping factors. Equations (8.29) and (8.30) compute the relationship:

$$\omega_n = \sqrt{K_{cco}K_i} \tag{8.29}$$

$$\zeta = \frac{\frac{K_p}{K_i}}{2}\omega_n \tag{8.30}$$

Equations (8.31) and (8.32) show the relationship of the charge pump PLL to the servo terminology for a transfer function with natural frequency and damping factor:

$$\omega_n = \sqrt{K_{vco}\frac{I_p}{2\pi C}} \tag{8.31}$$

$$\zeta = \frac{\left(CR_2 + \frac{2\pi}{\omega_i}\right)\omega_n}{2} \tag{8.32}$$

8.3.4 Unique Nonlinear Conditions Simulated by the Difference Equation

The difference equation can be modified to study several nonlinear conditions in a PLL with fast computing speed. The effects of amplitude quantization, phase wrapping, reacquisition of lock, drift on jitter, and negative DAC steps will be studied.

8.3.4.1 Amplitude Quantization

For instance, amplitude quantization can be added to the equation. In the example PLL, the integral filter uses a DAC to generate current into the CCO. The amplitude quantization is added to the difference equation by quantizing the current output of the integral filter in integer steps. Table 8.7 shows the design requirements for a high-gain and low-gain mode. The high-gain mode is used to lock the PLL, and the low-gain mode is initiated after lock. Table 8.8 shows the given values (the first six columns) and the calculated values (the last four columns) for the example PLL. Figure 8.26 shows the results of quantizing the amplitude. Current steps result

Table 8.7 Design Requirements for the Two Modes of Operation for the Example PLL

Low Gain (Narrow Bandwidth) f_n	Damping	High Gain (Wide Bandwidth) f_n	Damping
50–100 Hz	>3	4–8 kHz	>0.2

Table 8.8 Parameters for Frequency Step Response of Narrow Bandwidth Mode in Example PLL

I_{step}	N_{pdclk}	f_i	N_{div}	K_{cco}	I_p	K_i	K_p	f_n	Damping
500 na	1/32	32 kHz	512	7.8E11 rad/s/A	22 μa	81 μa/rad/s	3.18 μa/rad	56 Hz	>3

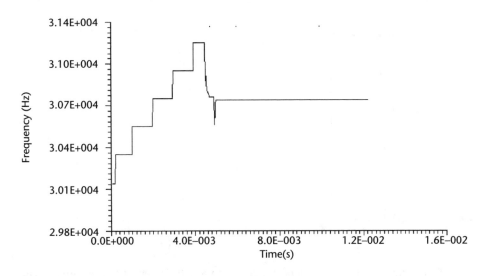

Figure 8.26 Difference-equation simulation to 600-Hz frequency step.

in discrete frequency steps in the transient solution. The equivalent charge pump PLL parameter values I_p = 22 μa, K_v = 7.8E11 rad/s/V, C = 31 μF, K_{ip} = 6E–4, and N_{div} = 512 for the simulation describe an equivalent PLL transfer function in servo terminology with a natural frequency of 74 Hz and damping greater than 3. A C of 31 μF would be difficult to integrate because of the large area that it would take. The area saved is the advantage of the digital part of the example PLL.

In Figure 8.26, a 600-Hz frequency step is large enough to toggle several DAC steps. The loop settles to a final result in 6 ms. SPICE simulations for this circuit at the transistor level took 2 days for each millisecond of computed time. The difference-equation solution took a few seconds.

Even though the integral filter is quantized, the proportional-filter output remains linear. Figure 8.27 shows that a smaller step size of 200-Hz change of the 32-kHz reference frequency does not toggle a DAC step. This response is only the proportional filter, which is a type 1 loop with an 8-kHz loop bandwidth. This figure shows a 240-μs switching-time response to switch from 118.9 to 119.6 ns at the output divided by two periods (8.4-MHz output). This response was obtained by modeling the multiplying effect of the CCO (256) on the feedback frequency to give an output at 8.4 MHz.

8.3.4.2 Phase Wrapping

Phase wrapping the output of the phase detector is another nonlinear condition that can be modeled in the difference equation. Phase wrapping can be added to the difference equation by conditionally checking for phase error greater than 2π or less than -2π, then subtracting 2π or adding 2π to generate the phase wrap. This models the performance of the phase/frequency detector as shown in Figures 8.28 and 8.29. Figure 8.28 shows the VCO frequency (47.5 MHz) less than the

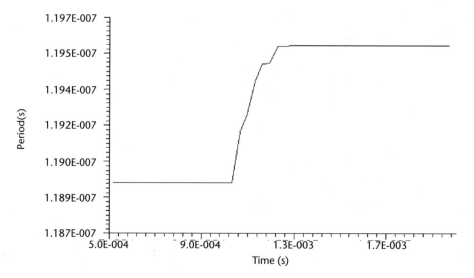

Figure 8.27 Difference-equation simulation to a 230-Hz frequency step, small phase error, and low gain.

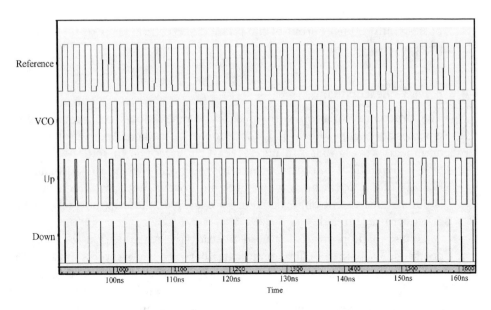

Figure 8.28 Phase/frequency detector phase-wrapping VCO less than reference.

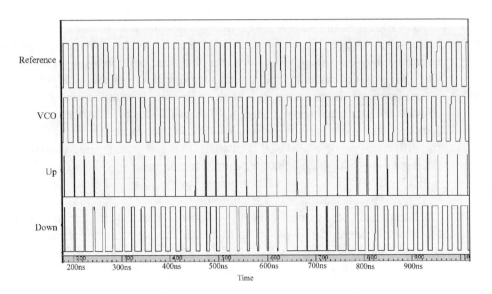

Figure 8.29 Phase/frequency detector phase-wrapping VCO frequency greater than the reference frequency.

reference frequency (50 MHz). High VCO and reference frequencies were selected to speed up SPICE simulation time. At time 1,350 ns, a phase wrap occurs. The maximum up-pulse-width (20 ns) output of the phase detector rolls over to the minimum up pulse width (<1 ns).

Figure 8.29 shows a VCO frequency (50 MHz) greater than the reference frequency (47.5 MHz). At time 630 ns, a phase wrap occurs. The maximum down-pulse-width (20 ns) output of the phase detector rolls over to the minimum down pulse width (<1 ns).

8.3.4.3 Reacquisition

The difference equation with phase wrapping allows the acquisition response, which is another nonlinear PLL condition to be studied. Acquisition is an extremely nonlinear process that can take several milliseconds to settle out, which makes SPICE simulations impractical. Acquisition occurs when a disturbance jumps the PLL out of lock to the point where cycle slips occur. Without a frequency acquisition circuit (exclusive-OR phase detector for example), the loop must rely on a nonlinear process to regain lock.

Phase-plane portraits are used to study the reacquisition response [6]. In the portraits, a separatrix curve demarcates between cycle-slip response and no cycle-slip response. A response with an initial condition inside the separatrix will not have a cycle slip and will follow the spiral track and lock at zero phase difference and zero frequency difference. A response with an initial condition outside the separatrix cycle slips in a damped sinusoidal waveform to the next 2π phase plane that is identical to the original plot but takes a slightly different path. After many cycle slips, the response path will cross the separatrix and follow a spiral track into lock.

Let's look at an example PLL with a 56-Hz natural frequency and a damping factor greater than 3 for this case. Table 8.9 shows the given values (the first six columns) and the calculated values (the last four columns) for the example PLL. The charge pump current is increased to make it find lock faster. Figure 8.30 shows a simulation of the example PLL with a difference equation that has many cycle slips before lock is achieved. Figure 8.30 shows 20 ms for relock time. The response required 40 ms of simulation time, which took 1 minute of computing time using the difference equation. SPICE transistor-level computing time is estimated to be 80 days. The equivalent charge pump PLL parameter values $I_p = 40$ μa, $K_v = 7.8E11$ rad/s/V, $C = 31$ μF, $K_{ip} = 6E{-}4$, and $N_{div} = 512$ describe the equivalent PLL transfer function in servo terminology.

Figure 8.31 shows measured data that compares well with the simulation in Figure 8.30. Here, the relock time is 40 ms for a 1-kHz initial disturbance. The differences between Figures 8.30 and 8.31 are due to mismatching of the initial conditions of phase and phase velocity.

8.3.4.4 Phase Wrapping and Quantization Effects at Wide Loop Bandwidths

The difference equation with phase wrapping and amplitude quantization gives a more accurate model of the actual circuit than the difference equation without

Table 8.9 Parameters for Reacquisition Simulation of Example PLL

I_{step}	N_{pdclk}	f_i	N_{div}	K_{cco}	I_p	K_i	K_p	f_n	Damping
500 na	1/32	32 kHz	512	7.8E11 rad/s/A	40 μa	81 μa/rad/s	6.3 μa/rad	56 Hz	>3

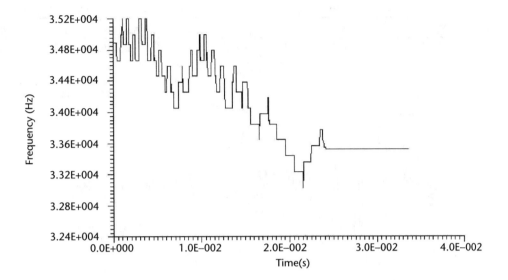

Figure 8.30   Outside the separatrix, difference-equation simulation.

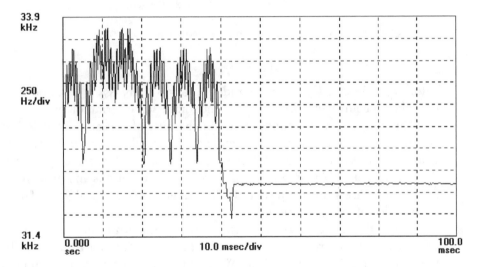

Figure 8.31   Outside the separatrix, measured response.

phase wrapping and amplitude quantization. The improved accuracy can be seen by studying the dramatic effect on PLL performance for wide loop bandwidths close to the reference frequency.

Let's look at an example PLL with a natural frequency of 8 kHz and a 0.13 damping factor for this case. A natural frequency of 8 kHz puts the loop close to oscillation because it is close to the reference frequency of 32 kHz. Table 8.10 shows the given values (the first six columns) and the calculated values (the last four columns) for the example PLL. Figure 8.32 shows the difference equation without phase-wrapping calculations. The equivalent charge pump PLL parameter

Table 8.10 Parameters for Phase-Wrapping and Amplitude-Quantization Simulations of the Example PLL

I_{step}	N_{pdclk}	f_i	N_{div}	K_{cco}	I_p	K_i	K_p	f_n	Damping
1.5 μa	128	32 kHz	512	1.3E12 rad/s/A	34 μa	0.33 a/rad/s	6.3 μa/rad	8 kHz	0.13

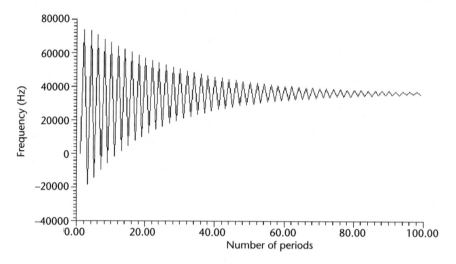

Figure 8.32 Unstable response from sampling-delay effects.

values $I_p = 34$ μa, $K_v = 1.3E12$ rad/s/V, $C = 31$ μF, $K_{ip} = 3.0$, and $N_{div} = 512$ describe an equivalent PLL transfer function in servo terminology.

This almost unstable condition occurs because of the sampling-delay effect. To achieve this result, the nominal integral gain was tripled from 0.3 to 1. The integral gain of 1 corresponds to a 1.5-μa-per-DAC step, which is three times larger than the nominal-case DAC step value. Consequently, conditions and process must shift in the same direction to cause a 200% change in the integral gain. This analysis shows a lot of design margin.

Next, adding the effects of phase wrapping, quantization of integral gain, and windowing of the proportional gain into the linear difference equation illustrates these modeling effects on the PLL. Windowing of the proportional gain minimizes the jitter of the loop by putting a limit on the pulse width out of the phase detector. In the program for the difference equation, a window is created by testing for error conditions greater than the window and setting the phase detector error output equal to the windowed value. For the example PLL, the window was set to 500-ns up and 500-ns down pulse widths.

Figure 8.33 shows that adding phase wrap, quantization of integral gain, and windowing of the proportional gain significantly reduces the damping factor and finds a stable operating point within 20 clock periods, while the linear simulation in Figure 8.32 continues to oscillate past 100 reference-clock periods. Here, the loop is operating at a point near the sampling frequency (unstable operation). Consequently, the effects of sampling delay are minimized because of phase wrapping and quantization.

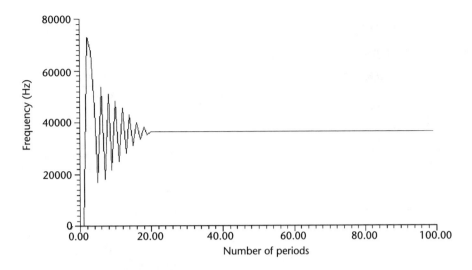

Figure 8.33 Nonlinear difference-equation response.

8.3.4.5 Negative DAC Step

Earlier we quantized DAC steps to equal current step sizes. Nonlinear effects of the DAC can also be added to the difference equation to study the effects on the response of the PLL. For example, the effect of a nonmonotonic DAC on the response of the loop can be studied. The nonlinearity of the DAC is mapped into a vector that becomes a look-up table in the computer program.

Let's look at an example PLL with a 56-Hz natural frequency and damping factor greater than 3 for this case. Table 8.11 shows the given values (the first six columns) and the calculated values (the last four columns) for the example PLL. The charge pump current is weakened to 10 μa to emphasize the negative DAC step effect. Figure 8.34 shows the calculated negative DAC step response. After the third step, a negative DAC step (−2 LSB) disturbs the normal transient response that was presented earlier in Figure 8.26. The equivalent charge pump PLL parameter values I_p = 10 μa, K_v = 7.8E11 rad/s/V, C = 31 μF, K_{ip} = 1.3E−3, and N_{div} = 512 for the simulation describe an equivalent PLL transfer function in servo terminology. The negative DAC step causes several-hundred-microsecond delay in the settling time.

8.3.4.6 Jitter from Drift

Another nonlinear condition occurs when the DAC steps from a slow drift in voltage or temperature, which causes jitter at the output of the loop. The difference equation can be programmed to model this condition. A ramped offset current is

Table 8.11 Parameters for Negative DAC Step Simulation of the Example PLL

I_{step}	N_{pdclk}	f_i	N_{div}	K_{cco}	I_p	K_i	K_p	f_n	Damping
500 na	1/32	32 kHz	512	7.8E11 rad/s/A	10 μa	81 μa/rad/s	1.6 μa/rad	56 Hz	>3

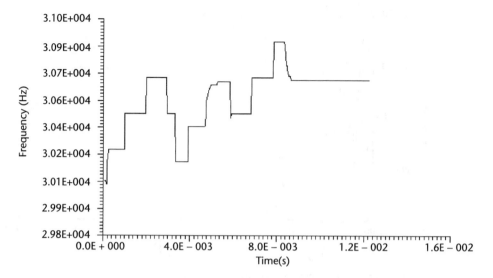

Figure 8.34   Response to 600-Hz frequency step with a –1 DAC step value.

added to the DAC current in order to cause a DAC step. This simulates the conditions of the loop during worst-case operation of the system with a drifting bias current caused by temperature or voltage drift. The DAC has to take a step in order to stay at the same output frequency.

Let's look at an example PLL with a 56-Hz natural frequency and damping factor greater than 3 for this case. Table 8.12 shows the given values (the first six columns) and the calculated values (the last four columns) for the example PLL. Figure 8.35 shows the VCO clock output divided by 61. The step change in the base period at 3.9 ms shows that a DAC step was taken. The offset current was adjusted so that the DAC steps from count 64 to count 63. The DAC step at 64 has a 3-LSB weighting and the proportional gain is at its maximum for a worst-case jitter condition.

The equivalent charge pump PLL parameter values $I_p = 40$ μa, $K_v = 7.8E11$ rad/s/V, $C = 31$ μF, $K_{ip} = 6E-4$, and $N_{div} = 512$ describe the equivalent PLL transfer function in servo terminology. The simulation time of 5 ms took one-quarter minute to execute with the difference equation. SPICE execution time was estimated to be 10 days. Figure 8.35 shows a proportional gain that has 250-ns peak period changes before the 3-LSB DAC step and 0-ns peak period changes after the DAC step.

Figure 8.36 shows the worst-case (largest DAC step and highest charge pump current) peak-to-peak jitter for varying the VCO divide ratio from 2 to 512 in log

Table 8.12   Parameters for DAC Step Jitter Simulation of the Example PLL

I_{step}	N_{pdclk}	f_i	N_{div}	K_{cco}	I_p	K_i	K_p	f_n	Damping
500 na	1/32	32 kHz	512	7.8E11 rad/s/A	40 μa	81 μa/rad/s	6.3 μa/rad	56 Hz	>3

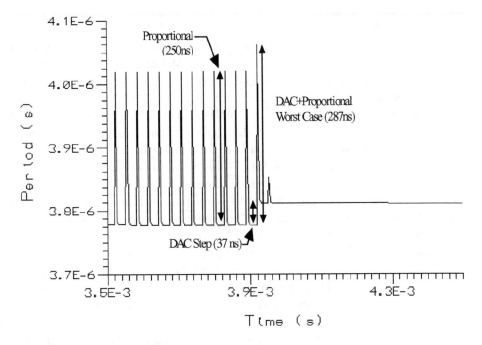

Figure 8.35 Blown-up view of DAC step with worst-case PLL parameters.

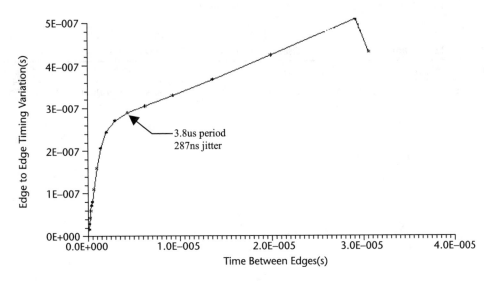

Figure 8.36 Worst-case jitter with a DAC step versus time between edges.

steps. Varying the divide ratio from 2 to 512 for a CCO output period of 59.5 ns varies the time interval between edges from 119 ns to 31.25 μs. The 220-ns peak-to-peak jitter in Figure 8.35 is plotted on Figure 8.36 at the 3.9-μs time between edges point. Execution time with the difference equation took 4 minutes (15 sec × 16 points). SPICE execution would take 160 days (10 days × 16 points).

8.3.4.7 Local Oscillation from Quantization Error

The ratio of coarse-tune, integral-filter, minimum step size to proportional-filter, fine-tune, maximum adjustment range provides another nonlinear condition from quantization that can be modeled with the difference equation. A quantization error occurs when DAC step sizes are larger than the fine-tune range adjustment of the proportional filter. This quantization error is programmed into the difference equation by increasing the DAC LSB size until toggling appears in the simulation.

Let's look at an example PLL with a 56-Hz natural frequency and damping factor greater than 3 for this case. Table 8.13 shows the given values (the first six columns) and the calculated values (the last four columns) for the example PLL. Figure 8.37 shows that, after 1 ms from the initial turn-on condition, the DAC step toggles with a 2-ms period. The 2-ms period occurs because 32 periods of the 32-kHz clock are blanked until a DAC step is allowed. The 32 periods of blanking produce a loop bandwidth of 56 Hz after initial lock is achieved. In Figure 8.37, the DAC minimum step size is greater than the proportional-filter maximum range, which causes the DAC to toggle between two values. This toggling is a local oscillation and increases the output jitter of the PLL.

To generate Figure 8.37, difference-equation execution time for a 4-ms simulation time took about 10 seconds. SPICE execution time would take 8 days. The equivalent charge pump PLL parameter values $I_p = 21\ \mu a$, $K_v = 7.8E11$ rad/s/V, $C = 31\ \mu F$, $K_{ip} = 9.2E{-}4$, and $N_{div} = 512$ for the simulation describe an equivalent PLL transfer function in servo terminology.

Table 8.13 Parameters for Localized Oscillation Simulation of the Example PLL

I_{step}	N_{pdclk}	f_i	N_{div}	K_{cco}	I_p	K_i	K_p	f_n	Damping
500 na	1/32	32 kHz	512	7.8E11 rad/s/A	10 μa	81 μa/rad/s	1.6 μa/rad	56 Hz	>3

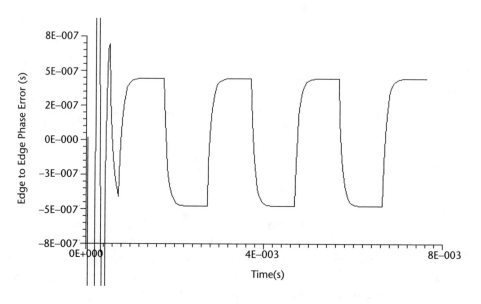

Figure 8.37 Localized oscillation from imbalance between integral and proportional gain.

8.3.4.8   Cycle Slip Reacquisition Time

Using a cycle-slip detection circuit to reacquire lock presents another nonlinear condition that can be modeled by the difference equation. A cycle-slip detector is modeled in the program by using the phase-wrap condition of $\pm 2\pi$ to sense a cycle slip. Frequency-step stimuli of 1, 3, and 10 kHz are large enough to cause cycle slips because the bandwidth of the loop is 300 Hz. Figure 8.38 shows 1-, 3-, and 10-kHz frequency step responses of the PLL reacquisition time with cycle-slip detection. For ± 1 kHz, there was no cycle slip. This small frequency step was used for a reference point for the larger steps. For a +10-kHz step, reacquisition was achieved in 20 ms. Execution time for this plot with the difference equation was 1.5 minutes. SPICE simulation at the transistor level would be 180 days.

Table 8.14 shows a comparison of execution time with simulation type and design requirement. Computers used for the simulations were a SUN Sparc Ultra 10 for level SPICE simulation and a PC 120-MHz Pentium for the other simulations.

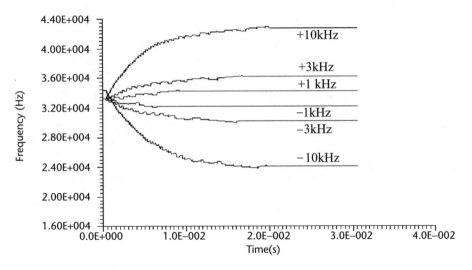

Figure 8.38   Family of curves for reacquisition time with cycle slip.

Table 8.14   Comparison of Execution Time with Simulation Type and Design Requirement

Simulation	Transistor-Level Execution Time/ Simulation Time	Behavior Model Execution Time/ Simulation Time	Difference-Equation Execution Time/ Simulation Time
Open-loop gain	N/A	5–10s/100 MHz	N/A
Power on	6 days/12 ms	5–10s/12 ms	5–10s/12 ms
Frequency step	1 day/2 ms	5–10s/2 ms	5–10s/2 ms
Reacquisition	80 days/40 ms	N/A	60s/40 ms
FM jitter from drift	160 days (5 ms * 16 pts)	N/A	240s/(5 ms * 16 pts)
Local quantization oscillation	8 days/4 ms	N/A	10s/4 ms
Cycle-slip reacquisition	180 days/(6 runs * 30 ms)	N/A	90s/(6 runs * 30 ms)

The leftmost column shows the PLL simulation that needs to be done. The top row shows the type of simulator is broken down into the three types of transistor level, behavioral model, and difference equation. The values in the table (i.e., 6 days, 12 ms) show execution time on the left of the slash and simulation range on the right of the slash. This table can help in making decisions about which simulator to use based on PLL design requirements.

Some general observations can also be made from experience in running simulations. There are a few net list characteristics that slow SPICE simulation time. First, a component inserted on the supply or ground lines, such as a voltage regulator, resistor, transmission line, or inductor, slows execution. Next, a high ratio of VCO frequency to phase detector input frequency slows execution. Finally, a narrow loop bandwidth (<1/1,000 of the reference frequency) slows down execution.

In conclusion, this section showed the versatility and accuracy of difference equations. A derivation of the difference equation was presented. An application to an example PLL was shown. Finally, several unique simulations were presented that demonstrate the accuracy of the difference-equation model, its versatility in computing a solution, and the resulting savings in computing time.

Table 8.15 shows a list of simulators and the types of simulations that can be done. This table, in combination with Table 8.14, provides a method for comparing computer programs to be used in analyzing and simulating PLLs, but it is not intended as an endorsement of one vendor over another. In addition, the accuracy of the programs and the simulations shown are not guaranteed. Please contact the vendors to obtain up-to-date and accurate information. Table 8.15 shows a long

Table 8.15   List of PLL Simulators and the Type Simulation That Is Done

Simulator	Transistor Level	Behavioral with Edges	Behavioral Without Edges	Difference Equation	Vendor
PSPICE	Yes	Yes	Yes	No	OrCAD
HSPICE	Yes	Yes	Yes	No	Synopsys
TI SPICE	Yes	Yes	Yes	No[a, d]	Texas Instruments
C programming language or other languages	—[c]	—	Yes[d]	Yes[d]	—
Powermill (nanosim)[e]	Yes	No	No	No	Synopsys
Saber	—	Yes	—	—	Synopsys
Verilog/VHDL	Yes	Yes	—	—	—
Matlab/Simulink	—	Yes	Yes[b, d]	Yes[b, d]	Mathworks
Mathcad	—	—	Yes[b, d]	Yes[b, d]	Mathsoft
Spectre	Yes	Yes	Yes	—	Cadence
HP EEsof (ADS)	Yes	Yes	—	—	Agilent
Eagleware = PLL =	—	Yes	Yes	—	Agilent

[a]Can compile C program and use in simulator
[b]Numerical errors because not double precision
[c]Blanks mean not evaluated
[d]Have to write code
[e]Faster SPICE simulator

list of simulators, and one can spend a lot of time learning about different simulators that do PLLs without improving accuracy or speed or on simulating characteristics that have not been previously simulated. Combining this table with Table 8.14 helps to sort out which of the three or four simulation tools are needed to make the required PLL simulations. This will help readers save money because they will not have to buy all the simulators on this list, hoping that one of them will solve their PLL problem.

To summarize, this chapter has covered simulation techniques for PLLs. The types of PLL simulators vary from SPICE transistor-level solutions, to behavioral-model solutions, to equation-based solutions. Our discussion of transistor-level simulations showed that many conditions can be simulated. Characteristics of supply-voltage sensitivity responses, pulse feed-through levels, slew rate, powering up, powering down, and loop recovery from supply-rail conditions were demonstrated. Depending on the circuitry, simulator, and computer, the biggest disadvantage is the computer execution time that some of these simulations can take.

Our discussion of the behavioral-level model showed that it can analyze open-loop gain, closed-loop gain, error response, and notch-filter responses. It has the advantages of fast execution, gives a time response, and runs on the low-cost evaluation version of PSPICE. The behavioral model has the disadvantages that it generates settling-time response inaccuracies, it maps frequency into voltages, the simple phase detector model will not show loss of lock response, and the phase detector model will not show cycle slips in the time domain. The control-systems analysis done by the behavioral model is the preferred method for adjusting and centering the compensation circuit. Next, our discussion of the difference-equation-level model showed increased computing speed, simulated noise effects, frequency step response, frequency ramp, quantization limits, nonlinear transfer functions, and acquisition responses. The speed of the behavioral and difference-equation models allowed many iterations of what-if questions to be studied.

The choice of simulation and simulator depends on the speed and accuracy of the results. For instance, accurate SPICE simulations can take several weeks to run when looking at acquisition times. This chapter showed a method to help the reader make an informed choice.

Questions

8.1 For a charge pump PLL with $K_v = 600E6$ rad/s/V, $K_d = 10$ μa/rad, $N = 500$, $F_n = 10$ kHz, and damping factor $= 0.6$, compute the compensation filter values. Then, simulate the transient response with SPICE for a small-signal frequency step response.

8.2 For a charge pump PLL with $K_v = 600E6$ rad/s/V, $K_d = 1$ μa/rad, $N = 4,000$, $F_n = 100$ Hz, and damping factor $= 0.6$, compute the compensation filter values. Then, simulate the transient response with SPICE for a small-signal frequency step response.

8.3 For a charge pump PLL with $C_1 = 10,000$ pF, $C_2 = 200$ pF, $R_1 = 6,000$, $N = 128$, $K_D = 13.5$ μa/rad, $K_V = 175E6$ rad/s/V, and reference frequency of 200 kHz, plot the open-loop gain and phase.

8.4 For a charge pump PLL with $C_1 = 70,000$ pF, $C_2 = 2,000$ pF, $R_1 = 3,000$, $N = 441$, $K_D = 13.5$ μa/rad, $K_V = 175E6$ rad//s/V, and reference frequency of 40 kHz, plot the open-loop gain and phase.

8.5 For a charge pump PLL with $C_1 = 40$ pF, $C_2 = 0.01$ pF, $R_1 = 25$ kΩ, $K_I = (2\pi*1E6*600E3)$, $g_{ma} = 1.5$ ma, $N = 8$, $K_D = 2$ μa/rad, and $K_V = (g_{ma}*K_I)$, plot the open-loop gain and phase.

8.6 For a charge pump PLL with $C_1 = 1,000$ pF, $C_2 = 100$ pF, $R_1 = 800\Omega$, $N = 8$, $K_D = 0.0002$ rad/V, $K_V = 300E6$ rad/s/V, and reference frequency of 20 MHz, plot the open-loop gain and phase.

8.7 For a charge pump PLL with $C_1 = 100$ pF, $C_2 = 40$ pF, $R_1 = 4,000\Omega$, $N = 6$, $K_D = 30$ μa/rad, $K_V = 940E6$ rad/s/V. and a reference frequency of 16 MHz, plot the open-loop gain and phase.

8.8 For a digital PLL with a feedback divide ratio of 24, a 10-MHz input frequency, 4.0 divide ratio for integral-filter clock, 8-bit DAC, an LSB of 1.4 μa, a DAC offset current of 0.136 μa, a charge pump current of 211.0E–6, a VCO gain of 1E11 Hz/A, a natural frequency of 300 kHz, a 0.2 damping factor, and a 1-MHz step frequency, simulate the transient response.

8.9 For a digital PLL with a feedback divide ratio of 40, a 10-MHz input frequency, 4.0 divide ratio for integral-filter clock, 12-bit DAC, an LSB of 24 na, a DAC offset current of 20 μa, a charge pump current of 80 μa, a VCO gain of 5E12 Hz/A, a natural frequency of 1 MHz, a 0.4 damping factor, and a 1-MHz frequency step, compute the missing parameters and simulate the transient response.

References

[1]   Tan, E., "Phase-Locked Loop Macromodels," University of California Berkley, Memo No. UCB/ERL M90/114, December 10, 1990.

[2]   Best, R., *Phase-Locked Loops: Design, Simulation, and Applications*, 3rd ed., New York: McGraw-Hill, 1997, pp. 229–249.

[3]   Gardner, F. M., "Charge Pump Phase Lock Loops," *IEEE Transactions on Communications*, Vol. COM-28, November 1980, pp. 1849–1858.

[4]   Martin, K., and D. Johns, *Analog Integrated Circuit Design*, New York: Wiley Interscience, 1997, pp. 681–688.

[5]   Richard, M., "Clock Module Stability Analysis Report," Quadic Systems Internal Technical Memo, February 21, 1995.

[6]   Gardner, F. M., *Phaselock Techniques*, New York: Wiley Interscience, 1979.

CHAPTER 9

Applications and Extensions

This chapter presents PLL applications and extensions. PLLs are used to generate frequencies, recover a clock from a signal, resynchronize signals, and help convert signals to a digital representation. First, PLLs synthesize frequencies to be used in computers (microprocessors and DSPs), test instruments, land-based communications (cell phones), space-based communications, radar, and electronic-warfare systems. The wide variety of applications with different requirements leads to different synthesizer configurations. We will study the various configurations to help in selecting the best one.

In clock recovery, many systems transmit or receive data without any additional timing reference. To an ever-increasing extent, communication links use digital formats to transmit information synchronously in a continuous uniform pulse stream. We will study clock-recovery techniques because every digital communication link uses a PLL in the reception of this information.

Communication systems, radar systems, data-acquisition systems, and test equipment convert analog signals to digital words for digital processing. Digital processing assumes equally spaced time samples; however, actual samples from analog-to-digital (A/D) converters have slightly unequal time spacing. Phase noise from the oscillator that generates the sampling clock produces this change in time spacing. In digital processing, equally spaced time samples are critical to system performance. The phase noise of the sampling clock affects the dynamic range of the A/D converter by adding noise. We will study the limitations of the conversion process so that adjustments can be made to PLL designs to overcome these limits.

As digital processes reduce the cost of circuit functions and design tools to handle large-scale digital circuits, design in the digital domain becomes easier. Consequently, it has created a demand for all-digital PLLs (ADPLL). Studying ADPLLs will improve our ability to recognize the correspondence between the fundamentals we have studied and structures that look much different.

This variety of design topics will be studied to illustrate solutions to particular design problems. This will familiarize readers with the various design approaches so that they can extrapolate these approaches to their particular solutions.

9.1 Design Trade-Offs in Frequency Generation with PLLs

Synthesizers are being used in computers (microprocessors and DSPs), test instruments, land-based communications (cell phones), space-based communications, radar, and electronic-warfare systems. Each application has a different set of requirements that leads to different synthesizer configurations. Consequently, studying various configurations can help in selecting the best configuration.

Frequency synthesizers have numerous topologies. The choices between topologies lead to numerous solutions to the same frequency-synthesis problem. The number of solutions to frequency generation is analogous to the number of ways that software can be written to solve a problem. There is no right or wrong solution, just a variety of ways to solve the problem. At first, the endless number of solutions seems overwhelming; however, the topologies can be separated into several categories, and the advantages and disadvantages of each can be analyzed. Comparing these advantages and disadvantages with a properly weighted specification for the synthesizer can enable selecting the optimum solution.

9.1.1 Classification

First, we will try to classify the many solutions. Frequency-generation techniques can be divided into two major categories, coherent and incoherent synthesis. Coherent synthesis is derived from one source, while incoherent synthesis is derived from multiple unrelated sources. The category of coherent synthesis can be even further subdivided into direct, indirect, or a hybrid of indirect and direct. Coherent direct synthesis uses a combination of multipliers, mixers, digital logic, and dividers to synthesize various frequencies from one source. Coherent indirect synthesis uses a control-systems feedback approach to stabilize one or many sources to a single source. A coherent indirect frequency synthesizer can be made with a single PLL or multiple PLLs. Hybrids use combinations of direct and indirect in a control-systems-feedback topology. Figure 9.1 organizes the various topologies into a tree that helps identify the nature of the topologies and shows the relationships between them.

The overall identification of the various types now allows us to study frequency synthesis in more detail. First, the various building-block topologies will be discussed. Then, the trade-offs between the topologies will be compared with specifications. This study of trade-offs will show specification priorities that have to be made in order to achieve the optimum synthesizer for the customer. Otherwise,

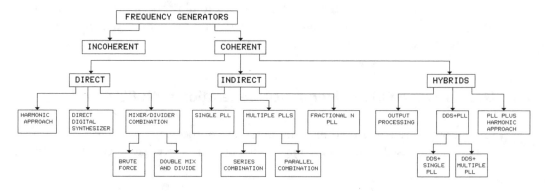

Figure 9.1 Classification of frequency generators.

equal weighting of each specification will make finding an optimum solution difficult.

9.1.2   Direct Synthesis

The study of each topology begins with the direct-synthesis method. This method uses combinations of mixers, multipliers, and dividers to synthesize frequencies [1, p. 7]. In general, this approach has the advantages of being more straightforward, having low phase-noise levels, and having the fastest switching time; however, it has the disadvantages of having high spurious outputs and a lot of components at high frequencies. In addition, increasing the number of frequency increments causes a geometric increase in circuit complexity.

9.1.2.1   Brute-Force Combinations of Mixers and Dividers

The brute-force method is used to generate simultaneous frequencies when an optimum topology is not clear-cut. Combinations of mixers, dividers, and multipliers are used to generate the required simultaneous frequency outputs. Figure 9.2 shows an example of the brute-force method. Branching off of the circuitry at key points in the frequency-generation process allows several simultaneous output frequencies to be generated. This example is from an HP8662 RF synthesizer.

9.1.2.2   Harmonic Synthesis

The harmonic-synthesis approach generates evenly spaced harmonics that are selected by a tunable filter. Unfortunately, this approach relies on the design of a narrowband tunable filter with a large number of poles for selectivity. This filter is costly to build and difficult to implement.

Comb Generator/Tunable Filter
The comb generator/tunable filter approach for harmonic synthesis is shown in Figure 9.3. As shown in the figure, a tunable filter selects the harmonic output of a comb generator. A comb generator/tunable filter has the advantages of low phase noise with high Q and wide tuning range; however, it has the disadvantages of large size, a low operating frequency, spurious signals, and high power consumption.

From studying this configuration, a few characteristics can be stated. First, selecting higher harmonics increases the rejection requirements for the filter. After filtering the harmonics, an amplifier and comparator are needed to convert to square wave. Equation (9.1) mathematically describes the relationship of the output frequency to the input frequency for this topology:

$$f_{syn} = nf_{in} \tag{9.1}$$

Comb Generator/Harmonic Canceller
For lower-frequency operation, a harmonic canceller functions like the comb generator/YIG filter combination does at microwave frequencies. Figure 9.4 shows an example of this circuit.

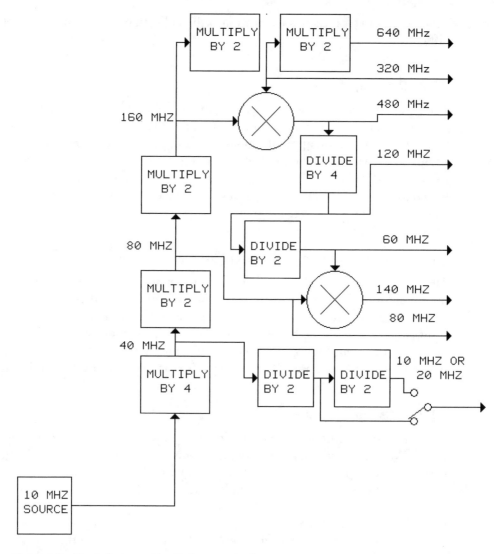

Figure 9.2   Block diagram of brute-force method.

This topology has the advantages of fast switching time and low phase-noise levels; however, it has the disadvantages of generating spurious signals and using a large number of high-frequency components. Equation (9.2) mathematically describes the output frequency for this topology:

$$f_{syn} = f_{vco} - [(f_{vco} - nf_{in}) - f_{in}] \tag{9.2}$$

where

$$f_{vco} = n_M f_{in} \tag{9.3}$$

$$n_M > n - 1 \tag{9.4}$$

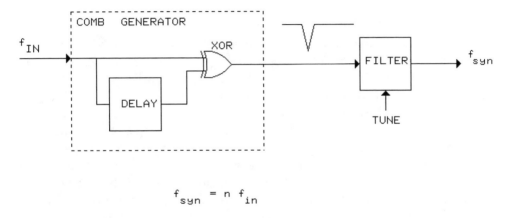

$$f_{syn} = n\, f_{in}$$

Figure 9.3   Block diagram of direct harmonic approach using a comb generator and tunable filter combination.

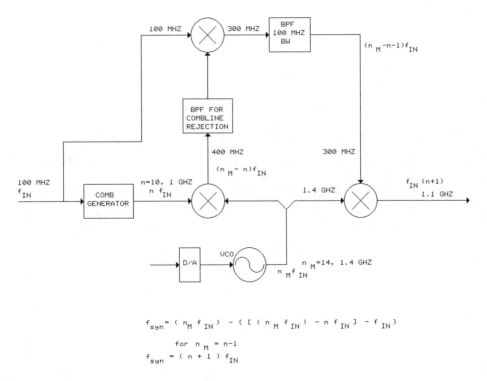

$$f_{syn} = (\,n_M\, f_{IN}\,) \; - \; (\,[\,(\,n_M\, f_{IN}\,) \; - \; n\, f_{IN}\,] \; - \; f_{IN}\,)$$

$$\text{for } n_M = n-1$$

$$f_{syn} = (\,n+1\,)\, f_{IN}$$

Figure 9.4   Block diagram of direct harmonic approach using the RF equivalent of a microwave comb generator and YIG filter combination.

Letting $n_M = n - 1$ and substituting (9.3) into (9.2) reduces (9.2) to (9.5):

$$f_{syn} = (n + 1)/f_{in} \qquad\qquad (9.5)$$

Equation (9.5) mathematically describes the relationship of the output frequency to the input frequency for this synthesizer. Furthermore, this equation shows the harmonic relationship that exists between the input and output frequencies.

9.1.2.3   Direct Digital Synthesis

Direct digital synthesis generates digital addresses that cycle through a look-up table for a sine wave. A digital-to-analog (D/A) converter converts these digital words to an analog sine wave. Filtering out the stair-step voltages at the output of the D/A converter produces a smooth sine wave at the output [1, p. 37]. Figure 9.5 shows an example of this circuit.

Direct digital synthesis has the advantage of having a phase-coherent switching response, small frequency steps, and a low phase-noise output; however, it has the disadvantages of being limited to a below-microwave frequency and of producing spurious signals. The D/A converter generates a major portion of the spurs. Consequently, selection of the D/A converter is critical to the performance of this topology. Equation (9.6) computes the relationship of the output frequency to the input frequency for the direct-digital-synthesis topology:

$$f_{syn} = f_r n_M / 2^{n_o} \qquad\qquad (9.6)$$

where

f_r = clock frequency (Hz);

n_o = total number of states;

n_M = input address of one of the states.

9.1.3   Indirect Synthesis

Indirect synthesis uses control systems to combine a frequency-controllable oscillator to a reference oscillator. The method of combination determines the relationship of the output frequency to the reference frequency.

9.1.3.1   Single PLL

First, a single PLL will be analyzed. Figure 9.6 shows a block diagram of a single PLL frequency synthesizer. A single PLL has several advantages. A single PLL is the simplest method of frequency synthesis and minimizes the number of high-frequency components. Minimizing the number of high-frequency components produces a circuit with small size and low power consumption. In addition, the maximum output frequency for this topology is only limited by the operating frequency of the dividers and the VCO.

A single PLL also has some disadvantages. A small frequency-step size in a single PLL requires a high divider ratio and a low reference frequency. Consequently, obtaining small frequency steps in a single PLL multiplies spurious signals and multiplies phase noise. In addition, the low reference frequency for obtaining small frequency steps produces a small bandwidth that causes a single PLL to have a slow switching speed. Equation (9.7) computes the relationship of the output frequency to the input frequency for a single loop synthesizer:

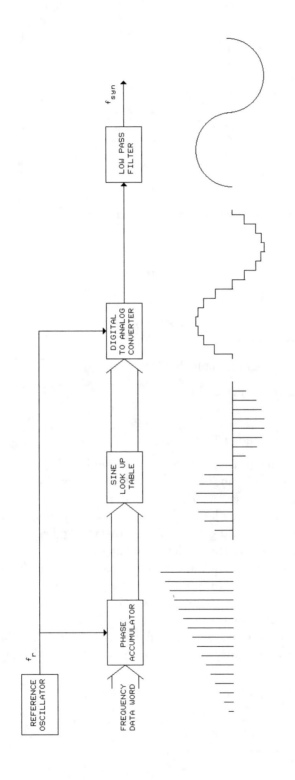

Figure 9.5   Block diagram of direct digital synthesizer.

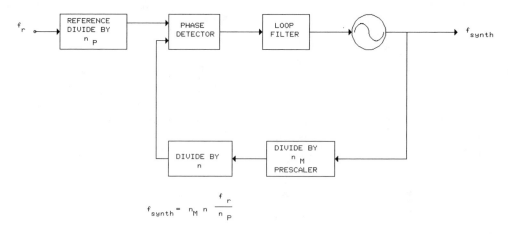

Figure 9.6   Block diagram of a conventional single-loop synthesizer.

$$f_{synth} = n_M n f_r / n_p \tag{9.7}$$

where

n_M = divide ratio of prescaler;

n = divide ratio of programmable divider;

n_p = divide ratio of reference divider.

9.1.3.2   Multiple PLLs, Cascaded Translation Loops

Next, multiple PLLs will be studied. Indirect synthesis with multiple PLLs distributes the problems of a single PLL among the many PLLs. It has cascaded translation loops that are combined through dividers at the output.

It has the advantages of smaller frequency steps and small size. Small size in this topology is achieved by using PLL chips. However, indirect synthesis with multiple PLLs has the disadvantages of slow switching speed, high output spurious levels, and narrow PLL bandwidths, and it is difficult to keep each looped phase locked. In addition, increasing the number of frequency increments with this multiple-PLL approach increases the complexity of the circuitry, as well as the number of high-frequency components.

Figure 9.7 shows an example of a multiple-PLL approach. The relationship of the output frequency to the input frequency for this method is calculated by (9.8):

$$f_{synth} = n_D f_{in} / n_{oc} + n_C f_{in} / n_{ob} + n_B f_{in} / n_{oa} + n_A f_{in} \tag{9.8}$$

where

n_A = divide ratio for PLL A;

n_{oa} = divide ratio for the reference check to PLL A.

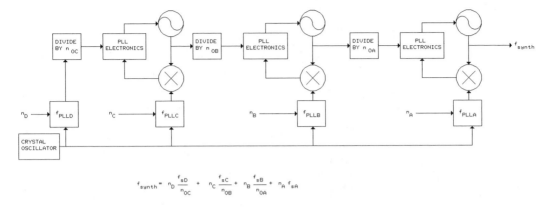

$$f_{synth} = n_D \frac{f_{sD}}{n_{OC}} \; + \; n_C \frac{f_{sC}}{n_{OB}} \; + \; n_B \frac{f_{sB}}{n_{OA}} \; + \; n_A \, f_{sA}$$

Figure 9.7 Block diagram of cascaded translation loops.

9.1.4 Direct-Indirect Hybrids

Hybrid topologies have come about because pure direct or indirect topologies do not satisfy enough performance requirements. Consequently, hybrid topologies attempt to combine the best of each topology to achieve a higher-performing synthesizer. Some of the more common hybrids will be discussed.

9.1.4.1 Microwave Single PLL with Sampling Downconversion

Now, the microwave single PLL with sampling downconversion will be studied. A very popular configuration in test equipment, it uses a sampler (narrow pulse generator) to sample the high-frequency waveform out of the YIG oscillator through a mixer. Then, a lowpass filter on the mixer IF reconstructs the sinusoidally down-converted frequency waveform. This behaves like a YIG filter that selects the harmonic out of a comb generator and downconverts the YIG oscillator output to a lower-frequency IF. After sine waves are converted to square waves, a divider divides the IF frequency to a feedback frequency input to the phase detector, which is compared to a divided-down reference frequency. The resulting phase error tunes the YIG oscillator to the correct output frequency.

This configuration has the advantages of a wide operating frequency and low far-from-the-carrier phase noise from the YIG VCO. However, the microwave single PLL with sampling downconversion has several disadvantages: It has slow switching speed and high current consumption, and the YIG is susceptible to ac voltage swings on the power-supply line. Furthermore, it has large size from the YIG, and the output operating frequency is restricted to greater than 500 MHz.

Figure 9.8 shows an example of this topology. Equation (9.9) computes the relationship of the output frequency to the input frequency for this topology:

$$f_{syn} = f_{ref} n_p / n_{ref} + n_{samp} f_{ref} \tag{9.9}$$

where

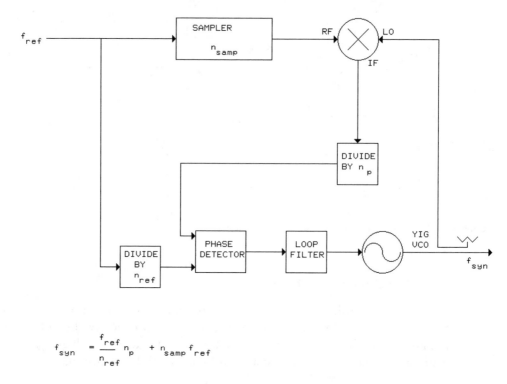

$$f_{syn} = \frac{f_{ref}}{n_{ref}} n_p + n_{samp} f_{ref}$$

Figure 9.8 Block diagram of microwave PLL and harmonic hybrid approach.

n_{comb} = multiplication factor of comb;

n_{ref} = divide ratio of reference clock;

n_p = divide ratio of programmable divider.

9.1.4.2   Direct Digital Synthesizers and PLLs

A more recent trend in synthesizers has combined direct digital synthesis with PLLs. This combination translates the small frequency-step size, low phase-noise performance, and fast switching speed of the direct digital synthesizer (DDS) to higher-frequency applications.

DDS Plus Single PLL

Figure 9.9 shows a simple approach to combining a direct digital oscillator with a PLL. The DDS connects to the input of the phase detector that is multiplied up to the VCO output frequency by the ratio set by the programmable divider.

This approach has the advantages of low phase noise, small step size, and fast switching time; however, it has the disadvantages of having high spurious outputs. Using the direct digital oscillator for the reference clock to the PLL multiplies the spurious output of the digital oscillator. The multiplication factor for the PLL determines the increased level of the spurious signals. Equation (9.10) mathematically describes the relationship of the output frequency to the input frequency for this topology:

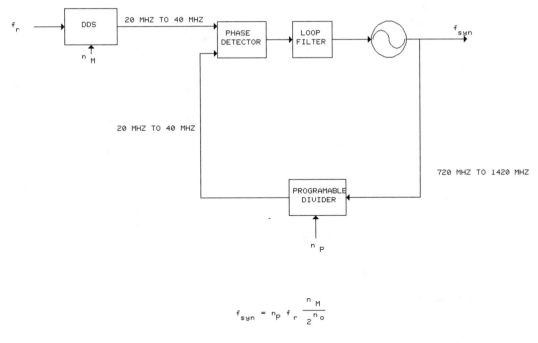

$$f_{syn} = n_p \, f_r \, \frac{n_M}{2^{n_o}}$$

Figure 9.9   Block diagram of DDS plus single PLL.

$$f_{syn} = n_p f_r n_M / 2^{n_o} \tag{9.10}$$

DDS Plus Multiple PLLs

The DDS plus multiple loops reduces the spurious levels of the previous topology by distributing the problems among several PLLs. This configuration combines a coarse PLL with a low multiplication ratio with a baseband DDS that gets up-converted by the translation loop to the output VCO frequency. The upconversion process in the PLL does not multiply the spurious signals in the DDS because the feedback ratio for the translation loop is unity.

This topology has the advantages of a fast switching time, low phase noise, a high output-frequency range, and wide PLL bandwidths; however, it has the disadvantages of a large number of spurious signals and high spurious-output levels and of having to find lock for each PLL. Figure 9.10 shows an example of this topology. Equation (9.11) computes the relationship of the output frequency to the input frequency for the multiloop synthesizer with DDS:

$$f_{synth} = f_r n_M / 2^{n_o} + n_p f_r \tag{9.11}$$

9.1.4.3   Single PLL with Fractional-*N* Divider

A fractional-*N* synthesizer (see Figure 9.11) uses a single PLL with direct digital synthesis to generate an output frequency that has an integer and fractional multiplication factor of the reference frequency. The divider has an integer part and a

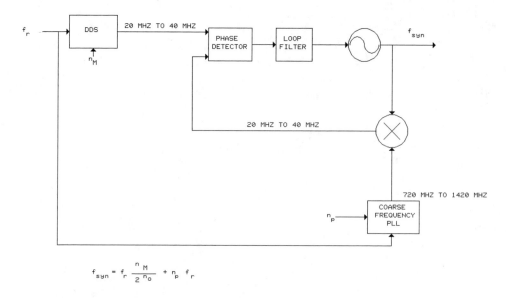

$$f_{syn} = f_r \frac{n_M}{2^{n_o}} + n_p\, f_r$$

Figure 9.10   Block diagram of DDS plus multiple PLLs.

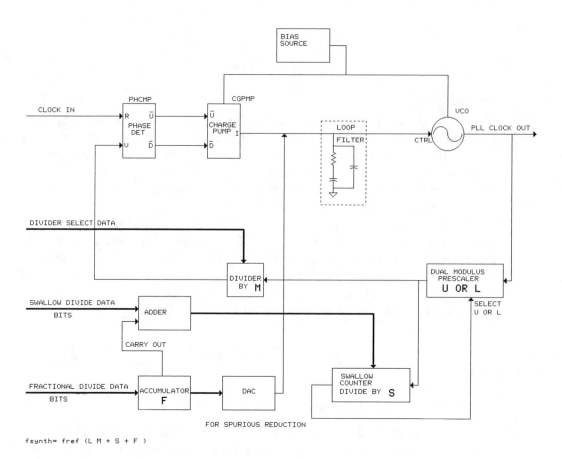

fsynth= fref (L M + S + F)

Figure 9.11   Block diagram of fractional-*N* synthesizer.

fractional part. The fractional part is obtained by modulating the dual modulus prescaler at the rate of the reference frequency. An accumulator, which is clocked at the reference frequency, calculates the moment when the divide ratio changes. The overflow bit of the adder determines the divide ratio. Some applications use a D/A converter to decode the accumulator address in order to the suppress reference sidebands also [1, pp. 43–48]. Equation (9.12) computes the relationship of the output frequency to the input frequency for this topology:

$$f_{synth} = f_r(n_L n_M + n_S + n_F/2^{n_o}) \tag{9.12}$$

where

n_L = lower divide ratio of the dual modulus prescaler;

n_M = programmable divide ratio;

n_S = swallow-counter divide ratio;

n_F = fractional amount;

n_o = number of bits in the accumulator.

A fractional-N synthesizer has several advantages. This method reduces the multiplication factor for small frequency-step sizes. This reduction in multiplication factor helps reduce phase noise inside the loop bandwidth.

A fractional-N synthesizer has several disadvantages. This method increases reference sideband levels because it adds phase-error pulses coming out of the phase detector that have to be filtered. The increase in phase-error pulse width requires a narrow loop bandwidth to filter the additional sidebands. In addition, to get smaller frequency steps requires that loop bandwidth narrow even further. Narrowing the loop bandwidth increases the phase noise inside the loop bandwidth, which negates the original advantage.

9.1.4.4   Multiple Output Processing

Multiple output processing uses the direct-synthesis approach to extend the output-frequency range of any synthesizer. Multiple output processing has the advantages of eliminating additional loops and having a lower input frequency; however, it has the disadvantages of having multiple spurious outputs and a high level of spurious outputs, as well as of increasing the number of high-frequency components.

Figure 9.12 shows an example of this topology. Equation (9.13) computes the relationship of the output frequency to the input frequency for the synthesizer:

$$f_{out} = f_{syn}(2S_1 + S_2 + 0.5S_3 + 0.25S_4) + S_5(f_{syn} - f_{lo}) \tag{9.13}$$

where

S_1 = logic state of switch 1 (1 or 0).

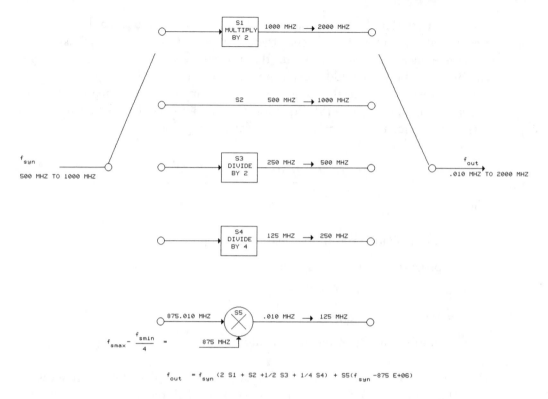

Figure 9.12   Block diagram of multiple output processing.

Equation (9.14) computes the frequency requirements for the LO in this topology:

$$f_{lo} = f_{smax} - f_{smin}/4 \qquad\qquad (9.14)$$

where

f_{smax} = maximum output frequency of synthesizer (Hz);

f_{smin} = minimum output frequency of synthesizer (Hz).

For example, consider a synthesizer that covers a frequency range of 500 MHz to 1 GHz. Equation (9.13) computes a 875-MHz LO and expands the synthesizer to cover an output-frequency range of 10 kHz to 2 GHz.

9.1.5   Application of Topologies

Now that the types of synthesizers have been classified, some example synthesizers (the HP8673 microwave synthesizer and HP8662 low phase-noise synthesizer) will show the applications of these topologies. Figure 9.13 shows a microwave synthesizer. This synthesizer uses a combination of harmonic synthesis and multiple PLLs to achieve a 2–18-GHz frequency range. In this synthesizer, multiple PLLs

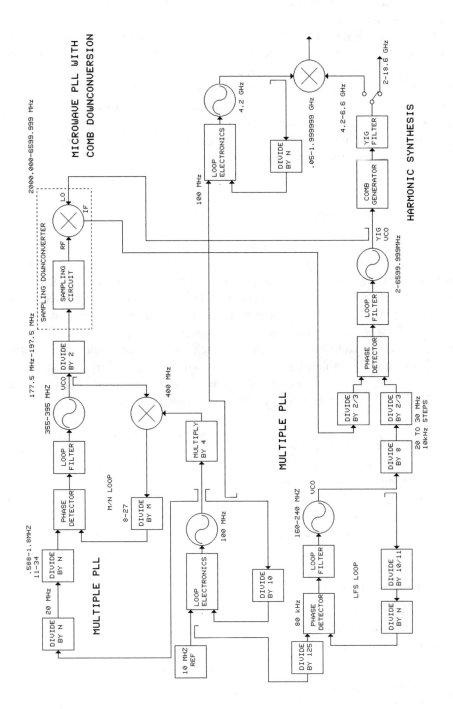

Figure 9.13 Block diagram of microwave synthesizer.

generate a very-high-frequency (VHF) signal for a harmonic downconversion in the main microwave PLL. Additional PLLs generate the reference clock for the main PLL. Finally, the main PLL down-converts the output of the YIG to the input of the phase detector.

Figure 9.14 shows the HP8662 RF low-noise synthesizer. This synthesizer uses a combination of brute-force direct synthesis, multiple PLLs, and output processing to generate a 0.01–1,280-MHz output-frequency range. First, multiple PLLs generate a 10–20-MHz reference for the main PLL. Combining this output with a harmonic downconversion PLL in a translation loop produces a 320–640-MHz output frequency. Finally, output processing expands the output range of the synthesizer to a final output of 0.01 to 1,280 MHz. These examples show the application of various topologies in existing hardware.

9.1.6 Design Trade-Offs

The numerous methods and the limitations of each method make choosing a topology very confusing. This chapter shows that one approach does not solve the problems of all synthesizers. Consequently, great care must be used to determine the importance of each requirement compared with the other requirements. Assigning proper weighting to each requirement leads to choosing the optimum synthesizer topology for the required application. Aaron notes [2], "As with all models of performance, the shoe has to be tried on each time an application comes along to see whether the fit is tolerable; but, it is well known, in the Military establishment for instance, that a lot of ground can be covered in shoes that do not fit properly."

Ideal PLL requirements for synthesizers include:

- Small frequency-step size;
- Low-noise and jitter:
  - Low close-in phase noise;
  - Low far-out phase noise;
  - Low-noise ear;
  - Low reference sidebands;
  - Low spurious levels;
- Fast time response;
- Wide tuning range;
- High stability;
- High output frequency;
- Fast power up;
- High noise immunity;
- Minimum tracking error;
  - No resets;
- Small size;
- Low power;
- Low cost.

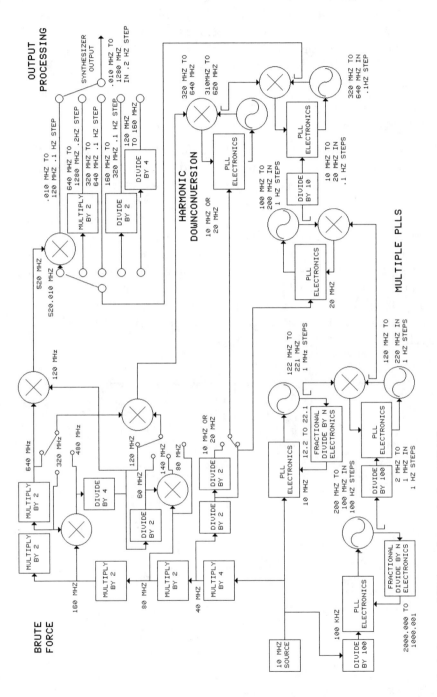

Figure 9.14 Block diagram of RF low-noise synthesizer HP8662.

9.1.6.1 Synthesizer Matrix

A decision matrix that compares requirements to topology performance can help in selecting a synthesizer approach. Figure 9.15 shows this decision matrix for determining an optimum synthesizer topology. This matrix shows that the optimum synthesizer topology depends on the performance requirements. Slight changes in performance requirements can require a complete topology change to achieve an optimum solution. In addition, this matrix shows that one topology does not neatly solve all the performance parameters.

9.1.6.2 PLL Matrix

If the selected topology contains a PLL, then this further complicates the selection issue. Another decision matrix must be used to determine the feasibility of the optimum PLL for this topology. Figure 9.16 shows this decision matrix for determining the optimum PLL parameters. From this figure, conflicting component selections and conflicting PLL design-parameter selections depend on performance requirements. Slight changes in performance requirements can require a complete redesign of the PLL. In addition, this matrix shows that one set of components and PLL parameters does not neatly satisfy all of the performance requirements.

Starting with the first PLL parameter and going across the requirements, we see the following: Increasing the divide ratio makes smaller frequency steps and increases the output frequency. Decreasing the divide ratio directly lowers the close-

(DESIRED) PERFORMANCE PARAMETERS	SYNTHESIZER TOPOLOGIES	INDIRECT	ANALOG SINGLE LOOP	ANALOG MULTIPLE LOOPS	SAMPLED LOOPS	FRACTIONAL N	DIRECT	DIGITAL OSC.	HARMONIC APPROACH	MIXER AND DIV.	DELAY LOCK LOOPS	HYBRID	DDS + PLL	SYNTHESIZER WITH OUTPUT PROCESSING	PLL + HARMONIC APPROACH
(SMALL) FREQ STEPS			▲	▼	▲	▼		▼	▲				▲	▼	
(LOW) CLOSE IN NOISE LEVEL			▲		▼	▼		▼	▼	▼	▼		▼		▼
(LOW) FAR OUT NOISE LEVEL			▼	▼	▲			▼	▲	▲	▼		▲		▼
(LOW) REF SIDEBAND LEVEL			▲		▲	▲		▼	▼	▼			▼		
(FAST) TIME RESPONSE			▼	▼		▼		▲	▲	▲	▲		▲		▼
(HIGH) OUTPUT FREQUENCY			▲	▲	▼	▲		▼	▲	▲	▼		▲		▲
(LOW) PHASE NOISE EAR LEVEL			▲	▼	▼			▼	▼	▼	▼		▼		
(WIDE) TUNING RANGE			▲	▲	▲	▲		▲	▼	▼	▼		▲	▲	▲
(SMALL) SIZE			▼					▼		▲	▲		▼	▲	
(LOW) COST			▼	▲	▼	▼		▼					▼	▲	▲
(LOW) SPURIOUS LEVELS			▼	▲	▲	▲		▲	▲	▲	▲		▲	▲	▲
(LOW) POWER CONSUMPTION LEVEL			▼	▲	▼	▼		▼	▲	▲	▼		▼	▲	▲
(SMALL) NO. OF HIGH FREQUENCY COMPONENTS			▼	▼	▲	▼		▲	▼	▲	▲		▼	▲	▲

Figure 9.15 Decision matrix of synthesizer performance specifications versus synthesizer topology.

PLL PARAMETERS	PERFORMANCE PARAMETERS (DESIRED)	(SMALL) FREQ STEPS	(LOW) PHASE NOISE CLOSE IN	(LOW) PHASE NOISE FAR OUT	(LOW) REFERENCE SIDEBANDS	(FAST) RESPONSE TIME	(HIGH) FREQUENCY OUTPUT	(LOW) NOISE PHASE NEAR	(LOW) POWER CONSUMPTION	(HIGH) STABILITY	(WIDE) TUNING RANGE	(SMALL) SIZE	(LOW) SPURIOUS	(LOW) JITTER
DIVIDE BY N		↑	↓		↓	↓	↑	↓	↓			↓	↓	↓
LOOP BANDWIDTH			↑	↓	↓	↑					↑			↑
DAMPING FACTOR			↑			↓					↑			↑
REFERENCE PHASE NOISE			↓				↑	↓	↑			↑		↓
VCO PHASE NOISE				↓			↑	↓	↑	↓		↑		↓
REFERENCE FREQUENCY		↓					↑	↑		↑	↑	↑	↑	↑
SPEED OF DIVIDERS		↑					↑	↑	↓		↑	↓		↑
SPEED OF P/F DETECTOR					↑	↑	↑		↓	↑	↑	↓		↑
SPEED OF CHARGE PUMP/OP AMP					↑	↑	↑		↓			↓		↑
REF FILTERING					↑				↓	↓			↑	↑
MULTIVIBRATOR/RING VCO			M		R			R			M	R	M	M
CHARGE PUMP/OP AMP		O	O		O	C		O	C	O	O	C	O	O
GAIN OF OP AMP			↑			↑		↑	↓	↑		↓		↑
OFFSET OF OP AMP					↓					↓		↑		↓

Figure 9.16 Decision matrix of monolithic PLL parameters versus PLL performance specifications.

in phase noise, reduces the reference sidebands, speeds up the frequency step response, reduces spurious signals, lowers jitter, and indirectly reduces power consumption and size. Next, increasing loop bandwidth directly lowers close-in phase noise, speeds up frequency step response, and lowers long-term jitter. Narrowing loop bandwidth decreases far-from-the-carrier noise and reduces reference sideband levels. Higher damping factor improves stability, reduces jitter, and lowers close-to-the-carrier phase noise. Low damping factor makes the PLL switch faster.

9.1.7 Architecture Design Example

A design example where the weighting has been changed will show the sensitivities of the topologies to seemingly small changes in requirements. For example, a synthesizer that has been produced by x company for several years has a frequency range of 500 MHz to 1 GHz with 1-kHz steps, small size, low power, low phase noise, and 1-ms switching time. Figure 9.17 shows the original topology.

This synthesizer uses five PLLs in the multiple-PLL topology. Now, after the synthesizer is in production, the customer comes back and says he wants 10-Hz steps instead of 1-kHz steps, and he does not want to change the design. Maintaining the same topology and keeping most of the existing circuitry limits your choices to adding only three more PLLs for a total of eight PLLs to get a 10-Hz step as shown in Figure 9.18.

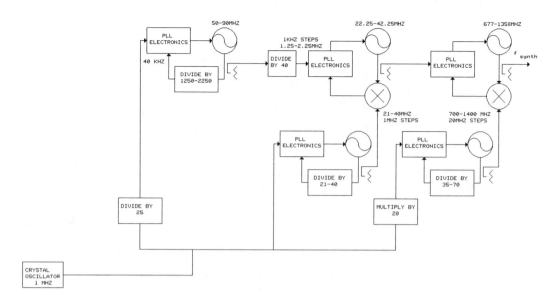

Figure 9.17   Block diagram of original example design for 1-kHz steps.

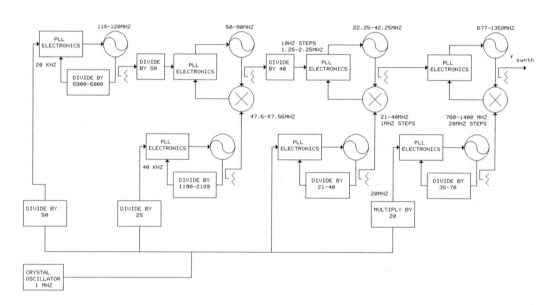

Figure 9.18   Block diagram of modified example design for 10-Hz steps.

However, after you study the problem further, a change in topology to multiple PLLs and a DDS (Figure 9.19) would have been more optimal had it been known at the beginning that small frequency steps were required. At the expense of higher harmonics, this topology would improve switching time, lower phase noise, reduce the number of high-frequency parts, and simplify the design. Consequently, the all-too-familiar choice of scraping the existing design and starting over or living with the old design and adding to it is presented. Joseph Petzval, a technical philosopher in the 1800s, commented about this predicament, "The optimal solution is the best one you have when the money runs out" [3].

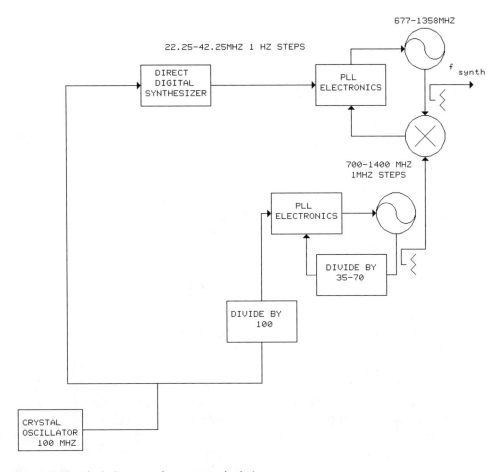

Figure 9.19 Block diagram of new example design.

In summary, this section has shown that studying synthesizer topologies can help with selecting the best configuration. First, synthesizer topologies were classified. Trade-offs between topologies were discussed. Applications of the topologies were shown. Decision matrices were presented to help in the selection process. And, finally, an example showed that slight changes in requirements can significantly change the design of a synthesizer.

9.1.8 Monolithic Synthesizer Example

An example monolithic design of a synthesizer will help show the trade-offs that have to be made in order to meet design requirements. Designing over a wide output-frequency range and wide input reference frequencies presents a different set of technical challenges than a fixed frequency design.

Figure 9.20 shows the schematic of the example synthesizer. The synthesizer has classic phase/frequency detector, a simple charge pump, and RC loop compensation. The VCO is a current-starved ring oscillator followed by a pulse-swallowing divider with a divide-by-4/5 dual modulus prescaler. The simulation test bench

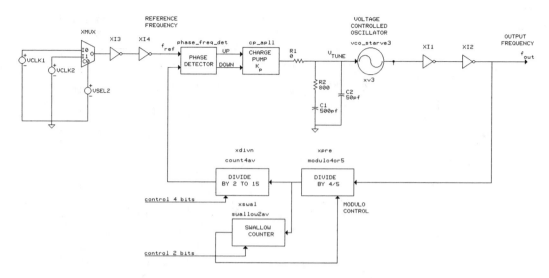

Figure 9.20 Example simulation test bench and functional block diagram for a synthesizer.

uses a multiplexer to switch between input frequencies in order to do frequency-step-response transient tests.

To characterize the example synthesizer, we will simulate the VCO tune corners, VCO gain corners, and phase detector charge pump corners; the large frequency step response with weak conditions and small and large divide ratios; the small frequency step response with weak conditions and small and large divide ratios, strong conditions, and a small divide ratio; and the power brown-out response.

First, let's look at the design parameters for the synthesizer for a 0.55-μm gate-length process with a 5-V supply. Table 9.1 shows the design parameters for the synthesizer. The VCO gain is given from an estimate of previous designs of a current-starved ring oscillator.

From the design parameters, we calculate the loop-compensation filter using the charge pump synthesis equations. The charge pump gain is an estimate from previous designs of a simple charge pump. A high current is desired in the charge pump to keep the SNR high. Table 9.2 shows the calculation for the loop compensation.

9.1.8.1 Component Simulations

Now we need to simulate the VCO and charge pump phase detector to verify our assumptions about the VCO gain and charge pump gain.

Table 9.1 Design Parameters for the Example Synthesizer

Design Parameter	Value
Reference frequency	1–30 MHz
K_v	500E6 rad/s/V
N	8–60
Loop bandwidth	0.85 MHz for $N = 8$
Damping factor	0.6

Table 9.2   Calculation Results for
the Loop Compensation

Component	Charge Pump
K_d	0.4 mA/rad
R_1	0
R_2	800Ω
C_1	500 pF
C_2	50 pF

Figure 9.21 shows the current-starved VCO tune curves for nominal and worst-case corner conditions. The weak condition has 10% low supply voltage, 125C temperature, and weak process. The strong condition has 10% high supply voltage, −40C temperature, and strong process. From these curves, we can find the operating frequency range of the VCO by looking for the overlapping range. The minimum operating frequency is determined by the lowest frequency for the strong conditions, which is 8 MHz at a tune voltage of 0.9V. The maximum operating frequency is determined by the highest frequency for the weak condition, which is 70 MHz at a tune voltage of 4V. Consequently, the working range of the VCO is 8 to 70 MHz for a tune voltage of 0.9V to 4V.

Figure 9.22 shows the current-starved VCO gain curves for nominal and worst-case corner conditions. From this curve, we can find the range of VCO gain so that we can adjust the design to operate under these conditions. This figure shows worst-case maximum gain of 96 MHz/V at a 2.2-V tune voltage for the strong condition and a worst-case minimum gain of approximately 12 MHz/V at 0.95V and 3.9V for the weak condition. This is an eightfold variation in the gain, which

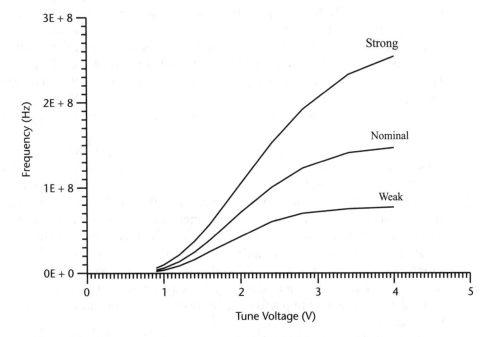

Figure 9.21   Nominal and worst-case corners for current-starved ring VCO frequency versus tune voltage.

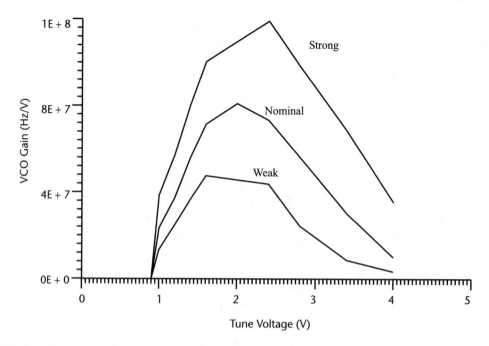

Figure 9.22   Nominal and worst-case curves for the gain of the current-starved ring VCO.

will change the bandwidth by a factor of the square root of 8 or a factor of 2.8 times. This variation needs to be accounted for during the design.

Figure 9.23 shows a phase detector and charge pump transfer function over nominal and the worst-case control voltages. A reference frequency and feedback divided-down VCO frequency of 10 MHz was used for these curves. From this curve, we can find the range of phase detector gain so that we can adjust the design to operate under these conditions. For a 3.5-V control voltage, the positive phase error has a 429-μa/rad gain, and the negative phase error has a 199-μa/rad gain. For a 1.0-V control voltage, the positive phase error has a 199-μa/rad gain, and the negative phase error has a 356-μa/rad gain. For a 2.5-V nominal control voltage, the positive phase error has a 390-μa/rad gain, and the negative phase error has a 295-μa/rad gain.

The 3.5-V and 1.0-V control voltages have gain slopes that vary by 2 to 1. This is enough variation for the loop to be able to oscillate about the zero phase error because of the difference in up and down gain. We will look for this characteristic in the system simulations. In addition, there is an offset of 50 μa at the zero-phase-error point. Consequently, the figure of merit for the phase detector reduces to approximately 4, which is poor and will cause high reference sidebands (jitter). Now, the VCO must operate under a range of less than 1V to 3.5V. For a 5-V supply, we are throwing away 50% of the range. At lower supply voltages for faster processes, this will not be acceptable. Other rail-to-rail architectures for the phase detector charge pump and the current-starved VCO will have to be used for these cases.

Figure 9.24 shows phase detector and charge pump transfer function curves for nominal and worst-case corners. For the weak condition, the positive phase

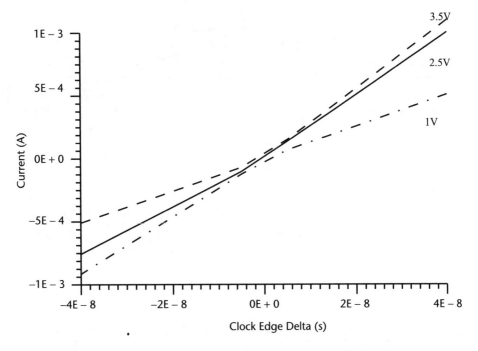

Figure 9.23 Variation of charge pump transfer function versus nominal and worst-case output tune voltages.

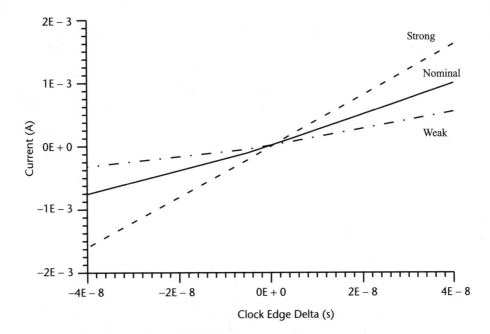

Figure 9.24 Nominal and worst-case curves of the charge pump transfer function.

error has a 219-μa/rad gain, and the negative phase error has a 127-μa/rad gain. For the strong condition, the positive phase error has a 624-μa/rad gain, and the negative phase error has a 644-μa/rad gain. For nominal conditions, the positive phase error has a 390-μa/rad gain, and the negative phase error has a 295-μa/rad gain. From all these conditions, the worst-case gains to be used in the analysis are 127 to 644 μa/rad. This is a 5-to-1 variation in gain, which will further complicate the synthesis of components. Table 9.3 summarizes the component simulations for the phase detector and VCO.

9.1.8.2 System Simulations

With the component gains determined, we can simulate the control system and adjust loop-compensation values to best fit the variation in gains and divide ratios. Figure 9.25 shows the open-loop Bode plot for nominal and worst-case conditions. Figure 9.25 shows that with 1-MHz reference, a divide ratio of 60, and weak conditions, the loop is close to oscillation as a result of the wide variation in gains

Table 9.3 Summary of Component Values for the Example Synthesizer

PLL Parameter	Minimum	Nominal	Maximum
Charge pump gain	127 μa/rad	—	644 μa/rad
Charge pump control range	1V	—	3.5V
VCO frequency range	8 MHz	40 MHz	70 MHz
VCO gain	12 MHz/V	72 MHz/V	96 MHz/V
VCO control range	0.9V	1.6V	4V

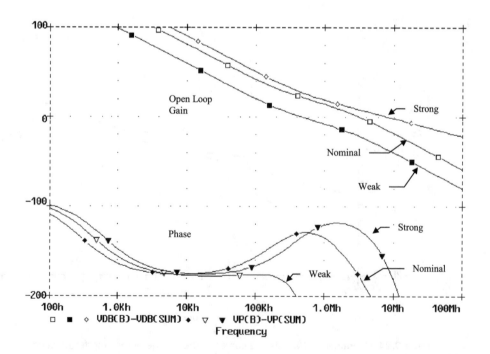

Figure 9.25 Nominal and worst-case magnitude and phase of the open-loop Bode plot of the example synthesizer.

and divide ratios. We will leave this close-to-oscillation condition and see how the simulations detect this condition. The 0-dB crossover point varies from 400 to 9,000 kHz. The nominal case with divide-by-8 and the strong case with divide-by-8 and a 20-MHz reference frequency have low phase margin (15° to 20°). Let's do some simulations to show the effects in transient analysis.

Figure 9.26 shows the control-voltage response to a small frequency step of 2.5 to 3.75 MHz with a feedback divide ratio of 8 and strong conditions (5.5V, −40C, and strong transistor models). The output frequency steps from 20 to 30 MHz. The rise time is 100 ns, and the overshoot is 3.8V/2.5V = 52%. Curve fitting, using (3.142), (2.25), and (2.24), gives a 25° phase margin and a 2.6-MHz 0-dB crossover frequency. This is the widest bandwidth condition for the circuit because of the strong conditions and the smallest divide ratio. This condition defines the maximum bandwidth side of achieving stability.

Figure 9.27 shows the response of the control voltage to a small frequency step at 15 μs from 1 to 1.06 MHz with a feedback divide ratio of 60 and weak

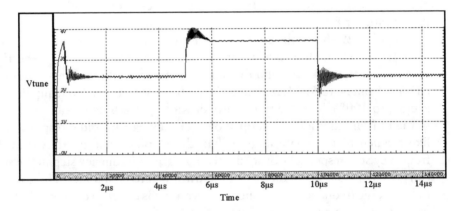

Figure 9.26 Small frequency step response for divide ratio of 8 and strong conditions of the example synthesizer.

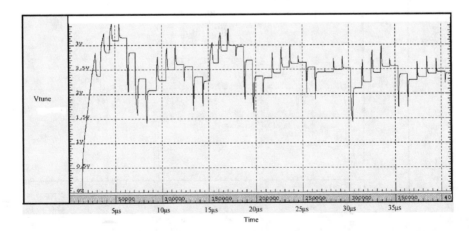

Figure 9.27 Small frequency step response for divide ratio of 60 and weak conditions of the example synthesizer.

conditions (4.5V, 125C, and weak transistor models. The x-axis scale is 0.1 ns per unit. The output frequency changes from 60 to 63.6 MHz. The rise time is 1 μs, and the overshoot is 100%, which makes the loop marginally stable. Curve fitting, using (3.142), (2.24), and (2.25), gives a 6° phase margin and a 0.3-MHz 0-dB crossover frequency. This is the narrowest small-signal bandwidth case because of the weak conditions and the highest divide ratio. This condition defines the minimum bandwidth side of achieving stability. In addition, notice how difficult it is to distinguish the step response at 15 μs from the ringing that occurs before and after the response. There is a tendency to increase the frequency-step size to make the step response more observable, but this can give false results because of the slew rate of the PLL.

Figure 9.28 shows the control-voltage response to a small frequency step from 4 to 8 MHz for a divide ratio of 8 and for weak conditions. The output frequency changes from 32 to 64 MHz. The 10% to 90% rise time is 400 ns, and the overshoot is 50%. Curve fitting, using (3.142), (2.25), and (2.24), gives a 20° phase margin and a 0.4-MHz 0-dB crossover frequency. Reducing the divide ratio and increasing the reference frequency produces a more stable response for the divide ratio of 8.

Figure 9.29 shows the control-system response to a large frequency step from 1 to 1.15 MHz with a divide ratio of 60 and weak conditions. The rising step response occurs at 15 μs, and the falling response at 30 μs. The output frequency changes from 60 to 69 MHz. The 10% to 90% rise time is 1.5 μs, and the overshoot is close to 100%. The rise time slows down by 50% because of slew rate limiting in the PLL. In addition, the loop shows a decaying sinusoidal response that looks like a much higher phase margin than the 6° phase margin shown in the small frequency step response. Consequently, too large a frequency step in the simulation can give erroneous results about the bandwidth and the stability. There is a tendency to have large frequency step responses because it is easier to distinguish the response from the noise artifacts of the simulation.

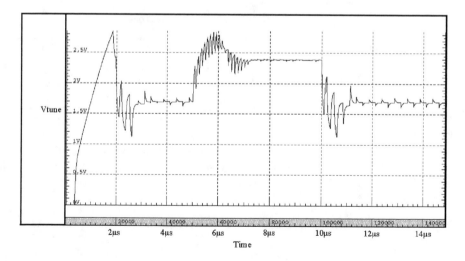

Figure 9.28   Small frequency step response for divide ratio of 8 and weak conditions of the example synthesizer.

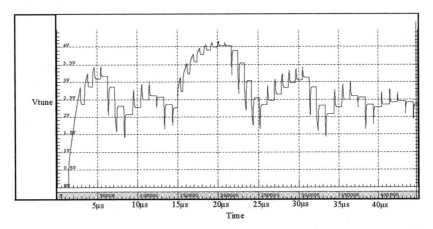

Figure 9.29   Large frequency step response for divide ratio of 60 and weak conditions of the example synthesizer.

Figure 9.30 shows the control-system response to a large frequency step from 4 to 10 MHz with a divide ratio of 8 and weak conditions. The x-axis has units of 0.1 ns. The output frequency steps from 32 to 80 MHz. The 10% to 90% rise time is 3 μs, and the overshoot is about 10%. In this case, the slew rate distorts the small-signal response even more.

Table 9.4 shows a summary of the hold-in ranges from the simulations that were done. This table shows a narrow hold-in range for the times-60 frequency multiplication.

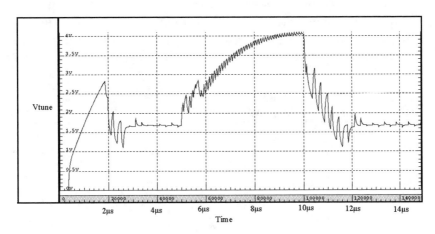

Figure 9.30   Large frequency step response for divide ratio of 8 and weak conditions of the example synthesizer.

Table 9.4   Summary of Hold-In Range Results

Parameter	Input Frequency (MHz)	VCO Output (MHz)	Control Voltage (V)
Hold-in range × 60	1–1.15	60–69	2–4.0
Hold-in range × 8	4–10	32–80	1.6–4.0

Table 9.5 shows a summary of the frequency step-response results. Simulation data of rise and fall times for each condition is converted to bandwidth and phase margin by fitting the response to the transient solutions discussed in earlier chapters. It shows a 300-kHz loop bandwidth for the times-60 multiplication factor and weak process conditions. Therefore, the narrow frequency range occurs because the 300-kHz bandwidth prevents the reference frequency from going below 1 MHz from the loss of phase shift from sampling delay, which will make the loop unstable. Consequently, wide output-frequency range with a high multiplication factor for a single loop PLL synthesizer requires that the loop bandwidth be narrow. To make it more stable requires reducing the phase detector gain or increasing the capacitor size.

From the simulations, the VCO must operate under a range of less than 1V to 3.5V. For a 5-V supply, we are throwing away 50% of the VCO range. At lower supply voltages for faster processes, this will not be acceptable. Other rail-to-rail architectures for the phase detector charge pump and the current-starved VCO will have to be used for these cases.

Table 9.6 shows a summary of the other parameters measured in the simulations. It shows that the lock time varies from 3 to 15 μs, and the loop bandwidth varies from 300 to 2.6 MHz. The variation in loop bandwidth accounts for the large variation in lock time. To speed up lock time in a PLL would require a wide loop bandwidth, which is a trade-off that would reduce output-frequency range. In most synthesizer applications, output-frequency range is more important than lock time.

Synthesizer design for a wide range of output and input frequencies produces a wide range in control-system variations. Methods to reduce this variation must be used to have the widest-possible-range synthesizer. Varying the phase detector gain (charge pump current) with divide ratios is one method to improve the output-frequency range. Switching loop compensation with divide ratios is another method to improve the output-frequency range. Finally, using a programmable output

Table 9.5   Summary of Frequency-Step-Response Data over Worst-Case Conditions

Parameter	Simulated Data
Frequency step response × 60 weak	1-μs rise time, 100% overshoot, 0.3-MHz f_x, 6° phase margin
Frequency step response × 8 weak	400-ns rise time, 50% overshoot, 0.4-MHz f_x, 20° phase margin
Frequency step response × 8 strong	100-ns rise time, 52% overshoot, 2.6-MHz f_x, and 25° phase margin

Table 9.6   Summary of the Other Parameters Measured in the Simulations

Parameters	Minimum	Maximum
Lock time	3 μs	15 μs
Loop bandwidth	300 kHz	2,600 kHz
Phase margin	>5°	250

divider to divide-down the output frequency is another method to improve output-frequency range.

9.2   Clock Recovery

To an ever-increasing extent, communication links use digital formats to synchronously transmit information in a continuous, uniform pulse stream. Many systems transmit or receive data without any additional timing reference. Almost every digital communication link uses a PLL in the transmission of this information. Digital data can be transmitted over serial or parallel data channels. A clock signal, which is normally not transmitted, synchronously clocks the data to be sent to the receiver. Thus, the receiver must extract the clock information out of the received data signal. A clock-recovery circuit extracts the clock information by producing a timing-clock signal from the stream of binary data. Most clock-recovery circuits use phase-locking techniques. Optimum detection of the data requires a local clock generator that has close alignment with the received pulse train. Clock-recovery processes the data stream to synchronize the frequency and phase of the clock. A PLL accomplishes this function. Clock-recovery applications include compact disk players, floppy disk readers, and satellite data links.

In general, transmitting digital data uses *baseband* and *carrier-bused* transmission methods. Baseband transmission directly sends the digital signal over the link. A carrier system uses the digital signal to modulate a carrier, which is usually a high-frequency signal. Amplitude modulation (AM), frequency modulation (FM), or phase modulation (PM) techniques can modulate the carrier, or different combinations can be used, such as AM and PM. In carrier systems, a number of digital signals can be modulated to different carriers. Consequently, the bandwidth of carrier systems can be much larger than the bandwidth of baseband systems. However, in this section we will deal only with baseband systems [4].

Digital communication systems convey information by a series of bits, 1s and 0s. Figure 9.31 shows a typical binary data signal. The data stream has a bit sequence of 0,1,1,0,1,0,0,1,1,0,1,0,1,0, where 1 represents a pulse, and 0 represents the absence of a pulse. The bit rate is called the baud, $f_b = 1/T_b$. To process the data correctly, the receiver synchronizes a clock to the data by making the clock frequency f_o equal to the bit rate f_b.

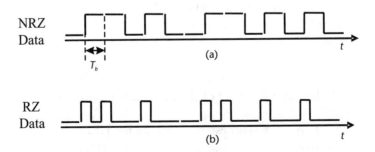

Figure 9.31   (a) Nonreturn-to-zero (NRZ) data, and (b) return-to-zero (RZ) data.

The return-to-zero (RZ) format goes to zero between consecutive bits, which can easily be confused with the nonreturn-to-zero (NRZ) format. For the same bit rate, RZ data has the advantage of more transitions then NRZ [5].

The application of PLLs to clock recovery has special design considerations. The random nature of the data stream restricts the choice of phase detectors. In particular, three-state phase detectors will not work, and other means must be used to aid acquisition. The random data also causes the PLL to add undesired phase variation (timing jitter) in the recovered clock. Proper design of the PLL can minimize this jitter. Trischitta and Varma [6] provide a comprehensive reference on jitter sources, effects, and standards.

Let's list the ideal requirements for a clock-recovery PLL. This will give us a clearer ability to adjust the architecture and parameters in the design:

- Generation of a positive clock edge centered between consecutive data-transition time points;
- Low additional jitter and noise to the input:
  - Low far-out phase noise (VCO);
  - Low-noise ear/peaking (control system);
  - High-order poles (improved jitter tolerance);
  - Low reference sidebands;
  - Low spurious levels;
  - Low multiplication factors to minimize accumulated jitter;
- Fast time response for initial lock;
- Zero phase start up;
- Slow time response to changing data;
- Wide frequency range;
- High stability;
- High output frequency;
- Fast power up;
- High noise immunity;
- Minimum tracking error:
  - No resets;
  - Minimum reference edge to output edge skew;
  - Tracking of low-frequency input modulation;
  - Rejection of high-frequency input modulation;
  - Minimum sensitivity to missing input edges;
  - Low phase detector offset;
- Small size;
- Low power;
- Low cost.

The following lists some of the multitude of clock-recovery architectures:

- PLL frequency synthesizer reference for DLL;
- PLL frequency synthesizer reference for ADPLL;
- Quadricorrellator;

- Early/late gate;
- Modified phase/frequency detector.

9.2.1 Properties of NRZ and RZ Data

Digital systems commonly transmit data in an NRZ format. Figure 9.32(a) shows the NRZ data stream V_i'. The bit sequence is 1,1,0,1,0,0,1,1. The NRZ format does not have the signal going to zero between adjacent pulses represented by 1s. Each NRZ bit has a bit time duration of $T_b = 1/f_b$, and a bit rate of $r_b = 1/T_b$ bits/s. Each bit has equal likelihood of being a 1 or 0 and has statistical independence from the other bits. Furthermore, the highest frequency (a repetitive 0,1 sequence) in the data stream is $f_b/2 = 1/(2/T_b)$ and equals half the input-generating synchronous clock $f_o/2$ [7].

NRZ data has two characteristics that make clock-recovery difficult. First, the data may have long sequences of 1s and 0s. With no transitions for the circuit to operate, the clock-recovery circuit must hold the clock frequency with negligible drift during these events. Second, the spectrum of NRZ has nulls at even-integer multiples of the bit rate. Consequently, there is no energy at the clock rate for the data.

In general, binary bit streams of data have random characteristics. Consequently, statistics can be used to describe them. From the autocorrelation function of a random binary sequence, (9.15) computes the power spectral density [8]

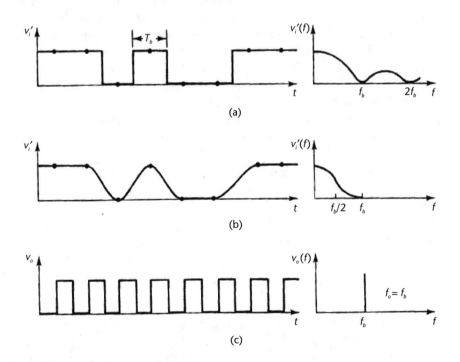

(a)

(b)

(c)

Figure 9.32 (a) NRZ data, (b) band-limited NRZ data, and (c) recovered clock. (*From:* [7]. © 1991 Prentice-Hall. Reprinted with permission.)

$$P_X(\omega) = T_b \left[\frac{\sin\left(\omega \frac{T_b}{2}\right)}{\omega \frac{T_b}{2}} \right]^2 \tag{9.15}$$

Figure 9.32(a) shows the shape of the power spectral density for NRZ data. The x-axis shows frequency at integer multiples of $1/T_b$. The maximum frequency of the data for a periodic 0,1 pattern is $1/T_b/2$. The recovered clock is at $1/T_b$, which is two times the maximum frequency of the data.

Figure 9.32(a) shows that the spectrum of NRZ has nulls at integer multiples of the bit rate. The band-limited form of the NRZ data [shown in Figure 9.32(b)] has most of the spectrum of this signal below $f_b/2$. With no spectral component at f_b, a PLL will not lock to the data to produce the clock signal.

An example shows the effect of nulls at integer multiplies of the bit rate. A 1 Gbps data stream with alternating 1s and 0s produces a 500-MHz square wave, which is the fastest waveform in the data stream. A signal at the 1-GHz clock rate does not exist. This characteristic makes PLLs lock to spurious signals or not lock at all.

To avoid false lock, the data stream goes through a nonlinear operation to create a frequency component at the clock frequency f_b. Then, a PLL can recover the desired clock signal V_o, as shown in Figure 9.32(c). For instance, the nonlinear operation (differentiator followed by a full wave rectifier) generates a pulse at each detected data transition.

When in lock, the PLL usually phases the clock so that its rising edges are centered on the data pulses (see the dots on the V_i' waveforms). If the PLL aligns the falling edge of V_o in the center of the V_i pulses, the complement of the clock can be used for data sampling.

Now. let's study the RZ format. Consider the data signal in Figure 9.31(b), where the second and third bits are adjacent 1s. The RZ format goes to zero between these consecutive bits of 1. The highest frequency in RZ data (a repetitive 0,1 sequence) in the data stream is f_b and equals the input-generating synchronous clock f_o; however, RZ uses more channel bandwidth than NRZ. Consequently, where larger channel bandwidths are costly, NRZ is preferred.

The RZ format produces a signal that has a periodic square wave with period T_b with some of the pulses missing. Therefore, the spectrum of this waveform has a line component at $f_b = 1/T_b$. In addition, the spectrum has a continuous component that extends beyond $f = 2f_b$. This portion of the spectrum corresponds to the random pattern of missing pulses. Communication applications lowpass-filter the RZ data to eliminate as much noise as possible. The filtering generates a rounded data-stream waveform in the time domain and a narrower spectrum in the frequency domain. The filter still has a broad-enough bandwidth to return the signal to zero between pulses in the time domain, and it still has a line component at f_b in the frequency domain. Clock-recovery extracts this spectral line as either a sine wave or a square wave.

Aligning the rising edges of the recovered clock at the center of the input data-stream pulses gives the optimum phase position. Then, the recovered clock samples the data at the optimum time to determine whether the bit is a 1 or a 0.

9.2.2 Edge Detection

One method to recover clock from NRZ data converts the NRZ data to an RZ-like data signal that has a line component at f_b. Then, the clock can be recovered from that RZ data with a PLL. Edge detection generates a waveform with an RZ-like format. Edge detection of a binary bit stream requires a method for sensing both positive and negative edges. Figure 9.33 shows NRZ data and the output of a positive and negative edge detection scheme [5].

Figure 9.34 illustrates the conversion process for band-limited NRZ data. The positive and negative transitions of the data stream V_i' (as shown in the top waveform in Figure 9.34) contain the phase information of the imbedded clock. Consequently, differentiating (d/dt) the data stream generates a pulse (either positive or negative as shown in the second waveform in Figure 9.34) corresponding

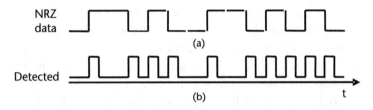

Figure 9.33 (a) NRZ data that is (b) edge-detected in a manner similar to RZ format.

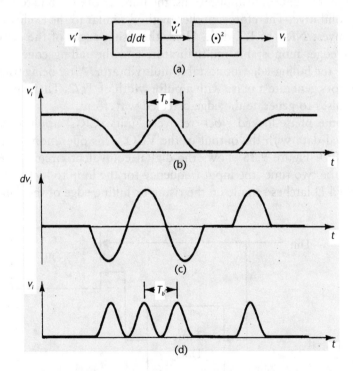

Figure 9.34 Resulting waveforms from converting band-limited NRZ data to RZ data: (a) functional diagram of the conversion process of (b) bandlimited NRZ data and the resulting waveforms (c) after differentiation and (d) after squaring to get RZ data. (*From:* [7]. © 1991 Prentice-Hall. Reprinted with permission.)

to each transition (from 0 to 1 or from 1 to 0). Squaring the differentiated signal dV_i' makes these pulses V_i all positive (as shown in the third waveform in Figure 9.34). Squaring produces fewer harmonics and reduces the in-band noise compared with full-wave rectification.

Signal V_i looks just like RZ data. The waveform has pulses spaced at intervals of T_b and some missing pulses. However, unlike standard RZ format, a pulse represents a transition rather than a 1 [7].

Low-noise applications use other options for converting NRZ data to RZ data. Figure 9.35 shows a method that is useful at high data rates ($f_b > 100$ Mbps). An exclusive-OR gate produces the desired edge detection by splitting the data stream into two paths with on of them delaying the data V_i' by $T_b/2$ and comparing the original data path with the delayed version of itself. After a data transition, the exclusive-OR gate senses the logical difference in $V_i'(t)$ and $V_i'(t - T_b/2)$ input and generates a pulse at the output.

For example, a data rate of $f_b = 100$ Mbps makes the bit spacing $T_b = 10$ ns and the desired delay $T_b/2 = 5$ ns. Let's assume signals propagate through an inverter delay in about 0.2 ns. Then, a 5-ns delay requires 26 inverters. Consequently, low signal-to-noise clock-recovery applications do not use this method because the long string of logic devices decreases the effective signal-to-noise ratio; however, the simplicity of this circuit in comparison with the circuit in Figure 9.34 makes it a good choice for the high-SNR application of recovering a clock from a logic data stream.

For low-data-rate applications, the number of inverters to realize T_b delay may be prohibitive. Therefore, another method similar to the exclusive-OR can be used to convert NRZ to RZ data. Here, the rising edges of the data stream trigger a rising-edge, monostable multivibrator, and the falling edges of the data stream trigger the falling-edge monostable multivibrator. After being triggered, both multivibrators generate a pulse with a pulse width of $T_b/2$. Then, an OR gate combines the pulses to generate the edge-detected waveform.

Some phase-locked clock recovery units (CRC) applications mix the edge-detected data with the output of the VCO. This operation samples the output of the VCO. Figure 9.36 shows the digital equivalent circuit. The VCO frequency has to be two times the input frequency for the loop to lock with this circuit. The clocked D latches sample on the rising or falling edge of the data and hold it. The

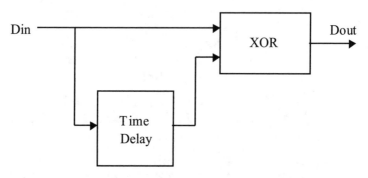

Figure 9.35 Edge detection of a data stream by comparing the original data stream to a delayed version of itself.

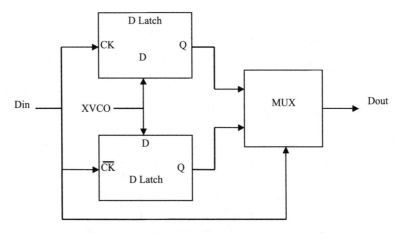

Figure 9.36 Diagram of double-edge-triggered flip-flop.

multiplexer selects the latch that is in the hold condition. This circuit is called a double-edge-triggered flip-flop [9].

9.2.3 Clock-Recovery Architectures

We will now discuss some of the clock-recovery architectures. Clock recovery consists of edge detection and a narrow-bandwidth tracking loop. The narrow-bandwidth tracking loop generates a periodic output that quickly settles to the input data rate and has negligible drift when input edges are missing. This operates like a sample-and-hold circuit. When there are edges, we want to correct the loop to the phase error that is detected, and when there are no edges, we want to hold the error value as long as possible until the next edge occurs. Figure 9.37 shows how a PLL can be used after an edge detector to recover a clock [5].

The output of the edge detector doubles the clock rate of the NRZ data into the mixer. The PLL locks to this clock rate. A missing edge does not cause the multiplier to add an additional signal to the lowpass filter, and the charge on the capacitor drifts over time. A narrow-bandwidth lowpass filter minimizes this drift; however, a narrow-bandwidth lowpass filter makes a narrow loop bandwidth. A narrow loop bandwidth produces a narrow capture range that can cause the loop to unlock and never recover. Consequently, most clock-recovery circuits use frequency

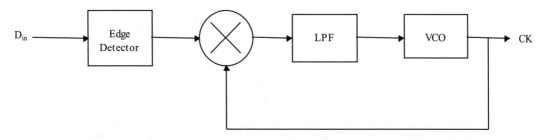

Figure 9.37 Edge detector followed by a PLL to recover the clock.

detection to steer the VCO frequency close enough (approximately within five loop bandwidths) to lock the loop.

The popularity and success of the three-state phase/frequency detector makes it the first phase detector to consider for any application. However, for clock-recovery applications, this detector is a poor choice. The three-state PFD produces false pulse widths for an NRZ bit stream.

Figure 9.38 shows the effect of pseudorandom code on a PLL with a phase frequency detector. The PLL was locked with a square wave data input, and then the pseudorandom code was applied. The top signal shows the input data to the edge detector, and the second signal shows the detected edges out of the edge detector. The third signal shows the modulation of the VCO output clock. The amount of modulation in frequency shows the difficulty the loop is having.

The fourth signal shows the tune voltage to the VCO. The up-and-down movement of the control line shows that the loop is having difficulty tracking the input. The next two signals show the up and down output signals out of the phase detector. The pulse-width modulation of the up and down signals show large phase-error variations in the loop. A lack of edges causes the up output to go high and cause an up error by driving the VCO to a higher frequency. Later, the loop corrects with some down pulses to try to regain lock.

This figure shows up pulses when there are no transitions, which will force the loop out of lock. The ideal would be a hold with no transitions. When the phase wraps in the comparison of the divided-down VCO output with the NRZ bit stream, the PFD falsely detects this as a high frequency and generates a wide down pulse to correct the PLL. Consequently, other choices for phase detector require careful examination, which will be discussed in Section 9.2.4.

9.2.3.1 Quadricorrelator

A quadricorrelator clock-recovery architecture combines frequency detection with a narrow-bandwidth PLL. Introduced by Richman in 1954, the analog version of the quadricorrelator has had a long history of use in the electronics industry [10]. The quadricorrelator has three loops, as shown in Figure 9.39. Loops I and II

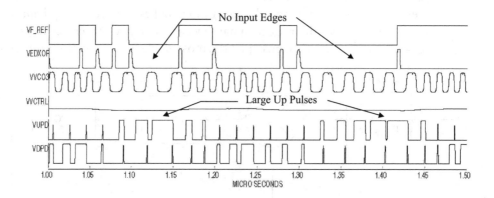

Figure 9.38   Poor clock-recovery detector response for a phase/frequency detector with pseudorandom input data.

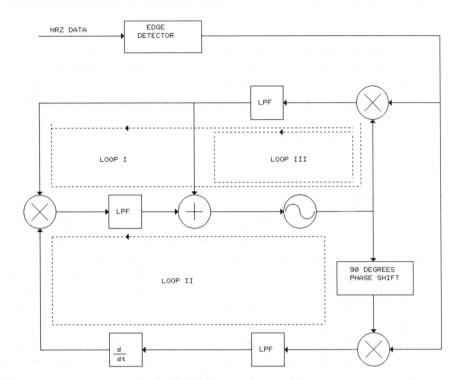

Figure 9.39 Block diagram of analog quadricorrelator.

perform frequency detection. Loop III does the phase detection. The frequency detection makes the capture range independent of the locked loop bandwidth. Consequently, the phase-detection loop can have a narrow loop bandwidth to minimize VCO drift between data transitions. This architecture requires the VCO to generate quadrature outputs. Consequently, a differential VCO would be preferred.

Mixing ω_1 for the output of the edge detector with ω_2 for the output of the VCO produces $\sin[(\omega_1 - \omega_2)t]$ at the output of the in-phase mixer and $\cos[(\omega_1 - \omega_2)t]$ at the output of the quadrature mixer. Differentiating the output of the quadrature mixer and mixing with the output of the in-phase mixer produces $(\omega_1 - \omega_2)\cos^2(\omega_1 - \omega_2)t$ at node P. Filtering the high-frequency component produces a voltage proportional to $(\omega_1 - \omega_2)$ for tuning the VCO. The negative feedback in the loop tunes the VCO until this is minimized to zero frequency error. As the loop approaches zero frequency error, the output of loop III begins to dominate, and the phase error begins to be minimized. After the loop achieves frequency lock, loops I and II should be disabled to prevent noise and spurious-signal disturbance paths.

A mixer (Gilbert cell) can electronically realize the multipliers in the circuit. The lowpass filters can be RC filters. A highpass filter can realize the differentiator. The 90° phase shifter can be a tap off of a differential ring oscillator.

Digital versions of the quadricorrelator replace the in-phase and quadrature mixers with the double-edge-triggered flip-flop, and this eliminates the need for an edge detector [9]. This produces quadrature beat notes at the difference frequency between the VCO and the bit rate into the circuit. From the outputs of the double-edge-triggered flip-flops, the signals can be processed by another double-edge-triggered flip-flop and some logic to generate an up and down frequency signal to control the frequency of the VCO. This circuit begins operation by minimizing the frequency difference. Then, the in-phase double-edge-triggered flip-flop is connected in summation with the frequency detector output and dominates the loop response. Other types of phase and frequency detectors are described in [10–13] and will be discussed in Section 9.2.4.

9.2.3.2 Hogge Clock Recovery

Another clock-recovery architecture uses a Hogge phase detector. Figure 9.40 shows a schematic for a clock-recovery circuit using a Hogge phase detector. The whole clock-recovery circuit has a coarse and a fine tune. In this figure, we are only concentrating on the fine-tune portion and model the coarse-tune portion with a dc supply voltage. The input data stream is fed to the input of the Hogge phase detector and compared with the VCO output. The Hogge detector generates and up and down signals that are used by a loop compensation to generate a fine-tune dc voltage VFINE_OUT. The fine tune voltage is summed together with the coarse tune voltage to adjust the VCO into the lock condition. The tune voltage adjusts the power-supply voltage in the VCO, which consists of a ring of RS latches. Adjusting the power-supply voltage changes the output frequency. Finally, a D flip-flop uses the recovered clock in order to clock the input data and produce the data output stream. This example minimizes the number of components and transistors needed to produce the fastest computer execution time.

Tuning of the coarse voltage is required in the computer simulation because of the small lock range (<5%) of the phase detector. The tuning is done by changing the coarse tune voltage until the VCO frequency is within 5% of the required lock frequency. The up and down signals will start to show a distinct trend once the frequency gets close to lock. Far away from the lock frequency, the up and down signals will have a nondistinct wave shape. Also, lock conditions to frequency harmonics and fractional harmonics of the VCO can be observed.

Figure 9.41 shows the key waveforms that are produced in the clock-recovery circuit when it is correctly locked to a pseudorandom data-stream input with 2^6 combinations. Studying these waveforms helps us understand clock-recovery circuits. The x-axis varies from 39 to 41 μs. The top waveform shows the input data stream at a 5-Mb rate. The fastest 1 and 0 square wave waveform is at 2.5 MHz. The data stream has a single transition at 39.5 μs, followed by multiple edges at 40.5 μs at the fastest bit rate. The next waveform shows the recovered data that was generated by the recovered clock. The recovered data is shifted from the input data by half a clock period of the recovered clock because of the 90° phase shift in the recovered clock. The next waveform is the recovered clock at 5 MHz. At 40.5 μs, the recovered clock waveform has a positive edge in the middle of the positive and negative pulse widths of the top waveform input data stream. This is the optimum position for the recovered data clock. The fourth and fifth waveforms are up and down outputs of the Hogge detector. The final waveform is the fine tune-voltage adjustment out of the clock-recovery filter.

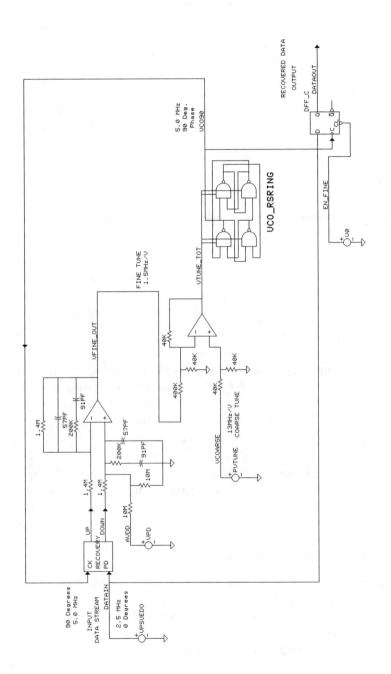

Figure 9.40 Functional schematic of the fine-tune portion of a clock-recovery circuit using a Hogge phase detector.

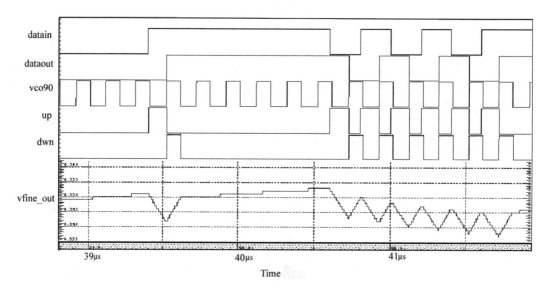

Figure 9.41   Waveform timing out of a clock-recovery circuit using a Hogge detector.

At 39.5 μs a single edge transition in the data produces an up and a down pulse out of the Hogge detector. This causes a ramp down followed by a ramp up in the fine tune voltage waveform. A balance in these pulse widths causes the net voltage to remain the same. An absence of transition edges causes the fine-tune output voltage also to remain the same. An up-signal pulse width larger than the down-signal pulse width causes the VCO frequency to increase, and a down-signal pulse width greater than the up-signal pulse width causes the VCO frequency to decrease. Consequently, a balance in the pulse widths keeps the clock-recovery circuit locked.

9.2.4   Phase Detectors for Clock Recovery

Studying how the various phase detectors work with random bits helps in understanding the trade-offs between them and in selecting the best detector for the application. Here, we will do simulations, measure the current output waveform for each phase detector with lag, zero, and lead phase errors, and look at the waveforms to understand phase detector operation.

9.2.4.1   Phase/Frequency Detector

The popularity and success of the three-state phase/frequency detector makes it the first phase detector to consider for any detector application. However, for clock recovery, this detector is a poor choice. The three-state PFD produces false up pulses for an NRZ bit stream.

Figure 9.42 shows what happens when trying to use a phase/frequency detector to recover a clock. The conditions for this case are different from those of the previous simulation with the phase/frequency detector. In this case, the loop is not closed, and the VCO frequency does not change with the resulting error signal. Instead, the VCO is aligned to have almost zero phase error when a data edge occurs. The top two inputs are used to generate the third from the top output, which is an edge-detected waveform.

The second from the top waveform is the data stream, NRZ. The top waveform is the ideally delayed input data stream. The third from the bottom waveform is the ideal recovered clock, which is twice the f_b. The bottom two waveforms are the up and down outputs out of the phase/frequency detector. With no clock edge, the phase detector generates an up signal with a wide pulse width equal to $T_b/2$ and nT_b for two or more consecutive 1s or 0s. Ideally, we want a phase detector to generate an error only when there is an edge and to hold the value until the next edge comparison. Clearly, this configuration will not work for clock-recovery applications; however, it does show that we can evaluate clock-recovery phase detectors with this test fixture configuration.

Figure 9.43 shows an eye diagram. This waveform is generated on an oscilloscope by displaying a pseudorandom data stream and using the generating clock or the recovered clock for a trigger. On a simulator, a triangle wave must be generated at the data clock rate; the pseudorandom data-stream magnitude is plotted on the y-axis, and the magnitude of the triangle wave is plotted on the x-axis. The sinusoidal shape of the logic levels shows that the minimum bandwidth was used on this data. A recovered clock at points B or D will generate bit errors. We want the ideal clock to have a positive edge at point C midway between the zero crossing times of B and D. This is optimum because it will take the largest level of FM on the clock to cause a bit error.

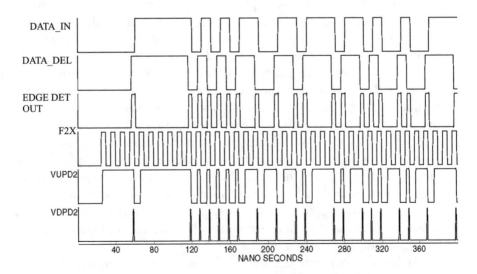

Figure 9.42 Phase/frequency detector with a zero-phase-error, ideal-clock-recovery waveform relationship.

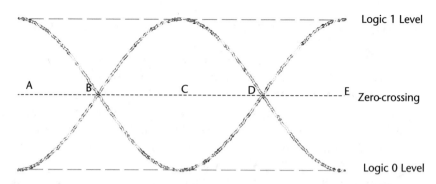

Figure 9.43   Eye diagram showing the optimum clock position to recover the data [2].

9.2.4.2   Exclusive-OR Edge Detector

Many of the clock-recovery phase detectors use exclusive-OR gates. Consequently, exclusive-OR gates are a basic building block, and understanding how it processes clock-recovery data is essential for understanding clock-recovery phase detectors. A simple exclusive-OR gate (multiplier) can be used as a phase detector for clock recovery. A simulation of the exclusive-OR gate with one input connected to a pseudorandom data stream that has been edge detected with another exclusive-OR gate and with the other input connected to a VCO clock that is set to the optimum position for zero phase error helps in understanding how the detector works. The output connects to a unity gain operational amplifier with some lowpass filtering. The current across the feedback resistor shows the equivalent charge pump current that would be generated.

Figure 9.44 shows the resulting waveforms for the exclusive-OR gate with a pseudorandom input data stream that has a 50-ns bit rate T_b and the following logic pattern: 0,0,0,0,1,1,1,1,1,0,1,0,1,0,1,1,0,0,1,1,0,1,1,1,0,1,1,0,1,0,0,1,0,0, 1,1. Studying the waveforms shows that the falling edges of the VCO clock (frequency of twice the data rate) splits the data pulses evenly in the zero-phase-error position. Figure 9.44 shows a zero average (V_{davg}) of the detector output V_d.

At this zero-phase-error condition, the product V_d consists of pulses with equal positive and negative areas, about the 2.5-V level for a 5-V supply, which is the zero-average phase-error point. The data stream has six consecutive edges between 0.5 and 0.9 μs in the data and six clocks at the output of the edge detector. This produces 12 edges in the V_d, which shows how the phase detector evenly distributes the area to obtain zero phase error. This operation looks like an early and late gate method. In an early and late gate method, a pulse before the detected edge is canceled by a negative pulse after the detected edge. Equal area alignment about the detected edge gives a net zero change. This balanced operation could make it sensitive to duty cycle. Later on in this section, we will look at this condition. Also, this method could generate a lot of spectral energy at various frequencies. This can make it difficult to filter out and will modulate VCO. If the modulation is large enough, the loop can lose lock.

Let's do a comparative study of the output current waveform when leading, in-phase, and lagging VCO edges are used. A bit rate of 50 ns was used for this simulation. A slightly delayed clock from the data produces a positive phase error

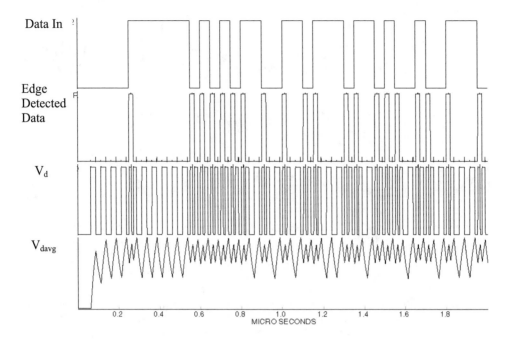

Figure 9.44 Exclusive-OR phase detector with a zero-phase-error, ideal-clock-recovery waveform relationship.

θ_e. For this condition, the V_d pulses have more positive than negative area, as shown in the bottom waveform in Figure 9.45. This positive area makes V_{davg} positive. A slightly leading clock to the data produces a negative phase error θ_e. For this condition, the V_d pulses have more negative than positive area as shown at the top waveform in Figure 9.45. This negative area makes V_{davg} negative.

Incrementally sweeping the phase error and calculating the average output voltage for each produces the transfer function for the phase detector. A continuous input data stream and a 32-bit random data stream were used to see the effect on the phase detector transfer function. Figure 9.46 shows the resulting phase detector characteristic of V_{davg} versus θ_e for 100% and 50% edge densities. The overlay of these waveforms shows that the triangular-shaped transfer function has a maximum and minimum value that depends on the density of data pulses.

This reduction in peak-to-peak value for the random data reduces the gain of the phase detector. A reduction in phase detector gain makes the loop narrower when compared with continuous data. This gain reduction needs to be taken into account when making bandwidth calculations for the PLL.

Let's develop an expression for the phase detector gain from the simulated transfer function. Studying Figure 9.44 shows V_d has an average of zero for bit times with no V_i pulses and a maximum or minimum level for the nonzero θ_e. The minimum and maximum voltages have half the 100% edge density value for the case with half the pulses missing on average. Then, for $\theta_e = \pi/2$ and a pulse density of 50%, V_d will be $V_h - V_l$ half the time and an average of zero the other

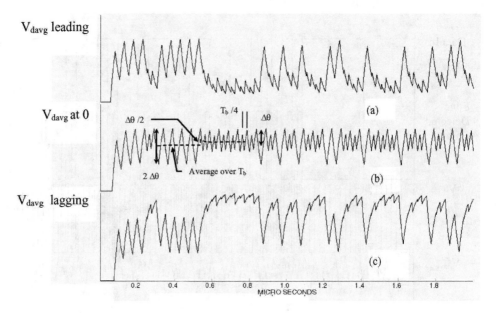

Figure 9.45 Waveforms for exclusive-OR with (a) a negative V_{davg} for lead phase shifts from the ideal clock, (b) a zero V_{davg} for zero phase shifts from the ideal clock, and (c) a positive V_{davg} for lag phase shifts from the ideal clock.

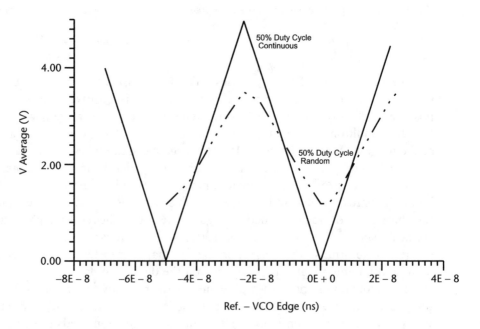

Figure 9.46 Exclusive-OR transfer functions with continuous and random data (50% density).

half the time. Consequently, $V_d = (V_h - V_l)/2$ corresponds to $\theta_e = \pi/2$ for a 50% pulse density as shown by the transfer function in Figure 9.46. Therefore, (9.16) computes the corresponding phase detector gain for a data stream with 50% pulse density:

$$K_{d50} = (V_h - V_l)\,\pi \tag{9.16}$$

where

V_h = logic high-level voltage;

V_l = logic low-level voltage.

Now, for the same example, let's vary the duty cycle on the 20-MHz clock frequency to see its effect on the transfer function. Figure 9.47 shows the effect of 25% duty cycle on the transfer function for the phase detector. The phase-error operating range is reduced by 40%, and time is shifted 5 ns, but the phase detector gain (slope) stays the same. Low duty cycles should be avoided because of the distortion they cause. To use the exclusive-OR phase detector, prior knowledge of the bit rate must be known so that the edge detector can adjust the delay to the exclusive-OR to have 50% duty cycle. In addition, we may want to study other phase detectors that are not sensitive to duty cycle.

9.2.4.3   Pattern-Dependent Jitter

In general, the lowpass nature of the PLL transfer function $H(s)$ makes the PLL respond to V_{davg}, the average of V_d. For the PLL bandwidth K much less than the baud f_b, V_d has a long-term average of zero in Figure 9.44 for $\theta_e = 0$. However, for a wideband PLL, the pattern of V_d, which depends on the data, does have an effect on the output phase of the loop θ_o.

The analysis here holds for an exclusive-OR phase detector. Consider the data pattern V_i shown in Figure 9.44 with the corresponding V_d pattern for $\theta_e = 0$.

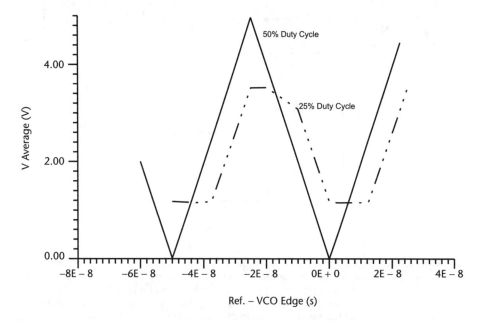

Figure 9.47   Exclusive-OR transfer function for 50% and 25% duty cycles.

Making V_d a phase-modulation input at $m(t)$ in Figure 9.48 allows us to analyze its effect. Figure 9.48(a) shows phase modulation of a PLL for a signal injected $m(t)$ after the phase detector.

Figure 9.48(b) shows a signal flow graph of a type 1 PLL with modulation $m(t)$. The transfer function from $m(s)$ to $\theta_o(s)$ has a flat frequency response up to the loop bandwidth. The forward gain from m to θ_o gives $A = F(s)K_v/s$. The feedback from θ_o to m gives $B = K_d$. From control theory, this produces (9.17):

$$\frac{\theta_o(s)}{m(s)} = \frac{A}{1 + AB} = \frac{F(s)K_v/s}{1 + K_d F(s)K_v/s} = \frac{G(s)/K_d}{1 + G(s)} = \frac{1}{K_d}H(s) \approx \frac{K/K_d}{s + K} \quad (9.17)$$

$$\frac{\theta_o(s)}{V_d(s)} \approx \frac{K/K_d}{s + K}$$

where

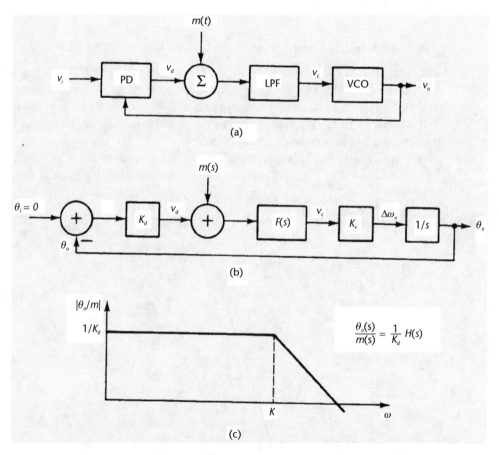

Figure 9.48 Signal flow graph of a type 1 PLL and the resulting transfer function [7]: (a) block diagram of a phase modulated PLL by a signal $m(t)$ injected after the phase detector in the time domain; (b) a signal flow graph of a type 1 PLL in the S domain with the injected signal; and (c) the resulting transfer function plot of output phase over input signal ratio versus radian frequency. (*From:* [7]. © 1991 Prentice-Hall. Reprinted with permission.)

$H = G/(1 + G)$;

$K = K_d K_v F(0) \approx$ loop bandwidth for Type I PLL.

The closed loop $H(s)$ has unity gain with a high-frequency cutoff at $\omega = K$. Figure 9.48(c) shows $\theta_o(s)/m(s)$ has a gain of $1/K_d$ with a high-frequency cutoff at $\omega = K$.

The $\theta_o(t)$ waveform in the middle of Figure 9.45 shows that the lowpass transfer function characteristic of Figure 9.49 shapes the waveform. For pulses on V_i, θ_o has a triangular wave shape with a peak-to-peak amplitude of $\Delta\theta$. For no pulses on V_i, θ_o has a triangular wave shape with a peak-to-peak amplitude of $2\Delta\theta$. For changes in the data pattern on V_i, θ_o has a decaying lowpass-filter exponential transient. This transient is the short-term average of θ_o. The transient has a $\Delta\theta/2$ amplitude.

The description of the waveform allows us to develop an expression for $\Delta\theta$ in terms of K and T_b. For frequencies on the order of f_b (therefore greater than K), (9.18) becomes approximately (9.17):

$$\frac{\theta_o(s)}{V_d(s)} \approx \frac{K/K_d}{s} \tag{9.18}$$

Rearranging and taking the inverse La Place transform produces (9.19) and (9.20):

$$s\theta_o(s) \approx V_{davg}(s)K/K_d \tag{9.19}$$

$$d\theta_o(t) \approx V_{davg}(s)K/K_d \tag{9.20}$$

Figure 9.45 shows that $V_d = V_h - V_l$ for an interval $T_b/4$ in the presence of a data pulse on V_i. Then, (9.21) describes the change in θ_o during this interval:

$$\Delta\theta \approx d\theta_o T_b/4 = (V_h - V_l)K/K_{d50}T_b/4 \tag{9.21}$$

Substituting the phase detector gain [$K_{d50} = (V_h - V_l)/\pi$] for a density of 50% 1s in the data in (9.16) produces (9.22):

$$\Delta\theta = (\pi/4)KT_b \tag{9.22}$$

The absence of a pulse in the data pattern V_i in Figure 9.45 makes the time interval $T_b/2$, during which there are θ_o ramps, and the change becomes $2\Delta\theta$. From the exponential-transient amplitudes of $+\Delta\theta/2$ and $-\Delta\theta/2$, (9.22) for $\Delta\theta$ computes the peak-to-peak phase jitter of θ_o.

The specific data pattern determines the specific form of the transient behavior. A pattern of alternate 1s and 0s gives a virtually nonexistent jitter; however, a pattern of long strings of 1s and 0s produces the jitter amplitude computed in (9.22).

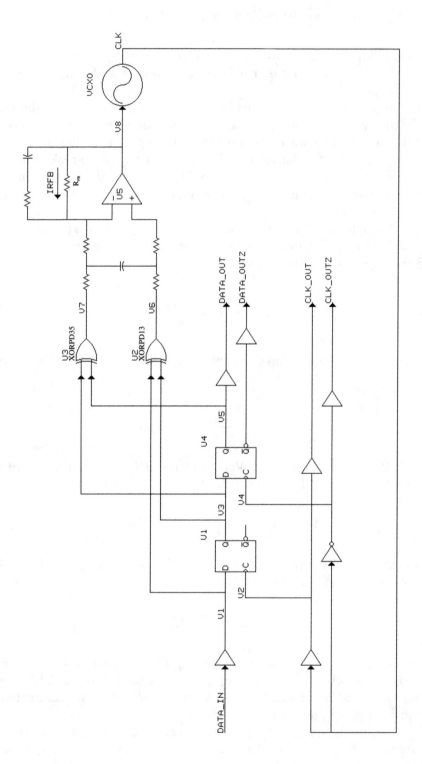

Figure 9.49   Schematic of a Hogge detector.

Example 9.1

A PLL with an exclusive-OR phase detector recovers a clock from RZ data. The PLL has a bandwidth of one-tenth the baud rate, which makes $K = 0.1 \times 2\pi f_b$. Find the pattern-dependent jitter. Then, find the pattern-dependent jitter for a bandwidth of 1/100th of the baud rate.

Substituting $f_b = 1/T_b$ into the PLL lowpass-filter expression gives $K = 0.2\pi/T_b$. Substituting the result into (9.22), $\Delta\theta = (\pi/4)KT_b$ gives $\Delta\theta = 0.2\pi^2/4 = 0.5$ rad. This shows significant pattern jitter of 8% [0.5 rad/(2π) 100] of a bit interval for a bandwidth of one-tenth the baud rate. For 1/100th the bandwidth, $K = 0.02\pi/T_b$, which gives a $\Delta\theta = 0.02\pi^2/4 = 0.05$ rad. This results in a 0.8% pattern jitter which is acceptable. Consequently, this gives us a rough estimate of the bandwidth that we need to meet the jitter requirements of a clock-recovery circuit.

9.2.4.4 Offset and Accumulated Jitter

The dc offset V_{do} of a phase detector produces another kind of pattern-dependent phase jitter. The offset voltage causes the clock phase to drift for the no-data-pulse condition (no phase information). The variations of the data pattern determine the length of the drift ([10] covers this in more detail).

Offset jitter may seldom be a problem for one PLL. However, for several PLLs in series, the offset can accumulate to a significant level. This condition occurs for a chain of data repeaters. After some distance of transmission, distortion and noise require that the signal be regenerated. Data stream V_i transmitted at the head end has the clock recovered by a PLL. The PLL cleans up the regenerated data, which we will call V_1. Repeated clock recovery can occur as many as 1,000 times in a long-distance transmission. Each clock-recovery circuit in the transmission path adds its own offset jitter to the total jitter ([10] covers this in more detail).

9.2.4.5 Hogge Detector

Up until now, our discussion has concentrated on a method to recover clock from NRZ data that required an edge detector to convert the NRZ data to RZ-like data, as shown in Figure 9.35. Now we will discuss an RZ phase detector that operates on the RZ data to recover the clock. Figure 9.49 shows a circuit to compare the phase of NRZ data directly with a clock that skips the need for an edge detector circuit. This circuit has similar characteristics to a two-state phase detector with a phase range from $-\pi$ to π, as shown in Figure 9.50.

The operation and connection of the circuit will be studied to get a deeper understanding of how this circuit works. The NRZ data signal connects to the D flip-flop U_1 and the exclusive-OR U_2. The noninverting CLK clocks D flip-flop U_1, and the inverting CLK clocks D flip-flop U_4. D flip-flop U_1 serves both the retiming decision circuit for the receiver and a part of the clock-recovery circuit. The rising clock edge V_o samples the data V_i. The retimed data from D flip-flop U_1 connects to exclusive-ORs U_2 and U_3 and the D flip-flop U_4. The exclusive-OR U_2 compares the sampled data Q_1 with the data.

The exclusive-OR outputs produce signals V_6 and V_7. The exclusive-OR output V_6 from U_2 produces a variable pulse width for each transition of the data signal.

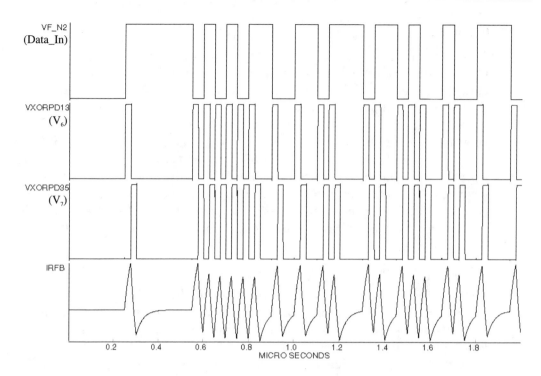

Figure 9.50 Comparison of EXOR and Hogge transfer functions with random data (50% density).

The pulse width depends on the position of the clock within the eye opening. For a clock that lags behind the ideal-clock position, the pulse widths vary from $T_b/2$ to T_b. The exclusive-OR output V_6 from U_2 also produces a variable pulse width for each transition of the data signal. For a clock that leads, the pulse widths vary from 0 to $T_b/2$. V_7 has a fixed-width square pulse with a pulse width of half a clock period for each transition of the retimed data. Connecting the outputs of the exclusive-OR gates to the differential active filter and filtering the two pulses out of the exclusive-OR gates produces the error voltage V_8 for controlling the VCO clock.

For θ_e going from $-\pi$ to π, the pulse width of V_6 goes from 0 to T_b. For $\theta_e = 0$, the pulse V_6 has a width of $T_b/2$, and the average of V_6 depends on the data-transition density (the number of V_6 pulses). Therefore, the average of V_6 alone does not give us the ideal 0V-average result. However, independently of θ_e, waveform V_7 maintains the same pulse width of V_6 for the $\theta_e = 0$ condition. Consequently, the reference waveform V_7 from the exclusive-OR output of U_3 has a fixed-width square pulse with a pulse width of half a clock period for each transition of the retimed data. The waveform at V_7 gives us a reference to compare against the varying pulse-width waveform V_6. Differentially comparing V_6 and reference V_7, then lowpass-filtering, produces the ideal 0V-average waveform output for $\theta_e = 0$ and a linearly varying transfer function from $-\pi$ to π. Consequently, properly centering the leading edge of the clock with the data produces a pulse width, V_6, out of the variable-pulse-width exclusive-OR U_2, that is identical to that of the reference fixed-width pulse V_7. Therefore, $V_8 = V_6 - V_7$ always has a zero average for $\theta_e = 0$.

One circuit method for differentially comparing the signals connects the outputs of the exclusive-OR gates to a differential active lowpass filter. The output of the differential active lowpass filter produces the error voltage for controlling the VCO clock. Figure 9.49 shows this connection.

Another circuit method for differentially comparing the signals connects the outputs of the exclusive-OR gates to charge pumps that differentially sum currents. An output capacitor lowpass-filters the output waveform to produce the error voltage for controlling the VCO clock.

Figure 9.50 shows the timing diagram for the response of a Hogge detector to a pseudorandom data stream. The top waveform of the data stream contains three logic 1s and seven logic 0s (0,0,0,0,1,1,1,1,1,0,1,0,1,0,1,1,0,0,1,1,0,1,1,1, 0,1,1,0,1,0,0,1,0,0,1,1) with the clock properly centered within the data-bit interval. The waveform numbers correspond with the node numbers in Figure 9.49. The pulse patterns from the outputs at 6 and 7 XORPD13 and XORPD35 have identical average values, which results in zero error voltage from the active loop filter at node 8 of Figure 9.49.

The active loop filter consists of differential amplifier U_5 and a feedback network. In conjunction with the integration contribution by the VCO, the active loop filter makes a second-order loop with a lowpass filter and phase-lead correction. The effect of combining the differencing and integrating functions into one circuit is the same as if we were to integrate the two pulse waveforms (nodes 6 and 7 XORPD13 and XORPD35) separately, then subtract the result to produce the correction current at IRFB.

Varying the phase error above and below the clock produces the waveforms in Figure 9.51. The same data-stream pattern was used (0,0,0,0,1,1,1,1,1,0, 1,0,1,0,1,1,0,0,1,1,0,1,1,1,0,1,1,0,1,0,0,1,0,0,1,1). The top waveform in Figure

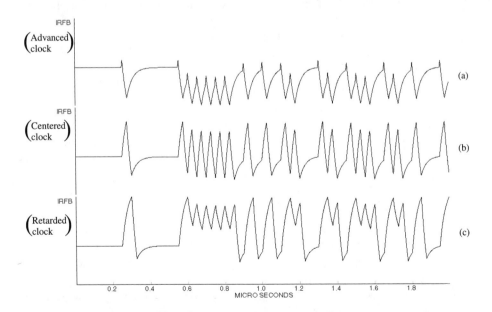

Figure 9.51   Hogge phase detector with (a) a negative I_{davg} for a leading phase shift from an ideal clock, (b) a zero I_{davg} for a zero phase shift from an ideal clock, and (c) a positive I_{davg} for a lagging phase shift from an ideal clock.

9.51 shows a timing diagram with the clock signal advanced relative to the center of the data-bit interval. This phase error causes the logic 1 pulses at exclusive-OR output node 6 to become narrower, while those at exclusive-OR output node 7 remain the same width. This produces a pulse pattern at node 6 that has an average value more negative than that at node 7. Filtering and taking the difference of these pulses produces a negative correction voltage at node 8.

Similarly, the bottom waveform in Figure 9.51 shows a timing diagram with the clock signal retarded relative to the center of the data-bit interval. This phase error causes the logic 1 pulses at exclusive-OR output node 6 to become wider, while those at exclusive-OR output node 7 remain the same width. This produces a pulse pattern at node 6 that has an average value more positive than at node 7. Filtering and taking the difference of these pulses produces a positive correction voltage at node 8.

A comparison of the exclusive-OR detector in Figure 9.45 with Figure 9.51 shows that the Hogge detector removes edges between 0.35 and 0.55 μs, which reduces spectral energy at $2f_b$. This reduction in energy should result in lower output jitter. Furthermore, the data pattern has six consecutive edges between 0.5 and 0.9 μs in the data. Studying this data pattern shows additional differences in the detectors. The exclusive-OR detector in Figure 9.45 has 12 edges in the feedback current between 0.5 and 0.9 μs. Figure 9.51 shows six edges in the feedback current between 0.5 and 0.9 μs. Consequently, this waveform will have reduced spectral energy at $2f_b$.

Stable loop operation does not require that the data have an equal number of logic 1s and 0s over any time interval. The loop only requires that a sufficient number of transitions (data activity) in either direction occur to keep the loop stable. The phase detector gain factor varies with this data activity. An example shows that the gain factor variation with data activity can easily be calculated.

Example 9.2

Assume a Hogge detector has square pulses at nodes 6 and 7. Next, assume a $1 - V$ logic swing and one data transition within five-bit intervals. Calculate the effect of a $\pi/8$ shift in the clock position.

A shift in the clock position of $\pi/8$ will shift the dc level at node 6 by 25 mV [$(1V/\pi$ detector gain$)(\pi/8$ shift$)(1/5$ density$)$]. This gives a 63.66 mV/rad [25 mV/($\pi/8$ shift)] gain factor for that transition density.

The decrease in the phase detector gain factor decreases the open-loop gain for the decreased transition density. This change decreases the closed-loop bandwidth for decreasing transition density. Decreasing loop bandwidth makes the circuit more selective to changes. Consequently, the circuit holds its value over an occasional long string of 1s or 0s.

At high frequencies the delay through the flip-flop U_1 becomes significant. At low frequencies, the circuit in Figure 9.49 relies on a negligible propagation delay through the flip-flop U_1 relative to a one-bit interval. For high-frequency cases (~140 MHz), comparable delay between the originating node and U_2 must be added to compensate for the delay through U_1. For even-higher-frequency cases (~565 MHz), a $T_b/2$ delay must be substituted for U_4 between the originating node and U_2 to compensate for the delay through U_1.

Figure 9.52 shows the transfer function of the Hogge detector for output average current versus phase error with a 20-MHz clock frequency. A zero average for $\theta_e = 0$ corresponds to $V_{davg} = 0$ for the phase detector characteristic as shown in the figure. The maximum value of the characteristic depends on the transition density. For a 50% density, the maximum value is $(V_h - V_l)/4$, where V_h is the logic high level, and V_l is the logic low level.

Comparison of Hogge and exclusive-OR phase detector characteristics in the figure show that the Hogge detector has twice as much phase-error range. In addition, this circuit has no duty-cycle sensitivity and lower reference sidebands.

The Hogge detector has several advantages. First, having the loop closed around the decision flip-flop self-corrects the clock position to the middle of the data changes. Other approaches to clock recovery that do not include this feature have to align the clock by other means. For the alternate approach, the extraction of the clock reference signal from the incoming data can have a long path before it produces a jitter-free clock that is applied to the decision flip-flop. The path to produce the jitter-free clock can consist of several stages of active, and one or more passive (high Q L C or SAW filter), circuits before it reaches the decision flip-flop. The timing relationship between the recovered clock and the data requires careful timing alignment. The alignment must be maintained over the life of the product. The effects of temperature and power-supply variations, as well as of component aging, on other approaches can cause significant misalignment, while the Hogge detector method would not suffer from these variations because the decision flip-flop is included in the loop.

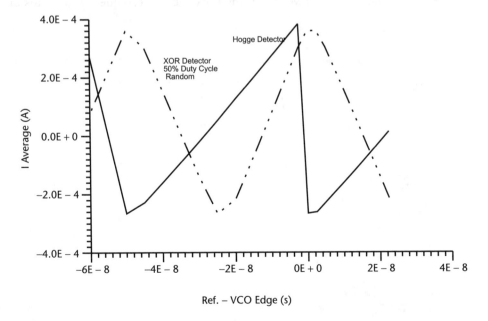

Ref. – VCO Edge (s)

Figure 9.52 For Hogge, negative, zero, and positive I_{davg} for lead, zero, and lag phase shifts from the ideal clock.

The Hogge detector has other advantages. The Hogge detector uses NRZ input data stream directly instead of RZ, which an exclusive-OR detector requires. Consequently, a Hogge detector does not need an additional edge detector. It has a large phase range and lower high-frequency energy and is duty-cycle insensitive.

Using this circuit requires one precautionary note. This circuit does not have a frequency-acquisition circuit like the phase/frequency detector has. Therefore, the clock source should either be crystal stabilized, or a sweep acquisition circuit should be included to acquire lock. Considering all of these factors, the simple and self-correcting Hogge circuit has both performance and cost advantages.

9.2.5   Clock-Recovery Tests

Clock-recovery circuits have their own set of measurements that have very little meaning to other PLLs. This section discusses these measurements.

9.2.5.1   Jitter-Tolerance Measurement

Figure 9.53 shows the test setup used to measure the jitter tolerance of a clock-recovery circuit. The test setup in Figure 9.53 shows a typical bit-error-rate (BER) test, except the transmitter uses an external clock. The external clock uses frequency modulation to produce jitter. Increasing the amount of jitter at a fixed frequency and monitoring the BER until it passes a BER limit of 10^{-10} makes a jitter-tolerance measurement. A BER of 10^{-10} is called the limit of the jitter tolerance.

Figure 9.54 shows the measured jitter tolerance with a comparison to the SONET specification. The y-axis is expressed in unit intervals (UIs). What is a UI? Studying narrowband FM theory and the concept of frequency as the instantaneous

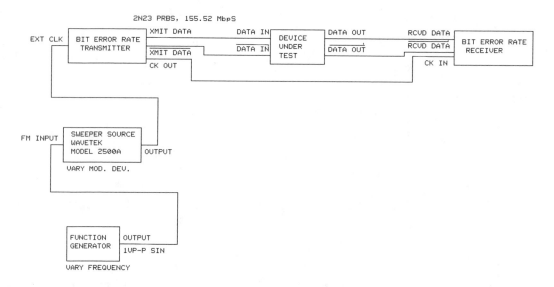

Figure 9.53   Test setup for jitter tolerance of a clock-recovery circuit.

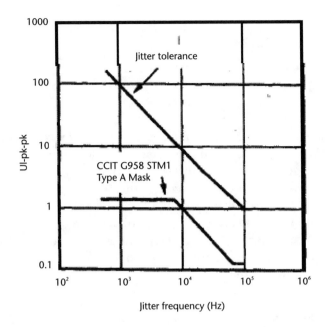

Figure 9.54 Jitter-tolerance measurement compared to SONET specification [5].

derivative of phase makes understanding this unit of measure easier. A frequency-modulated clock can be represented by (9.23):

$$V(t) = \cos[2\pi f_c t + \Delta\omega/\omega_m \sin(2\pi f_m t)] \tag{9.23}$$

where

f_c = clock rate;

f_m = jitter modulation frequency;

$\Delta\omega$ = zero-to-peak frequency deviation.

Equation (9.24) shows that the derivative of the argument of the cosine gives the instantaneous frequency of this modulated signal:

$$\omega(t) = 2\pi f_c + \Delta\omega \cos(2\pi f_m t) \tag{9.24}$$

Thus, (9.24) shows that the peak frequency deviation is Δf, and $\Delta\omega/\Delta\omega_m$ is recognized as the modulation index. From (9.23), doubling the amplitude of the sinusoidal phase term gives (9.25) for the peak-to-peak phase deviation in radians:

$$\Delta\theta_{pp} = 2\Delta\omega/\omega_m \text{ rad} \tag{9.25}$$

Finally, dividing this phase deviation in (9.25) by 2π (to normalize to one cycle) gives jitter in UI_{pp}. Equation (9.26) shows the result of this operation:

$$\text{UI}_{pp} = \Delta\omega/\omega_m/\pi = \Delta f/f_m/\pi \tag{9.26}$$

Equations (9.27) and (9.28) show the UI relationship to modulation index and deviation ratio (DevRat):

$$\Delta f / f_m = \mathrm{UI}_{pp}\,\pi \qquad (9.27)$$

$$\mathrm{DevRat} = \mathrm{UI}_{pp}\,\pi \qquad (9.28)$$

Signal generators capable of external frequency modulation commonly have a modulation range of zero-to-peak frequency deviation for a $1 - V$ zero-to-peak modulation input. The ratio of the peak frequency deviation to the modulation frequency is known in radio jargon as the deviation ratio, or DevRat. The peak-to-peak UI of jitter is then given by (9.28).

The measurement setup for jitter tolerance requires a frequency-modulated carrier test generator that can generate enough jitter at low frequencies to test the device fully. Most jitter generators will not create a test signal with a large number of unit intervals of jitter.

9.2.5.2   Jitter Transfer Function Measurement

Maximum bandwidth requirements for a regenerator necessitate measurement of the jitter transfer function of a clock-recovery device. Also, jitter transfer measurements aid jitter-tolerance characterization.

Figure 9.55 shows a test fixture to phase-modulate a reference carrier with the excitation source from a network analyzer. Then, the modulated carrier clocks a data generator to drive the device under test (DUT). A comparison of the phase

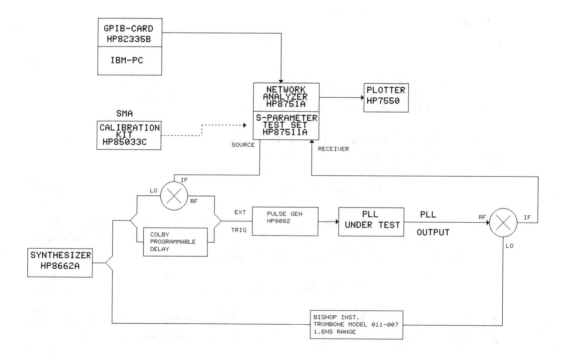

Figure 9.55 Jitter transfer function test setup.

of the recovered clock with the original unmodulated reference carrier produces a signal that is received and detected by the network analyzer. Then, the network analyzer displays the resulting transfer function.

As an example, the jitter transfer function of a clock-recovery device with various loop-damping ratios has been measured. For damping ratios of 1, 2, 5, and 10, the circuit exhibits jitter peaking of 2, 0.5, 0.08, and 0.02, dB, respectively. In addition, these conditions produces measured −3-dB bandwidths of 130, 100, 90, and 90 kHz, respectively.

The SONET specification for OC3 has a maximum bandwidth requirement of 130 kHz and maximum jitter peaking of 0.1 dB. Consequently, we are required to use a damping of 5 or greater. Figure 9.56 shows a measured jitter transfer function for another clock-recovery device with different damping capacitors. The family of curves here agrees remarkably well with theoretical results.

9.2.5.3 Bit-Error Rate Versus Signal-to-Noise Ratio

Figure 9.57 shows the test setup for measuring BER versus SNR. The Tektronix CSA 907T and 907R generate data and measure the BER. Attenuating the data output from the generator and mixing with noise creates any desired SNR at the input to the DUT. The hybrid combiners have two purposes in this test setup. First, they provide isolation between the A and B input ports. Second, one of the combiners inverts the phase of the noise signal to be added to the already differential data from the generator. Also, the inputs of the DUT have 50Ω to terminate each of the combiners separately.

The noise in this test must be lowpass filtered with a corner frequency at 0.7 times the bit rate to simulate the dynamics of a fiber-optic receiver. These fiber-optic receivers generally have an analog bandwidth of 0.7 of the bit rate. This

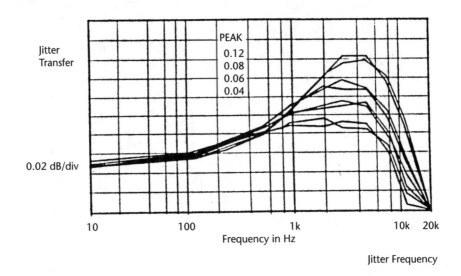

Figure 9.56 Measured data of jitter transfer functions. (*From:* [14]. © 1991 IEEE. Reprinted with permission.)

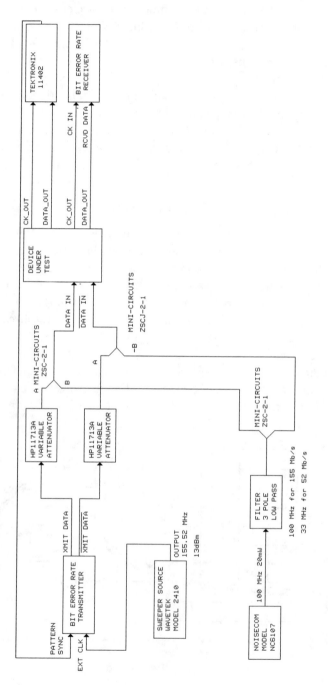

Figure 9.57 Test setup for BER versus signal-to-noise ratio.

bandwidth compromises between intersymbol interference and noise. Figure 9.58 shows the schematic details of the 33-MHz filter used in Figure 9.57. Measurements at 155 MHz do not use the filter because the band limit of the noise source itself is 100 MHz. Averaging of the noise signal by the 33-MHz filter makes the noise more Gaussian. The clipping of extreme peaks causes the unexpected slope of the 155-MHz data in Figure 9.59(b). But the 52-MHz data in Figure 9.59(a) lies more closely to theoretical expectations.

The data of Figure 9.59 has been plotted on special graph paper that plots the theoretical result as a straight line. This line has been labeled the complementary error function. All of the data fall close to the line. However, more importantly, the slope of the data line matches the theoretical result. Data plots of BERs that are better than theoretical indicate an oversimplified (i.e., wrong) theory.

The data at 155 MHz does not match the ideal slope as well as the data at 52 MHz. This mismatch results from a noise generator that does not give truly Gaussian noise. In fact, the generators have been known to clip extreme noise peaks. Consider the situation of a high signal-to-noise ratio and low BERs. In this case, extreme noise peaks cause errors. Consequently, knowing the exact characteristics of the noise peaks supplied by the generator helps determine the sensitivity of the measurements to these characteristics.

To begin the theoretical calculation of BER versus SNR for a binary signal with additive white Gaussian noise, we make the amplitude of the binary signal equal to A. Then, we set the comparator threshold at half the amplitude, $A/2$, to decide between a logic 0 and a 1. Consequently, a noise peak greater than $A/2$ makes an error.

Two types of errors can occur. A 0 can be mistaken for a 1, and a 1 for a 0. Then, summing the areas in the tails of the two Gaussian distribution curves calculates the total error rate. Equation (9.29) gives the result of this calculation:

$$\text{BER} = 1/2 \; \text{erfc}\left[1/\left(2\sqrt{2}\right) S/N\right] \qquad (9.29)$$

where

erfc( ) = complementary error function;

S/N = raw signal-to-noise ratio.

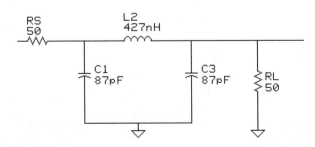

Figure 9.58   33-MHz LPF used in BER versus signal-to-noise ratio test. (*From:* [14]. © 1991 IEEE. Reprinted with permission.)

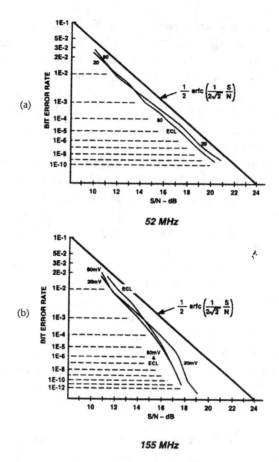

Figure 9.59 Measured and theoretical BER versus signal-to-noise ratio for (a) 52 MHz, and (b) 155 MHz. (*From:* [14]. © 1991 IEEE. Reprinted with permission.)

9.3 Effect of Phase Noise on A/D Converters

Many systems, such as communication systems, radar systems, data-acquisition systems, test equipment, and electronic-warfare receivers, convert analog signals to digital words for digital processing. Digital processing assumes equally spaced time samples; however, actual samples from A/D converters have slightly unequal time spacing. Phase noise from the oscillator that generates the sampling clock produces this change in time spacing. In some digital processing, equally spaced time samples are critical to system performance. Dynamic range is the system performance requirement that is most affected. Understanding the limitations of the conversion process allows adjustments to be made to system configurations that can overcome these limits.

The phase noise of the sampling clock affects the dynamic range of the A/D converter by adding noise. Reduction in the number of effective bits measures the reduction of the A/D converter's dynamic range. Current systems require an increasing amount of dynamic range to satisfy increasing customer demand for improved systems performance. Increasing the number of A/D bits and increasing the number of processed samples are two ways of increasing the dynamic range. An increase in the number of A/D converter bits requires a corresponding increase in input analog dynamic range. Lowering the input noise to the A/D converter increases the input analog dynamic range. The phase noise of the sampling clock adds to the input noise of the A/D converter. Therefore, the importance of phase noise in the sampling clock increases with the number of A/D bits. Consequently, a lower phase noise for the sampling clock may be required.

Processing more samples is the second way of increasing the dynamic range of the system; however, an increase in the number of A/D samples without changing the sampling rate increases measurement time. Oscillator stability exponentially decreases over a longer measurement period. This decrease in stability results in an increase in noise from the oscillator that is greater than the noise-reduction effects of processing a greater number of samples. Consequently, more A/D samples can further limit the dynamic range of the system.

These limitations to A/D converter dynamic range by the sampling clock can be quantified by studying the relationship of oscillator phase noise to the number of effective A/D bits. Studying the effects of phase noise on A/D converters requires studying equations that relate the measurement of frequency in the time and frequency domains. This is accomplished in two steps. First, an equation for converting phase noise to time jitter is studied. Then, the relationship of time jitter to A/D converter dynamic range is studied. Finally, applications of the derived relationships to example systems and an FM sideband illustrate the usefulness of these equations.

9.3.1   Conversion of Phase Noise to Jitter

First, let's study the conversion of phase noise to time jitter. An A/D converter samples an input waveform in the time domain. Phase noise measures oscillator stability in the frequency domain. Allan variance measures oscillator stability in the time domain. Let's begin to develop equations that allow us to convert phase noise to time jitter by studying the Allan variance measurement. Later in the development at (9.33), we will study equations in the frequency domain. Equation (9.30) computes the Allan variance, which is the variance of fractional frequency fluctuations [15]:

$$\sigma_y(T)^2 = \frac{1}{f_c^2 2(m-1)} \sum_{k=1}^{m-1} (f_{k+1} - f_k)^2 \tag{9.30}$$

where

T = gate time of measurement (sec);

$\sigma_y(T)^2$ = Allan variance of fractional frequency fluctuations;

m = total number of measurements;

f_c = frequency of the carrier (Hz);

f_k = array of frequency measurements with index k (Hz).

Rearranging (9.30) produces (9.31) for computing the standard deviation of frequency fluctuations:

$$\sigma_f(T) = \sqrt{\frac{1}{2(m-1)} \sum_{k=1}^{m-1} (f_{k+1} - f_k)} \tag{9.31}$$

where

$\sigma_f(T)$ = standard deviation of frequency fluctuations (Hz).

Inverting (9.31) for the standard deviation of frequency fluctuations produces (9.32) for the standard deviation of time fluctuations:

$$\sigma_t(T) = 1/\sigma_f(T) \tag{9.32}$$

where

$\sigma_t(T)$ = standard deviation of time fluctuations.

Equations (9.30) through (9.32) are the time-domain relationships that will be needed in the development of equations to convert phase noise to time jitter.

Now, to convert phase noise to standard deviation of time calculations requires the development of equations in the frequency domain [16]. Let (9.33) be the instantaneous frequency of a sum of narrow-banded noise components:

$$f = f_c + \sqrt{2} \sum_{n=1}^{n_{max}} \delta_n \cos(2\pi f_n t + \alpha_n) \tag{9.33}$$

where

δ_n = frequency deviation, RMS, of noise component n (Hz);

f_n = modulation frequency of noise component n (Hz);

α_n = phase angle of noise component n (rad);

n_{max} = total number of noise components.

Next, let's use (9.34) to compute the average frequency of the first measurement:

$$f_{av1} = \frac{1}{T_a} \int_0^{T_a} f \, dt \tag{9.34}$$

where

T_a = time duration of first measurement (sec).

Substituting (9.33) into (9.34) and calculating the integral produces (9.35):

$$f_{av1} = f_c + \frac{1}{\sqrt{2}\pi T_a} \sum_{n=1}^{n_{max}} \frac{\delta_n}{f_n} [\sin(2\pi f_n T_a + \alpha_n) - \sin(\alpha_n)] \qquad (9.35)$$

For the first measurement, (9.35) computes the average frequency of the instantaneous frequency of a narrow-banded noise component. Likewise, for the second measurement, (9.36) computes the average frequency:

$$f_{av2} = \frac{1}{T_a} \int_{T_s}^{T_a + T_s} f \, dt \qquad (9.36)$$

where

   T_s = amount of time from the start of the first measurement until the start of the second measurement (sec).

Substituting (9.33) into (9.36) and calculating the integral produces (9.37):

$$f_{av2} = f_c + \frac{1}{\sqrt{2}\pi T_a} \sum_{n=1}^{n_{max}} \frac{\delta_n}{f_n} \{\sin[2\pi f_n (T_a + T_s) + \alpha_n] - \sin(2\pi f_n T_s + \alpha_n)\}$$
$$(9.37)$$

Equation (9.37) calculates the average frequency of the second measurement. Subtracting the two frequency averages produces (9.38):

$$f_\Delta = \frac{1}{\sqrt{2}\pi T_a} \sum_{n=1}^{n_{max}} \frac{\delta_n}{f_n} \qquad (9.38)$$

$$\times \{\sin[2\pi f_n (T_a + T_s) + \alpha_n] - \sin(2\pi f_n T_s + \alpha_n) - [\sin(2\pi f_n T_a) - \sin(\alpha_n)]\}$$

where

   f_Δ = frequency difference of the two measurements (Hz).

Applying identities for the sum and difference of trigonometric functions to (9.38) produces (9.39):

$$f_\Delta = \frac{2\sqrt{2}}{\pi T_a} \sum_{n=1}^{n_{max}} \frac{\delta_n}{f_n} \sin(\pi f_n T_a) \sin(\pi f_n T_s) \sin[\pi f_n (T_a + T_s) + \alpha_n] \qquad (9.39)$$

Computing the average squared of the phase angle α over 2π produces (9.40):

$$f_{avg\Delta}^2 = \frac{4}{\pi^3 T_a^2} \sum_{n=1}^{n_{max}} \frac{(\delta_n)^2}{f_n^2} [\sin(\pi f_n T_a)]^2 [\sin(\pi f_n T_s)]^2 \qquad (9.40)$$

$$\times \int_0^{2\pi} \{\sin[\pi f_n(T_a + T_s) + \alpha_n]\}^2 \, d\alpha_n$$

where

$f_{avg\Delta}$ = average frequency difference of the two measurements (Hz).

Integrating (9.40) and simplifying the result produces (9.41):

$$f_{avg\Delta}^2 = \frac{4}{\pi^2 T_a^2} \sum_{n=1}^{n_{max}} \frac{(\delta_n)^2}{f_n^2} [\sin(\pi f_n T_a)]^2 [\sin(\pi f_n T_s)]^2 \qquad (9.41)$$

Taking the limit in (9.41) produces (9.42):

$$\sigma_f(T)^2 = \frac{4}{\pi^2 T^2} \int_0^{f_{max}} S_\phi(f_n) [\sin(\pi f_n T)]^4 \, df_n \qquad (9.42)$$

where

$S_\phi(f_n)$ = spectral density of phase fluctuations (rad^2/Hz);

f_{max} = maximum measurement frequency, A/D bandwidth (Hz).

Equation (9.42) converts phase noise to Allan variance by integrating over the measurement bandwidth. For an A/D converter, we will integrate over the analog input bandwidth. Combining (9.32) and (9.42) allows the standard deviation of time period fluctuations, which is time jitter, to be computed from phase noise.

9.3.2 Relationship of Time Jitter to Dynamic Range

With developed equations relating phase noise to time jitter, we can now study the relationship of time jitter to dynamic range. We begin with some definitions. Fluctuations in the sampling clock edge produce aperture jitter. Aperture jitter is the time variation at the point that the sample is taken. Figure 9.60 shows that the worst-case aperture jitter occurs with the highest frequency into the A/D converter and at the zero crossing of the sine wave, which has the fastest slope. The amount of aperture jitter is one of the major components in determining the dynamic range of an A/D converter.

Equation (9.43) computes the effect time jitter has on the dynamic range of an A/D converter [17]:

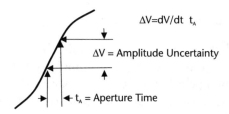

$\Delta V = dV/dt \; t_A$

ΔV = Amplitude Uncertainty

t_A = Aperture Time

Figure 9.60 Relationship of time uncertainty (aperture time) to amplitude uncertainty.

$$V_{ad} = 2\pi f_s / \sqrt{2}\, \sigma_t(1/f_s) \tag{9.43}$$

where

V_{ad} = ratio of A/D dynamic range;

$\sigma_t(1/f_s)$ = aperture time jitter of the clock;

f_s = highest frequency of sine wave into the A/D (Hz).

Equation (9.43) is based on the worst-case error for a Nyquist-Rate A/D converter, which occurs at the zero-crossing point for a sine wave. The signal-to-noise ratio is 20 log (V_{ad}).

Equation (9.43) computes the dynamic range for the analog input. The number of effective bits is the figure of merit for the dynamic range at the digital output. Equation (9.44) calculates the number of effective bits by converting dynamic range:

$$n_{eb} = [10 \log(V_{ad}) - 1.9]/6.02 \tag{9.44}$$

where

n_{eb} = number of effective bits.

Consequently, the effect of time jitter on the number of effective bits can be computed by using (9.32), (9.42), and (9.43). Calculating the maximum allowable jitter allows maximum phase-noise specifications to be computed. Rearranging (9.43) produces (9.45) for time jitter:

$$\sigma_{tmax}(1/f_s) = P_{ad}/(2\pi f_s) \tag{9.45}$$

where

σ_{tmax} = maximum allowable time jitter (sec).

Equation (9.45) computes the maximum allowable jitter to maintain the required dynamic range. This can be used to determine the maximum amount of phase noise.

9.3.3   Phase Noise Versus Effective Bits

To understand the effect that phase noise has on effective bits, let's study an example A/D converter system. The example system has an A/D converter with a 150-MHz sampling clock and 100-MHz bandwidth. Now, let's vary the close-to-the-carrier phase noise and far-from-the-carrier phase noise on the example A/D system to see the changes in the number of effective bits.

Figure 9.61 shows the variation of far-from-the-carrier phase noise to be used in the analysis. Figure 9.62 shows the variation of close-to-the-carrier phase noise to be used in the analysis. Goldman [15] provides formulas to model these variations in the phase noise of the sampling clock, and substituting the results into the derived equations computes the number of effective bits.

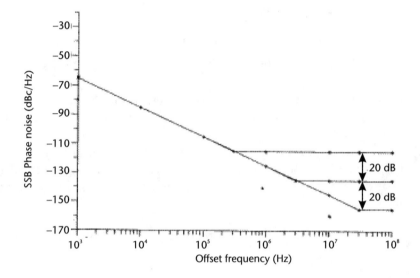

Figure 9.61   Variation of far-from-the-carrier phase noise to be used in the effective-bits analysis.

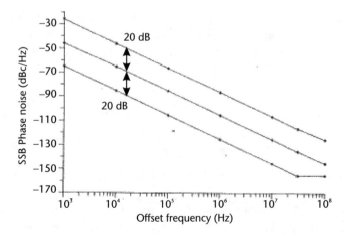

Figure 9.62   Variation of close-to-the-carrier phase noise to be used in the effective-bits analysis.

Figure 9.63 shows the effect of variation of far-from-the-carrier phase noise on the number of effective bits. This figure shows that an increase of flat, far-from-the-carrier phase noise causes a loss in effective bits that flattens out with faster measurement times.

Figure 9.64 shows the effect of variation of close-to-the-carrier phase noise on the number of effective bits. This figure shows that an increase of close-to-the-carrier phase noise causes an equal loss in the number of effective bits with measurement time. This analysis section shows the different effects that far-from-the-carrier and close-to-the-carrier phase noise have on the number of effective bits. In general, close-to-the-carrier phase noise has an equal effect on long- and short-term measurement times and far-from-the-carrier phase noise has a greater effect on short-term measurement times. With this knowledge, the requirements for performance improvements in systems can be addressed.

9.3.4   Effective Bits at High Frequencies

Another example shows the effect of increasing operating frequency on effective bits for an A/D converter. For this case, two example A/D converter systems will be considered. First, a low-frequency example system has an A/D converter with a 10-MHz crystal sampling clock and a 10-MHz bandwidth. Second, a high-frequency example system has an A/D converter with a 10-GHz YIG sampling clock and a 10-GHz bandwidth.

Figure 9.65 shows a comparison plot of the effective bits for the high- and low-frequency example systems. This figure shows significant performance limitation for the higher-frequency example system. Consequently, this analysis indicates that

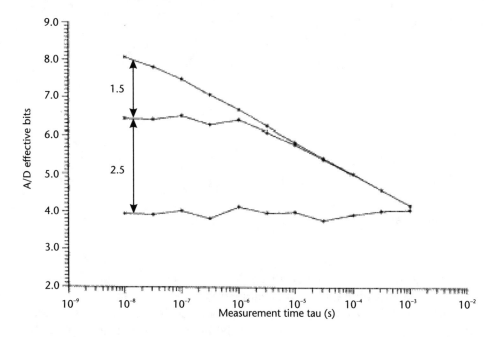

Figure 9.63   Variation of number of effective bits from variation of far-from-the-carrier phase noise.

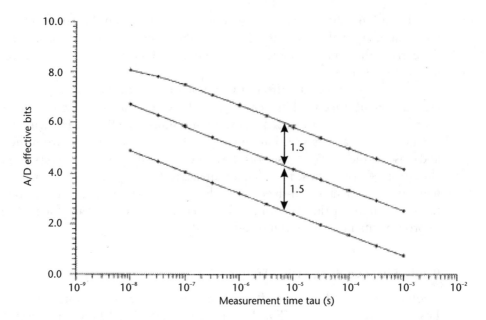

Figure 9.64   Variation of number of effective bits from variation of close-to-the-carrier noise.

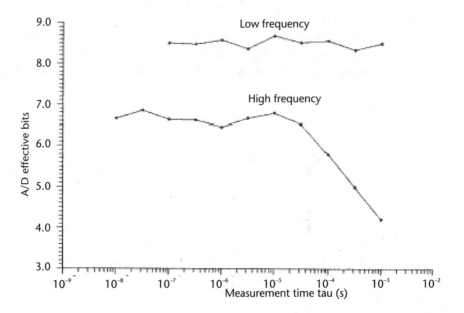

Figure 9.65   Effective-bits comparison plot for the high- and low-frequency examples.

a high-frequency system with an A/D converter front end will have significant performance limitations. This degradation in performance is a result of wider input bandwidth and less oscillator stability at high frequencies. In general, oscillator stability is less at high frequencies because of the difficulty in achieving high resonator Qs at high frequencies.

9.3.5 Effect of FM Sideband on Effective Bits

Not only does phase noise limit the number of effective bits, but spurious signals also limit the number of effective bits. From (9.41), (9.46) computes the variance of frequency fluctuations from a spurious discrete signal:

$$\sigma_f(T)^2 = \frac{4}{\pi^2 T^2} \int\limits_0^{f_{max}} [\eta(f_n)]^2 \, [\sin(\pi f_n T)]^4 \, df_n \qquad (9.46)$$

where

η = modulation index.

Combining (9.32), (9.43), (9.44), and (9.46) computes the number of effective bits. Consequently, the effects of sideband level on the number of effective bits in the system can be evaluated.

For instance, let's introduce a 60-dBc, 2.5-MHz sideband into the example system with a 150-MHz clock. Substituting these parameters into the equations shows that this sideband limits the effective bits, as shown in Figure 9.66. Consequently, the maximum allowable sideband level in a system can be computed. In addition, Figure 9.66 shows that a single sideband in the frequency domain does not produce a single time perturbation. This effect results because time and frequency have a transform relationship.

In summary, this section developed a method for calculating the effects of sampling clock phase noise on A/D dynamic range. In addition, this section showed

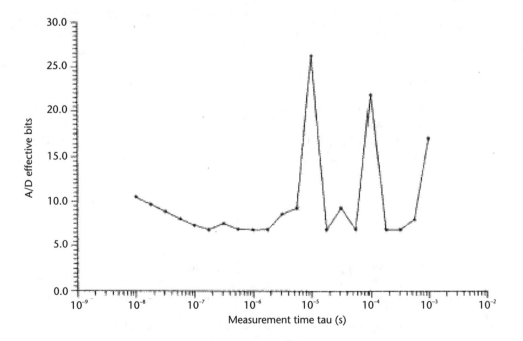

Figure 9.66 Effect of sideband modulation on the number of effective bits.

increasing A/D dynamic range requires a more stable source. Increasing the dynamic range by increasing the number of processed samples or increasing the number of A/D converter bits requires two different phase-noise stability improvements. First, increasing the number of samples to improve dynamic range requires lower close-to-the-carrier phase-noise levels. Second, increasing A/D input analog bandwidths requires lower far-from-the-carrier phase-noise levels from the sampling clock source. The future is clear. More advanced A/D converters and signal-processing requirements will require more stable sources to maintain dynamic range for high-frequency input signals.

9.4   All-Digital PLLs

As digital processes reduce the cost of circuit functions and as design tools to handle large-scale digital circuits make it easier to design in the digital domain, demand is created for all-digital PLLs. All-digital PLLs have the advantages of entailing lower cost by using standard digital CMOS process, requiring less training on design tools because they follow ASIC flow tools, and remembering their last state after power-down for fast reacquisition. They also have several disadvantages, including high jitter with the minimum jitter equal to the clock frequency (>2 ns), higher power consumption as more functions run at maximum clock frequency, and higher maintenance in troubleshooting and testing because errors are in windows and are not linearly extrapolated between results. Other disadvantages are that well-behaved circuits require clocks much higher than natural frequency (>100), the ADPLL has a maximum frequency determined by the ripple carry out in counters (<200 MHz for 0.8-μm gate lengths), and it is harder to reuse cells because each design has its own unique solution. Because of all these disadvantages, ADPLLs have limited application.

9.4.1   Operation of a Simple ADPLL

Figure 9.67 shows a simple version of an ADPLL. This configuration helps us understand the operation of an ADPLL. However, it has a narrow frequency range. Later, the circuit will be modified to work over a wider range. Numbers represent all of the electronic signals. The circuit has a clock input of frequency f_{in} and a

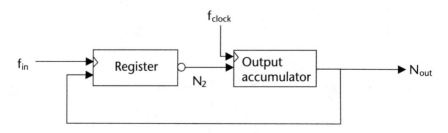

Figure 9.67   Simple ADPLL. (*From:* [18]. © 1998 Wiley Interscience. Reprinted with permission.)

number output that has an average repetition rate of f_{in}. The loop parameters determine how closely the number output follows the input.

The VCO has been replaced by an output accumulator (OA). Each clock cycle changes the output of the OA by an amount equal to its input, N_2. The OA recycles to zero ($N_v - 1$ being the highest number limit) each time the output accumulator reaches capacity, N_v. Thus, N_v represents one output cycle. Consequently, (9.47) models the output phase of the OA cycles:

$$\theta_{out} = \left(\frac{N_{out}}{N_v}\right) \text{cycles} \qquad (9.47)$$

Equation (9.48) defines the relationship of output phase to output frequency:

$$f_{out} = \frac{\theta_{out}}{\Delta t} \qquad (9.48)$$

Equation (9.49) computes the output frequency:

$$f_{out} = \frac{N_2}{N_v} f_{clock} \qquad (9.49)$$

where

f_{clock} = frequency of clock to OA.

For example, if f_{clock} = 8.192 MHz, N_v = 4,096, and N_2 = 512, then the output frequency would be 1.024 MHz, or one-eighth of f_{clock}.

Each cycle ($T = 1$ cycle/f_{clock}) of the OA clock increments the output by N_2. If we consider N_2 to be the tuning signal, we can obtain the OA gain constant. Equation (9.50) computes the change in output frequency with change in tune variable N_2:

$$K_v = \frac{d}{dN_2} f_{out} \qquad (9.50)$$

Substituting (9.49) into (9.50) produces (9.51):

$$K_v = \frac{f_{clock}}{N_v} \qquad (9.51)$$

From the previous example, the K_v would be equal to 2,000 Hz per unit N_2.

The register stores the value of N_{out}. This value represents the output phase of the loop for each cycle of the input signal. Thus, the stored number represents the output phase at the last transition of f_{in}. This can be considered the phase difference between the output and input signals. Consequently, the register functions as a phase detector (and zero-order hold), but we need a phase inversion. Making

N_2 the complement of the sampled value of N_{out} gives us a simple way to invert the phase. In the figure, the small circle at the output of the register represents this inversion. Equation (9.52) describes the value of N_2 during the nth input cycle (following the nth input clock transition):

$$N_2 = -N_{out} \ (\theta_{in} = 2\pi n) \tag{9.52}$$

Equations (9.53) and (9.54) compute the phase detector gain constant:

$$K_d = -\frac{\partial}{\partial \theta_{out}} N_2 = -\frac{\partial}{\partial N_{out}} N_2 \frac{\partial}{\partial \theta_{out}} N_{out} \tag{9.53}$$

$$K_d = -(-1) \frac{N_v}{1 \ \text{cycle}} = N_v \tag{9.54}$$

Consequently, (9.55) computes the forward gain constant:

$$G_{fwd} = K_d K_v = f_{clock} \tag{9.55}$$

This gives us the main components of a PLL. The output from a (phase detector) register produces a control signal that advances the output phase of an OA (controlled oscillator) at a rate proportional to the control signal. This is a type 1 loop.

Let's study the output for the frequency changes from low to high because sampling effects will make it change. For small values of N_2, the output frequency increases with increasing N_2, beginning with $f_{out} = f_{clock}/N_v$ when $N_2 = 1$. For small N_2, the output has a smoothly increasing sawtooth shape. Continuing to increase N_2 reduces the number of output changes between overflows. Eventually, the output no longer resembles a smoothly increasing phase. At $N_2 = N_v/2$, the output changes by half of the capacity of the output accumulator for each input period. This makes a square wave output.

At even higher values of N_2 ($>N_v/2$), the output frequency decreases. The slope of the phase change with time becomes negative. Further increases in value of N_2 decreases the output frequency until the maximum value of $N_2 = N_v - 1$. At this point, the output frequency returns to where we started at a frequency of f_{clock}/N_v. However, it will have a negative sign. Sampling the advancing value of N_{out} at a frequency of f_{clock} causes this behavior.

9.4.2 Sampling and Stability

The simple loop has two sampling processes. The first occurs in the OA at f_{clock}. The second occurs in the register at f_{in}, which is the frequency of the output from the OA during lock. To make the OA operate like a VCO with an approximately continuous phase change at its output, the OA clock f_{clock} has to be much greater than the input clock. Equation (9.56) defines this condition:

$$f_{clock} \gg f_{in} \tag{9.56}$$

Not meeting this inequality will make only a few steps at the output rather than a ramp of phase.

Let's begin to look at stability by studying the change in phase at the next sample. To accomplish this calculation, let the input frequency f_{in} have a discrete phase change of $\Delta\theta$ between two of the transitions that clock the register. At the first input sample after this step, (9.57) computes the frequency change that will occur at N_{out}:

$$\Delta f_{out} = G_{fwd} \Delta\theta = f_{clock} \Delta\theta \tag{9.57}$$

Equation (9.58) computes the resulting phase change at the next input sample, which occurs 1 cycle/f_{in} later:

$$\Delta\theta_2 = -\Delta f_{out} \text{ cycle}/f_{in} = -\Delta\theta f_{clock}/f_{in} \tag{9.58}$$

The phase error grows by $-f_{clock}/f_{in}$ for each input sample. According to (9.56), a substantial increase in phase will occur each time. The sampling process makes this first-order loop unstable. Multiple overshoots of the final value by an amount larger than the pervious error causes a growing oscillation. Dividing N_2 by 2^q solves this problem. Shifting bits between the register and the OA and connecting 2^i from the register to 2^{i-q} at the OA input performs the division. This reduces the gain by 2^q; however, (9.59) shows that the maximum value of N_2 has also been reduced by the same amount:

$$N_2 \leq (N_v - 1) 2^{-q} \tag{9.59}$$

Equation (9.60) computes the new highest frequency that can be achieved by dividing by N_2 and shifting bits:

$$f_{in} = f_{out} \leq \frac{N_v - 1}{N_v} 2^{-q} \cong 2^{-q} f_{clock} \tag{9.60}$$

The gain change of 2^{-q} alters the incremental phase change calculation in (9.58). Adjusting for the gain change produces (9.61) for computing the new ratio of phase change magnitudes between adjacent samples:

$$\frac{-\Delta\theta_2}{\Delta\theta} = 2^{-q} \frac{f_{clock}}{f_{in}} > 1 \tag{9.61}$$

Equation (9.61) approaches unity at the maximum input frequency, f_{in}. Unity in (9.61) gives a desirable fast response because the output phase follows the input step in one input sample period. However, at $-\Delta\theta_2/\Delta\theta = 2$, the circuit becomes unstable. The mechanism for this instability can be explained as follows: For a value of 2, the magnitude of the error in $\Delta\theta$ increases, instead of decreasing, for each input sample. Consequently, the error grows until some limit is reached. This value-of-2 stability limit constrains operation to a narrow range of frequencies that depend on the value of output accumulator f_{clock} and q.

Adding a constant N_{off} to the phase detector output N_2 reduces the constraint. Now, the value of N_2 is not tied so directly to f_{in}. Then, (9.59) and (9.60) do not hold, which allows us to lower the gain by operating at higher input frequencies than previously allowed by (9.60). Incorporating all of these changes into Figure 9.67 gives us the new block diagram shown in Figure 9.68.

The Z-transform representation of the loop shows the stability problems more clearly. Again, we will assume the conditions of (9.56), in which the clock frequency of the output accumulator is much higher than the input clock frequency, so that the analysis can be simplified. Ignoring the faster sampling in the OA simplifies the analysis, which means we have to reassert (9.56). Figure 9.69 shows the mathematical block diagram with the sampler in terms of La Place transforms. Now, however, the phase detector gain in (9.53) has changed to (9.62).

$$K_d = N_v 2^{-q} \tag{9.62}$$

Equation (9.63) shows the open-loop La Place transfer function:

$$G(s)H(s) = \frac{1 - e^{Ts}}{s^2} G_{fwd} = \frac{1 - e^{Ts}}{s^2} f_{clock} 2^{-q} \tag{9.63}$$

Converting (9.63) for the open-loop transfer function to the Z-transform produces (9.64):

$$G(z)H(z) = \frac{(1 - z^{-1})Tz^{-1}}{(1 - z^{-1})^2} f_{clock} 2^{-q} = 2^{-q} \frac{f_{clock}}{f_{in}} \frac{1}{z - 1} \tag{9.64}$$

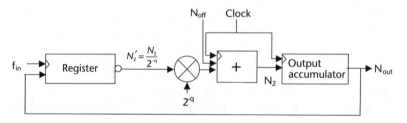

Figure 9.68 Workable simple ADPLL. (*From:* [18]. © 1998 Wiley Interscience. Reprinted with permission.)

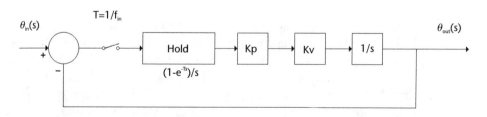

Figure 9.69 ADPLL mathematical model with sampler. (*From:* [18]. © 1998 Wiley Interscience. Reprinted with permission.)

Substituting (9.64) into the control-system definition of a closed-loop response produces (9.65):

$$\frac{G(z)}{1 + G(z)H(z)} = \frac{2^{-q}\dfrac{f_{clock}}{f_{in}}}{z - 1 + 2^{-q}\dfrac{f_{clock}}{f_{in}}} \tag{9.65}$$

Setting the denominator in (9.65) to zero sets up the equation for determining the stability limit. Equation (9.66) shows the denominator set to the zero condition:

$$z - 1 + 2^{-q}\frac{f_{clock}}{f_{in}} = 0 \tag{9.66}$$

Setting $z = -1$ in (9.66) gives the Z-transform stability limit, which gives us (9.67):

$$-1 - 1 + 2^{-q}\frac{f_{clock}}{f_{in}} = 0 \tag{9.67}$$

Rearranging and combining terms in (9.67) gives us (9.68) for the stability limit:

$$2^{-q}\frac{f_{clock}}{f_{in}} = 2 \tag{9.68}$$

The left side of the equation must be less than 2 for stable loop operation.

Figure 9.70 shows a z-plane plot of the open-loop and closed-loop poles for this loop. This graphically describes the stability issues that were previously discussed. The closed-loop pole locus begins at the open-loop pole at $2^{-q}f_{clock}/f_{in} = 0$

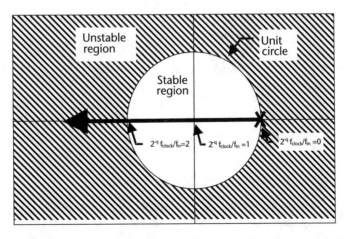

Figure 9.70   Z-plane stability plot of the ADPLL. *(From:* [18]. © 1998 Wiley Interscience. Reprinted with permission.)

(then, $z = +1$). Increasing $2^{-q} f_{clock}/f_{in}$ moves the locus along the real axis to the left on the z-plane. The best response occurs in the center of the unit circle at $2^{-q} f_{clock}/f_{in} = 1$ (then, $z = 0$). Continuing to increase $2^{-q} f_{clock}/f_{in}$ causes the locus to pass the unit circle at $2^{-q} f_{clock}/f_{in} = 2$ (then, $z = -1$) and enter the region of instability, which is outside the unit circle. At this point and beyond, the loop oscillates. This verifies the simple analysis we did earlier. The absence of an offset added to the phase detector output restricts operation to $2^{-q} f_{clock}/f_{in} < 1$ in the left half-plane only. The offset doubles the stable range of operation from $2^{-q} f_{clock}/f_{in} = 1$ to $2^{-q} f_{clock}/f_{in} = 2$.

9.4.2.1 Choice of Values

Example 9.3

A first-order ADPLL has $N_v = 2^{12} = 4,096$; $f_{clock} = 8.192$ MHz; and $f_{in} = 400$ kHz. Find the fastest switching condition value given by (9.61) for q, N_2, N_{off} (if required), ω_x, and OA output frequency range.

  a. With $2^{-q} f_{clock}/f_{in} = 1$ giving the fastest switching response and $f_{clock}/f_{in} = 21$, we will try $q = 4$ and 5 for $2^{-q} = 1/16$ and $1/32$ while using $1/32$ gives $2^{-q} f_{clock}/f_{in} = 0.64$, which is less than 1 and not optimum. Using $1/16$ gives $2^{-q} f_{clock}/f_{in} = 1.31$, which operates close to the optimum. From (9.48) and (9.49) and substituting the example values, $f_{in} = f_{out} = 4 \times 10^5$ Hz $= (N_2/4,096) 8.192 \times 10^6$ Hz $= N_2 \times 2$ kHz. Then, solving for $N_2 = 200$.
  b. $N'_2 = N_2/2^{-q} = 16 \times N_2 = 3,200$. No offset N_{off} is required since $N'_2 = 3,200 < N_v$.
  c. For this first-order loop, $\omega_x = G_{fwd} = 2^{-4} f_{clock} = 5.12 \times 10^5$ rad/s and $\omega_x/2\pi = f_x = 81.49$ kHz.
  d. N_2 can vary from 0 to INT$(4,095/16) = 255$, (9.49) shows that the possible OA output range of f_{out} is from 0 to $f_{out\_max} = N_{2\_max}/N_v \times f_{clock} = (255/4,096) 8.192$ MHz $= 510$ kHz.

Also, in the example, N_2 can be constant if f_{in} is a power-of-two multiple of the step size, 2 kHz. However, for other frequencies, N_2 will cycle between two numbers such that the average output frequency, which cycles between two adjacent multiples of 2 kHz, equals the input frequency. This is an example of jitter or ripple. Quantization causes the jitter because the signals are quantized rather than continuously variable. Sampled systems can have the unwelcome possibility of a false lock. A small numbered relationship by ratio of output over input frequencies gives the conditions for the possibility of a false lock [18, pp. 228–230].

9.4.3 ADPLL by Pulse Addition and Removal

Figure 9.71 shows a second ADPLL implementation. An available integrated circuit uses this implementation. The TI SN54LS297 IC has all the blocks in Figure 9.70, except the divide-by-N divider. The divider must be supplied externally. The IC also has a flip-flop phase detector. See the data sheet [19], but notice the differences

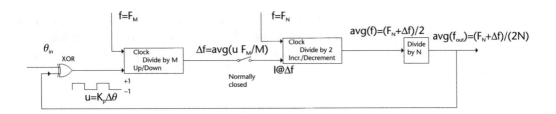

Figure 9.71 Block diagram of an ADPLL using pulse addition and removal.

in symbols. Studying this variation of the ADPLL will improve our ability to recognize the correspondence between the fundamentals we have studied and structures that look much different. More information can be obtained from the IC data sheet and from an application note [19].

Removing pulses from, or adding pulses to, a fixed-frequency clock and dividing the result by a large number to generate low jitter produces the VCO function. Removing a small number of pulses, relative to the total, gives adjacent cycles approximately the same duration.

Let's begin our study at the switch, which is open for purposes of discussion only. The signal to the right of the switch we will call I. This series stream of pulses has a value of +1 or −1 and an average net frequency of Δf (the number of positive pulses less the number of negative pulses divided by time). The following divide-by-2 circuit divides the clock frequency, F_N, by 2. However, receiving an increment (+1) or decrement (−1) command at I adds one clock edge or subtracts one clock edge, respectively. Thus, (9.69) describes the average frequency at the output of the divide-by-2 circuit:

$$f_{avg} = (F_N + \Delta f)/2 \tag{9.69}$$

Then, (9.70) describes the average output frequency by dividing (9.69) by N:

$$f_{oavg} = (F_N + \Delta f)/(2N) \tag{9.70}$$

9.4.3.1 Transfer Function

This ADPLL implementation uses an exclusive-OR gate phase detector. Let's define the output as u, with true = +1 and false = −1 as its value. For exclusive-OR inputs in phase, the output has a value of −1. Shifting the relative phase of the inputs in the positive direction produces a more positive average output until it equals +1 for the out-of-phase condition. During the half-cycle change in the relative phase of its inputs, the average output of the exclusive-OR changes proportionately from −1 to +1 over a π change in input phase error (0.5 cycles). The transfer function has a triangle shape, as shown in Figure 5.13, with the limits of −1 and +1. Thus, (9.71) gives the phase detector gain constant:

$$K_p = 2/(0.5 \text{ cycle}) = 4/\text{cycle} \tag{9.71}$$

A phase detector output of +1 makes the divide-by-M circuit count up. A phase detector output of −1 makes the divide-by-M circuit count down. Therefore, the output frequency from the divide-by-M counter has an average value described by (9.72):

$$\Delta f = u F_M / M \tag{9.72}$$

Equation (9.73) gives the gain of the VCO block by substituting (9.72) into (9.70) and taking the derivative:

$$K_v = df_{out}/du = F_M/(2MN) \tag{9.73}$$

Figure 9.72 shows a block diagram using these constants. The l/s block converts frequency to phase. Figure 9.72 models a first-order loop. Equation (9.74) computes the unity-gain bandwidth from multiplying the blocks in the figure:

$$\omega_x = K = [2F_M/(MN)] \text{ rad/cycle} \tag{9.74}$$

9.4.3.2 Tuning Range

Studying Figure 9.70, (9.70) and (9.72), and the +1 or −1 phase detector limits produces (9.75) for the hold-in range:

$$F_H = \pm F_M/(2MN) \tag{9.75}$$

Studying (9.70) produces (9.76) for the center frequency:

$$F_c = F_N/(2N) \tag{9.76}$$

Consequently, from (9.75) and (9.76), changing the external clock frequencies (F_M and F_N) or setting divider ratios (M and N) controls both the center frequency and the bandwidth.

9.4.3.3 Stability

Continuous time control theory shows first-order and type 1 loops to be inherently stable. Digital implementation of the loop produces samples of the phase-error

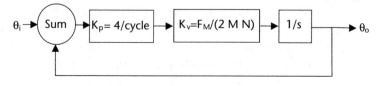

Figure 9.72 Mathematical block diagram for the ADPLL.

signal. Consequently, let's look at the effect of sampling on stability. These effects will be similar to the ones we found with the sample loop of Section 6.4.

The ADPLL has three sampling processes that occur from each input clock. The clock at frequency F_M, the clock at frequency F_N, and the reference input clock at the phase detector do the sampling. Two transitions of the output waveform per cycle sample the phase information with the exclusive-OR phase detector. The phase information does not change between transitions. Consequently, this defines a sampling process for the phase detector.

The sampling process at frequency f_s will not appreciably affect stability under the condition shown in (9.77):

$$f_s \gg \omega_x = K \tag{9.77}$$

Substituting (9.74) for bandwidth into (9.77) for stability gives (9.78) and (9.79) for stable conditions with $F_N \gg \omega_x$ rad/s and $F_M \gg \omega_x$ rad/s, respectively:

$$MN \gg 2 \tag{9.78}$$

$$MN \gg 2F_M/F_N \tag{9.79}$$

Rearranging and substituting (9.74) to (9.76) for bandwidth, tuning range, and center frequency for the lowest input sampling frequency gives (9.80):

$$2f_{in} \geq 2(f_c - F_H) = F_N/N - F_M(MN) = \omega_L/2(F_N/F_M M - 1) \tag{9.80}$$

Rearranging and inspecting (9.80) gives (9.81):

$$M \gg 2F_M/F_N \tag{9.81}$$

Meeting the condition shown in (9.81) will produce an input frequency much greater than ω_x.

Thus, satisfying (9.78), (9.79), and (9.81) guarantees stability. Furthermore, these conditions can be easily met. First, we make M and N large numbers in order to reduce ripple. Then, we make clock F_M not too much greater than clock F_N.

9.4.3.4 Ripple Control

Ripple refers to jitter on the loop output. Variations in instantaneous frequency cause this jitter. The loop controls average frequency by removing pulses. Depending on the sequence, different cycles may have different lengths; however, the average frequency remains correct.

From the loop design, the jitter can be limited to no more than $1/N$ cycles at the output. Initially, let's set the output frequency to the center at f_c. In the middle of phase detector range, it has a 50% duty cycle. From Figure 5.11, the time duration at one output state of the phase detector equals one-quarter of an output cycle at the zero-phase-error operating point. Substituting (9.76) for the center frequency gives (9.82):

$$T_{PD} = 1/(4f_c) = N/(2F_N) \tag{9.82}$$

Making the duration of the M count at least as long as T_{PD} prevents an increment or decrement pulse at I greater than 1. Equation (9.83) shows the required condition to make this happen:

$$M/F_M \geq T_{PD} \tag{9.83}$$

Combining these last two equations gives

$$M \geq NF_M/(2F_N) \tag{9.84}$$

Meeting this restriction limits the unnecessary outputs from the M divider to 1 in each direction during lock at center frequency. Even though frequency offsets can be much greater than one increment or decrement from the center, meeting the condition in (9.84) will limit the number of divide-by-M outputs in the "wrong" direction to one. This occurs because the output width of the phase detector is narrower than (9.82). Therefore, the conditions in (9.83) are still met. This illustrates one method of reducing ripple in an ADPLL.

9.4.3.5 Section Summary

To summarize, the principles discussed in this section can be applied to many other circuits and software functions. Some types of PLLs may require further considerations (e.g., control of jitter) peculiar to those types; however, this section illustrated a loop theory that can be used to obtain a basic understanding of diverse circuits that employ the common properties of PLLs.

9.5 Summary

This chapter presented PLL applications and extensions. First, trade-offs between frequency synthesizer topologies were discussed. Applications of the topologies were shown. A change in ranking requirements can change the whole synthesizer architecture. In addition, one set of components and PLL parameters does not neatly satisfy all of the performance requirements. Finally, studying example synthesizer topologies can help select the best configuration.

For clock-recovery applications, it was shown that aligning the rising edges of the recovered clock at the center of the input data-stream pulses gave the optimum phase position. Characteristics of different digital data-stream formats were studied. Several different clock-recovery architectures were presented. The application of PLLs to clock recovery had special design considerations. The random nature of the data stream restricted the choice of phase detectors. The popular three-state PFD produced false pulse widths for an NRZ bit stream. Consequently, design trade-offs were discussed for several different phase detector techniques.

It was shown that the phase noise of the sampling clock affects the dynamic range of the A/D converter by adding noise. Equations were developed that related phase noise to time jitter and related time jitter to dynamic range. It was shown that increased A/D dynamic range required a more stable source.

A simple version of an ADPLL was studied. This configuration helped us understand the operation of an ADPLL, even though it had a narrow frequency range. In the ADPLL, numbers represented all of the electronic signals. An OA was used as a VCO. To make the OA operate with an approximately continuous phase change at its output, the OA clock had to be much greater than the input clock. A register functioned as a phase detector. The stored number represented the output phase at the last transition of the input clock. This was shown to be the phase difference between the output and input signals. Having studied this variation of the ADPLL helped us recognize the correspondence between the fundamentals we have studied and structures that look much different.

Questions

9.1 In Figure 9.2 for the brute-force method, what would be the biggest obstacle in integrating this solution?

9.2 What are the two main categories of frequency synthesis?

9.3 What are the differences between a coherent signal versus a synchronous signal?

9.4 A single PLL synthesizer with an input reference divider has a 12-MHz crystal clock available, and an audio customer has audio data sampled at 441 kHz. For his CODEC, he requires an output frequency that is 35.28 MHz. Calculate the highest phase detector reference frequency, the reference divide ratio, and the feedback divide ratio.

9.5 A single PLL synthesizer with an input reference divider has a 32.768-kHz crystal clock available, and an audio customer has audio data sampled at 441 kHz. For his CODEC, he requires an output frequency that is 35.28 MHz. Calculate the highest phase detector reference frequency, the reference divide ratio, and the feedback divide ratio.

9.6 Using a single PLL synthesizer, a customer wants 16.75–24.75-MHz PLL output. Calculate the highest phase detector reference frequency and the minimum and maximum feedback divide ratios.

9.7 Using a single PLL synthesizer, a customer wants 101–847-MHz PLL output with spacings of 31.25 kHz. Calculate the highest phase detector reference frequency and the minimum and maximum feedback divide ratios.

9.8 Name some of the different types of phase detectors for clock recovery.

9.9 Can you use a phase/frequency detector for clock recovery?

9.10 Name some bit-stream formats?

9.11 Why does an all-digital PLL have more jitter than an analog PLL?

9.12 What is the minimum jitter for an all-digital PLL?

References

[1]   Manassewitsch, V., *Frequency Synthesizers: Theory and Design*, New York: John Wiley and Sons, 1987.

[2]   Aaron, M. R., "The Use of Least Squares in System Design," *IEEE Transactions on Circuit Theory*, December 1956, pp. 224–231.

[3]   Cuthbert, T. R., *Circuit Design Using Personal Computers*, New York: John Wiley and Sons, 1983.

[4]   Best, R. E., *Phase-Locked Loops: Design, Simulation, and Applications*, 3rd ed., New York: McGraw-Hill, 1997, pp. 156–163.

[5]   Razavi, B., *Monolithic Phase-Locked Loops and Clock Recovery Circuits*, New York: IEEE Press, 1996, pp. 33–36.

[6]   Trischitta, P. R., and E. L. Varma, *Jitter in Digital Transmission Systems*, Norwood, MA: Artech House, 1989.

[7]   Wolaver, D. H., *Phase-Locked Loop Circuit Design*, Upper Saddle River, NJ: Prentice-Hall, 1991, pp. 211–237.

[8]   Shanmugam, S. K., *Digital and Analog Communication Systems*, New York: John Wiley and Sons, 1979.

[9]   Pottbacker, A., U. Langmann, and H. U. Schreiber, "A Si Bipolar Phase and Frequency Detector IC for Clock Extraction up to 8 Gb/s," *IEEE Journal of Solid State Circuits*, Vol. 27, December 1992, pp. 1747–1751.

[10] Richman, D., "Color-Carrier Reference Phase Synchronization Accuracy in NTSC Color Television," *Proc. IRE*, Vol. 42, January 1954, pp. 106–133.

[11] Gardner, F. M., "Properties of Frequency Difference Detectors," *IEEE Transactions on Communications*, Vol. COM-33, February 1985, pp. 131–138.

[12] Messerschmitt, D. G., "Frequency Detectors for PLL Acquisition in Timing and Carrier Recovery," *IEEE Transactions on Communications*, Vol. COM-27, September 1979, pp. 1736–1746.

[13] Hogge, C. R., "A Self-Correcting Clock Recovery Circuit," *IEEE J. Lightwave Technology*, Vol. LT-3, December 1985, pp. 1312–1314.

[14] DeVito, L., et al., "A 52-MHz and 155-MHz Clock Recovery PLL," *ISSCC Dig. Tech. Papers*, 1991, p. 142.

[15] Goldman, S. J., *Phase Noise Analysis in Radar Systems Using Personal Computers*, New York: John Wiley and Sons, 1989.

[16] Egan, W. F., *Frequency Synthesis by Phase Lock*, New York: John Wiley and Sons, 1981.

[17] Martin, K., and D. Johns, *Analog Integrated Circuit Design*, New York: Wiley Interscience, 1997, p. 458.

[18] Egan, W. F., *Phase-Lock Basics*, New York: Wiley Interscience, 1998, pp. 252–263.

[19] Troha, D. G., "Digital Phase Locked Loop Design Using SN54/74LS297," Texas Instruments Application Note sdla005a.

Letter Symbols

A_v	opamp gain
B	3-dB bandwidth
B_{FFT}	equivalent FFT bandwidth
B_{pll}	3-dB bandwidth of the closed-loop PLL
B_{rcv}	receiver bandwidth
B_{sa}	spectrum analyzer bandwidth
c	speed of light
C_o	controlled output in control-systems theory
C_o/R_i	closed-loop transfer function used in feedback and control systems
f	frequency
F	noise figure
F_{rcv}	receiver noise figure
f_c	frequency of the carrier
f_d	maximum frequency deviation in frequency-modulation theory
f_{di}	deviation of the instantaneous frequency
f_{IF}	intermediate frequency
f_{in}	input frequency
f_{int}	frequency of intercept with white phase noise.
f_l	corner frequency of flicker noise
f_m	frequency of modulation (sideband, offset, Fourier, baseband)
f_n	natural frequency
f_{op}	operating frequency of a radar system
f_{out}	output frequency of the PLL
f_{3p}	frequency of 3-dB point
f_{ref}	reference frequency in a PLL
f_{syn}	output frequency of the synthesizer
f_v	frequency value at the relative power level $L(f_i)$ used in Bode plot models of phase noise
f_x	0-dB magnitude crossover frequency
f_z	frequency location of the zero in a PLL
G	gain
$G(s)H(s)$	open-loop transfer function used in feedback and control systems
$G(f)$	forward transfer function used in feedback and control systems

G_n	amplifier gain
G_{pll}	gain constant of a PLL
$H(f)$	feedback transfer function used in feedback and control systems
I_P	charge pump current
k	Boltzmann's constant
K_{cco}	transfer function of a current controlled oscillator
K_d	phase detector's gain in a PLL
K_v	transfer function of the VCO in a PLL
L	insertion loss
$\mathcal{L}(f_m)$	ratio of single-sideband phase noise to total signal power in a 1-Hz bandwidth f_m Hz from the carrier
$L_A(f)$	filter transfer function
L_{mc}	mixer conversion loss
$L_{pll}(f)$	PLL filter transfer function
L_{res}	insertion loss of a resonator
n_{mf}	frequency multiplication factor
n_{FFT}	number of FFT points
N	noise power
N_o	noise-power density
N_L	total noise-level thermal plus phase noise
$N_{mf}(f)$	measurement noise floor
N_{sa}	spectrum-analyzer noise floor
N_T	thermal noise level
P	power
P_{cal}	phase-noise measurement calibration factor
P_{sbc}	calibration factor that converts measurements to single-sideband phase noise, which has a value of 3 dB
P_i	input power
$P_{IF}(f_m)$	power measured at the IF port of a mixer
P_{ps}	power level of beat note, which equals the power level of the mixer's phase slope
P_o	output power
$P_m(f_m)$	power level at offset frequencies from the carrier
P_{mW}	1,000 mW/W to convert watts to milliwatts
P_c	power level of the carrier
P_{LO}	power level at the local oscillator port of a mixer
P_{os}	1-dB output power saturation
P_s	signal power
P_{ssb}	power of a single sideband
Q	quality factor
Q_l	loaded quality factor of a resonator
Q_{ul}	unloaded quality factor of a resonator

R_1, R_2	resistors of filter in PLL
R_L	load resistance, which is usually 50Ω
R_i	reference input in a control system
$S_{\Delta f}(f)$	spectral density of frequency fluctuations
S/N	signal to noise
$S_\phi(f)$	spectral density of phase noise
$S_y(f)$	fractional spectral density of frequency fluctuations
t	time variable
t_A	aperture time
T_0	temperature, Kelvin
T_c	time period of the fundamental frequency
V	voltage
V_{im}	imaginary part of voltage
V_{rl}	real part of voltage
V_p	peak voltage of sinusoidal waveform
α	exponential growth factor in crystal oscillator startup
ϕ, θ, β	phase angle variables
θ_m	phase margin
Δf_{res}	residual frequency modulation
η	modulation index
ζ	damping factor in a control system
τ	delay time
τ_{sd}	sampling-delay time
ω	angular frequency
ω_n	angular natural frequency
ω_c	angular frequency of the carrier
ω_{beat}	angular frequency difference between two carriers
ω_{lo}	angular frequency at the local oscillator port of a mixer
ω_m	angular modulation frequency
ω_{rf}	angular frequency at the radio frequency port of a mixer

Glossary

Terms

Aliasing	Improper sampling of a function causes high-frequency responses to be mixed with low-frequency responses. This mixing is called aliasing. Aliasing limits a system's ability to measure input signals.
Baseband bandwidth	See *Video bandwidth*.
Carrier synchronizer	Circuit that generates the received carrier.
Cascode	An arrangement of electronic active devices that combines two amplifier stages for increased output resistance.
Charge pump	A current source that is controlled by the phase error (e.g., a pulsed current source).
Clock synchronizer	A circuit that generates the received clock waveform.
Coherent	Correlated waveforms that are characterized by a fixed phase relationship between points on the electromagnetic wave. Synchronous signals are a subset of this definition with a 0° phase relationship.
Coherent integration	A signal-processing technique that uses a signal's magnitude and phase information.
Correlated (deterministic/ systematic) jitter	See *Spurious jitter*.
Cycle-to-cycle jitter	Time variation from adjacent period measurements. Also known as edge-to-edge jitter. The difference between successive periods (Nth period—$N - 1$ period).
Cycle slip	The event when phase comparison of the VCO with the input reference oscillator slips one or more cycles.
Denominator	The expression written below the line in a network function.
Dynamic range	In general, the power range from the minimum detection level to the saturation point. In regards to a spe-

cific receiver, it can have many meanings, depending on the performance specification (probability of detection or probability of an error) of the system. For example, it could be defined as anywhere from the 0.1-dB saturation point to a signal-to-noise level of 6 dB.

Focused ion beam (FIB)	An analytical instrument used in the semiconductor industry.
Fractional-*N* loop	PLL that uses a fractional divider in its feedback.
Gain margin	The increase in gain to cause the unity gain crossover frequency to move the open-loop phase angle to 180° (unstable point).
Groupe Spécial Mobile (GSM)	The European group set up to establish European mobile telephony protocols and which was later anglicized to *Global System for Mobiles* to preserve the acronym.
Hold-in	Frequency range over which the loop will hold lock.
Intermediate frequency bandwidth	The frequency difference between the frequency of the −3-dB point below the carrier and the frequency of the −3-dB point above the carrier.
Intermodulation products	Modulation of the components of complex waves, which produces frequencies equal to the sums and differences of integral multiples of those components.
Injection lock	The product of adding a signal directly into the tune circuit of an oscillator.
Jitter	See *Time jitter*.
Jitter, frequency	Frequency-related, abrupt, spurious variations in the frequency of successive pulses.
Jitter, phase	Phase-related, abrupt, spurious variations in the phase of successive pulses referenced to the phase of a continuous oscillator.
Jitter, timing, tolerance, transfer function	Telecommunications terms relating to the transmission and reception of digital data.
Lock-in range	With the loop initially unlocked, the range within which the input frequency is tuned and the loop locks up without a cycle slip.
Long-term jitter or accumulated jitter	See *Time-interval jitter*.
Loss of lock	Repeated cycle slips between the VCO and the input reference oscillator at the phase detector inputs.
Loop bandwidth	The loop bandwidth of a PLL is that range of frequency modulation over which the PLL will respond

satisfactorily. A 3-dB decrease in the response of a PLL to frequency modulation usually defines the loop bandwidth.

Low voltage differential signaling (LVDS)

An electrical signaling system that can run at very high speeds over cheap, twisted-pair copper cables.

Metal Oxide Semiconductor Implementation Service (MOSIS)

Probably the oldest integrated circuit (IC) foundry service.

Modulation bandwidth, −3 dB

The modulating frequency applied to the tune line of the VCO at which the output-frequency deviation decreases to 0.707 of its dc value.

Natural frequency

The frequency of free oscillations with a zero-damping factor (no loss).

Noncoherent integration

A signal-processing technique that only uses a signal's magnitude information.

Numerator

The expression written above the line in a network function.

OC3

Optical carrier levels describe a range of digital signals that can be carried on SONET fiber optic network.

Open-loop gain

The gain of the system without feedback.

Open-loop transfer function

The complex quantity that is measured through the forward and feedback functions of the system.

Order of the loop

The loop order determined by the highest power of the s polynomial in the denominator of the transfer function for the loop.

Pacing

An HP1743 modulation analyzer term for the number of edges allowed to pass between time stamps.

Period jitter

Variation of period during the measurement time (nonsuccessive periods).

Phase jitter

See *Jitter, phase*.

Phase margin

180° minus the absolute value of the loop phase angle at a frequency where the loop gain is unity; the number of degrees away from the unstable, open-loop, 180° phase angle.

Phase noise

A randomly occurring signal in the time domain displayed as time jitter on an oscilloscope. Phase noise and time are related by convolution. Phase noise in the frequency domain is the ratio of a single-sideband phase-noise power level to the carrier power level in a 1-Hz bandwidth, f_m Hz away from the carrier frequency.

Pole	Any real or complex value substituted for s in a network function $[N(s) = (s - z)/(s - p)]$ that causes the network function to be infinite.
Posttuning drift	The frequency error compared to a final stabilized value, at a specified time after application of a step change in tuning voltage to the VCO.
PSPICE	SPICE analog circuit simulation software developed by MicroSim (bought by OrCAD) that runs on personal computers, hence the first letter "P" in its name.
Pull-in	Self-acquisition of the loop without frequency-acquisition aids.
Pull-out	Frequency-step limit that causes the loop to skip cycles.
Pulling, frequency	Maximum peak-to-peak frequency variation observed while an output load of specified resistance is varied (50Ω load phase varied with line stretcher from $0°$ to $360°$); expressed in frequency variation peak to peak.
Pushing, frequency	The change in output frequency corresponding to the variation of the supply voltage about the quiescent operating point; expressed in frequency change per volt.
Sampling delay	A delay in the time behavioral model in a PLL due to the sampling characteristics of a phase detector.
Sampling interval	The time between samples in a periodic sampling system.
Sampling rate	The inverse of the sampling interval.
Saturation point or limit	In general, the point at which a change in input power produces no change in output power. For a specific receiver, the saturation point depends on the receiver's accuracy requirements. Usually, the receiver's saturation point is at the receiver's 1-dB compression point.
SONET	Synchronous optical networking is a method for communicating digital information using lasers.
SPICE (Simulation Program with Integrated Circuits Emphasis)	A general-purpose analog circuit simulator.
Spurious jitter	Time jitter caused by a dominant spurious signal; also sometimes referred to as correlated (deterministic/systematic) jitter.
Spurious signal	Any signal other than the desired signal that produces a spectral line in the frequency domain or the modulation domain.

Static random access memory (SRAM)	A type of semiconductor memory that retains its contents as long as power remains applied.
Synchronous	Signals or events produced as a result of a clock edge.
Time stamp	An HP1743 modulation analyzer term for the time of occurrence of an edge.
Time interval	An HP1743 modulation analyzer term for the time between time stamps; the time interval from one edge to the nth edge. If the number of skipped edges equals 1, then the interval is equivalent to a period.
Time-interval jitter	Variation of time interval during the measurement time (nonsuccessive periods); also referred to as long-term jitter or accumulated jitter.
Time jitter	Time-related variations in the duration of any specified, related interval. Quantitative use of time jitter requires specifying the measure of time with average, root-mean square, or peak to peak. Jitter can be caused by noise or spurious signals.
Type of loop	Classification number for the number of perfect integrators in the loop (number of poles at the origin).
Unit interval	The part of a signal that occupies the shortest interval of signal coding.
Video	In receiving systems, the signal after envelope or phase detection. Video contains the relevant information after the removal of the carrier frequency (baseband).
Video bandwidth	The bandwidth of the signal after video detection. It equals one-half of the intermediate frequency bandwidth when the received signal is down-converted to baseband through a mixer.
Zero	In a network function $[N(s) = (s - z)/(s - p)]$, any real or complex value substituted for s that causes the network function to be zero.
Zero-decibel magnitude crossover frequency	The frequency point on a graph of the open-loop gain where the magnitude equals 0 dB.
Zero, frequency location of	The frequency location that causes a zero in a network function to make the network function equal to zero.

Acronyms

A/D	Analog-to-digital converter (also ADC, A/D, or A to D)
ADPLL	All-digital phase lock loop
AM	Amplitude modulation

ASIC	Application-specific integrated circuit
BER	Bit-error rate
BW	Bandwidth
CCO	Current-controlled oscillator
CML	Current mode logic
CMOS	Complementary metal oxide semiconductor
CMRR	Common mode rejection ratio
CRC	Clock recovery circuit
CSL	Current switch logic
DAC	Digital-to-analog converter
DDS	Direct digital synthesizer
DevRat	Deviation ratio from radio terminology
DPLL	Digital phase lock loop
DRO	Dielectric resonator oscillator
DUT	Device under test
ECL	Emitter coupled logic
EMI	Electron magnetic interference
ESD	Electrostatic discharge
FET	Field effect transistor
FFT	Fast Fourier transform
FIB	Focused ion beam, an analytical IC instrument
FM	Frequency modulation
FSCL	Folded source coupled logic
GPIB	General-purpose interface bus
GSM	Groupe Spécial Mobile the European mobile telephony protocols
HPF	Highpass filter
IC	Integrated circuit
IF	Intermediate frequency
I/O	Input-output device
JFET	Junction field effect transistor
LO	Local oscillator
LPF	Lowpass filter
LSB	Least-significant bit
LVDS	Low voltage differential signaling
MOS	Metal oxide semiconductor
MOSFET	Metal oxide semiconductor field effect transistor
MOSIS	Metal Oxide Semiconductor Implementation Service
MSB	Most significant bit
MUX	Multiplexer
NF	Noise figure
NMOS	N-type MOS
NRZ	Nonreturn to zero

OC3	Optical Carrier levels for SONET fiber optic network
PCB	Printed circuit board
PFD	Phase frequency detector
PLL	Phase lock loop
PM	Phase modulation
PMOS	P-type MOS
PSPICE	SPICE program that runs on personal computers
PSRR	Power supply rejection ratio
PWB	Printed wiring board
RC	Resistance-capacitance circuit
RF	Radio frequency
RMS	Root mean square
RSFF	Reset and set flip flop
RZ	Return to zero
SAW	Surface acoustic wave
SCL	Source coupled logic
SNR	Signal-to-noise ratio
SONET	Synchronous optical networking using lasers
SPICE	Simulation Program with Integrated Circuits Emphasis
SRAM	Static random access memory
SSO	Simultaneous switching outputs
TTL	Transistor-transistor logic
UI	Unit interval
VCO	Voltage-controlled oscillator
VHF	Very high frequency
YIG	Yttrium iron garnet

About the Author

Stanley Goldman received a B.S. in electrical engineering from Carnegie Mellon University in 1975. For more than 25 years he has had a career designing and developing distinctive phase lock loops at Scientific Communications as a senior engineer in the Receiver Group on Electronic Warfare Receivers; at E-Systems as an engineering specialist in the Special Projects Group; at Rockwell International as a technical consultant on microwave satellite communications; at Reliance Electric as a principal engineer on broadband communications; and at Texas Instruments, Inc., in the Wireless Terminal Business Unit. Mr. Goldman has developed and taught class sessions on phase lock loops at the University of Texas at Dallas and given several presentations at technical conferences and IEEE meetings.

For over 20 years at Texas Instruments (TI), Mr. Goldman has been applying phase lock loop solutions to computer timing, radar transmission and reception, automotive electronic control, and communication systems. His most recent work has been designing low-power, small-area, and low-noise phase lock loops as a distinguished member of the technical staff for the Wireless Analog Technology Center at TI.

Mr. Goldman has been a Senior Member of the IEEE and a registered professional engineer for more than 20 years. He was selected to be in Marquis' *Who's Who in Engineering and Science* in 1997. Mr. Goldman has published and presented more than 24 papers and has 7 U.S. patents on the subjects of frequency synthesizers, oscillators, multipliers, phase noise, phase lock loops, receivers, and signal processing. He is the author of *Phase Noise Analysis in Radar Systems* (Wiley, 1989).

Index

Recent Titles in the Artech House Microwave Library

RF and Microwave Coupled-Line Circuits, Rajesh Mongia, Inder Bahl, and
  Prakash Bhartia

RF and Microwave Oscillator Design, Michal Odyniec, editor

RF Power Amplifiers for Wireless Communications, Second Edition, Steve C. Cripps

RF Systems, Components, and Circuits Handbook, Ferril A. Losee

Stability Analysis of Nonlinear Microwave Circuits, Almudena Suárez and
  Raymond Quéré

System-in-Package RF Design and Applications, Michael P. Gaynor

*TRAVIS 2.0: Transmission Line Visualization Software and User's Guide, Version
  2.0,* Robert G. Kaires and Barton T. Hickman

Understanding Microwave Heating Cavities, Tse V. Chow Ting Chan
  and Howard C. Reader

For further information on these and other Artech House titles, including previously
considered out-of-print books now available through our In-Print-Forever® (IPF®)
program, contact:

Artech House	Artech House
685 Canton Street	46 Gillingham Street
Norwood, MA 02062	London SW1V 1AH UK
Phone: 781-769-9750	Phone: +44 (0)20 7596-8750
Fax: 781-769-6334	Fax: +44 (0)20 7630 0166
e-mail: artech@artechhouse.com	e-mail: artech-uk@artechhouse.com

Find us on the World Wide Web at: www.artechhouse.com